Rubber Technology and Manufacture

Rubber Technology and Manufacture

Edited by C. M. BLOW BSC, PHD, FRIC, FIRI

*Lecturer, Institute of Polymer Technology,
University of Technology, Loughborough*

*Formerly, Technical Director, Precision Rubbers Division,
The Dunlop Co. Ltd*

Published for the
Institution of the Rubber Industry

BUTTERWORTHS LONDON

THE BUTTERWORTH GROUP

ENGLAND
Butterworth & Co (Publishers) Ltd
London: 88 Kingsway, WC2B 6AB

AUSTRALIA
Butterworth & Co (Australia) Ltd
Sydney: 586 Pacific Highway Chatswood, NSW 2067
Melbourne: 343 Little Collins Street, 3000
Brisbane: 240 Queen Street, 4000

CANADA
Butterworth & Co (Canada) Ltd
Toronto: 14 Curity Avenue, 374

NEW ZEALAND
Butterworth & Co (New Zealand) Ltd
Wellington: 26–28 Waring Taylor Street, 1
Auckland: 35 High Street, 1

SOUTH AFRICA
Butterworth & Co (South Africa) (Pty) Ltd
Durban: 152–154 Gale Street

First published 1971
© Institution of the Rubber Industry, 1971
ISBN 0 408 70165 X

Filmset by Filmtype Services Ltd, Scarborough, Yorkshire

Printed in England by Cox & Wyman, Fakenham, Norfolk

Contributors

G. F. BLOOMFIELD, PhD, FRIC, FIRI
Formerly, International Co-ordinator, Technical Advisory Service,
Natural Rubber Producers' Research Association

C. M. BLOW, BSc, PhD, FRIC, FIRI
Lecturer, Institute of Polymer Technology,
University of Technology, Loughborough
Formerly, Technical Director, Precision Rubbers Division,
The Dunlop Co. Ltd

B. B. BOONSTRA, PhD
Associate Director of Rubber & Plastics Research,
Cabot Corpn, USA

C. L. BRYANT, AIRI
Assistant Technical Manager, Revertex Ltd

W. COOPER, MSc, PhD, FRIC
Research Manager, Materials, Dunlop Research Centre

B. G. CROWTHER, AIRI
Process Control & Technical Manager, Horninglow Factory,
BTR-Leyland Industries Ltd

H. M. EDMONDSON, BSc
Technical Manager, Services Division, BTR-Leyland Industries Ltd

C. W. EVANS, FIRI, ARTCS, AFInstPet
Group Technical Manager, Dunlop-Angus Hose Group,
The Dunlop Co. Ltd

W. P. FLETCHER, BSc, FInstP, FIRI
Manager, Elastomers Research Laboratory,
Elastomer Chemicals Department, Du Pont Co. (UK) Ltd

5

R. C. HAINES, BSc, ACT(Birm), AIRI
Group Technical Manager, International Sports Co. Ltd

J. A. C. HARWOOD, MSc, PhD
Group Research Physicist, Avon Rubber Co. Ltd

G. HOPKINS, MSc, ARIC, AIRI
Senior Technical Superintendent, Bonding,
Polymer Engineering Division, The Dunlop Co. Ltd

J. B. HORN, BSc, ARIC, AIRI
Technical Manager, Cabot Carbon Ltd

J. E. JACQUES, BSc, ARIC
Technical Controller, Supplies Control Department,
Avon Rubber Co. Ltd

D. G. JONES, AIRI
Technical Manager, Dunlop Footwear Ltd

T. J. MEYRICK, BA, ARIC, FIRI
Manager, Application Research and Technical Service Department,
Rubber Chemicals Group, Dyestuffs Division,
Imperial Chemical Industries Ltd

S. H. MORRELL, MA, DPhil, FIRI
Manager, Rubber Technology,
Rubber and Plastics Research Association

G. F. MORTON, BSc, AInstP, FIRI
Development Director—Tyres, The Dunlop Co. Ltd

W. C. NELLER
Technical Manager, Belting Division, The Dunlop Co. Ltd

B. PICKUP, BSc, MA, AInstP, FIRI
Director, Anchor Chemical Co. Ltd

D. POLLARD, CEng, MIEE
Technical Manager, Wiring & General Cables Division,
Leigh Works,
British Insulated Callender's Cables Ltd

J. F. POWELL, MA, FRIC, AIRI
Metalastik Development Manager, Polymer Engineering Division,
The Dunlop Co. Ltd

G. B. QUINTON
Formerly, Manager, Factory Technical Operations, Fort Dunlop,
The Dunlop Co. Ltd

J. M. RELLAGE
   Manager, Research & Development, Elastomers Division,
   Royal Dutch Shell Plastics Laboratory,
   Delft, Holland

J. R. SCOTT, PhD, FRIC, FInstP, FIRI
   Formerly, Director, Rubber and Plastics Research Association

D. N. SIMMONS, BSc
   Formerly, Scientist,
   Central Research and Development Division,
   The Dunlop Co. Ltd

R. SINGLETON, ARIC, AIRI
   Technical Manager, United Reclaim Ltd

H. J. STERN, PhD, FRIC, FIRI
   Consultant

G. J. VAN DER BIE, dr. ir., FIRI
   Manager, Elástomers Division,
   Royal Dutch Shell Plastics Laboratory,
   Delft, Holland

C. VERVLOET, ir.
   Senior Research Scientist, Elastomers Division,
   Royal Dutch Shell Plastics Laboratory,
   Delft, Holland

J. G. WEBSTER, MSc, ARIC
   Technical Manager, Expanded Rubber & Plastics Ltd

J. O. WOOD, BSc, AInstP
   Manager, Basic Research Section, Textile Research Department,
   Central Research & Development Division, The Dunlop Co. Ltd

A short time before publication, the Editor learned with regret of the deaths, in retirement, of Mr D. N. Simmons and Mr G. B. Quinton.

Foreword

BY J. M. BUIST
Chairman of Council, Institution of the Rubber Industry, 1969–1970

The Institution of the Rubber Industry, since its formation in 1921, has promoted a better understanding of the science and technology of natural rubber and synthetic polymers. By lecture programmes, symposia, publications, and international conferences, the IRI has encouraged and stimulated free interchange of ideas and discussion of methods of solving the many interesting problems which arise in any industry where the technology has often been in advance of the scientific explanations and theories which enable the technology to advance further. In the early 1950s the IRI sponsored and published a series of Monographs designed to assist the growing number of scientists and technologists working with polymers. The IRI has held its examinations for the Associateship since 1925 and the Licenciateship since 1932, and has co-operated fully with educational bodies to provide the necessary preparatory facilities for studying and maintaining a supply of adequately trained technologists for an expanding industry. From time to time the IRI has also supported and encouraged the publication of books on various aspects of rubber and polymer technology in order to disseminate knowledge and encourage students.

With this background, it is highly commendable, but not surprising, that the IRI has sponsored and organised the publication of *Rubber Technology and Manufacture*. In the brief prepared by the Editor, Dr C. M. Blow, it was stated that 'the book is intended for graduates in chemistry, physics and engineering wishing to acquire a sound practical knowledge of rubber technology with emphasis on the processing, compounding and manufacturing of rubber products rather than on the chemistry and physics of rubbers, although the basic facts and theories of the latter must be adequately, though briefly, covered. It is further intended to be a "guide" to undergraduates reading for the AIRI and degrees in polymer science or technology.'

The different chapters of the book have been written by authors who are specialists with considerable familiarity and knowledge of their subject, and Dr Blow is to be congratulated on the team of experts he chose and encouraged to work with him on this project. Personally, I am well aware of the additional effort these authors have had to make over and above their normal business duties, and they deserve our thanks and congratulations. Dr Blow has spent his working life in the polymer industry in a variety of roles, and we are fortunate that a man of his very wide experience has been willing and able to undertake the task of being Editor with such gratifying results.

Although significant advances are made each year in the science and technology of polymers, I believe this book will serve both as a landmark in the recording of the progress of the technology of the polymer industry and as a stimulus to further investigation and research to advance the frontiers of our knowledge in the future.

It has been a privilege and honour to encourage the preparation of this publication during my term of office as Chairman of Council.

J. M. Buist

Preface

The aim of the book is to provide an up-to-date guide for students, for entrants to the rubber manufacturing and associated supplying industries, and for the users of rubber products in other industries, public corporations, and government departments. Emphasis has been placed on the main scientific facts, technological methods, and manufacturing techniques, with comprehensive coverage of both raw materials and testing methods. Each of the authors has gone straight to the subject of his particular section and written an authoritative contribution. No attempt has been made to eliminate repetition and overlapping completely nor to blend the sections together apart from presenting them in a logical order. Instead, Chapter 2 has been written, as it is entitled, to *outline* the breadth and width of rubber technology and to provide connecting links for the subsequent chapters.

The preparation of a 500 page book on rubber technology and manufacture has been possible only by omitting detailed discussion of certain branches of the subject. Latex, ebonite, and adhesive technologies may be said nowadays to be distinct and separate and have, therefore, been considered outside the scope of this book; short notes on them appear, for reference, in Chapter 2. Because, in recent years, there have been many conferences held and several books published on the use and application of elastomers in engineering, it was decided that this subject should also receive only scant treatment. The third and fourth omissions are the specialised field of chemical analysis in connection with rubbers and rubber products, and the general subject of instrumentation of machines and equipment in such respects as temperature control, or thickness measurement at the calender and spreading plant. Finally, the book is concerned essentially with rubbers; discussion of thermoplastics is in general omitted. It is, of course, realised that the latter materials are processed by rubber manufacturers alongside rubber in certain sections of the industry, notably belting, footwear, and cables, and the authors refer to this in Chapter 10.

The lengthy contents list and indexes, and the abundant cross-references, it is hoped, will be useful in meeting the requirements of those who use the book for reference. For those new to the industry, attention has been paid to nomenclature and to the terms, abbreviations, trade names, etc., in common use. In particular, the code letters for rubbers (ASTM D1418) have been adopted to save repeated use of the full names. Throughout the text,

Registered Trade Names are printed with an initial capital letter. Abbreviations are listed on pp. *20–21*, and the symbols used in the book appear on pp. *21–22*.

The glossary of terms published by the British Standards Institution (BS 3558, 1968) not only records the terms used in the industry but suggests reforms by deprecating the use of certain words. Of the deprecated terms, 'compound' is to be found frequently in this book because it is in very common usage to denote a rubber to which have been added all the ingredients necessary for processing and vulcanisate properties. 'Stock', which means much the same and is also deprecated by the BSI, occurs occasionally. The other terms used in this book have the meaning given in the BSI glossary. I have commented in Chapter 2 on the terms 'rubber' and 'elastomer', which, if I interpret the definitions in BS 3558 correctly, are not synonymous. Rubber is 'macromolecular material which *has, or can be given,* properties of (1) at room temperature returning rapidly to the approximate shape from which it has been substantially distorted by a weak stress, and (2) not being easily remoulded to a permanent shape by the application of heat and moderate pressure'; whereas elastomer is 'macromolecular material which can return rapidly to the approximate shape from which it has been substantially distorted by a weak stress'.

These definitions suggest to me that many (rubber) polymers need to be crosslinked to be elastomers and are my justification for confining the term 'elastomer' to vulcanisates only. 'Polymer' and 'rubber' are used for the raw materials, and the latter additionally for mixes, vulcanisates, and products. It is hoped that the context removes any ambiguity.

There has never been any doubt in my mind, or in that of anyone I consulted, that metric units should be used throughout with preference for SI units where these are accepted. The industry, at the time of writing, has not, it appears, made up its mind regarding the use of the newton; furthermore, if the newton is adopted, there seems reluctance to use the recommended meganewton per square metre (MN/m^2) for the tensile strength and stress unit; therefore, kilograms-force per square centimetre (kgf/cm^2) is used. For those who wish to convert to British units, a table is given on p. *22*.

Some explanation is perhaps necessary of the slightly unorthodox way in which the references to the literature and patent specifications have been arranged. Instead of suffixed numbers in the text, the Harvard system has been adopted of identifying in the text the author(s) and the year of publication, with numeral 1, 2, etc., added if there is more than one publication in one year by an individual author. The references are not listed at the end of each chapter, but are assembled in alphabetical order of first author at the end of the book in what is, in fact, a name index.

As a supplement to the concise Chapter 11 on physical testing, a full list of standards is included. For follow-up reading, a Bibliography, grouped into general and special categories, corresponding to chapter subjects, will be found on pp. 479–485.

Editor's Acknowledgements

Professor R. J. W. Reynolds, Director of the Institute of Polymer Technology, suggested that the IRI should sponsor a textbook on rubber technology, and I am indebted to him not only for being instrumental in my undertaking the editorship but also for much practical assistance and encouragement. Without his help and that of the small sub-committee in planning the book and suggesting authors, I would never have started. Mr E. W. Madge and Dr D. G. Marshall of The Dunlop Co. enlisted several contributors from their organisation; for that assistance and interest I am grateful. Also I must thank Mr J. M. Buist for suggesting contributors and for so willingly agreeing to write the Foreword.

I thank all contributors for the effort they put into the preparation of their articles and for so readily agreeing to my requests for amendments and additions. I must place on record my gratitude to several of my former colleagues in The Dunlop Co., including Messrs R. N. Thomson and D. W. Southwart for assistance, and to Mr Claude Hepburn who read the text and whose comments enabled me to correct errors and omissions.

Finally, my thanks are due to Miss Mary Lamb and Miss Janet Arnold who worked hard for me at their typewriters, and to my wife who spent many hours checking scripts with me.

Loughborough   C.M.B.

Contents

Useful Information

Abbreviations

AC	azodicarbonamide	EPM	ethylene–propylene copolymer rubber
ADS	air-dried sheet		
AHEM	Association of Hydraulic Equipment Manufacturers	ETU	ethylene thiourea
		EU	polyurethane (ether) rubber
APF	all purpose furnace (black)	EV	efficient vulcanisation
ACM	polyacrylic rubber	EVA	ethylene vinyl acetate
ASTM	American Society for Testing and Materials	FEF	fast extrusion furnace (black)
		FF	fine furnace (black)
AU	polyurethane (ester) rubber	FPM	fluorocarbon rubber
AZDN	azoisobutyronitrile	FSI*	fluorosilicone rubber
		FT	fine thermal (black)
BASRM	British Association of Synthetic Rubber Manufacturers	FVSI*	fluorosilicone rubber
BR	polybutadiene rubber	GMF	quinone dioxime
BS	British Standard	GPF	general purpose furnace (black)
BSH	benzene sulphonyl hydrazide	GR-S	Government rubber-styrene (SBR)
BSI	British Standards Institution		
		HAF	high abrasion furnace black
CBS	cyclohexyl benzthiazyl sulphenamide	HBS	cyclohexyl benzthiazyl sulphenamide
CCV	catenary continuous vulcanisation	HCV	horizontal continuous vulcanisation
CF	conductive furnace (black)		
CLIIR	chlorinated butyl rubber	HMDA	hexamethylene diamine
COD	cyclooctadiene	HMF	high modulus furnace (black)
CR	polychloroprene rubber	HMT	hexamethylene tetramine
CSM	chlorosulphonated polyethylene	HOFR	heat resisting, oil resisting, and flame retardant
CV	continuous vulcanisation		
CVNR	constant viscosity natural rubber	HPC	hard processing channel (black)
		HS	high structure (black)
DBP	dibutyl phthalate	HSMB	hydrosolution masterbatch
DCPD	dicyclopentadiene		
DNPT	dinitrosopentamethylene tetramine	IEC	International Electrotechnical Committee
DOTG	di-o-tolylguanidine		
DPG	diphenyl guanidine	IIR	isobutylene–isoprene (butyl) rubber
DPTS	dipentamethylene tetrasulphide	IISRP	International Institute of Synthetic Rubber Producers
EDTA	ethylene diamine tetraacetic acid	IR	polyisoprene rubber (synthetic)
ENB	ethylidene–norbornene	IRHD	international rubber hardness degree
EP	ethylene–propylene rubber		
EPC	easy processing channel (black)	IRI	Institution of the Rubber Industry
EPDM	ethylene–propylene terpolymer rubber	ISAF	intermediate super abrasion furnace (black)

ISO	International Organisation for Standardisation		PVA	polyvinylalcohol
			PVC	polyvinylchloride
LM	low modulus (black)		PVSI*	phenyl vinyl methyl silicone rubber
LS	low structure (black)		RAPRA	Rubber and Plastics Research Association
LVN	limiting viscosity number			
LVNR	low-viscosity natural rubber		RFL	resorcinol–formaldehyde–latex
			r.h.	relative humidity
MB	masterbatch		RRIM	Rubber Research Institute of Malaya
MBI	mercaptobenzimidazole			
MBT	mercaptobenzthiazole		RSS	ribbed smoked sheets
MBTS	dibenzthiazyl disulphide			
MDI	diphenylmethane-4,4′-diisocyanate		SAF	super abrasion furnace (black)
MOCA	4,4′-methyl-bis-2-chloroaniline		SATRA	Shoe and Allied Trades Research Association
MPC	medium processing channel (black)			
MPF	medium processing furnace (black)		SBR	styrene–butadiene rubber
MRFB	Malayan Rubber Fund Board		SC	slow-curing (black)
MT	medium thermal (black)		SCF	super conductive furnace (black)
			SCI	seal compatibility index
NBR	acrylonitrile–butadiene copolymer (nitrile) rubber		SI*	dimethyl silicone rubber
			SMR	standard Malaysian rubber
NDI	naphthalene-1,5-diisocyanate		SP	superior processing (NR)
NR	natural rubber		SPF	super processing furnace (black)
NRPRA	Natural Rubber Producers' Research Association		SRF	semi-reinforcing furnace (black)
NS	non-staining (black)		TCR	technically classified rubber
			TDI	toluene diisocyanate
OB	*pp′*-oxy-bis-benzene sulphonylhydrazide		TDM	tertiary dodecyl mercaptan
			TETD	tetraethyl thiuram disulphide
OENR	oil-extended natural rubber		TMTD	tetramethyl thiuram disulphide
OEP	oil-extended polymer		TMTM	tetramethyl thiuram monosulphide
OESBR	oil-extended styrene–butadiene rubber		TR	polysulphide rubbers
			VCV	vertical continuous vulcanisation
PAUS	pale amber unsmoked sheet		VGC	viscosity gravity constant
PBN	phenyl-β-naphthylamine		VP	vinyl pyridine
PMMA	polymethylmethacrylate		VSI*	vinyl methyl silicone rubber
PP	partially purified			
p.p.h.r.	parts per hundred of rubber		WLF	Williams–Landel–Ferry (equation)
p.p.m.	parts per million		XCF	extra conductive furnace black
PRI	plasticity retention index			
PSI*	phenyl methyl silicone rubber		ZDC	zinc diethyl dithiocarbamate
PTFE	polytetrafluorethylene		ZDMDC	zinc dimethyl dithiocarbamate

Symbols

A	area	ΔH	latent heat of vaporisation
B	bulkiness (of black)	k	Boltzmann's constant
C	driving torque, rate of creep, length of crack, electrical conductivity	L	length, torque
		ΔL	maximum change in torque during vulcanisation
c	volume fraction of filler	M	molecular weight
D	diffusion coefficient	M_c	number average molecular weight between crosslinks
E	Young's modulus, extrusion shrinkage		
F	force	N	number of chains per unit volume
f	shape factor	n	flow index
G'	elastic shear modulus	P	penetration rate
G''	viscous shear modulus	Q	anisometry (of black), permeation coefficient
H	hysteresis		

* ASTM D1418 gives Si as the symbol for silicone rubbers; SI is more commonly used in the UK (BS 3502, 1967).

Q_s	differential heat of adsorption	$\dot{\gamma}$	rate of strain
R	$\begin{cases}\text{rate of recovery,} \\ \text{gas constant}\end{cases}$	δ	$\begin{cases}\text{phase angle,} \\ \text{solubility parameter}\end{cases}$
r	radius	ε	strain (tension)
S	rate of stress relaxation, solubility	$\dot{\varepsilon}$	rate of strain (tension)
T	$\begin{cases}\text{absolute temperature,} \\ \text{structure factor (of black)}\end{cases}$	η	viscosity
		λ	extension ratio
T_g	glass transition temperature	ν	number of crosslinks (per cubic centi-
t	time		metre)
v_r	volume fraction of rubber in swollen gel	ρ	density
V_s	molecular volume of the solvent	σ	stress
V	volume adsorbed	$\mathcal{T}$	characteristic energy (per unit area of
V_m	volume adsorbed to form monolayer		crack)
W	$\begin{cases}\text{stored energy density,} \\ \text{work per unit volume}\end{cases}$	τ	relaxation time
		χ	interaction constant (Flory–Rehner
γ	$\begin{cases}\text{strain,} \\ \text{surface energy}\end{cases}$		equation)
		Ω	angular velocity

Conversion Factors

To convert	Multiply by
Millimetres (mm) to inches	0·039
Metres (m) to feet	3·28
Metres (m) to yards	1·09
Square centimetres (cm^2) to square inches	0·155
Cubic centimetres (cm^3) to cubic inches	0·061
Cubic centimetres (cm^3) to pints	0·0018
Grams (g) to ounces	0·035
Kilograms (kg) to pounds	2·205
Kilograms-force per square centimetre (kgf/cm^2) to pounds-force per square inch	14·3
Degrees Centigrade (°C) to degrees Fahrenheit	1·8, and add 32

1 megagram (Mg) = 1 ton approx.
1 gigagram (Gg) = 1000 tons approx.

Multiples and Sub-Multiples of Metric Units

Multiple	Prefix	Symbol	Sub-multiple	Prefix	Symbol
10^{12}	tera	T	10^{-1}	deci	d
10^9	giga	G	10^{-2}	centi	c
10^6	mega	M	10^{-3}	milli	m
10^3	kilo	k	10^{-6}	micro	μ
10^2	hecto	h	10^{-9}	nano	n
10	deca	da	10^{-12}	pico	p
			10^{-15}	femto	f
			10^{-18}	atto	a

1 millimicron (mμ) = 1 nanometre (nm).

ONE

History

BY H. J. STERN

1.1 Introduction

In this historical review it is necessary to consider far more than the history of rubber itself, because the material is seldom, if ever, used in its original state. Rubber manufacture, that is to say the manufacture of articles from rubber, involves the addition to the rubber of many other materials. The nature of these, the means by which they are added, and the subsequent treatment of the mixture are all of equal importance. This history, therefore, is not the history of a single product but of a series of parallel developments and of the interaction between the various lines of progress. The *History of the Rubber Industry**, published by the IRI, covered more than 400 pages when it appeared in 1952, and much has happened since then. Omissions in the present chapter are therefore inevitable, but many references to relevant patent specifications and articles are included.

1.2 The Era of Natural Rubber

Until the synthetic material became a reality (Section 1.4.2), rubber was obtained from trees. There are a number of species of trees yielding rubber which are native to many parts of Africa and the Far East, but there is no record of these, or of the rubber from them, prior to the discovery of America where, as excavations have shown, it was known as early as the sixth century. The first mention of rubber trees is to be found in the eighth 'decade' of *De Orbe Novo* by Pietro Martire d'Anghiera (Peter Martyr), published in Latin in 1516. The outstanding source of natural rubber, the large forest tree *Hevea brasiliensis*, occurs in the southern equatorial region of America. The use of rubber for waterproofing garments was apparently first noted by Pizarro in 1540, but the Spaniards did not bring the process to Europe. This was done much later as a result of the travels of Charles de la Condamine. By the end of the eighteenth century the properties of rubber as obtained

*Several items of information from this book have been utilised in the present chapter.

1

from the tree were known throughout Europe. The name was derived from the use of the material as an eraser, an application popularised by Joseph Priestley in 1770. The original South American name of *caoutchouc* and its variations are still used to denote rubber in many languages other than English, where it is reserved for the pure rubber hydrocarbon.

In Europe and North America progress was long delayed by lack of a satisfactory solvent until in 1823 Charles Macintosh of Manchester patented the use of coal tar naphtha. In its original condition, rubber was thermoplastic, becoming soft and sticky in summer and hard in winter. It was placed between two textile fabrics to minimise these defects, and from this material the so-called double-texture waterproof garments, or macintoshes, were made. This susceptibility of rubber to changes of temperature was overcome by vulcanisation (Section 1.2.2).

1.2.1   MASTICATION

Although naphtha enabled rubber to be dissolved and the solutions to be spread, these solutions were of high viscosity and low solids content; there was also no satisfactory means of shaping rubber in solid form. About 1830

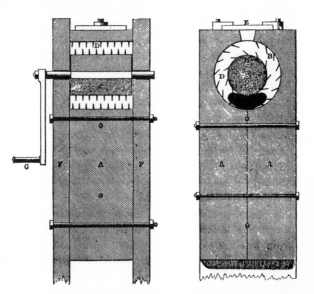

Fig. 1.1   Hancock's pickle (1820), the forerunner of the internal mixer: AA, two pieces of wood bolted together; B, a hollow cylinder cut out of AA and studded with teeth; C, a cylinder of wood studded with teeth and having a spindle passed through it; D, space between the two cylinders B and C; E, an opening with a cover; FF, two pieces of wood bolted on both sides of AA and enclosing the space D and cylinder C; G, a winch; the darkened spot in space D represents the charge of rubber. From Stern (1967), courtesy Elsevier

Thomas Hancock, working in London, discovered the phenomenon now known as mastication. He used a simple machine (Fig. 1.1) which he did not patent but, in order to keep secret, called by the misleading name of a 'pickle'. Later, Hancock masticated rubber on a horizontal two-roll mill, but the principle of the pickle was sound and was revived over a hundred years later as the internal mixer.

Rubber which has been masticated is softer and flows more easily than in its original state, and solutions of high solids content can be obtained. Another important effect is that rubber is rendered tacky, so articles can be built up from several pieces, or layers, of masticated rubber or rubberised fabric without the use of solvent. Further developments in machinery are described in Section 1.8.

1.2.2 VULCANISATION

The main defect of raw rubber—susceptibility to temperature change—is removed by vulcanisation, for which the popular name 'cure' is therefore appropriate. It is achieved by combining rubber chemically with sulphur. Credit for the original discovery, which was made about 1839, must go to Charles Goodyear of Woburn, Mass., USA, but, doubtless because of trade depression, he made little commercial progress. Thomas Hancock of London, who, in 1843, took out a patent on the same lines as that of Goodyear, was more successful in exploiting the commercial possibilities. He found that if he mixed sulphur with rubber and heated the mixture, a 'change' occurred to which the name 'vulcanisation' (from Vulcan, the God of Fire) was given by his friend, the artist William Brockeden (Stern, 1945-1). In 1843 Hancock found that thin sheets of rubber could conveniently be vulcanised by immersion in a bath of molten sulphur, and the use of this process lingered on until about 1965. The rubber which was immersed was usually in the form of cut sheet, produced as follows. Masticated rubber was pressed to form either a block from which sheets were cut by a reciprocating knife, or a cylinder from which a continuous length could be made. Cutting was facilitated by freezing the mass.

The discovery of vulcanisation meant that by about 1850 a whole range of rubber articles was available, and many are illustrated in the books written by Goodyear (1800–1860) and Hancock (1785–1865).

The so-called cold cure, namely treatment with sulphur chloride, was discovered by Alexander Parkes (1840). It was extensively used for proofing and dipped goods but is now virtually obsolete.

1.2.3 WILD RUBBER

During the period of fundamental invention just described, the raw material came almost entirely from South America, where the rubber was collected from the forest in the Amazon valley by crude methods which had changed little over the centuries. Rubber occurs in the Hevea tree as latex, a milky liquid which is exuded when the tree is wounded or cut. The collectors in the Amazon valley would go out in the morning with a hatchet or machete and make cuts in the tree trunk, fastening below them a small cup or calabash into which the latex flowed. They would return later and collect the latex, which they then coagulated by dipping in a wooden paddle and rotating it in the smoke from burning palm tree nuts. Repetition of this operation gave a wet smoked coagulum which was removed from the paddle and dried in the sun. Wild rubber thus produced was variable in quality and subject to wide fluctuations in price. In 1910, because the plantations were not yet ready to meet the sudden demand for tyres for the then new motor car, the price of rubber rose to a record figure of 153*d.*/lb. Despite this, the lot of the collectors was little removed from that of slavery.

1.2.4   PLANTATION RUBBER

It was Hancock who foresaw the benefit of a regular uninterrupted supply of rubber, and in 1876 the India Office in London commissioned Sir Henry Wickham to collect Hevea seeds. He obtained some 70 000, which were planted at Kew Gardens, but few survived and another batch of seeds was brought back in 1877. The seedlings were sent to a number of equatorial parts, including Java, Borneo, and Ceylon, but it is in what was the Netherlands East Indies and more particularly in Malaya that the greatest success in the cultivation of rubber trees was achieved. This success was largely due to the work of H. N. Ridley (1859–1956), who in 1888 took charge of the Botanical Gardens in Singapore and devised proper methods of tapping, instituted the use of acetic acid to coagulate the latex, and produced rubber in thin sheets which would dry rapidly instead of the irregular shapes previously made.

The plantations made a very slow start. The total production in the year 1907 was only 1000 tons (1 Gg) of rubber, but, as the number of motor vehicles increased and the price of rubber rose, more plantation rubber was produced. The consumption of wild rubber declined and became virtually obsolete by about 1930 (Table 1.1). Attempts at various periods to obtain rubber from plants other than the Hevea tree have been shortlived. Amongst the plants may be mentioned the Ceara tree. Guayule (a shrub cultivated in North Mexico) and the Russian Dandelion rubbers were both expedients of

Table 1.1   WORLD PRODUCTION OF PLANTATION AND WILD RUBBER

Year	Plantation rubber (1000 tons) (Gg)	Wild rubber (1000 tons) (Gg)	Total	Plantation rubber (%)	Wild rubber (%)
1905	0·2	59·3	59·5	0·3	99·7
1914	73·2	48·1	121·2	60·4	39·6
1920	316·6	37·0	353·6	89·5	10·5
1933	839·3	11·8	851·1	98·6	1·4

World War II, and production ceased when synthetic rubber became available.

Although rubber cultivation falls outside the scope of this chapter, mention must be made of developments which, over a period of years, have enabled the plantation industry to survive and which may be described as selective breeding. By bud grafting and the use of selected seeds, the annual yield has been increased more than fivefold to a present-day figure approaching 2000 lb per acre (2 Mg per hectare) per year.

Mention must also be made of improvements in the form in which plantation rubber is delivered to the user. For many years there have been two main types; crepe prepared by coagulating the latex and washing, rolling, and drying the coagulum; and smoked sheet prepared on similar lines but by drying the coagulum in the presence of smoke. Within each grade—crepe and smoked sheet—are a number of subgrades, and these, together with types prepared from various residues, resulted in no fewer than 31 standard grades (Table 4.1) classified entirely on appearance without regard to the properties most important to the manufacturer, such as plasticity and speed of vulcanisation.

In addition to variability, the presence of foreign matter (dirt) has been another disadvantage of natural rubber. Efforts to introduce technically classified rubber have always been retarded by the reluctance of buyers to pay the slightly higher price required. The introduction of synthetic rubber, marketed as a clean uniform product, obtained by pressing together small particles to form bales which could be suitably wrapped, could not fail to influence the natural rubber producers, some of whom now produce, pack, and market it, in the same form as synthetic, under the description SMR.

The improvements described above have enabled natural rubber to meet, at least in part, the competition of synthetic.

1.3 Direct Use of Latex

The direct use of latex, practised in South America, lay dormant for many years. The revival began in the 1920s when 4 gal. sealed tins became available in Europe and elsewhere. Preserved with ammonia, the use of which for this purpose had been known since 1791 (de Fourcroy), this latex was of 30% solids; it was therefore necessary to transport some 7 parts of water for every 3 parts of rubber. Concentration to 60% achieved by either creaming, centrifuging, evaporation, or electrodecantation, was a step forward.

Creamed latex became available about 1925, and some is still produced (Traube, 1924). Centrifuged latex, dating from about 1927 (Utermark, 1923), is now the standard latex of commerce, available in bulk. Much was made of electrodecantation in the 1940s, but production was shortlived (Semperit, 1937). Evaporated latex (Revertex, 1931) dates from 1930 and is still in production (Section 2.4.1).

The use of latex instead of rubber solution for making dipped articles, such as gloves, began about 1924 and still continues. Latex consumption remained small until the introduction about 1930 of latex foam by the Dunlop Rubber Co. (1929) using the silicofluoride process (Fig. 1.2). Earlier, Schidrowitz (1914) had made sponges from latex.

Foam production has expanded rapidly, being today one of the largest latex-consuming industries. The process devised by Talalay (1934) and depending on freezing is now widely used (Section 2.4.3.4).

Early disadvantages in the use of SBR latex, such as low solids content and unpleasant odour, were eventually overcome (Bennett and Burridge, 1962; Jones, 1962). This led to a fall in the price of natural latex and to a situation where either natural or SBR or a mixture of these latices may be used.

It was discovered by Schidrowitz (1921, 1922) that latex could be vulcanised.

Fig. 1.2   The original beater (cake mixer) used at the Dunlop laboratories for the development of latex foam in 1928. Courtesy The Dunlop Co. Ltd

The electrodeposition or anode process for latex was shortlived (Klein, 1923) but within the last few years has been revived on a large scale for use with emulsion paints, which are essentially latices.

1.4   Synthetic Rubber

1.4.1   THE COMPOSITION OF NATURAL RUBBER

As early as 1826 Faraday found an empirical formula of C_5H_8 for natural rubber, and Greville Williams (1860) recognised that rubber was a polymer of isoprene. Following the work of Harries, Pickles (1910) put forward the now universally accepted straight-chain polymer structure for the rubber hydrocarbon:

$$\sim\sim CH_2-\underset{\underset{CH_3}{|}}{C}=CH-(CH_2)_2-\underset{\underset{CH_3}{|}}{C}=CH-(CH_2)_2-\underset{\underset{CH_3}{|}}{C}=CH-CH_2\sim\sim$$

The elucidation of the structure of natural rubber paved the way for the development of synthetic rubbers.

Because of the presence of an asymmetric carbon atom, *cis* and *trans* forms are possible. In fact, rubber is the *cis* and gutta the *trans* form of polyisoprene

(Meyer and Mark, 1928). These findings have also been of importance in work on synthetic rubber and gutta.

1.4.2 THE QUEST FOR SYNTHETIC RUBBER

The composition of the rubber hydrocarbon having been established, it was possible to start work on its synthesis, and methods for the preparation of the dienes were worked out at a comparatively early date in England and in Russia. In 1909–1910, Prof. W. H. Perkin Jr and his colleagues, F. E. Matthews, E. H. Strange, Sir Wm Ramsay, Dr Ch. Weizman, and Prof. Fernbach, devised and patented methods for making butadiene which could be polymerised to yield rubber (Perkin *et al.*, 1910). It was not until much later that a synthetic polyisoprene was produced (Section 4.3). Russian chemists were active in the years prior to 1914, in particular Kondakoff, Butlerow, Ostromislensky, and Lebedev, who was making butadiene as early as 1910 and whose work played an important part in the development of the USSR synthetic rubber industry.

Most synthetic rubbers are produced in two main stages: first, the production of the monomer or monomers, then the polymerisation to form a rubber. Although alcohol and also acetylene have been used in the past as starting points for monomer preparation, this has now become part of petroleum technology and hence falls outside the scope of this review. The most significant early discovery in the polymerisation process was achieved when Matthews and Strange (1910) patented the use of metallic sodium. Their application was made on 25th October, forestalling by only three days the application of Harries and The Bayer Co. for the same invention. Litigation ensued, and the claims of Matthews and Strange were upheld.

During the war of 1914–1918 methyl rubber was made in Germany on a commercial scale. It was a polymer of 2,3-dimethylbutadiene which had been discovered by Kondakoff. After an interval following the war, work in Germany began again. At first the rubbers (of two types) were polymers of butadiene obtained by the action of sodium—hence the name 'Buna' from butadiene and natrium, the German name for sodium. Similar rubbers were made in the USSR (Talalay, 1942), and production of these continued. After the Germans had decided to use a copolymer prepared from butadiene and styrene (I.G. Farbenindustrie, 1930), a further advance was the introduction of emulsion polymerisation, which brought many advantages, including greater speed, little change in viscosity, and a more homogeneous polymer. Small amounts of peroxide were used as catalyst. The idea arose from the fact that rubber is formed in the plant as a latex, and patents had been taken out by The Bayer Co. as early as 1912.

Production in Germany of the copolymer of butadiene and styrene (Section 4.2.1) increased from 2000 tons (2 Gg) in 1937 to 40 000 tons (40 Gg) in 1940. The process was licensed to the Standard Oil Co. in the USA, and, when war broke out between Germany and the USA, developments proceeded independently but on similar lines in the two countries. In Germany, production rose from 70 000 tons (70 Gg) in 1941 to 103 000 tons (103 Gg) in 1944. Production in the USA was negligible at the outbreak of war but, between 1942 and 1944, 87 factories with a total annual output of 1 000 000 tons (1 Tg) were erected. The rubber, at first called Buna, then GR-S, and later SBR, was (and is) a copolymer of butadiene (about 77 %) and styrene (about 23 %), made by emulsion polymerisation (Section 4.2.1).

After the ending of the war in 1945, production in Germany ceased but started again in 1959 at the point of technical advance reached in the meantime in the USA.

As usual, the old single type of SBR was replaced by a number of specialised types (Section 4.2.1.3).

1.4.3   IMPROVEMENTS IN SBR

The original rubbers (up to about 1940) were extremely difficult to process. This was mainly because their molecular structure, unlike that of natural

Table 1.2   RECIPES FOR THE PRODUCTION OF STYRENE–BUTADIENE RUBBERS

Ingredient	Buna S (about 1936)	Buna S3 (about 1943)	'Cold' poly- merised SBR (1967)	Nature of ingredient
Butadiene	68	68	70	Main monomer
Styrene	32	32	30	Additional monomer
Sodium salt of naphthalene sulphonic acid (Nekal for Buna, Orotan SN for SBR)	2·85	2·85	0·15	Emulsifier
Linoleic acid	2·0	—	—	Part of emulsifier system
Paraffinic fatty acid	—	0·5	—	Part of emulsifier system
Sodium hydroxide	0·5	0·5	—	Part of emulsifier system
Potassium stearate	—	—	2·5	Part of emulsifier system
Rosin soap (Dresinate 214)	—	—	2·5	Emulsifier
Diproxid	—	0·08	—	Modifier
TDM	—	—	0·2	Modifier
EDTA	—	—	0·035	Sequestering agent
Potassium persulphate	0·45	0·45	—	Catalyst
Paramenthane hydroperoxide	—	—	0·05 ⎫	
Rongalite	—	—	0·05 ⎬ Redox catalyst system	
Ferrous sulphate	—	—	0·02 ⎭	
Potassium sulphate	—	—	0·5	Viscosity regulator
Sodium dimethyl dithiocarbamate	—	—	0·1 ⎫ 'Shortstop'—to stop	
Polyamine H	—	—	0·015 ⎬ further polymerisation	
Polygard	—	—	1·25	Antioxidant—added as a dispersion after completing polymerisation
Water	105	105	200	

rubber which consists of straight chains, was branched and irregular, and the molecular weight could not be properly controlled. A significant improvement came with the use of the so-called modifiers which favoured straight-chain formation. The Germans used Diproxid (diisopropyl xanthogen disulphide), and the Americans TDM. It was also found that polymerisation at a low temperature, which favoured straight-chain formation, could be achieved by the use of a reducing agent together with the oxidising catalyst, thus forming a Redox system. The addition to the Redox system of a sequestering agent was found to accelerate the polymerisation still further. These developments may be seen in Table 1.2 and apply equally to other rubbers produced by emulsion polymerisation, e.g. nitrile rubbers (Sections 1.4.8 and 4.10).

1.4.4 OIL EXTENSION

Improvements, described above, in the polymerisation process resulted in a reduction in the price of the product. An important further reduction came when oil extension was introduced (The General Tire & Rubber Co., 1955). To rubbers of high molecular weight substantial amounts—around 50%—of a suitable mineral oil may be added without appreciable detriment to the properties of the vulcanisate, provided a sufficient quantity of carbon black is used.

The practice of oil extension has been adopted also with IIR, BR, EPM and EPDM, and NR.

1.4.5 POLYCHLOROPRENE (NEOPRENE) RUBBER

The first synthetic rubber to achieve true commercial success was one which, unlike natural rubber, was oil resistant after vulcanisation. This was Neoprene, first marketed in 1932 under the name of Duprene, the name being changed in 1936. Its production was based on the researches of Father Nieuwland, who found (1921–1923) that vinyl acetylene combined with

$$CH_2{=}CH{-}C{\equiv}CH + HCl \longrightarrow CH_2{=}\underset{\underset{Cl}{|}}{C}{-}CH{=}CH_2$$

hydrochloric acid to yield 2-chloro-1,3-butadiene to which the name 'chloroprene' was given (Du Pont Co., 1933-1).

Chloroprene was readily polymerised, especially in emulsion form (Du Pont Co., 1933-2), to give a rubberlike polymer which, unlike natural rubber, could be vulcanised not by sulphur but by zinc oxide to give an oil-resisting product. Many different types of Neoprene have now been produced, differing mainly as to whether or not a modifier is used in the polymerisation process and in the nature of the modifier—sulphur or a mercaptan (Table 4.16) (Du Pont Co., 1961). Both solid rubbers and latices are available.

1.4.6 POLYSULPHIDE RUBBERS

Ethanite was produced in Belgium about 1934, and Thiokol in the USA some two years earlier. These oil- and solvent-resisting rubbers filled a gap during the 1939–1945 war. Because of their unpleasant odour and poor physical properties, they were later largely replaced by nitrile rubbers when these became available. Liquid Thiokol rubbers have, however, been developed in recent years for use as sealants (Section 4.16) (Fettes and Jorczak, 1950).

1.4.7   BUTYL RUBBER

The polymerisation of isobutylene to give a rubbery but non-vulcanisable polymer was patented (Standard Oil Co., 1935; I.G. Farbenindustrie, 1940) some years before Thomas and Sparks (1939, 1941) discovered that, by the addition of a small amount of an unsaturated hydrocarbon containing conjugated double bonds, a vulcanisable rubber with valuable properties could be produced at a low price.

At first, butadiene was the added monomer, and in the early days the rubber was difficult to vulcanise because of the low degree of unsaturation. The use of a purer isobutylene enabled more of the unsaturated component to be used and this, together with the substitution of butadiene by isoprene, brought about a great improvement, and a wide range of butyl rubbers is now available (Section 4.5).

The manufacturing process is unique in that polymerisation occurs with great rapidity at about $-65°C$, aluminium chloride being used as catalyst.

1.4.8   NITRILE RUBBERS

Nitrile rubbers are characterised by their excellent resistance to solvents. They are copolymers of butadiene and acrylonitrile, first produced in Germany about 1935. Since that date improvements have been made by the same methods as those used for SBR. The rubbers are now available in a number of types, differing mainly in the relative proportion of the two monomers (Section 4.10).

1.4.9   STEREOREGULAR POLYMERS

As explained already, natural rubber is essentially a *cis* polymer of isoprene with a straight-chain structure, to which it owes its excellent properties, particularly in processing. Efforts to make a synthetic rubber of the same structure were unsuccessful until about 1956, when two catalyst systems were invented independently.

In that year, Staveley and his co-workers (Staveley *et al.*, 1956; Firestone Tire & Rubber Co., 1957) discovered that finely divided lithium would give the desired result. A rubber, first called 'Coral' from its appearance, was produced with a 94·3% *cis*-1,4 structure, compared with 97·1% for natural rubber (Section 4.3). This may be considered as a highly refined development of the original sodium process. Overcoming the difficulties of handling the spontaneously inflammable lithium represented a considerable technical achievement.

Similar results are obtainable with butyl lithium, and this is used in making a *cis*-polybutadiene (Diene rubber) (Section 4.4).

A second way of producing stereoregular polymers arose from the use of the special catalysts which were discovered in 1953 (Horne *et al.*, 1956) by Prof. Karl Ziegler (1959) and which bear his name. These are mixtures of organo-metallic compounds, such as aluminium alkyls, with heavy-metal compounds such as titanium chloride. These catalysts were first used to produce polymers of ethylene at low pressures, but it was also found that, with the correct catalyst, isoprene polymers which were essentially either *cis*, similar to natural rubber (Natsyn: Goodyear Tire & Rubber Co.), or *trans*, similar to gutta percha (Trans Pip: The Polymer Corpn), could be obtained at will.

The above polymers are mass or solution polymers and represent the closest approximation, so far, to a true synthetic rubber (or gutta).

Ethylene propylene rubbers were originated about 1955 by the Montecatini Co. (Mazzanti, 1969), who found that, despite their saturated nature, vulcanisable polymers could be obtained by using a Ziegler catalyst, e.g. alkyl aluminium with vanadium oxychloride (Montecatini Co., 1960).

It was later found that sulphur vulcanisable polymers could be obtained by the use of a third component, e.g. 1,4-hexadiene. This was claimed to give mainly straight-chain molecular structure and hence improved processing characteristics, and is believed to be used in Nordel (Du Pont). A complicated patent situation resulted (Section 4.6) (Union Carbide, 1960; Esso Research, 1960, 1961; Shell International Chemical Co., 1961, 1962; Hercules Powder Co., 1961, 1962).

An interesting development in the field of styrene–butadiene polymers was the production by Shell (1965) of polymers of which the molecular chain consists of alternate 'blocks' of polymerised styrene, which is rigid, and polymerised butadiene, which is flexible. This gives an elastic polymer which may be injection moulded in the same way as conventional thermoplastics, and used without vulcanisation. Because they remain thermoplastic, the rubbers are not suitable for use at elevated temperatures (Section 4.7).

1.4.10 THE NEWER SPECIALITY RUBBERS

1.4.10.1 *Fluorocarbon Rubbers*

Fluorocarbon rubbers of a highly specialised nature are distinguished by their outstanding resistance to heat and chemical attack, and arose from the work carried out by Du Pont (Dixon, Rexford, and Rigg, 1957). A fabric coated with a rubber of this kind can have a useful life of more than 100 hours at 290°C. The rubbers, which are essentially copolymers of vinylidene fluoride and hexafluoropropylene, are sold under the name of Viton and are cured by special reagents (Section 4.12) (Du Pont Co., 1960).

1.4.10.2 *Silicone Rubbers*

The organic chemistry of silicon has been studied since an early date, silicone tetrachloride having been prepared by Berzelius as early as 1823. Friedel and Crafts prepared tetramethylsilane, $Si(CH_3)_4$, in 1863, but present-day industrial applications had their origin in the investigations of F. S. Kipping (Kipping *et al.*, 1923), who, as Professor of Chemistry at University College, Nottingham, made extensive researches over a period of some 30 years. It was Kipping who coined the name 'silicone' and applied it to the silicon analogues of the ketones. However, in present-day nomenclature, the term silicone refers to any organo-silicon compound based on a silicon–oxygen backbone with hydrocarbon radicals attached to the silicon atoms. Thus a typical silicone is represented by the formula

$$
R-\underset{\underset{R}{|}}{\overset{\overset{R}{|}}{Si}}-O-\left[\underset{\underset{R}{|}}{\overset{\overset{R}{|}}{Si}}-O-\right]_n\underset{\underset{R}{|}}{\overset{\overset{R}{|}}{Si}}-R
$$

The group inside the brackets is called 'siloxane'. In the case of the poly-dimethylsiloxanes (i.e. where R is a methyl group), n may vary from zero to several thousands. These compounds are amongst the commonest types of silicones.

Kipping (1937) did not foresee any practical value in the compounds which he investigated, but researches were begun at Corning Glass Works in the USA about 1932. The first commercial silicone of sealing compound for aircraft engines was introduced in 1942, and in 1943 Dow Corning Corpn was formed to manufacture and market silicones, Dow Chemical Co. having co-operated in the research work. Other investigations were made by the General Electric Co. of the USA.

Silicone rubbers were introduced to British industry soon after World War II when Dr Shaylor L. Bass (1947) came on a visit from the USA and read papers to a number of societies. For some time silicone rubbers were imported from the USA, but in 1952 Midland Silicones Ltd started manu-facture in South Wales and soon afterwards ICI, in association with GEC, began production in Scotland.

Silicone rubbers are distinguished by their exceptional thermal stability, showing little change in chemical and electrical properties between about $-80°C$ and $250°C$ (Section 4.13).

1.5   Carbon Black

The reinforcement of rubber by carbon black was discovered in January 1904 at Silvertown (London) by S. C. Mote, F. C. Mathews, and others. They were investigating the effects of numerous ingredients on physical properties. Their experiments were controlled in a manner which would do credit even to present-day workers. All the rubber was taken from a specially blended reserve (of fine hard Para); curing times were carefully controlled and re-corded, and properties were measured on special testing machines developed and made at Silvertown. Curing was carried out in a glycerine bath for tensile specimens and in an autoclave for the specimens used to measure hardness and elasticity.

With carbon black it was found impossible to use the standard amount of 1 part to 2 parts rubber, so a special mix was made consisting of 100 of rubber, 10 of sulphur, and 30 of the black. This was the Eclipse brand of Chance & Hunt. Its nature is no longer known to the firm, but subsequent work (Stern, 1945-2) showed it to correspond to HPC and confirmed the amount present—greater than anything used before. The Silvertown workers found a tensile strength of 4175 lbf/in^2 (293 kgf/cm^2) after a cure of 3 hours at 280°F (138°C) and at an age of 10 weeks. Natural ageing was the rule in those days, tests being made after 10, 20, 30, 40, and 80 weeks. The Silvertown tester operated at constant rate of load.

The use of carbon black to obtain high tensile strength and other valuable properties lay dormant until 1910, when further tests confirmed the earlier results. Until then tyre failure had occurred in the fabric before the tread had been completely worn away. With the introduction of the Palmer cord, the fabric began to outlast the tread and the knowledge of carbon black re-inforcement, acquired some six years earlier, was utilised.

About 1912 the US rights for the Palmer cord tyre were acquired by the Diamond Rubber Co., later absorbed by B.F. Goodrich, their negotiator being J. D. Tew. The mixes were included in the deal. They were first seen in the USA by Gammeter, a fertile inventor of rubber machinery, who was quick to recognise the importance of the then novel feature—a high loading of carbon black. The new treads, in which carbon black replaced the zinc oxide previously used, were soon in production.

The discovery of the reinforcing effect of carbon black had a profound effect, and the amount used has shown a consistent increase, not only absolutely but in relation to the total consumption of rubber. This is largely because general purpose synthetic rubbers need carbon black (or another reinforcing filler) to develop adequate physical properties. In its absence, vulcanised SBR has a tensile strength of some 250 lbf/in^2 (17·5 kgf/cm^2) only, less than a tenth of that obtained when carbon black is used. Oil exten-sion has further augmented the amount of carbon black employed.

The first 'dustless' black, sold as 'Micronex beads' (Wiegand and Venuto, 1929, 1930, 1932) was followed by many others (Section 6.1.2.1).

For many years the black was made almost exclusively in the USA by burning natural gas to give channel black, so called from the iron girders, or channels, on which it was formed. When the demand for natural gas increased and the price rose, furnace blacks came into use. At first they did not equal channel blacks in their properties, but subsequent modifications (e.g. Cabot Corpn, 1953) have now enabled them to replace channel black almost completely (Section 6.1.7). Because furnace blacks are made by burning oil, which can readily be transported, factories now exist in many parts of the world, even those remote from oil fields.

1.6 Accelerators

Although the accelerating action of ammonia was observed by Thomas Rowley (1881), the first important advance came when Wo. and Wa. Ostwald

(1908) patented the use of aniline. It was especially valuable with methyl rubber, where it acted not only as an accelerator but as an antioxidant and an 'elasticator'. The Bayer Co. (1911) filed their application for organic bases, and the claims were extended by Hofmann and Gottlob (1914). Ostromislensky (1915) described the use of zinc alkyl xanthates, too fast for many purposes but applied years later to latex. In 1914 Boggs (1919) found that amines such as β-naphthylamine and p-phenylenediamine acted as both accelerators and antioxidants.

Other work was concentrated on aniline, which is extremely toxic, with the object of stopping up, so to speak, the toxic NH_2 group. In this way three accelerators were discovered. Thiocarbanilide was extremely slow, but diphenyl guanidine, patented by M. L. Weiss (1922), was probably the first accelerator to be generally used in Great Britain. Then came mercapto-benzthiazol, the accelerator which has been used in greater quantity than any other and which was patented at about the same time by Bruni and Romani (1920) in Italy, and Sebrell and Bedford (1925) in the USA. A much later patent of The Bayer Co. (1963) is also of interest.

The discovery of the accelerating action of TMTD by Molony (1920) was the work of a lone investigator following the traditions of the early pioneers. He made the accelerator in the kitchen of his house at Wellesely Hills, Mass., USA (Davis, 1951) and was the first to use the low sulphur compounds which became general many years later. It was, however, left to Bruni and Romani to find that TMTD would bring about the vulcanisation without the addition of elemental sulphur (Anon, 1921).

During the 1930s many delayed action accelerators were introduced, and for a time it was thought that further development had virtually ceased. However, with the change from channel to furnace black and the coming of the synthetic general purpose sulphur vulcanisable rubbers, the need arose for a more powerful accelerator with delayed action and this was met by the sulphenamides (Monsanto Co., 1940, 1951; Firestone Tire & Rubber Co., 1942, 1947, 1949; Banks and Wiseman, 1968).

1.7   Antioxidants

Despite the unsaturated nature of the rubber hydrocarbon, the vulcanisates obtained by the earlier workers aged well. This was partly because of the presence of natural antioxidants in rubber, but mainly because of the large amount of free sulphur which remained and acted as an excellent antioxidant. Aniline was not only one of the earliest accelerators but was also an antioxidant. Indeed, it has been stated that its use owes its inception to the observation that its toxic effect was due to prevention of the normal oxidation of blood.

By condensing the poisonous NH_2 group of aniline with an aldehyde, antioxidants of much reduced toxicity, such as butyraldehyde aniline (Calcott and Douglass, 1932) and aldol-α-naphthylamine (Winkelmann and Gray, 1924) (Agerite resin), were obtained and widely used. So too were the phenyl naphthylamines, alpha (Robinson, 1928) and beta (Chesnokov, 1933), but the former was later found liable to cause cancer of the bladder (Parkes, 1969).

Early antioxidants, such as the above, caused staining in light. Recent work has had the object of minimising this staining without reducing the antioxidant effect. This has led to the production of non-staining anti-oxidants (Section 6.5.4) (US Rubber Co., 1954; Imperial Chemical Industries, 1955). Unlike natural rubber (see above), an antioxidant is an essential addition to most synthetic rubbers. A paper by Naunton (1930) covers the history, types of antioxidant, and theories of their action at that time.

1.8   Machinery

1.8.1   MIXERS

Reference has already been made to Hancock's pickle (Section 1.2.1) and his use of two-roll mills. These were the standard machines, not only for masticating rubber but also for adding compounding ingredients, until about 1916. In that year, Fernley H. Banbury (1881–1963), an English engineer who had emigrated to the USA, took out his first patent (Banbury, 1916) for an internal mixer. The invention was opportune because two-roll mills, even in large numbers, were unable to deal with the then unprecedented demand for compounded rubber during World War I. The numbering of the Banbury machines is of interest, being in terms of 60 in (1·5 m) two-roll mills. Thus, the no. 3 had three times the capacity of a 60 in mill, whilst the no. 27 could do the work of no fewer than twenty-seven 60 in mills, a performance at first regarded with scepticism (Killeffer, 1962).

Other internal mixers followed, e.g. the Shaw Intermix, first marketed about 1934.

1.8.2   CALENDERS

The spreading machine was invented by Hancock in 1837, and the calender by Chaffee in the USA at about the same time. Subsequent developments in calenders were largely in the nature of refinements to secure more accurate and rapid working, and these facilitated the development in due course of calenders for plastic materials. In fact, early PVC sheeting was made on rubber calenders—even on one which was very old.

1.8.3   EXTRUDERS

Although the 1846 machine of Werner von Siemens was described as a screw press, in this small hand-operated machine the screw operated a piston or ram. Larger and more powerful ram extruders for gutta were patented by R. A. Brooman (1845) and by Henry Bewley (1845). Bewley's machine operated downwards (Fig. 1.3) and was used by Charles Hancock for the covering of wire. It may be considered as the forerunner of the most modern type of cable-covering extruder operating in conjunction with a continuous vulcanisation process (Fig. 1.4).

These piston or ram extruders have the disadvantage of operating dis-continuously. Continuous screw extruders for rubber are said to have been devised as early as about 1866, but the earliest description, including all the important features of the present-day rubber extruder, was that given in the patent of Matthew Gray (1879). The Gray family for many years owned and

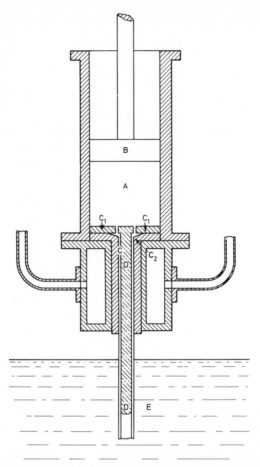

Fig. 1.3   The extruder of Henry Bewley of Dublin, chemist (1845): A is a cylinder containing the gutta percha, B is a piston, and C is a die box heated by steam (through the two pipes shown); C consists of a disc C_1 having a number of holes through which the material is forced into a cup C_2 then round the core D, where it forms a tube that passes into the cold water E below. Bewley's disc C_1 is still a feature of extruders and, further, he says that the gutta percha in the cylinder may be 'in a granular state'

managed their factory at Silvertown. Francis Shaw of Manchester in-dependently made a screw extruder at about the same time as Gray, and a little later, John Royle of Petersen, New Jersey, started to build screw extruders in the USA.

Iddon of Leyland, Lancs., devised a continuous extruder which did not depend upon the action of a screw. It was, in effect, a small vertical two-roll calender. The rubber was fed between the rolls and passed on the opposite side into a cone-shaped tube, the end of which held a die. Extruders of this type were in use until at least about 1930 for the extrusion of treads for aeroplane tyres.

As with other machines, further developments involved detailed improvements, the basic principles remaining the same. Effective cooling enabled fast vulcanising compounds to be extruded without the danger of scorching or precure in the extruder, and the use of special steels reduced wear and increased the accuracy of the sections extruded. For the covering of wires, normal (straight-line) extruders with a hole through the screw were first used

Fig. 1.4   Dual extruder combination, Davy Plastics Machinery (1969): this shows the main units in a vertical continuous extruding and vulcanising system for the simultaneous application of the semiconducting layer and the main insulation in high-voltage cables; the two extruders feed into a common crosshead. Courtesy Davy Plastics Machinery Ltd

but later the now well-known right-angle types (T-head) were developed (Section 10.8.3.2).

In the meantime, the piston or ram extruder was still in use for inflammable substances, such as celluloid and cordite, where the danger of heat generated by a screw could not be tolerated*. The use of the ram extruder for rubber was revived in 1954 by Barwell, who recognised its advantages in those cases where continuous large-scale production was not necessary.

1.9   Tyres

Early tyre development took place mainly in Great Britain. Hancock made rubber tyres, saying: 'These tires are about an inch and a half wide and one-quarter thick. Wheels shod with them make no noise and they greatly relieve concussion on pavements and rough road.' R. W. Thompson (1845), a civil

*When the Nobel Division of ICI started silicone production at Ardeer, use was made of the cordite ram extruders, available there, to strain and refine the silicone rubber stocks. [Ed.]

engineer of Adam Street, Adelphi, London, patented 'a hollow belt ... of caoutchouc ... inflating it with air'. His further description of the pneumatic tyre holds good today for he refers to 'the comparatively small amount of power required ... the absence of all jolting ... nearly all noise, the high speed that may be safely attained'. However, the first inflated tyre to be commercially successful was the invention of J. B. Dunlop (1888, 1889) and was entitled 'An Improvement in Tyres of Wheels for Bicycles, Tricycles, or other Road Cars', the last two words being truly prophetic (Fig. 1.5).

A satisfactory means of fixing the tyre to the rim of the wheel was the main problem in the early days. This, like the means adopted later for changing

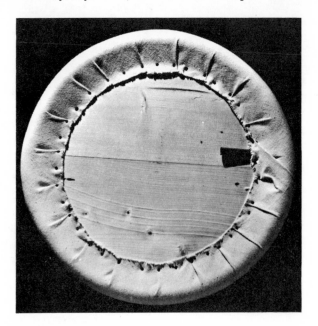

Fig. 1.5   Replica of J. B. Dunlop's first experimental tyre: the tube, filled with air, was covered with canvas and tacked to a wooden disc. Courtesy The Dunlop Co. Ltd

tyres, or wheels, does not form part of rubber technology, but mention must be made of the outstanding patents. The idea of a rigid bead containing a wire was conceived by C. K. Welch (1890), and W. E. Bartlett (1889, 1890) took matters a stage further with his clincher edge where the bead fitted into the wheel base rim—a system which has continued ever since and which later, about 1951, enabled tyres to be used without tubes.

It must be remembered that early developments were concerned only with bicycle tyres.

Changes in the fabric reinforcement gave great improvement in life and reliability. Being a Belfast man, Dunlop used linen, but this was soon replaced for general use by cotton of the normal square-woven type.

The first successful cord-type tyre—a bicycle tyre—was invented in the USA by J. F. Palmer (1893). Following this, the famous Palmer cord tyres for cars were developed at Silvertown by C. H. Gray and T. Sloper (1903). Their cord has been the basis of all subsequent tyre construction. It was used for the first aeroplane tyres in 1910, and the original cord method continued in use for these tyres until as late as 1933, the rubber-impregnated cords being laid by hand around pins on a wooden former.

Cord fabric (i.e. fabric consisting mainly of cord-like warp threads but with a few cross or weft threads to hold the warps together during handling) is now in general use. Cotton is now virtually obsolete, having been replaced mainly by rayon and nylon, and several new synthetic fibres are being tried. Metallic wire reinforcement seems first to have been thought of in the USA (Gautier, 1910; Blackwelder, 1924) but was mainly developed by the Michelin Co. in France.

Until about 1923 vehicle tyres were of comparatively small section and operated at comparatively high pressures. About 1923 balloon low-pressure tyres were first marketed by Firestone, and their use soon became general. Tubeless tyres date from about 1951.

Until recently all tyre cords were arranged at an angle (usually about 45°) to the bead in accordance with the original Palmer–Gray method. About 1939, however, Michelin produced large tyres in which the 'cords' were arranged either at right angles to the bead or parallel to it (Goodyear Tire & Rubber Co., 1961), combining this method of construction with the use of metallic wire. Since about 1964 this arrangement of the cords has been increasingly adopted for passenger car tyres, giving the 'radial' tyres. Modifications resulting in the 'banded' construction have also been introduced, and it is anticipated that, with the general increase in road speeds, radial tyres will be widely used.

At all stages of its history, the tyre industry has been marked by much patent litigation. It is, however, of interest to note that the patenting of balloon tyres, and even an original patent for radial tyres, were not possible because of the early work of C. H. Gray (Gray and Sloper, 1913) and his co-workers at Silvertown. On the other hand, the Palmer patents were subsequently largely upset by the Flexifort fabric of Moseley, although this had never been used in tyres*.

Parallel with the improvements in mechanical construction came improvements in the rubber compounds. The introduction of carbon black as a reinforcing agent gave greatly increased tread wear, and the use of accelerators avoided exposure of the textile to lengthy periods at high temperatures and hence an increased useful life.

Originally tyres were built entirely by hand. Much manual work is still involved, but the amount has been progressively diminished by the invention of tyre-building machines (Section 10.1). Early methods of vulcanisation were carried out in vertical autoclave presses, which are still in use to some extent. The collapsible core was later replaced by the so-called air bags (Johnson and Gammeter, 1916) in individual vulcanisers.

---

*The statement on p.113 of the *History of the Rubber Industry* that Dunlop used Flexifort fabric is entirely incorrect.

TWO

An Outline of Rubber Technology

BY C. M. BLOW

2.1 Polymers

Rubber belongs to the class of substances termed 'polymers': high molecular weight compounds, predominantly organic, consisting of long-chain molecules made up of repeating units usually on a backbone of carbon atoms. These high molecular weight polymers have a lower temperature limit to their rubbery state. At the so-called glass transition temperature T_g, there is a fairly abrupt change to a glassy state. Materials in the class of polymers which are, at normal temperatures, plastics, become rubberlike as the temperature is raised above their T_g.

In their rubbery state, polymers behave in many ways like viscous liquids, because the links in the long chain are freely rotating and enable flow and distortion of the material to occur under stress. Because of the chain length and the presence of side-groups on the chain, their molecular freedom is restricted and they show both time-dependent viscous and elastic properties, and are said to be viscoelastic.

It must be emphasised that, because T_g is often difficult to measure precisely and because its value depends on the method of observation, the T_g of rubbers reported in the literature is subject to considerable variation; this variation is further increased because nominally identical polymers show differences in their chemical make-up and structure sufficient to affect the T_g value (Shen and Eisenberg, 1970).

Mention must be made of the phenomenon of crystallisation, which is more complex in rubbers than in ordinary low molecular weight substances. Crystallisation of rubbers takes place by local rearrangement of portions of molecules to form crystallites. Some portions of the chains will not fit into a regular lattice and remain amorphous; in other words, crystallisation is never complete and the crystallites formed are bound up in the structure and are not, therefore, separable.

These crystallites have a melting point; below this temperature crystallisation takes place, but it is a time-dependent process with a maximum, usually, in the rate of crystallisation–temperature relationship. At temperatures

20

below the melting point, crystallisation can be produced rapidly by stretching the rubber (Fig. 2.1). During elongation, rearrangement and orientation occur to form crystallites—a valuable phenomenon, though not essential for

(a)   (b)

Fig. 2.1   Diagram representing the molecular structure of crystalline rubber: (a) unstretched, and (b) stretched before crystallising; the parallel bundles represent crystallites. From Treloar (1958), courtesy The Clarendon Press

rubberlike properties, that gives natural rubber its high strength in the un-compounded state. As is discussed later, irregularity in the polymer molecular structure may mean that crystallisation cannot occur. Many synthetic rubbers fall into this class, and their strength is low in the pure gum state.

Polymers, in their rubberlike state above T_g, however, are far from perfectly elastic, as mentioned above, owing to the mobility of the molecules of which they are composed.

2.2   Natural Rubber, The Only Polymer for Elastomer Production for a Century

The raw polymer, natural rubber, while at room temperature having considerable strength and appreciable elasticity and resilience, is sensitive to hot and cold and is liable to oxidise to a sticky product. These factors imposed a severe limitation on the usefulness of the material, although the natives of South America made toys, balls, bottles, and footwear direct from the latex, the milky aqueous dispersion of rubber which exudes from the bark of the tree when cut. Limited quantities of these products were sold in North America and Europe in the early part of the nineteenth century. Charles Macintosh in Britain made a double-texture fabric with a rubber interlayer to provide waterproofing. The temperature sensitiveness of the rubber was inconsequential to a large extent in this product. Today, apart from adhesive solutions based on raw rubber, probably the only product made and used in the raw unvulcanised state is the crepe sole for footwear (Section 4.1.2.1).

2.2.1   VULCANISATION

Vulcanisation is the chemical reaction which, as is now known, brings about the formation of crosslinks between the long polymer chains. The three-dimensional structure so produced restricts the free mobility of the molecules and gives a product having reduced tendency to crystallise, improved elasticity, and substantially constant modulus and hardness characteristics over a wide temperature range. In fact, vulcanisation enables useful materials and products to be produced from the rubbery polymer. The initial discovery was the effect of heating the rubber in sulphur, but many improvements were to follow. Inorganic metal oxides, and then organic chemicals increasing in complexity and discovered and developed largely on an empirical basis, were found to accelerate the reaction and process of crosslinking and also to alter and improve the properties of the end product (Sections 1.2.2, 1.6, and 5.1).

The industry owes a debt to both the chemical manufacturers who produced these chemicals and the research workers in the laboratories of NRPRA and many other organisations for working out so thoroughly the mechanisms of vulcanisation and giving the compounder an understanding of the way particular crosslinks are formed and how the properties of the vulcanisate depend on the constitution of the structures produced. The elucidation of the chemistry of vulcanisation has had an important impact on technology over the last two decades. Many properties of the vulcanisate can be substantially modified by variation of the proportion of sulphur and type of accelerator (Section 5.6). Furthermore, production methods are more under control because prevulcanisation can be avoided by correct choice of curing agents.

Although sulphur still remains the basis of the most widely used crosslinking systems, new polymers have required other chemicals, e.g. peroxides, metal oxides, and amines, as described in Chapter 5. Many types of rubber can be cured by high-energy radiation, but very limited, if any, commercial application has as yet been found for this method. Selenium and tellurium can replace and supplement sulphur and have found some use in vulcanising systems for particular purposes. Although outside the scope of this book, ultra-accelerators, too fast for solid-rubber work, find application in latex and in room-temperature vulcanising rubbers made up in solution in solvent and used mainly for making joints of one rubber to another.

2.2.2   FILLER INCORPORATION

There is a limit to the hardness and modulus range that can be achieved by vulcanisation alone. The addition of fillers is almost as old as the industry itself. As soon as Hancock discovered that mechanical working softened natural rubber, a process termed mastication, the mixing of other materials with it was possible; the ability of the raw rubber to absorb quite substantial quantities of mineral fillers in powder form was no doubt soon apparent. This ability dictated the second direction in which the technology was to develop. Initially, with an already strong material, the interest lay in adding fillers to harden and cheapen the product. As time went on, the significance of compounding, as it is termed, to impart other benefits became apparent; these included improved processibility, resistance to abrasion and tearing, and pigmentation.

The developing rubber industry provided, therefore, a market for a wide variety of materials, details of which are given in Chapter 6. The suppliers of these materials, in order the better to sell their products, embarked on their evaluation and improvement. A large part of the compounding technology has been contributed by the supplying industry.

2.2.3 THE IMPORTANCE OF COMPOUNDING

As will be apparent, the first reason for compounding is to incorporate the ingredients and ancillary substances necessary for vulcanisation. The second reason is to adjust the hardness and modulus of the vulcanised product to the values required. Closely associated with these processes is the addition of active fillers which impart high tear strength and high abrasion resistance. By far the most important of these is carbon black, the producers of which now supply a range varying in particle size from 20 nm to 500 nm, in surface, physical, and chemical features, and in tendency to form structure (Sections 1.5 and 6.1). The value of the fine particle blacks in enabling high-strength elastomers to be produced from non-crystallising polymers can hardly be overemphasised (Section 9.4.3). Broadly, the finer the particle size the greater the increase of modulus and strength. The phenomena associated with the addition of these active fillers to rubber are referred to later (Section 2.6.2) and discussed in detail in Chapter 7.

The viscoelastic material, now consisting of polymer and fillers, has to be processed and formed to shape; its rheological properties may need to be adjusted by the addition of oils, esters, waxes, factice, and other substances as discussed in Section 6.3. To prevent or delay degradative failure of the rubber by oxygen, ozone, light, heat, and other influences, additives are available (Section 6.5). To make cellular products, to impart flame resistance, to ensure adhesion to textiles and metals, and to develop electrical conductivity, special compounding techniques are necessary (Sections 6.6 and 9.4). Further introductory discussion of these aspects of rubber technology is to be found later.

2.2.4 THE VERSATILITY OF NATURAL RUBBER

The rubber industry, by which is meant the industry that produces the finished rubber products, really began with the discovery of vulcanisation in 1839. For nearly 100 years its raw polymeric material consisted only of natural rubber. Initially natural rubber came from South America, where the tree that produces it, *Hevea brasiliensis*, grew wild. Not until the first decade of this century, however, did plantation-grown rubber become available in substantial quantities for the western world (Table 1.1). Like most natural products, it was subject to considerable variability, certainly in appearance but also in technical properties.

Nevertheless, during that hundred years a considerable part of the technology of the industry was developed, and it is not easy to think of rubber articles which were not originally made from the natural polymer. The performance and serviceability as regards temperature range and fluid resistance of many products have been changed by the advent of synthetic rubbers, but the technology was already there. It is worth recalling that oil seals for the first automobile hydraulic system were made from natural rubber; the oil sealed was castor oil selected to be suitable for natural rubber.

At one end of the hardness scale of rubber materials there is ebonite, originally essentially a natural rubber product, and at the other the cellular latex foam, which is produced by what can be regarded as a separate section of the industry based on latex technology. A digression will now be made to discuss briefly both ebonite and latex technologies.

2.3   Ebonite

2.3.1   INTRODUCTION

If rubber is heated for lengthy periods of time with larger amounts of sulphur than are used for vulcanisation, the product is, at first, very cheesy and weak—the so-called rotten rubber stage. On further heating, however, more sulphur combines and a hard product is formed, termed variously 'ebonite', 'vulcanite', or 'hard rubber'.

Ebonite has been known and products made from it since the middle of the last century. For many years it was thought that chemically it was $(C_5H_8S)_n$, representing a fully saturated material with 32% combined sulphur. Work has shown, however, that as much as 43% sulphur can be combined with rubber, from which it must be assumed that it is attached in both polysulphide crosslinks and some intramolecular cyclic structures.

2.3.2   COMPOUNDING AND MANUFACTURE

Heating times required to achieve an ebonite vary up to 10 hours at $150°C$; some accelerators, in particular diphenyl guanidine, are effective in reducing the cure time. The reaction is strongly exothermic, and this fact must be taken into account in moulding and forming operations. The volume loss on vulcanisation is high, approximately 6%.

Ebonites can be obtained from BR, SBR, and NBR, in addition to NR. Unlike their behaviour in soft rubber, carbon black and mineral fillers reduce the strength of the material and the only purpose in their incorporation is to minimise softening, or more exactly the deformation under load at elevated temperatures to which reference is made later. Reclaim, factice, bitumen, and especially finely ground ebonite dust are, however, used as fillers to aid processing and to vary final properties. Polychloroprene imparts flexibility to ebonite.

2.3.3   PROPERTIES

While ebonite is resistant to many liquids, including alcohol, glycerol, acetone, and aliphatic hydrocarbons, it is far from immune to swelling by aromatic and chlorinated solvents, in which volume increases of from 50% to 100% are reported. It is particularly resistant to water and shows outstanding stability dimensionally in moist conditions.

Ebonite is commonly reported as having a plastic yield temperature, whereas, in fact, it is a transition from a rigid to a viscoelastic state which occurs; its elastic recovery in the softened state is very time dependent. For ebonites based on natural rubber this transition temperature is 80°C; with polybutadiene and styrene–butadiene ebonites the values are 85°C and 90°–110°C, depending on the polymer and compounding. This transition property limits to some extent the applications of ebonites, some of which have been replaced by plastics.

The most important applications of ebonite are still electrical because of its high resistivity, strength, rigidity, and resistance to impact. Light causes some deterioration of the surface, leading to the development of acidity and a reduction of the electrical insulation properties.

The good stability of ebonite in moist conditions makes it the preferred material for water-meter components, pipe stems, and textile machinery parts. Its chemical resistance dictates its choice for such items as battery boxes, chemical pipe work, and tank and vat lining. Reference to its use as a bonding layer for adhesion of soft rubber to metal is made in Section 8.7.1. Microporous ebonites can be made and are used for battery separators and filter elements.

As will be apparent from the above, a special technology has grown up and, although some rubber manufacturers handle ebonite production, it has tended to become a speciality of a few firms. For this reason, more detailed discussion is considered to be outside the scope of this book.

2.4   Latex Technology

2.4.1   INTRODUCTION

Natural rubber, on which the technology of the industry was built, is obtained by the coagulation of the milky aqueous dispersion of rubber—called latex—produced by the tree *Hevea brasiliensis.*

As methods were discovered of preserving and transporting the latex from the tropical areas, in which the trees grow, to the USA and Europe, a separate technology for the manufacture of products direct from latex developed during the 1920s. The several methods of concentrating the latex, by heat, by creaming, centrifugally or chemically, and later by electrodecantation (Section 1.3), encouraged these developments because less water was transported and because higher solids content increased still further its advantage over solutions of rubber in organic solvents for some manufacturing operations.

This technology, being distinct from dry-rubber technology, is adequately dealt with in separate books and numerous publications in the literature. It is not proposed, therefore, to devote more than a short space to the subject in this book.

2.4.2   COMPOUNDING

The addition of any filler reduces the strength of the rubber obtained from the latex. Nevertheless, for certain applications and because of the high

C

strength of the natural latex rubber, substantial quantities of clay and other cheap fillers, suitably wetted out and dispersed in water, are added. Some resins, however, reinforce (Section 7.1). Compounding with curing agents is necessary so that a vulcanised product can be obtained. Because of the ease of incorporation as aqueous dispersions or solutions and the absence of any risk of prevulcanisation by heat, very rapid accelerators can be employed. In this connection, mention should be made of vulcanised or prevulcanised latex; this is latex in which the particles are vulcanised, and the product obtained from it does not require any heat treatment apart from that to remove moisture.

In addition to antioxidants and pigments and dyes, surface-active agents are incorporated to control stability and impart the necessary processing properties, e.g. fabric wetting and penetration. Gelling and heat-sensitising agents are also added (Sections 2.4.3.2 and 2.4.3.4).

2.4.3   MANUFACTURING TECHNIQUES

The principal manufacturing methods adopted are based on either drying, coagulation, or gelling of the latex; the last two accelerate the separation of water and therefore reduce drying time.

2.4.3.1   *Dipping and Coating*

Dipping a suitable former into the latex repeatedly with intermediate drying will produce a thin-walled article such as a glove. Impregnation and coating of fabrics and fibre bats are carried out by dipping, spreading, or spraying techniques, the rubber film being obtained by drying off the water by heat and/or air currents. Bonded hair products for upholstery and packaging purposes come into this category. Carpet backing, where rubber from latex is used to anchor the pile and simplify the weaving structure, at the same time providing a non-slip finish, is one of the larger uses of latex (Meyrick, Vautier, and Lombardi, 1969).

2.4.3.2   *Coagulation*

Thicker dipped deposits can be achieved if the former is first immersed in a solution of a coagulant, such as acetic acid or calcium chloride. Dipping the latex-coated former into a coagulant solution is also advantageous in preventing the flow of the latex and speeding the drying process.

Coagulation is used in the extrusion process for producing rubber thread. The latex is forced through fine jets into a bath of coagulant, the thread of coagulum being strong enough to be drawn off and dried.

2.4.3.3   *Slush Moulding and Rotational Casting*

Latex film can be deposited on the inside of a mould to produce hollow articles by the well-known slush moulding and rotational casting techniques. Several alternative methods are available. Moulds can be of metal or plaster. If the mould is of metal, coagulant can be applied to the surface prior to the latex, or latex containing a heat-sensitising agent can be used with a heated mould. Heat-sensitising agents have an inverse solubility–temperature coefficient and so cause the latex to gel in contact with a heated surface. Either the coagulant or the gelling agent leads to a thicker deposit of rubber on the wall of the mould.

Plaster moulds can be used with normal latex, the plaster absorbing some of the water and filtering out a film of rubber on its surface. In all cases, the technique is either to fill the mould completely and allow it to stand for coagulation, heat gelling, or water absorption to occur, the excess latex being then poured out for re-use; or a volume of latex containing the amount of rubber to form the required thickness of deposit is poured in, and the mould is rotated to spread it evenly until all is deposited or gelled.

2.4.3.4   *Latex Foam Rubber*

The largest use of latex is for cellular products. The basic latex foam rubber process (Dunlop Rubber Co., 1929) consists of three stages. The first is the foaming of the latex containing a surface-active agent by whipping it and/or blowing air through it; considerable research has gone into the technique of achieving a fine-textured uniform froth. The second stage is the gelling of the foam by adding a delayed-action gelling agent, sodium silicofluoride. Alternative gelling systems, including heat-sensitive latex as mentioned above, have been tried but are not favoured. The third stage is the filling of the moulds, usually of light aluminium construction, which are heated to bring about vulcanisation of the rubber. After cooling, the product is removed from the mould, washed to dispose of undesirable soluble materials, and dried.

A recent development is the adoption of the Talalay process (1946). Partially foamed latex is poured into a mould which is sealed and evacuated so that the foam expands to fill the mould completely. The next stage is to cool the mould to $-35°C$ in order to freeze the foam. Carbon dioxide is admitted; this penetrates the structure and, owing to the pH change, causes gelling. The final stage is heating of the mould to vulcanising temperature to complete the cure. In spite of the high capital cost of the equipment needed to produce the low and high temperatures and the vacuum, this process is currently used because of the excellent quality of the product and the low rejection rate.

2.4.4   SYNTHETIC LATICES

The majority of the processes described above were developed in the 1920s and 1930s using the available natural rubber latex. Since World War II, most of the widely used types of synthetic rubber have become available in latex form. Styrene–butadiene copolymer latices, with styrene contents in the range 25–50% and total solids contents of 40–65%, are supplied by a number of producers. Chloroprene and nitrile rubber latices are available with solids contents up to 50%. These have supplemented and replaced the natural product in a number of applications. The CR and NBR have oil and solvent resistance, which is taken advantage of in gloves and bonded fabrics designed to withstand dry-cleaning. The fire-retardant property of CR is also exploited.

2.5   Thirty-Five Years of Synthetic Rubber

2.5.1   PRE-1940 RUBBERS

So far little mention has been made of any raw polymer other than natural rubber, because for so long it was the only polymer available for elastomer production and much of the technology referred to above was developed for this polymer.

During the past 35 years, however, the technology has become more complex because of the availability of a variety of synthetic rubbers. In the early 1930s, two materials were offered commercially as synthetic rubbers: polychloroprene by the Du Pont Co. under the name 'Duprene', later changed to Neoprene; and polysulphide rubber, Thiokol in the USA and Ethanite of Belgian origin. From these two types of raw polymer, rubberlike products were obtainable having resistance to oils and solvents not achievable with natural rubber. These synthetic rubbers established usages and were not at that time in competition with natural rubber because they met applications which the latter could not meet.

Shortly afterwards these two types of synthetic rubber were followed by a third—copolymers of butadiene and acrylonitrile—also with the feature of oil resistance.

2.5.2   GENERAL PURPOSE SYNTHETIC RUBBERS

Prior to World War II, developments were being actively pursued in Germany in the production of a polymer as a replacement for the natural product, i.e. for general purpose application. Through commercial contacts between German and American manufacturers, much detail of these materials and their manufacture was known in the USA. Because of these circumstances, when Japan occupied Malaya in 1942 and the need arose for the Allies to have a raw material to make up for the deficiency of natural rubber supplies, large-scale manufacture of the styrene–butadiene polymers in the USA was possible.

The introduction of these general purpose rubbers was not a commercial enterprise but a wartime necessity, and represented a considerable achievement by both the chemical industry of the USA in producing it and the manufacturing industries in that country and Great Britain in mastering the technology to handle it and produce acceptable vulcanised products. Close co-operation in this project was established between producers and suppliers of ingredients and compounders in the rubber industries, as well as university and government laboratories.

Not only had adequate properties to be developed in the vulcanised product, but also it was expected that the raw polymer would be capable of being handled by the same basic processing technology as natural rubber. It is of interest and significance that the four types of rubber available during the immediate post-war period all had this feature. In detail, however, there were differences, and occasion will arise later to refer to this fact. The raw rubbers were millable and mixable on rubber machinery, and reinforcing fillers, softeners, curing ingredients, and antidegradants conferred on them the required properties both for subsequent processing operations and in the vulcanisate.

The term 'rubber' is used to convey several meanings, and it is unfortunate that the terminology of the industry regarding this one word was not clarified many decades ago. Rubber, apart from a turkish-bath attendant, means the raw material produced by the tree, a product for erasing pencil marks from paper, and the generic term for products from rubber; in other words, it covers both the raw unvulcanised material and the finished product. Designers specify their material requirement as 'rubber'. No word has ever been coined for synthetic rubber, but some trade names for particular or general classes of synthetic rubber have become part of the language: Neoprene, Hycar, Perbunan, Thiokol, Viton, for example. For others, the code letters or shorthand names have become fairly widely adopted: SBR, nitrile, EP, urethane are examples.

There is a great deal to be said for restricting the use of the term 'rubber' to the raw polymer and to refer to natural rubber, chloroprene rubber, polybutadiene rubber, styrene–butadiene rubber, and so on, adopting for brevity the agreed codes for these rubbers as listed in Tables 9.1 and 9.2. The term 'elastomer' would then be used for the vulcanised product. As has already been pointed out, the raw polymers are generally speaking useless, varying from soft plastic materials to tough gristly substances. They do not, as a rule, develop high elasticity and stability to temperature changes until they are vulcanised.

The multiplicity of raw polymers for elastomer production only emphasises the necessity for greater clarity and precision in discussing and specifying elastomer properties. The complexity of the situation is eased somewhat by the division into general purpose and special purpose categories. Although this division is not clear-cut and opinions differ as to where to draw the line, in this section and in the chapters that follow general purpose rubbers are those which are produced in large quantities and supplement and replace natural rubber with which they are comparable in their non-oil-resistant properties. Special purpose rubbers are those produced in much smaller quantities and having a different degree of oil and solvent resistance and/or heat resistance from those in the general purpose class.

The end of World War II left the US chemical and rubber industries with a large productive capacity for GR-S (government rubber-styrene), the styrene–butadiene copolymer rubber, and the prospect of a return of supplies of natural rubber. Competition between the two inevitably developed, and the balance sheet presented a variety of factors on each side. Natural rubber has high resilience and elasticity, easy processing, tack, and high strength without the addition of black; also, the industry has long experience of its handling qualities. But its presentation was as a commodity at a price subject to wide market fluctuations; it was often contaminated with foreign matter and shipped in 100 kg cubes which arrived at the factory in a very distorted condition, difficult to handle and stack. Variability in its technical properties, viscosity, rate of vulcanisation, and oxidation resistance, was well known.

The synthetic competitor GR-S, later to be renamed SBR, is supplied to a technical specification in well-shaped and compact bales of 32 kg, easily stacked and handled and at a stable price. Its viscosity is controlled and its milling properties are more uniform than, if not as satisfactory as, those of natural rubber. It lacks natural tack. The vulcanised product has low strength unless reinforced with carbon black, and the resilience of all mixes is considerably lower than that of natural rubber mixes.

As a study of this balance sheet will show, there was a place for each rubber and, indeed, the availability of the synthetic polymers was necessary to meet the substantially increased demand which natural rubber suppliers could not have met.

Supported by the vigorous research organisation NRPRA formed in 1938, natural rubber producers have met the challenge presented by the competition of synthetic rubber; in Section 4.1 the considerable changes and achievements in the technical marketing of natural rubber are fully described. Not only is classification into grades, meeting tight specifications for, in particular, contamination and oxidation resistance, proceeding very rapidly, but minor modifications to methods of production have resulted in special type rubbers with stable viscosity, easy processing characteristics, and so on.

Sections 4.2–4.7 deal fully with the general purpose and non-oil-resistant synthetic rubbers. In the immediate post-war period, important and significant changes were made in the polymerisation recipes for the styrene–butadiene rubbers, which improved both the processing and vulcanisate properties; at the same time, the variety of grades within this one type increased.

Before referring to the other important developments, mention should be made at this point of butyl rubber, the copolymer of isobutylene and a small percentage of isoprene (Section 4.5). This rubber became available in the early 1940s and found immediate application as an inner-tube rubber because of its special property of low permeability to gases. At that time, it tended to be classed as a special purpose rubber. With the change to tubeless tyres, its main outlet was substantially reduced and, in some ways, it can now be considered a general purpose rubber. Its good resistance to ozone, oxidation, and chemicals has led to its adoption for a number of applications as a replacement for natural rubber. Butyl rubber is now used for engine mounts, cable sheathing, tank and reservoir lining, and for airbags and similar components in tyre manufacture.

The discovery of the new stereospecific catalyst systems, associated with the names of Ziegler, Natta, and others, by which polymers of more regular formation similar to the natural product are produced, led several of the USA synthetic-rubber manufacturers, between 1955 and 1956, to embark on the production of polybutadiene (BR) and polyisoprene (IR) by solution processes. These two rubbers have made a considerable impact on the general purpose rubber market and have established themselves in tyre compounds (Sections 4.3 and 4.4).

About 90% of the BR production is used in tyres in admixture with SBR, in particular oil-extended SBR, because BR is capable of accepting large amounts of oil and black. The poor road-holding quality of BR restricts its use unblended, but it confers improved wear resistance and groove cracking resistance. A small usage of BR is found in conveyor belts, footwear, and shock absorbers.

IR also has become established in tyres, but its non-tyre use is increasing for footwear and for the production of chemical derivatives (Bonner, 1969).

The ethylene–propylene copolymers and terpolymers, discovered in 1956 and made using the Ziegler-type catalysts, are on the borderline between general and special purpose rubbers. Their production at present is small, and the copolymer is certainly only finding speciality uses because, having no unsaturation, it requires peroxide vulcanisation. The fact appears that the rubber industry is geared to sulphur vulcanisation and that section of it using the general purpose rubbers is reluctant to accept rubbers not capable of this type of cure. The future of the EP polymers lies then in the terpolymers, in which a third monomer provides unsaturation for sulphur crosslinking and the possibility of blending and covulcanising with NR, SBR, BR, and IR. The cost situation is complex (Section 4.6.3). The terpolymers seem likely to take some of the CR market for extrusions.

The newest materials, the so-called thermoplastic rubbers or thermo-elastomers, are on the boundary between thermoplastics and rubbers; they are not crosslinked but are rubberlike and elastic over a relatively narrow temperature range (Section 4.7). Brydson (1969, 1970) has drawn attention to the several ways in which materials display elastomeric properties over a restricted temperature range, without having chemical crosslinks of the type introduced by normal vulcanisation methods. The unvulcanised crepe sole rubber owes its elastic properties to entanglements of high molecular weight material; plasticised PVC and polyurethanes owe the properties to hydrogen bonding, either directly between polymer molecules or via plasticiser and filler. The thermoplastic rubbers are block copolymers in which domains of a polymer of high T_g are linked by rubbery polymer molecules of low T_g. The only available thermoplastic rubber to date is the block copolymer of styrene, as the reinforcing high T_g polymer, and butadiene, the flexible rubbery low T_g polymer. At the T_g of styrene, the material becomes plastic and its elastomeric properties disappear to a large extent.

It is very speculative to forecast the future of these materials. Other 'rubbers' of this type are likely to be produced for either general or special purposes; e.g. boron crosslinked natural rubber and ethylene–vinyl acetate copolymers are under development as thermoplastic rubbers.

2.5.3 SPECIAL PURPOSE RUBBERS

The chloroprene and acrylonitrile–butadiene rubbers remain the workhorses in the field of special purpose rubbers, because of their cost and their oil resistance. Both types are now produced in a range of grades (Sections 4.8 and 4.10). The market for chloroprene rubbers has been much widened by the exploitation of their excellent resistance to ozone and weather, and by their use in fire-hazardous situations, cable sheathing, and conveyor belting for mines. The largest outlets for nitrile rubbers are in the engineering industries for oil seals, O-rings, gaskets, and fuel and oil hose.

The last 25 years have seen the development and establishment of chloro-sulphonated polyethylene rubbers (Section 4.9), which find application where solvent, chemical, ozone, and weathering resistance are required, polyacrylic rubbers (Section 4.11), and fluorocarbon rubbers (Section 4.12). The latter two, with inferior low-temperature properties to the nitrile rubbers but superior oil and temperature resistance, represent improvements which have been very acceptable in the aircraft and automobile industries for applications where the requirements have become more severe. Poly-acrylic rubbers are particularly valuable for resisting sulphur-containing lubricants and greases. The high price of fluorocarbon rubbers is a limitation on their use, and the same can be said of silicone rubbers, which are unique in their wide serviceable temperature range.

Whereas all the rubbers so far referred to have followed the pattern of being offered to the manufacturing industry as raw polymers to be com-pounded, processed, and vulcanised, the silicones are a special case. A certain amount of special technology was required to produce satisfactory finished products, and this was developed largely by the producers of the polymers. Other reasons for the course of this development are probably the high price of the polymer and the consequent smaller market which did not attract the compounding ingredient suppliers to direct their attention to the technology and requirements of these polymers (Section 4.13).

Polyurethanes and their properties are only briefly described (Section 4.14) because two comprehensive books on this subject have been published recently. Other speciality rubbers are dealt with in Sections 4.15–4.18; many of these are in the early stages of development, and in some cases better vulcanising systems are required before they can become widely used.

By way of summary, therefore, it is apparent that the rubber compounder has now a wide spectrum of polymers from which to produce a range of elastomers, each capable of meeting one or more of the requirements ex-pected of the products today: abrasion resistance, tear strength, low-temperature flexibility, high-temperature resistance, fluid resistance, and high elasticity.

2.6   Compounding Technology

2.6.1   GENERAL PRINCIPLES

The reasons for and the broad principles of compounding have already been referred to (Section 2.2.3) and are fully dealt with in Chapter 9. In-evitably, in discussion of these principles, reference will have to be restricted to the most widely used general purpose and special purpose rubbers.

The wide range of polymer types and grades and the almost infinite variety of compounding ingredients can be bewildering to the technologist new to the industry. The general principles discussed in Chapter 9 must be looked upon as a guide. Supporting literature is that issued by the producers of the polymers and compounding ingredients, and the evaluations reported by research associations and other organisations. The value of these technical data can hardly be overstressed as a source to direct the course of develop-ment of new mixes.

Equally it can be said that the general principles apply to all polymers, with the exception of the vulcanising systems, which vary with the polymer (Sections 5.2–5.4). A further exception concerns silicone rubber (Section 4.13).

It must be emphasised, further, that in the formulation of a mix not only must the final purpose and the properties that the application demands be considered but also the method and process of manufacture; in addition, the cost must be kept low consistent with processing and product performance.

2.6.2 REINFORCEMENT

Certain fine particle carbon blacks act as 'reinforcing agents', while others of larger particle size are valuable fillers for the adjustment of hardness and processing characteristics; other fine particle fillers, notably silica, also reinforce under certain circumstances. The term 'reinforcing' is ill defined, unfortunately, in the rubber industry; Dr Boonstra gives his definition (Section 7.2); the technologist recognises reinforcement when he encounters it.

By way of introduction to the subject of particulate reinforcement, it may not be out of place to highlight the main phenomena and observations. There are a bewilderingly large number of papers, reviews, and other literature on the subject dating back 30 or 40 years.

2.6.2.1 *Reinforcement Phenomena in Unvulcanised Systems*

For want of a better description it is helpful to refer to the 'interaction' of fine particle carbon black with rubber, an interaction that manifests itself, in the unvulcanised mixture of the two materials, as an increase of viscosity, both 'dry' and in solution in, for example, toluene, greater than that postulated by the simple Einstein equation (Section 7.8). A freshly mixed piece of well-masticated natural rubber and channel black dissolves and disperses readily; on standing, however, the mix becomes more difficult to dissolve, and the resultant solutions have higher viscosity than those prepared from the freshly mixed batches (Blow, 1929). This phenomenon shows itself at quite low proportions, less than 15%, of black in rubber. As the proportion of black is increased and the time of standing after mixing is prolonged, there comes a point when the rubber has hardened considerably and solution without severe mechanical working becomes impossible. The mixture behaves as if it were vulcanised and in a solvent swells to a jelly without loss of shape; the supernatant liquid remains clear, but after some time is found to contain rubber. An apparent equilibrium is established, and a certain proportion only of the rubber is found to have dissolved. The remainder is held by the mix and is termed 'bound rubber' (Section 7.8) (Twiss, 1925; Stamberger, 1929).

Recently, Southwart and Hunt (1968, 1969, 1970) have reported similar phenomena in the case of pyrogenic silica and silicone rubber gum, in which systems the interaction of filler and polymer appears even stronger than that observed between carbon black and organic rubbers.

Similar observation of these phenomena can be made with the Mooney viscometer (Section 11.2.1.1). The mechanical shearing of an aged black–rubber mix in the instrument lowers its viscosity value, and some recovery towards the higher figure can be observed on standing, suggesting a thixotropic structure (Mullins and Whorlow, 1951).

2.6.2.2   *Reinforcement in Vulcanisates*

There is an inclination to associate the phenomena observed in the un-vulcanised mix with the higher modulus, tensile strength, and abrasion resistance of vulcanisates containing these reinforcing particulate fillers. The complexity of abrasion resistance is considered later (Section 2.9), and it is probably true to say that the improved abrasion resistance of carbon black filled rubber, as manifested in longer wearing tyre tread rubber, arises from a combination or interaction of several other properties imparted by black.

The phenomena that are relevant and on which many papers have been written can be placed in three groups. First, at strains of 10% or less, often below 1%, the modulus has been studied and the elastomer is found to be truly elastic; at least, over a period of 3–30 min, no creep occurs and the original length is recovered on removal of the stress. Secondly, at higher strains, the material is not elastic and the original length is not recovered on removal of the stress, i.e. there is considerable hystersis. Furthermore, stress softening occurs, i.e. the second and subsequent stress–strain cycles are very different from the first, showing a reducing modulus but tending to approach a limit. On standing or after certain treatments, a return towards the original stress–strain curve takes place. Thirdly, there is the frequency effect in the determination of modulus at very low strains. The dynamic modulus may well be two or three times that of the static modulus.

The wide departure from the Einstein relationship as the particle size of the filler is decreased suggests that, because particle size is having an effect, the equation is not applicable to such systems. The equation has been corrected to take account of particle aggregation into rod-like structures and of particle–particle interaction (Section 7.8). Another approach is to include in the volume fraction of filler the bound rubber from estimates of the value on unvulcanised mixes. It is, however, difficult to imagine two particles, one of solid filler and the other partly of filler and partly of rubber, as an outer covering, behaving in the same way when the system is strained. It may be that the effect of particle size is in reducing interparticle distance (from the surface of one particle to the surface of its nearest neighbour) and so increasing the likelihood of bridging this distance by one or more rubber molecules. Is the most operative factor filler–filler cohesion or rubber–filler adhesion? The bound rubber phenomenon suggests the latter but, on the other hand, if filler–filler structure develops a network of high rigidity and strength, the rubber molecules will have reduced mobility and be unable to diffuse out.

Perhaps too little attention has been given to the non-linearity of the stress–strain curve as a source of explanation and to the possibility that many structures are present each of which operates in a different region of the stress–strain–time relationship.

The practical facts are that fine particle carbon blacks enable highly abrasion-resistant rubbers to be achieved, even from non-crystallising low-strength rubbers. Black provides a ready means of increasing modulus and hardness, but at the expense of elastic properties. For many years, attention has been paid to resin and organic reinforcing agents, but these either impart some thermoplasticity to the system or lower the elastic properties. Work is going ahead on this subject and it is possible that some form of block copolymer may form the basis of an all-organic rubber of high elasticity and capable of production in a range of hardnesses, all with acceptable temperature stability of modulus and resilience.

2.6.3 OTHER COMPOUNDING INGREDIENTS

The variety of other fillers is amply covered in Section 6.2, and the plasticisers, softeners, and other processing aids are dealt with in Section 6.3. It is one of the characteristics of rubber that it will absorb not only quantities of particulate fillers but also oils, waxes, fatty acids, and other organic substances which are, in practice, added for several purposes. Stearic acid or other fatty acid or soap, e.g. zinc laurate, activates the vulcanisation system, aids dispersion of carbon black and other fillers, and improves the flow behaviour in processing. Mineral oils may be required to cheapen the compound, in association with a high filler content in order to give the correct vulcanisate hardness, as well as to aid processing. General purpose non-oil-resistant polymers will absorb substantial amounts of oil, but selection of the type must be made to achieve the best results (Section 6.3.2).

Oil-resistant polymers, by their very nature, will not absorb much mineral oil, but improvement in their processing is often required and this is achieved by incorporation of organic esters of high boiling point and low volatility— typically adipates, phthalates, and sebacates of octyl and higher alcohols, as well as phosphates. These same liquids, having low freeze or melting points, are valuable additions to make these oil-resistant elastomers, which stiffen at temperatures unacceptably high for some purposes, flexible at lower temperatures.

Waxes, like other products referred to in this group, serve the dual purpose of improving processing, particularly extrusion, and of imparting to the final product improved water resistance and electrical properties, and, if blooming to the surface, resistance to light and ozone.

2.6.4 RE-USE OF WASTE

Reclaiming of waste rubber in the form of spew from moulding, faulty products, and waste rubber articles, has been practised for a very long time (Section 6.4). Whereas its incorporation is most often with the object of reducing cost and results in some inevitable deterioration of 'quality' if proportions in excess of about 10% are added, the importance of reclaimed rubber in improving processing and handling must not be ignored. Mention should be made also of the use of finely ground scrap rubber, both as a cheapener and, in small amounts, to improve, in particular, the moulding behaviour of mixes that tend to be excessively soft in the uncured form.

2.6.5 ANTIDEGRADANTS

Attention must now be turned to the special additives used to achieve specific properties, structures, or effects. First among these are the chemicals which improve the resistance of elastomers to oxidation by delaying its onset: antioxidants and antiozonants.

Familiar to all is the liability of rubber to 'perish', to harden and crack or soften to a sticky residue. Natural rubber is particularly sensitive in this respect, and much successful work was carried out from the early days in the selection of materials to improve it. There is available today a wide range of these chemical additives, which, in many cases, are specific in their action in delaying deterioration by a particular influence, such as oxygen, ozone, mechanical flexing, light, poisons, or catalysts of the oxidation reaction e.g. copper and manganese (Section 6.5).

The oxidation of rubbers is important technologically, but the understanding of its mechanism has not made as much impact on technology as that made by the understanding of the vulcanisation reactions.

2.6.6 MISCELLANEOUS ADDITIVES

Finally there is a group of miscellaneous materials (Section 6.7): blowing agents for producing cellular products, flame retardants, antistatic agents, and stiffening agents (for unvulcanised rubber).

Fibrous fillers, such as flock, asbestos, and woodflour, are on occasion added to harden, toughen, and cheapen the mix. Several commercially successful proprietary products based on cork and rubber are offered for gasketing and similar purposes.

Whilst the majority of products are black because of the necessity to use black as a filler, the compounding of white rubbers and, by the use of pigments and dyes (Section 6.2), of attractive coloured rubbers must not be overlooked.

2.7 The Importance of Textile Reinforcement

Carbon black and other particulate fillers play a vital role in producing elastomers with properties to meet the many and varied applications. Equally important to the performance of many products made from these elastomeric materials is the incorporation of textile or, more generally, fibrous reinforcement. Tyres, belting, hose, footwear, flexible containers, dinghies, and many other everyday domestic and industrial products do not require and, indeed, cannot function if the full extensibility of several hundred percent, of which elastomers are capable, is present. Plies of low-extensible and high-modulus textile or metallic cords and fabrics restrain the stretch of rubber and yet the essential flexibility is not lost. A section on textiles in Chapter 6 deals with the types used and the requirements to be met for various articles. Whilst textile technologists are employed in the rubber industry and the producers of fibres and textile manufacturers provide technical data on these materials, it is desirable that the rubber technologist should have a working knowledge both of the terminology of the textile industry and of the basic properties, behaviour, and processing of these materials (Chapman, 1968).

The change from staple cotton to continuous filament man-made fibres, such as rayon, nylon, and polyester, gave rise to an adhesion problem. The preparation and use of adhesives is now part of the technology of the rubber industry (Section 8.5.2).

Extensibility can be varied by choice of fibre, yarn twist, and fabric construction, and in the laying up of the fabric the direction of maximum stretch can be controlled. There are products which require that, under stress, some stretch occurs in the textile reinforcement; others must show a minimum stretch, at least in certain directions; the belts of the belted bias and radial ply tyres, diaphragms, and hose are examples. The most inextensible fibre and the most stable to temperature changes is glass fibre, and there is evidence of an increasing use of this in a variety of products as the problems of adhesion, brittleness, and interfibre chafing during flexing in service are overcome (Clayton and Kolek, 1965).

2.8 Manufacturing Processes

The production sequence in the rubber manufacturing industry can be fairly sharply and clearly defined into three stages: mixing, forming, and vulcanising.

2.8.1 MIXING

The polymers, fillers, processing aids, vulcanising ingredients, and other additives, which the compounder has decided on, have to be mixed together. The two basic machines for this process are the two-roll mill and the internal mixer (Section 8.2.1); the latter is widely used, particularly for the larger quantity requirements, but, in many ways, it only distributes the ingredients in the polymer and the important dispersion of the fine particle fillers is often of necessity completed by repeated milling on two-roll mills or by passage through one or other of the newly developed continuous mixing machines (Section 8.2.7).

Prior to mixing natural rubber, there is the mastication step, previously referred to. The reduction of viscosity and increase of plasticity of natural rubber and some synthetic rubbers, brought about by mechanical milling and working, is a striking phenomenon, which brings with it both advantages and disadvantages. The advantages lie in the ability to reduce the molecular weight and viscosity so as to render the material more easily processable;

the disadvantage is that processing involving mechanical working of the mix reduces molecular weight and hence the modulus of the vulcanisate.

The mixing of the ingredients into rubber often involves a compromise. High viscosity promotes high shear, which is required to break up filler aggregates; low viscosity assists wetting of the particles, which is essential to achieve uniform modulus and other properties.

To avoid weighing small quantities of powders, to assist dispersion, and generally to aid production flow, masterbatching is widely practised; sulphur, zinc oxide, accelerators, and antidegradants are mixed in separate batches, in proportions of from 1 : 1 to 1 : 10 with the appropriate rubber. Carbon black masterbatching is the general procedure for tyre tread compounds (Section 8.2.5).

It has been said that the quality of the final product is made in the mill room or mixing department because it is there that a uniform and high level of dispersion and consistent rheological properties must be produced in the batches of mixed compounds.

2.8.2 FORMING

The second production stage is the forming of the material into the shape required. The variety of methods is wide, but basically there are only four: spreading on fabric from solution, extrusion through a die, calendering between rolls or bowls, and moulding. Many products require all four. At one extreme, only one of these basic techniques is needed: spreading will produce waterproof sheetings; extrusion will produce a tube, cord, or profiled shape; calendering a sheet or rubber-coated fabric; moulding a finished article or component from a rough piece of raw mixed stock. At the other extreme, might be instanced a large radial ply pneumatic tyre: into its formation go extrusions for the beads, sidewalls, and tread; calendered sheet for the various filling pieces; textile cord and other fabrics, adhesive treated and rubber coated on the calender; and brass-plated wire coiled and embedded in ebonite. From these, partly by hand-building and partly by automatic assembly, a composite is produced which requires final shaping during a moulding stage to produce the finished tyre.

Processing the viscoelastic raw stock through extruders and calenders introduces structures and strains which must be adequately relaxed immediately after processing or during the initial stages of vulcanisation if gross dimensional variation and poor surface appearance are to be avoided.

Associated with the basic forming methods are ancillary processes, such as the coating of fabric with adhesive and rubber from solution to ensure good adhesion. Another important ancillary technique is the preparation of metal for bonding to rubber to produce such products as engineering components, mountings, diaphragms, and seals.

Although not crosslinked, the raw polymers are viscoelastic, and the elastic component shows itself in what is commonly termed 'nerve' (Section 9.3.2) or recovery from deformation when the stress is removed. Much is known about, and there are practical methods for studying, the rheological properties of mixes (Sections 3.2 and 11.2). The compounder in adjusting his mixes for satisfactory processing is assisted by an understanding and appreciation of the phenomena.

In emphasising the importance of the compounder's art and science, the skill of the machine and equipment designers and of the operatives must not be minimised. Whereas an understanding of rubber technology has led to improved machine design, the skill of the experienced calender and extruder operatives and moulders is considerable in ensuring that the products obtained are of satisfactory dimensional consistency, appearance, and performance, often from material varying in rheological properties from batch to batch.

2.8.3   VULCANISING

Overriding and predominant in the technology of the industry, particularly with the emphasis on higher production rates and new techniques, is the control of the chemical reactions leading to crosslinking, to bonding to metal and fabrics, and to polymer breakdown.

The final step in the manufacturing sequence is the vulcanisation or curing of the formed product. Basically, the process is that of applying heat at a certain temperature for a certain time. Of the many methods (Section 8.9), vulcanisation is most often combined with the forming method of moulding.

For many years, simple compression moulding was universal: a blank of the mix was placed in the cavity of the mould, and the two or more pieces of the mould were closed together under pressure between heated platens. Of recent years, several alternative methods which may be grouped under the terms 'transfer' and 'injection' moulding, have occupied the efforts of polymer producers, equipment manufacturers, and the rubber manufacturers. It is the necessity to vulcanise rubber and the consequent reactivity of the material that distinguish the forming of thermoplastics by injection moulding from the forming and curing of elastomers. One can discern the increasing complexity of the operation and the greater 'strain' on the rubber prevulcanisation tendency in passing from simple transfer, where a blank is transferred from one part of the mould to the cavity, to that of forcing the rubber into the cavity from a screw or screw–ram extruder.

It is as well to consider the aims in going to injection moulding: to eliminate blank preparation, to pre-heat the rubber to near vulcanising temperature so that it arrives in the cavity at this temperature, to reduce flash and spew to a near-flashless moulding, and to shorten cure time and thereby increase mould productivity. The simplest transfer process achieves simplification of blank preparation and to some extent warms the blank. A single-shot ram injection shortens the cure time and reduces the flash in addition. Extruder filling reduces the risk of scorch by keeping the extruder working and the material moving.

Whereas sheet rubber is likely to remain for some time as the form in which unvulcanised rubber is fed to the forming process, pellets or powders have long been considered as an alternative. For many years, natural rubber powders have been available (Section 4.1.4.3), but, because they contained large amounts of antiagglomerating substances or were lightly vulcanised, they have been little used in rubber manufacture. Dicing and pelletting of mixed compounds is practised and appears a useful technique, particularly for feeding extruders and injection moulding machines; silicone rubber compounds are offered in this form by the polymer manufacturer. On the other hand, greater storage space is required and care must be taken that massing does not occur in storage and handling. Calenders and transfer and compression moulds are more readily fed by solid strip or blank.

The other alternative is rubber in liquid form, and the production and use of a fluid rubber also goes back many years (Stevens, 1931). Suitable mixes of natural rubber can be dissolved in mineral oil to give pourable liquids that can be cast into moulds and vulcanised at moderate temperatures by the use of ultra-accelerators. The vulcanisate is soft but of reasonable strength, and there were several applications for the process, including printers' rollers. Further improvements in this type of product can be brought about by the use of a depolymerised natural rubber.

More recently, liquid polymers of nitrile, fluorocarbon, and butyl (DelGatto, 1969) rubbers have been offered, and for some years production of some of the urethane rubbers has been by a liquid casting technique. Liquid polymers of butadiene and butadiene–acrylonitrile with hydroxyl-, carboxyl-, or mercaptan-terminated chains have been reported; the steps of making the short chains into long molecules and then crosslinking them to give a solid elastomer are brought about after casting into the mould (Drake and McGarthy, 1968; Duck, 1968; Moore, Kuncl, and Gower, 1969). The attraction of these materials lies in avoiding the necessity for heavy mixing machinery and moulds, but the entrainment of air bubbles often necessitates vacuum-degassing and mould-filling techniques.

2.9   The Properties of Polymers, Mixes, and Vulcanisates

2.9.1   POLYMERS

As already discussed, certain vulcanisate properties, notably low- and high-temperature performance, volume change in liquids, and chemical resistance are to a large extent inherent in the polymer; Chapter 4 gives considerable detail regarding the available polymers and their characteristic properties, and there is more to be found in the technical literature that issues in profusion from the suppliers.

Many of the effects and defects experienced in processing unvulcanised rubber mixes are likewise attributable to the base polymer; to ease matters for the manufacturer, grades of synthetic rubbers are offered with varying viscosity, nerve, and other rheological properties.

2.9.2   UNVULCANISED MIXES

Polymers, fillers, the vulcanising system, the antidegradant, processing aids—each contribute to the properties of the unvulcanised mix from which products have to be formed. The relative magnitude of the viscous and elastic components of its viscoelastic behaviour will be of paramount interest in many processes (Sections 2.11.2 and 11.2). Other even less well defined and less measurable rheological properties are important: e.g. tack and green strength. 'Tack' is a term used in several industries: in the rubber industry it means the property of uncured rubber to stick to itself; rubber can have tack without being sticky to other surfaces. It is important in building up products where one ply of rubber has to be added and must adhere to other plies. By 'green strength' is meant the strength of the unvulcanised rubber or, more usually, of the unvulcanised rubber mix. This property varies greatly from one rubber to another; natural rubber is superior to the synthetic polyisoprene in this property (Fig. 4.6). Of all the characteristics of the unvulcanised mix, its tendency to prevulcanise or scorch has received, during recent years, most attention. The various instruments developed and used for determining scorch time and rate of cure are fully described in Sections 5.5.2 and 11.2.2.

2.9.3   VULCANISATES

2.9.3.1   *Hardness and Modulus*

Hardness measurements on rubber are in the nature of a modulus determination at low strains and are different from hardness measurements on metals: the indentation of a ball-point produced by a load is measured, and not the permanent indentation made by a point (Section 11.3.5). Hardness is usually the first requirement to be met in developing a mix.

2.9.3.2   *Tension Stress–Strain Relationships*

Controversy centres on the subject of the value of tensile strength and elongation measurements. In so many products, they play no part: the rubber is never extended to more than a fraction of its ultimate elongation and tensile strength because it is restrained by the presence of textile or metal reinforcement. In other unreinforced products, the rubber is strained in compression and not in tension. Nevertheless, there are still two reasons put forward for retaining tensile strength and elongation determinations: first, the quality aspect—contamination, degradation, and adulteration—can be detected by these measurements. Secondly, a high tensile strength is usually an indication of good tear and abrasion resistance. Tensile strength falls sharply, for most elastomers, with rise of temperature; silicone rubber is an exception in this respect (Fig. 4.29).

2.9.3.3   *Abrasion and Tear*

No property of rubber presents more difficulty in its evaluation than abrasion. Both in laboratory instruments, in which a sample is worn away against an abrasive surface, and in actual products, the conditions of shear, rate of removal of rubber, temperature, and so on affect both the result and the comparisons between one material and another. The abrasion resistance of mixes for tyre treads must finally be evaluated by testing tyres on the road, and manufacturers spend substantial sums of money on such testing under very carefully controlled conditions and procedures. It is well known that vehicle, driver, road surface and condition, speed, route, and weather influence not only the wear of a particular tyre but also the relative performance of one tyre against another. The relative movement of tread rubber surface and road surface, which produces abrasion, depends on acceleration and braking, the deflection of the tyre and the tread under load, and the hardness of the rubber. Dynamically, tread-type mixes are much harder than their static values would suggest. As the wheel rotates and each part of the tread meets the road at high frequency, the deflection is less than that observed under static load. The writer recalls an experiment carried out many years ago in which small pellets of non-black pure gum rubber compound were vulcanised into the tread of a tyre; although this compound had very low wear resistance, the areas of it in the tread did not show any more wear, after a lengthy road trial, than the surrounding black tread compound, which had supported it and prevented its movement relative to the road (Section 7.10.2).

2.9.3.4   *Elasticity*

For perhaps the majority of products, it is the elastic property of rubber which is most important. It is as well, therefore, to stress the main features and the departure from perfect elasticity which shows itself in creep, stress relaxation, and set measurements. Time dependence is common to them all and complicates an interpretation of the results. These imperfections tend to arise from the base polymer but are modified to a greater or lesser extent by filler and plasticiser loading, the curing system, and the processing treatment. Compression set is still the most commonly used test; it is easy to carry out and does give some relative indication of the success achieved in producing an elastic material. In fact, it may serve as an indication of state of vulcanisation. It is a test, moreover, carried out at various temperatures (Fig. 4.28) (Section 10.9.5.4). Creep and stress relaxation (the increase of strain under constant stress, and the decrease of stress under constant strain respectively) may be more informative but are more tedious and time consuming to carry out.

2.9.3.5   *Liquid Resistance*

The interest in the resistance of elastomers to liquids lies in the volume changes that occur with time and temperature, the change in physical properties as the rubber absorbs the liquid, and the low-temperature flexibility.

The solubility parameter concept provides a useful approach to decide on the compatibility of elastomers with liquids (Section 9.2.3.1). However, this is by no means the whole story because chemical changes, chemical reactions, and hydrogen bonding in the elastomer or liquid can override the simple solubility parameter value. Copolymers will often show two values owing to the make-up of the monomers or polymer molecule. Solubility parameter will not give the rate of penetration of a liquid into an elastomer or the final absorption; crosslinked vulcanised materials will swell, will not dissolve, as a rule, and the rigidity of the three-dimensional structure will play a part in the amount absorbed.

The penetration rate varies considerably with the rubber type and with the liquid type: the higher the viscosity, the lower the penetration rate. Silicone and fluorocarbon can be cited as extremes; liquids generally have high penetration rate into the former and a low rate into the latter.

The acrylonitrile content of NBR affects very considerably the volume swell in mineral oils (Section 4.10.4). Mineral oils vary, moreover, in their aromatic content and therefore in their effect on nitrile rubbers. Work reported by the Association of Hydraulic Equipment Manufacturers (AHEM, 1968) has shown that, for a series of mineral oils in which were immersed a standard nitrile rubber vulcanisate and several commercial rubbers, there is a linear relation between the volume changes in any two elastomers. This fact forms the basis of the proposed 'Seal Compatibility Index' (Section 10.9.5.4).

2.9.3.6   *Ageing, Heat Resistance, and Flame Resistance*

For many years heating in air at 70°C has been used as an accelerated ageing test for natural rubber compounds, with changes in tensile strength, elongation at break, and modulus, or more correctly stress at an arbitrary elongation, being used as a measure of deterioration. Evidence is clear that such a test, the Geer oven test, provides a comparative evaluation of the ageing characteristics of elastomers, and it was used in the development of anti-oxidants. With the arrival of more heat-resistant and oxidation-resistant rubbers, test temperatures have been raised and it is difficult to distinguish tests to study the heat resistance of elastomers from those to estimate their ageing, oxidation, resistance. Interpretation of tests carried out at temperatures at which crosslinking reactions can continue to occur, even in vulcanisates, is often difficult; the tendency is to follow not only changes in tensile stress–strain properties but also to include compression set, stress relaxation, and liquid absorption measurements.

Several changes in the chemical network structure can occur during heat ageing: scission of the main chains and of the crosslinks, the formation of more crosslinks of the same type as those already present or of a different type which may be immune to further scission. Observation of the rate of stress relaxation in a rubber, either continuously or intermittently under strain, provides much information as to the changes taking place in the network structure, but is not wholly reliable as an ageing test because of the disturbing effect of fillers present in commercial products.

Many elastomers burn readily but can be compounded to have a measure of flame resistance (Section 9.4.6.2) by the use of additives (Section 6.6.2). Halogen-containing rubbers, CR, CSM, and FPM, are inherently moderately non-inflammable and are selected for use where this property is required.

2.9.3.7   *Fatigue*

In many products failure can occur by cracking promoted at the surface of a flexed rubber or by excessive heat generated within the material by cyclic stressing owing to the hysteresis loss. Tests have been developed for these properties (Section 11.3.9), and principles of compounding and manufacture must be observed to achieve satisfactory performance (Section 9.4.5).

2.9.3.8   *Low-Temperature Behaviour*

Whereas the properties of a rubber are changed both reversibly and irreversibly by high-temperature treatment, at sub-ambient temperatures the first order transition (crystallisation) and the second order glass transition produce only reversible stiffening and hardening (Section 9.4.7).

2.9.3.9   *Radiation*

Engineering requirements nowadays occasionally include resistance to high-energy radiation. All vulcanised rubbers deteriorate under prolonged exposure, most hardening, but butyl and polysulphide rubbers soften, eventually to a tarry residue. Harrington (1957, 1958) has classed isocyanate urethanes as the most resistant, followed by natural, styrene–butadiene, and polyacrylate elastomers. Chloroprene and nitrile are decidedly inferior, and silicone, fluorosilicone, and fluorocarbon the least resistant. Certain *p*-phenylenediamine antioxidants and antiozonants are found to improve performance somewhat.

2.9.3.10   *Gas and Vapour Solubility and Permeability*

Table 2.1 gives the solubility and permeability data for a number of elastomers vis-a-vis a number of gases and water vapour (Amerongen, 1964). IIR shows exceptionally low, and SI exceptionally high, permeability to gases. In many applications these factors have to be taken into account apart from the loss of gas from containers under pressure. In the high-pressure sealing of carbon dioxide, for example, 'expansion' of the rubber seal can occur when the pressure is lowered as the dissolved gas comes out of solution and forms cells. Solubility determines the amount absorbed and thrown out, and diffusion the rate at which the gas will leave the rubber. It is found advantageous to select an elastomer with a high diffusion rate (Anon, 1968-6).

2.9.3.11   *Bulk Modulus and Volume Compressibility*

Although for most practical purposes elastomers can be considered as incompressible, the values reported by Scott (1935) and by Wood and Martin (1964) show that the bulk compressibility of NR at 25°C and 1 kgf/cm^2 is

Table 2.1 SOLUBILITIES OF GASES IN ELASTOMERS, AND PERMEABILITY OF ELASTOMERS TO GASES AND MOISTURE. FROM AMERONGEN (1964), COURTESY AMERICAN CHEMICAL SOCIETY

Elastomer	Temperature (°C)	He S	He Q	H₂ S	H₂ Q	N₂ S	N₂ Q	O₂ S	O₂ Q	CO₂ S	CO₂ Q	Moisture at 39°C Q*
NR	25	0·011	23·7	0·037	37·4	0·055	6·1	0·112	17·7	0·90	99·6	100
	50	0·014	52·3	0·041	90·8	0·057	19·4	0·10	47·0	0·63	221·0	
SBR	25	—	17·5	0·031	30·5	0·048	4·8	0·094	13·0	0·92	94·0	118
	50	—	42·0	0·037	74·0	0·052	14·5	0·10	34·5	0·67	195·0	
NBR (18%)	25	—	—	—	—	0·038	1·9	—	—	1·13	48·0	78
	50	—	—	—	—	0·045	7·0	—	—	0·85	120·0	
NBR (27%)	25	0·008	9·3	0·027	12·1	0·032	0·8	0·068	2·9	1·24	23·5	
	50	0·010	23·4	0·030	33·7	0·037	3·6	0·073	10·5	0·88	67·9	
NBR (39%)	25	0·007	5·2	0·022	5·4	0·028	0·2	0·054	0·7	1·49	5·7	
	50	0·009	14·2	0·026	17·0	0·032	1·1	0·062	3·5	1·01	22·4	
CR	25	—	—	0·026	10·3	0·036	0·9	0·075	3·0	0·83	19·5	80
	50	—	—	0·030	28·5	0·038	3·6	0·078	10·1	0·62	56·5	
IIR	25	0·011	6·4	0·036	5·5	0·055	0·2	0·122	1·0	0·68	3·9	7
	50	0·014	17·3	0·039	17·2	0·057	1·2	0·105	4·0	0·52	14·3	
EPDM	25	—	—	—	—	0·08	6·4	0·13	19·0	0·71	82·0	—
FPM	30	—	—	—	—	0·084	0·3	—	—	1·77	14·5	—
	60	—	—	—	—	0·089	2·9	—	—	1·77	116·0	
AU/EU	25	—	—	—	—	0·026	0·4	—	—	1·50	13·5	1500
	50	—	—	—	—	0·027	1·8	—	—	1·10	48·4	
SI	20	—	—	—	570	—	200	—	400	0·43	1600	—
	50	—	—	—		—	280	—	500	—	1550	
(PTFE)	—	—	—	—	—	—	—	—	—	—	—	2

S = solubility, in cm³ of gas at n.t.p. per cm³ at 1 atm Q = permeation coefficient, in 10^{-8} cm² s⁻¹ atm⁻¹ D = diffusivity $Q = D \times S$ Q^* = Q relative to NR = 100

approximately $50 \times 10^{-6}\ \mathrm{kgf^{-1}\ cm^2}$, falling to $40 \times 10^{-6}\ \mathrm{kgf^{-1}\ cm^2}$ at higher pressures.

2.9.3.12   *Thermal and Electrical Properties*

Table 2.2 gives data on thermal conductivity of some elastomers as an indication of the values in comparison with other materials. The electrical properties of elastomers are referred to in Section 9.4.8, and some values typical of

Table 2.2   THERMAL CONDUCTIVITY

Material	Thermal conductivity (cal cm^{-1} s^{-1} °C^{-1})
Air at 0°C	5×10^{-5}
Cork	10×10^{-5}
NR	35×10^{-5}
NR (tread stock)	46×10^{-5}
SBR	60×10^{-5}
SI	50×10^{-5}
Sponge rubber	12×10^{-5}

the various elastomers appear in Tables 4.27, 4.30, 9.31, and 9.33. For conductive and antistatic rubbers, reference should be made to Section 6.6.3 also.

2.10   The Rubber Industry's Products

2.10.1   THE PNEUMATIC TYRE

In quantity consumption of raw polymers, carbon blacks, accelerators, and antioxidants, the pneumatic tyre manufacture dominates the industry.

The function of the tyre, as an essential component of the road vehicle, is to cushion the ride and to transmit the tractive effort and braking force from the vehicle to the road. Particularly of recent years, it has played a greater and greater part as a component of the suspension system contributing to steering control, stability on cornering, and road holding. Its design is an engineering function and, in the final product, are incorporated many, if not all, of the branches of rubber technology and considerable textile technology and metallurgy. The different pneumatic tyre constructions are described in detail in Section 10.1 along with the manufacturing techniques involved.

The range of tyre sizes produced today is quite remarkable, and in each case the product has to be built from numerous items by hand, aided by semi-automatic machines. The selection of the rubber for each part of the tyre is a complicated matter in which the blending of one or more polymers is common practice to achieve the compromise so necessary, because no one rubbery polymer is ideally satisfactory in every respect. The compounder has to consider the road grip, tendency to crack under flexure and ozone attack, resilience, heat build-up, and the wear properties of his mixes. The polymer to a large extent determines these properties (Table 10.1).

Tyre manufacture provides an excellent illustration of the need to compromise in the achievement of economically satisfactory products and of how no new development is as good as it seems at first sight. The radial tyre gives improved tread life, greater steering control, lower fuel consumption, and higher load-carrying ability at the price of a less smooth ride in some cases, more noise, and more steering effort with a near-stationary vehicle. For the fibrous reinforcement, cotton presents no significant adhesion problems but is low in strength; the moisture content of rayon has to be controlled, involving handling problems; 'flat-spotting' in tyres built of nylon is common, and the inherent stretch of this fibre entails extra processing steps; polyester fibre has presented an adhesion to rubber difficulty, which delayed its adoption; glass fibre with very high modulus is liable to chafing and brittle fatigue; and steel cords have brought complex production handling procedures.

2.10.2 ENGINEERING PRODUCTS

As indicated above, the tyre is a product specified more and more in terms of engineering requirements. Other products of the rubber industry required by the engineer in large quantities to meet precise specifications are belting, both transmission and conveyor, and hose for carrying fluids and solids under a wide range of pressures and temperatures (Sections 10.2 and 10.3).

The engineering industries require, in addition, a variety of smaller specialised items. Over the last 15 years the subject of the use of rubber as an engineering material and in engineering applications has been discussed at a great many conferences and symposia, and there are many papers in the literature as well as books on the subject. All the engineering disciplines have an interest in elastomer properties, and the rubber industry has set out to furnish the correct data to enable the engineer to design in rubber as he designs in metal.

Rubber is selected by the engineer for its flexibility, elasticity, and impermeability to gases and liquids. It is used to isolate equipment against shock and vibration, as a seal for anything from a watch to a container door, as a diaphragm to separate two fluids and transmit pressure differences, for drive rollers on business machines, for paper-making machinery, and so on.

While the techniques for the manufacture of these products are covered in general terms, with some detail in specific cases, in Chapters 8 and 10, it is outside the scope of this book to discuss their design; readers are referred to the literature.

2.10.3 FOOTWEAR, CELLULAR, AND OTHER PRODUCTS

Footwear is an increasingly large user of rubbers; the waterproof and wear properties are exploited to the full. Plasticised PVC has become a competitor to rubber in this field because of the rapid economical production methods of which it is capable. The range of footwear is wide, and a number of manufacturing techniques are employed depending on the product; at one extreme is the all-rubber product with fabric reinforcement, at the other is the microcellular soleing sheeting for attachment to traditional footwear (Section 10.4).

Cellular rubber has a long history, and its production has divided into several branches. Initially Hancock had patents for sponge rubber made by blowing with ammonium carbonate. The next development in the early part of the century was an expansion process based on nitrogen which gave a closed cell structure. In the last 30 years considerable application has been made of nitrogen-generating organic chemicals capable of producing either sponge or expanded rubber according to the production technique (Section 10.6).

Reference has already been made to the cellular products made from liquid latex by foaming and blowing (Section 2.4.3.4). Finally, the most recent product is polyurethane foam, which represents another cellular product of extremely low density produced from liquid base materials.

The successful expansion of practically all rubber and rubberlike materials has been achieved by one or other of the above-mentioned techniques, and the cellular products have an important market in the furnishing and packing industries and for insulation. It is remarkable how diverse are the products and for how long the majority of them has been made. This fact draws attention to the general purpose character of natural rubber, which is not approached by any other polymer.

The electrical industry is a large user of rubber as a flexible insulating material, although some uses have been replaced by plastics which are superior in some respects and more easily and cheaply processed, not requiring the vulcanisation step (Section 10.8). Conducting rubbers can be produced at two levels, conducting and antistatic, and are used to reduce the hazards of static build-up (Section 9.4.8) (Brokenbrow, Sims, and Stokoe, 1969).

Its sound-absorbing quality, durability, and insensitivity to temperature commend rubber as a flooring material, but it has never been able to compete very successfully either technically or economically with linoleum, a product that improves on oxidation, or with the newer plastics floor-coverings. It has been suggested that flooring of high natural rubber content with chopped nylon fibre to prevent stretching and rippling may make a comeback (Anon, 1969-2).

The versatility of rubber and the variety of products that are made from it have already been referred to and are covered in a later section (Section 10.9).

2.10.4   ADHESIVES

Adhesives are important materials in many industries and services. The rubber industry use adhesives to obtain a bond between rubber and textiles and between rubber and metals (Sections 8.5.2 and 8.7). There is also an offshoot of the industry engaged in developing and supplying a range of adhesives for many purposes. A number of these are based on natural rubber, chemically modified natural rubber, or synthetic polymers. Indeed several chloroprene rubbers, for example, are specially produced for use in adhesive preparation.

The technology is very specialised and finds no place in a textbook devoted to the technology of solid rubber (Wragg, 1967).

2.11 The Evaluation of Materials: Testing and Specifications

As in most industries, one of the occupations of the chemists, physicists, and technologists is the evaluation of new materials and the control of the quality of incoming supplies, materials in process, and final products.

The rubber industry is fortunate in having an active standards group within the British Standards Institution (BSI) and the International Organisation for Standardisation (ISO), and, as will be apparent from a study of Chapter 11 and the Bibliography, particular emphasis has been placed on the standardisation of test methods for unvulcanised rubbers, vulcanisates, and finished products.

2.11.1 CHEMICAL TESTING ·AND ANALYSIS

No more than passing reference is made in this book to the subjects of chemical analysis and testing. In addition to physical methods, polymers and compounding ingredients are commonly subjected to chemical analysis to identify the type and to control quality. Similarly, for control purposes and, on occasion, for matching or imitation, finished products need to be analysed. Some of the methods are routine, others are highly specialised; books are listed in the Bibliography and attention is also drawn to a number of British Standards.

2.11.2 PHYSICAL TESTING OF RAW RUBBER AND UNVULCANISED MIXES

In Chapter 11, the physical test methods are divided into those applied to raw rubbers and unvulcanised rubbers and those applicable to vulcanisates. The former are directed to assessing: (a) the rheological properties for processing—extruding, calendering, and moulding—usually by means of a viscosity or plasticity measurement; and (b) the vulcanisation characteristics of the material including the tendency to prevulcanise or scorch.

It is unfortunate that both 'viscosity' and 'plasticity' are widely used in the industry; Dr Scott has chosen to call the instruments 'plastimeters' and the property measured 'plasticity', except in the case of the rotation instruments —the Mooney—where the term 'viscosity' is almost universally used. The important point to remember is that high-viscosity material has a low plasticity, and vice versa. Several instruments give plasticity values which increase as the material increases in viscosity; these values are preferably called plasticity numbers (Section 11.2.1.2).

The flow properties of mixes are usually further assessed by trials using an extruder of standardised dimensions and performance with a special die (the Garvey die), which shows the ability to produce an unbroken feather edge and a smooth finish. Other factors measured are die swell, shrinkage, and rate of extrusion (ASTM D2230).

The avoidance of prevulcanisation is no new problem, but of recent years, with higher temperatures of processing and faster rates of cure being demanded, it has achieved increased importance (Section 5.6.2). Likewise, quick and easy methods of estimating rate of cure have been developed (Sections 2.9.2, 5.5.2, and 11.2.2).

2.11.3   PHYSICAL TESTING OF VULCANISATES

Tests on vulcanised rubbers covering a large number of properties are standardised in many countries. A number of these tests are designed to be carried out on specially prepared test-pieces, and this raises two problems. First, in developing a material to meet a specification calling for, say, hardness, tensile strength, and compression set, one requires different thicknesses of sheet. Secondly, few of the tests can be carried out on finished products. The latter may not be a very valid or important criticism, because, for each product, the supplier and user can usually agree on how test-pieces are to be prepared for a few critical tests which the customer wishes to carry out in his own laboratory as a control check on the finished product. The diversity of test-pieces required, however, for some of the commonly specified tests, necessitates both the equipment for their production and several moulding operations involving press-curing time and labour. Furthermore, there may be difficulty or dispute in arriving at equivalent curing conditions applicable to all samples.

The adoption by some laboratories of the two-thickness sheet is one solution. The moulded test sheet has an area of thickness suitable for tests on tensile strength, fluid immersion, etc., and another area of a thickness suitable for compression set; the two different thicknesses are plied together for hardness tests.

It cannot be too strongly emphasised that many of the laboratory physical tests applied to elastomers are very sensitive to degree of vulcanisation, test-piece size and shape, and procedure—hence the importance of standardisation. Nevertheless, there is all too often an unexpectedly large intra- and inter-laboratory lack of reproducibility. Many of the British Standards give data on reproducibility, but it may not always be reliable or sufficiently closely defined. A separate BSI subcommittee is preparing a document on the subject.

2.11.4   SPECIFICATION PHILOSOPHY

The value of standardised methods of testing vulcanisates lies in their forming a basis for agreed specifications between supplier and user. The laboratory of the rubber manufacturer, by adopting the standard procedures, has data on his mixes in a form which can be compared with the customer's requirements, if they too are stated in terms of standard methods. Each rubber manufacturer builds up a series of compounds covering a wide range of hardnesses, based on each of the polymer types which his business demands.

The manner in which the user presents his requirements will vary from the word(s) 'rubber' or 'synthetic rubber' on a drawing, to a detailed specification of properties and performance. In between these two extremes, there will be the situation in which a customer, with a product under development, approaches a rubber manufacturer for an elastomer for a particular application and set of environmental conditions. In these circumstances, the rubber supplier will offer a grade from his range which his experience suggests will meet the requirement; or he may modify one, or develop a completely new mix, depending on circumstances. The customer accepts components or test-sheets and evaluates them, partly by laboratory test and partly by rig or service trials, and reports back. The grade found satisfactory is adopted. The relationship may be such that the user relies on the supplier's assurance that the mix will be kept of constant composition and controlled by his well-established procedures. Alternatively, the user may wish to agree on certain values as representative of the properties of the particular rubber mix, and a specification, which is essentially a control specification, is drawn up. At a future date, a request may be originated by the buying department for another source of supply of the components in question, or for a competitive quotation. Often the user will adopt the above-mentioned specification as a performance specification and will seek from other suppliers a rubber grade meeting the same values for properties stated in that specification, not all of which are relevant to the performance required of the components to be produced from it.

In this way the proliferation of specifications starts, and the rubber compounder in the industry becomes engaged in developing and providing rubber materials to meet the user's requirements which are specified not in terms of the application and environmental conditions, but in terms of properties of rubber on which no work has been carried out to determine their relevance to the service conditions. The proliferation of specifications is therefore followed by the proliferation of mixes in the manufacturer's factory, reducing the efficiency of his production.

Included in the proliferation of rubbers are the general specifications drawn up by large organisations and official bodies to give designers a range of good-quality materials to quote on drawings and orders. A limited number of this latter class of specifications is desirable and acceptable, but this policy has to be kept in perspective.

A buying specification should state the requirements, in general terms, of the application and environment and, in more precise terms, the properties relevant to the use. Such a specification can indicate what properties are to be kept consistent and at what values and tolerances. First and foremost, it must be realised that the variation in the quality and properties of a rubber mix that can be tolerated will depend very much on the particular product and its application. Furthermore, the property which is important will be closely related to its use. A general purpose specification which calls up values for hardness, tensile strength, compression set, fluid swell, heat resistance, etc., with often some definition of the base polymer to be used and permitted ingredients, may set too high a quality and cost for some uses. For other applications, such a specification may permit a number of compositions differing quite substantially in basic make-up but not all capable of performing equally well in an exacting function in, say, an aeroengine component.

The desirable specification philosophy is that which recognises that, for the less exacting application, the requirements should be stated in general terms and, for the more exacting applications, each elastomer will need to be evaluated. The usual standard laboratory tests favoured by specification writers are insufficient, in the latter case, to ensure that two mixes meeting them will behave equally well in service. Proved rubbers need a control specification to control production quality within certain limits.

Much specification philosophy arises from a mistaken idea that there is no need for the multiplicity of rubber mixes, for which there are probably three valid reasons. First, in spite of the many polymers and compounding ingredients available today, it is not possible to optimise, in one mix, all the properties that are demanded—heat resistance, low-temperature flexibility, resistance to all liquids, elasticity, fatigue resistance, etc.; a range, therefore, is built up for various applications in which different compromises have been adopted to suit each exacting demand. Secondly, design and material interact. It is the common experience of suppliers of rubber components to find that a particular rubber mix which suits one customer's equipment will not suit another's which is made essentially for the same purpose, e.g. a piece of hydraulic equipment. It must perhaps be accepted that individualism, so valuable and essential for research and development, lies at the root of this situation. Thirdly, as already stated, mixes proliferate because specifications proliferate.

2.12   Statistics of Production and Usage

To conclude this introductory chapter, Tables 2.3–2.5 give the most recent available figures relating to the production and consumption of rubber polymers and other materials used by the rubber manufacturing industry.

The usage of the different synthetic polymers in various products, for the USA, is given in Table 2.6 (Anderson, 1969); the synthetic rubber usage in tyres and non-tyres by countries is given in Table 2.7. All data are for the Free World and are stated in Gg (1000 metric tons) per annum.

Table 2.3   PRODUCTION FIGURES—POLYMERS

NR		Synthetic rubbers	
Malaysia	1258	USA and Canada	2327
Other Far East countries	1407	UK	235
Africa	162	Europe excluding UK	791
		Japan	375
Total	2827	Total	3890

Table 2.4   CONSUMPTION FIGURES—POLYMERS

	NR	All synthetic rubber	SBR	BR	IR
USA and Canada	625	2105	1430	260	80
UK	190	230		80	45
Europe excluding UK	592	860			
Japan	250	342			
Total	2790	3760			

Table 2.5 CONSUMPTION FIGURES—OTHER MATERIALS

	USA	UK	Rest of Free World
All fillers	1960		
Carbon black	1350	150	1500
Reclaimed rubber	250	34	91
Tyre cord			
Rayon	50		
Nylon	150		
Polyester	75		
Glass fibre	10		

Table 2.6 USAGE OF POLYMERS IN THE USA

Application	SBR	BR	IR	IIR	EPM/EPDM	CR	NBR
Tyres	900	245	52	70	12	4	—
Mechanical	70	—	15	12	—	50	—
Footwear	50	—	8	—	—	—	3
Hose	—	—	—	—	4	—	20
Electrical	—	—	—	3	—	—	—
Wire and cable	12	—	—	—	4	18	—
Foamed	40	—	—	—	—	—	—
Seals	—	—	—	—	—	—	11
Rollers	—	—	—	—	—	—	3
Coated fabrics	—	—	—	—	2	—	5
Blends	—	—	—	—	—	—	4
Automobile	—	—	—	—	20	22	—
Adhesives	—	—	—	—	—	20	6
Appliances	—	—	—	—	5	—	—
Miscellaneous	248	15	5	10	13	6	13
Total	1320	260	80	95	60	120	65

Table 2.7 SYNTHETIC RUBBER USAGE AS PERCENTAGE OF TOTAL RUBBER USAGE

Country	Tyres	Non-tyres
USA	75	69
UK	58	49
France	56	65
W. Germany	63	56
Japan	58	57

THREE

The Physics of Raw and Vulcanised Rubbers

BY J. A. C. HARWOOD

3.1 Introduction

This chapter outlines the basic concepts of the rheological behaviour of raw and vulcanised rubbers, and describes the practical significance of various rheological measurements.

Rheology is concerned with the flow and deformation of matter and so, in theory, covers a very wide field indeed. However, in practice, the term has come to refer to the study of those materials which show a behaviour inter-mediate between those of liquids and solids. The greater part of rheology is concerned with the interrelation of stress, strain, and time, and their depend-ence upon such factors as temperature and chemical constitution; the strain γ is a displacement defined as a relative change of length, and the stress σ is the force acting per unit area.

Completely elastic materials require the same stress, which was applied during the last moment of deformation, to maintain the deformation in-definitely. Ideal liquids require no perceptible force to maintain the state at which the process of deformation ceased. Moreover, completely elastic materials do not show any dependence on the rate of deformation; they follow Hooke's law, i.e.

$$\sigma = E\gamma \qquad (3.1)$$

where E is the static modulus of elasticity. Furthermore, it is inherent in ideally elastic behaviour that the deformation process is reversible.

In the case of liquids, deformation in the static sense does not exist; pro-gressive deformations are not recovered, and energy cannot be stored in the liquid or regained. Instead, the process of deformation requires force, and the work done in this irreversible process is quantitatively dissipated as heat. For an ideal liquid, the stress is proportional to the rate of deformation $\dot{\gamma}$, i.e.

$$\sigma = \eta\dot{\gamma} \qquad (3.2)$$

and the dissipated energy is proportional to the square of the rate of deformation; η is the coefficient of viscosity in macroscopic flow.

Real materials exhibit a whole spectrum of behaviour from liquidlike to solidlike. Any material may be caused to flow, i.e. become fluid, by variation of the temperature and force field. In rubbers, the elastic element forms the continuous phase but encompasses frictional viscous elements. Such materials are termed viscoelastic. When they are deformed, the viscous elements consume energy and retard the elastic deformation; similarly, energy is dissipated when the elastic phase returns in the process of strain recovery and releases its stored energy. The viscous elements or internal friction are thus responsible for the energy difference, or hysteresis, H between work recovered and expended.

The superimposed elastic and viscous behaviour of rubbers is clearly demonstrated during oscillatory deformations. Consider a rubber specimen subjected to a sinusoidally varying deformation. During a deformation half-cycle—from zero deformation, through a maximum, to zero deformation—the rate of deformation, and hence the viscous force, is a maximum at zero deformation and zero at maximum deformation. Therefore, deformation and rate are 90° out of phase. The forces due to such cycling may be elastic owing to the deformation or viscous owing to the rate of deformation. Each of these force components will be in phase with its causative process and, therefore, 90° out of phase with each other in the cycle; their vector sum is the resultant force which is parallel to the rate and extent of the deformation. In an ideally elastic material, all stress is due to strain and is in phase with the deformation; in an ideally viscous material, the stress is in phase with the rate of deformation. In viscoelastic materials, the resultant stress magnitude lags behind the deformation by a phase angle δ; the more viscous the material, the greater the phase angle.

It can be shown that the response of such materials to imposed stresses is governed by a characteristic parameter τ (or distribution of them) with the dimension of time. This is the so-called relaxation or retardation time, which itself can be understood as the ratio of a viscosity to an elastic modulus.

The phase angle δ must be a function of frequency; the elastic component of a viscoelastic material responds instantaneously and does not dissipate energy, but viscous flow or relaxation requires time in dissipating energy proportional to the square of the deformation rate. At low frequencies (small compared with $1/\tau$), the viscous elements will operate but contribute little and the stress will be almost in phase with the deformation. As the frequency is raised, the dissipative effort will rise sharply and the stress will become increasingly out of phase with the deformation, but, as the frequency approaches and exceeds the relaxation times, the viscous mechanisms become more and more incapable of following. As a result, a maximum loss is observed.

The above description of viscoelastic behaviour is essentially phenomenological. The industrial rubber rheologist and physicist is concerned with translating rheological principles into practical tools with which he can facilitate processing and can assess the behaviour and the quality of rubber products.

The rest of this chapter is divided into two parts. The first deals with the rheology of raw (unvulcanised) elastomers, and the second with the viscoelastic phenomena observable in vulcanised rubbers. In both parts, those fundamental measurements which can be measured in the laboratory and their relevance to processing and mechanical performance will be discussed.

3.2   Rheological Measurements on Raw Rubbers

Students of rubber rheology may conveniently be divided into two groups according to their spheres of interest: (a) the fundamentalists, who are concerned with the measurement of rational property characteristics of a constitutive equation (e.g. a relaxation function), or of a particular class of deformations, (e.g. viscosity as a function of shear rate); and (b) the industrial investigator who, by rheological experiment, attempts to characterise the materials used in production processes over the relevant conditions of deformation rate, temperature, etc., obtaining in the process. Efficient factory processing demands that tests be made on raw and partly processed materials to anticipate possible processing problems. In the plastics industry, materials change little in properties from the raw material to the final product and often only one processing step is employed. In the rubber industry, however, production materials are compounds usually containing two or more different raw polymers and large quantities of other materials, such as fillers and processing oils. As a result, the demands made on rubber rheologists are difficult.

It is obviously a difficult problem to maintain adequate processing controls when handling materials which are non-Newtonian, elastic, and thixotropic, and which undergo molecular changes during processing. A large number of 'rheological tests' have been developed as aids, but the majority only provide variables for comparative measurements (i.e. plasticity); there are very few ways of rheologically characterising a rubber or rubber compound.

The flow behaviour of gases and low molecular weight liquids is generally Newtonian: i.e. their viscosity depends solely on the temperature and their composition; it is independent of the rate of deformation. The viscosity of rubbers can vary dramatically with deformation rate, and thus their rheological properties cannot be defined by measuring viscosity at one rate. Usually rheological properties are presented in the form of a flow diagram which plots shear stress as a function of rate of deformation.

The most basic deformation which can be applied to a fluid is constant motion of infinite duration. When the velocity field of such fluid motions is essentially one of simple shear, then the motions are known as laminar shear flows. The stress σ causing the motion is the 'shear stress', and the gradient of the velocity of the flowing fluid normal to the direction of flow is the 'shear rate' $\dot{\gamma}$. For non-Newtonian fluids these parameters are related by

$$\sigma = f(\dot{\gamma}) \tag{3.3}$$

whereas for Newtonian fluids the following expression is sufficient:

$$\sigma = \eta\dot{\gamma} \tag{3.4}$$

where η is the viscosity.

For many polymers it has been found that the flow curves—shear stress against shear rate—can be described by a simple power law of the form

$$\sigma = k(\dot{\gamma})^n \tag{3.5}$$

where k and n are constants.

The index of equation 3.5 is called the 'flow index', and its numerical value is a measure of the deviation from Newtonian behaviour. The more n differs from unity, the more the system is non-Newtonian. For $n < 1$, a fluid is called 'pseudoplastic', and the viscosity decreases with increasing shear rate; for $n > 1$, a fluid is called 'dilatant', and the viscosity increases with increasing shear rate. Rubbers mostly exhibit pseudoplastic characteristics.

3.2.1   CAPILLARY RHEOMETERS

The capillary rheometer has been widely used for measuring the viscosity of fluids. It is described in works by Bingham and by van Wazer *et al.* The instruments have been used for bulk polymers since the work of Marzetti (1923) and of Dillon and Johnston (1933). A useful experimental study for rubbers is given by Einhorn and Turetzky (1964).

Measurements obtained from a capillary rheometer are generally interpreted in terms of the analysis of Eisenschitz, Rabinowitsch, and Weissenberg (1929). This permits the evaluation of a shear stress–shear rate function from capillary pressure drop–flow rate data without assuming a non-Newtonian viscosity function. A more general solution was developed by Mooney (1931).

The advantage of the capillary rheometer lies in the wide range of shear rate over which it can be used. For rubbers, viscosities can be measured over the shear rate range from $1\ \mathrm{s}^{-1}$ to more than $1000\ \mathrm{s}^{-1}$, which makes it suitable for characterising injection moulding compounds. However, many problems involving the capillary rheometry of rubbers remain unsolved, e.g. the detailed development of velocity fields at the die entrance, and the use of die swell measurements to measure the normal forces (Weissenberg, 1947) exerted by the rubber compound.

3.2.2   CONCENTRIC-CYLINDER VISCOMETERS

Couette flow between concentric cylinders has proved one of the foremost methods of measurement for the rheological properties of liquids but, although significant studies have been made by Mooney (1936) and by Philipoff and Gaskins (1956), its use with bulk polymers has been limited.

Viscosity is given by

$$\eta = \frac{\sigma}{\dot{\gamma}} = \frac{C}{4\pi L\Omega}\left(\frac{1}{r_1^2} - \frac{1}{r_2^2}\right) \tag{3.6}$$

where r_1 and r_2 are the radii of the inner and outer cylinders respectively, L is their length, Ω is the angular rotor velocity, and C is the driving torque.

It is claimed (Mooney, 1958) that when this machine is used to obtain elastic recovery curves—by disengaging the drive—the dynamic shear modulus and dynamic viscosity can be calculated from the few rapidly damped oscillations occurring at the beginning of the torque–time curve.

3.2.3   THE MOONEY SHEARING DISC AND BICONICAL ROTOR VISCOMETERS

One of the most important instruments in the rubber industry is the shearing disc viscometer first introduced by Mooney (1934). A knurled disc rotates

D

within a serrated cavity, and measurements of torque are recorded as a function of time.

In his theory of the shearing disc and biconical rotor viscometer, Mooney assumed the Porter–Rao (1926) power law of fluidity, i.e.

$$\dot{\gamma} = a\sigma^n \tag{3.7}$$

where $\dot{\gamma}$ is the steady rate of shear, σ is the shearing stress, and a and n are the material constants. On this basis Piper and Scott (1945) analysed the *biconical rotor viscometer*, obtaining the equation

$$C = \frac{4\pi\rho^3}{3}\left(\frac{\Omega}{\alpha a}\right)^{1/n} + 2\pi^2\rho^2 r_1\left[\frac{\rho\Omega(n-1)}{ar_1(1-r_1/r_2)}\right]^{1/n} \tag{3.8}$$

where C is the driving torque, Ω is the angular velocity of the rotor, and α, ρ, r_1, and r_2 are dimensions characterising the rotor. If it is found experimentally that C is proportional to $\Omega^{1/n}$, the above equation serves to determine a and n. However, if the logarithmic plot of C against Ω is not strictly linear, a more elaborate procedure is required to obtain the true shear stress–shear rate relationship.

If $\dot{\gamma}$ is identified as Ω/α, then the above equation is equivalent to

$$\sigma = \frac{C}{\dfrac{4\pi\rho^3}{3} + 2\pi^2\rho^2 r_1\left[\dfrac{\rho\alpha(n-1)}{r_1(1-r_1/r_2)}\right]^{1/n}} \tag{3.9}$$

For any chosen value of Ω, the local values of n can be determined graphically from the slope of the logarithmic C–Ω plot. The corresponding stress σ can then be obtained from equation 3.9, and a point on the rheological curve is thus determined. The process can be repeated for other values of Ω, and the complete experimental shear stress–shear rate curve is thus obtained.

An analysis of the *shearing disc viscometer* by the Piper and Scott method leads to the following equations for the shear rate and shear stress:

$$\dot{\gamma} = \frac{r_1\Omega}{h} \tag{3.10}$$

and

$$\sigma = \frac{C}{\dfrac{4\pi r_1^3}{3+1/n} + \dfrac{4\pi nbr_1 h}{1-(r_1/r_2)^{2n}}} \tag{3.11}$$

where r_1 is the rotor radius, r_2 is the stator radius, b is the rotor thickness, and h is the clearance, top and bottom, between rotor and stator (Section 11.2.1.1).

3.2.4   PARALLEL-PLATE PLASTIMETERS

In the parallel-plate type of viscometer, a sample, initially in the form of a small pellet, is pressed between flat plates under a given load. The thickness of the sample following compression for a given time is generally taken as a measure of its 'plasticity'.

The first instrument of this type to be used extensively was the Williams plastimeter (1924). Williams used a 2 cm³ sample which was pre-heated to 100°C and then compressed for 5 min with a 5 kg load. The elastic recovery is measured by the height of the sample a fixed time after removal from the plastimeter.

Many variations of this principle have been developed, especially the Rapid Plastimeter (H. W. Wallace & Co.) and the Defo tests (Section 11.2.1.2).

3.2.5 DISCUSSION

The above techniques describe how rheological measurements can be made on materials exhibiting laminar shear flow. However, under severe deformation, smooth laminar flow breaks down and the unstable conditions of tearing and fracture occur. For example, at a certain deformation rate, an extrudate can emerge rough and jagged from the extruder die. Similarly, rubber on a mill can crumble.

Perhaps the major criticism of standard methods of viscosity measurements in the rubber industry, such as Mooney viscosity, is that the range of shear rates available to the instrument is considerably lower than that experienced in processing machinery. For example, the shear rate in the Mooney shearing disc viscometer is only $1-2$ s^{-1}, whereas in milling and calendering shear rates of $10-100$ s^{-1} are used; in extrusion shear rates of $100-1000$ s^{-1} are the rule, and in injection moulding the shear rate often exceeds 1000 s^{-1}.

The capillary rheometer is capable of measuring viscosity over this range of shear rate, and indeed its use is becoming more widespread (Hopper, 1967; White, 1969). Another instrument capable of high shear rate measurement, up to 300 s^{-1}, is the Brabender Plastograph (Kraus, 1965) which, in fact, is a minute internal mixer. The author has found this a useful instrument in differentiating between rubbers where lower shear rate instruments have failed. For example, torque measured on a Brabender Plastograph is sufficiently sensitive to detect a 1% change in the level of oil in oil-extended SBR; this sensitivity was not possible in the Mooney viscometer. However, although the Brabender is a convenient instrument for comparing rubbers and mixing procedures, it does not measure rational rheological parameters, chiefly because the rubber is not deformed at any one shear rate but at a spectrum of shear rates. Indeed, this criticism can also be levelled at the Mooney shearing disc viscometer; it is avoided in the biconical rotor machine, where the design of rotor ensures a uniform shear rate.

At present there is no commercially available rheological instrument suitable for rubber which accurately measures the elastic energy stored in a raw rubber. It is believed that such an instrument must be forthcoming before meaningful predictions of processing characteristics can be made.

3.3 The Physics of Vulcanised Rubbers

The long-chain polymeric nature of rubber molecules was finally established by the classic work of Staudinger (1925). Meyer, Susich, and Valko (1932) recognised that rubber molecules under thermal activation take up statistically random configurations. When the molecules are deformed and straightened, there is an associated decrease in entropy and a retractive force exerted on their ends. It was shown that this retractive force increased nearly proportionately to the absolute temperature and that the force is associated primarily with a change in entropy on stretching, the internal energy changes being very small. This concept was developed by Wall (1942), Guth and James (1943), and Flory and Rehner (1943), and was applied to a crosslinked rubber network; it provides the basis of the statistical theory of rubberlike elasticity. This work has enabled many large-deformation problems to be solved assuming the materials to be perfectly elastic and isotropic (Rivlin, 1948). However, for problems involving non-elastic rubber properties, such as energy dissipation at large strains, recourse must be made to non-linear viscoelastic concepts.

3.3.1　STATISTICAL THEORY OF RUBBERLIKE ELASTICITY

Raw rubber consists of a large number of flexible long-chain molecules possessing a structure which permits free rotation about certain chemical bonds along the chain. The crosslinks introduced randomly during vulcanisation build up a three-dimensional network which resists flow and partly overcomes irreversible deformation. The basic element of such a network is the portion of the molecule between two adjacent crosslinks and is known as a network chain. To develop a quantitative analysis of the changes of entropy with deformation and network structure, Treloar put forward a model wherein the real molecular chains are replaced by equivalent hypothetical chains comprising many freely jointed links about which there is complete freedom of rotation. Apart from the crosslinks, it is assumed that no interactions between chains occur and, furthermore, it is assumed that the crosslinks move affinely when the rubber is deformed.

Treloar derived the following expressions relating the forces f_1, f_2, and f_3 applied to a unit cube of rubber, and the resulting strain ratios λ_1, λ_2, and λ_3. Assuming rubber to be incompressible (i.e. $\lambda_1\lambda_2\lambda_3 = 1$):

$$\lambda_1 f_1 - \lambda_3 f_3 = NkT(\lambda_1^2 - \lambda_3^2)$$
$$\lambda_2 f_2 - \lambda_3 f_3 = NkT(\lambda_2^2 - \lambda_3^2) \tag{3.12}$$

In simple extension when $f_2 = f_3 = 0$ and $\lambda_2 = \lambda_3 = \lambda_1^{-1/2}$, the tensile force is

$$f_1 = G\left(\lambda_1 - \frac{1}{\lambda_1^2}\right) \tag{3.13}$$

where $G = NkT$; N is the number of chains per unit volume, k is Boltzmann's constant, and T is the absolute temperature.

A comparison between theory and experiment is shown in Fig. 3.1, where the parameter G was chosen to give the best fit at small strains. At a strain of about $\lambda_1 = 3$ the measured forces fall below the predicted curve, but at higher extensions the experimental stresses are considerably higher than

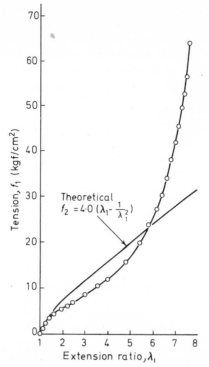

Fig. 3.1   Comparison of theory and experiment for simple extension. From Treloar (1958), courtesy The Clarendon Press

predicted. However, these discrepancies can be satisfactorily explained by later theories (Treloar, 1954).

3.3.2   NON-LINEAR VISCOELASTICITY

Except at very small strains, the stress–strain behaviour of vulcanised rubbers, whether filled or unfilled, is non-linear. A typical stress–strain extension–retraction cycle is shown in Fig. 3.2.

Whereas Hooke's law describes the stress–strain behaviour of a perfectly elastic material, an equation of the following type is required for non-linear viscoelastic materials (Landel and Stedry, 1960):

$$\sigma(\varepsilon, t) = f(\varepsilon)\,\phi(t) \tag{3.14}$$

where $\phi(t)$ is the time-dependent stress relaxation modulus, and $f(\varepsilon)$ is some function of the strain ε and approaches zero as ε approaches zero.

The above equation assumes that the time-dependent variable and the strain-dependent variable are separable. This has indeed been shown to be the case for many gum rubbers, and there is some evidence that it is also true of filled rubber (Harwood, Payne, and Smith, 1969). In extension from $\varepsilon = 0$ at a constant rate of strain, the imposed deformation time function is of constant form, i.e. ramp as opposed to, say, a step or sinusoidal deformation.

Thus the function $\phi(t)$ depends only on the duration of the ramp strain, i.e. on t. When different strains at the end of the ramp are reached in the same time t, i.e. when the strain rate is changed, these strains are said to be isochronous. It follows that $f(\varepsilon)$ can be described as the stress–strain form function for isochronous ramp strain, and $\phi(t)$ is the Young's modulus

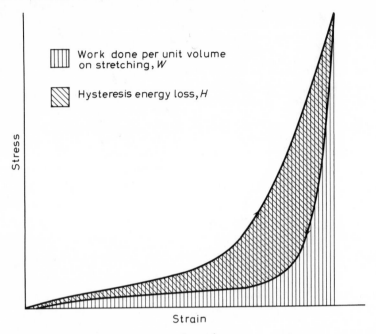

Fig. 3.2   *Typical stress–strain extension–retraction cycle*

(at $\varepsilon = 0$) for an isochronous ramp strain. The function $f(\varepsilon)$ takes account of the departure of the form of the isochronous stress–strain relation from Hookean behaviour and can be written as

$$f(\varepsilon) = \varepsilon\, g(\varepsilon) \tag{3.15}$$

where $g(\varepsilon)$ is the stress–strain non-linearity factor, and ε is the strain at the end of the ramp. Thus equation 3.14 can be rewritten as

$$\frac{\sigma(\varepsilon, t)}{g(\varepsilon)} = \varepsilon\, \phi(t) \tag{3.16}$$

The left-hand side of equation 3.16 is now the linearised stress at the strain ε.

In principle, both $f(\varepsilon)$ and $\phi(t)$ can be derived directly from the experimental data, but the determination of $\phi(t)$, which governs the dynamic behaviour of the rubber, is considerably facilitated if an analytical expression is available for $f(\varepsilon)$. A widely used expression to describe tensile stress–strain curves of rubber, up to moderate strains, is that due to Mooney (1940) and modified by Rivlin and Saunders (1951):

$$\sigma = 2A_0\left(C_1 + \frac{C_2}{\lambda}\right)\left(\lambda - \frac{1}{\lambda^2}\right) \tag{3.17}$$

where $\lambda = 1+\varepsilon$, A_0 is the area of cross-section of the unstrained test-piece, and C_1 and C_2 are constants characteristic of the rubber. However, Smith (1962) and Landel and Stedry (1960) have found that the strain function of SBR can be represented to very high extensions by an empirical equation that was originally proposed by Martin, Roth, and Stiehler (1956) for 'static' (i.e. relatively slow strain rate) stress–strain curves. This equation, the MRS equation, states that

$$\sigma = E\left(\frac{1}{\lambda}-\frac{1}{\lambda^2}\right) \exp A\left(\lambda-\frac{1}{\lambda}\right) \tag{3.18}$$

where A is a constant, assumed independent for isochronous ramp strains, and E is the time-dependent Young's modulus.

Equation 3.18 can be rewritten as

$$\log_{10}\frac{\sigma\lambda^2}{\lambda-1} = \log_{10} E+0.434A\left(\lambda-\frac{1}{\lambda}\right) \tag{3.19}$$

and, by plotting $\log_{10}[\sigma\lambda^2/(\lambda-1)]$ as a function of $\lambda-1/\lambda$, a straight line of

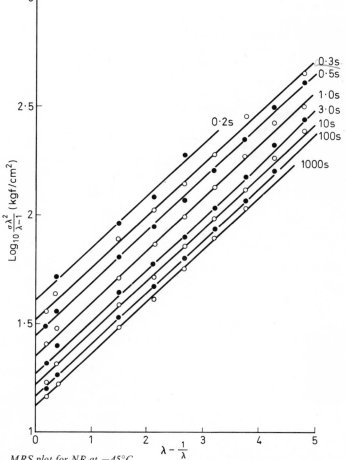

Fig. 3.3   *MRS plot for NR at* $-45°C$

slope $0.434A$ should be obtained, the intercept on the ordinate axis at $\lambda - 1/\lambda = 0$ being $\log_{10} E$.

The validity of equation 3.18 for relatively rapid rates of strain was tested (Harwood and Schallamach, 1967) by constructing isochronous stress–strain curves for various times t (equal to $\varepsilon/\dot{\varepsilon}$, where $\dot{\varepsilon}$ is the strain rate) from the experimental stress–strain curves. The isochronous data were plotted after equation 3.19 in Fig. 3.3. The straight lines shown in Fig. 3.3 were obtained from NR at $-45°$C. Three conclusions can be drawn from this plot:

1. The linearity of the lines proves the stress–strain curves obey the MRS expression (equation 3.18).
2. The constant slope shows that A is independent of time. Results at other temperatures gave the same value of A, showing that A is also independent of temperature.
3. The parallelism establishes the validity of the factorisation of the stress into strain-dependent and time-dependent functions.

The time-dependent modulus E can thus be determined directly from equation 3.19 or from the intercept at $\lambda - 1/\lambda = 0$ of the plot shown in Fig. 3.3. The importance of E lies in the fact that predictions can be made from it as to the stresses during retraction, and hence to the hysteresis energy losses during a tensile stress–strain cycle. The hysteresis is the energy lost during deformation and is given quantitatively by the area of the loop shown in Fig. 3.2. The significance of hysteresis to strength and fatigue characteristics is discussed in Section 3.3.6.

3.3.3   STRESS RELAXATION, CREEP, AND RECOVERY

A practical example demonstrating that three apparently different viscoelastic phenomena depend upon a single relaxation function was given by Gent (1962), who interrelated creep, stress relaxation, and recovery. Creep is deformation under constant load, stress relaxation is the observed reduction in the load required to maintain a constant extension, and recovery is the returning to the unstressed dimensions of a test-piece once a deforming load has been removed.

The rate of creep C may be formulated as

$$C = \frac{1}{\varepsilon}\left(\frac{\partial \varepsilon}{\partial t}\right)_{\sigma} \tag{3.20}$$

where ε is the deformation at a time t after the stress has been imposed, and σ is the constant applied stress. The rate of stress relaxation S at a constant deformation may be similarly formulated:

$$S = -\frac{1}{\sigma}\left(\frac{\partial \sigma}{\partial t}\right)_{\varepsilon} \tag{3.21}$$

and hence

$$C = -S\frac{\sigma}{\varepsilon}\left(\frac{\partial \varepsilon}{\partial t}\right)_{\sigma}\left(\frac{\partial t}{\partial \sigma}\right)_{\varepsilon} \tag{3.22}$$

Now consider the moment in time in both experiments when the initial deformation (or stress) has been reached and the process of relaxation (or creep) is about to begin. At this point one can write

$$C = -S\frac{\sigma}{\varepsilon}\left(\frac{\partial\varepsilon}{\partial\sigma}\right)_t \qquad (3.23)$$

The rate of recovery R (when the deforming stress is removed) is given by

$$R = \frac{1}{\varepsilon}\left(\frac{\partial\varepsilon}{\partial t}\right)_{\sigma=0} \qquad (3.24)$$

By a reasoning similar to that leading to equation 3.23,

$$R = -S\frac{\sigma}{\varepsilon}\left(\frac{\partial\varepsilon}{\partial\sigma}\right)_{t,\sigma=0} \qquad (3.25)$$

The factors of S in equations 3.23 and 3.25 are properties of the relation obtaining between the nominal applied stress σ and the ensuing strain ε. One such equation due to Mooney and Rivlin (equation 3.17) has been found to describe well the stress–strain behaviour of gum rubbers up to moderate strains (less than 200%). Gent found that the rates of creep and stress relaxation calculated from the stress–strain relationship (equation 3.17) were in good agreement with the experimentally determined values.

3.3.4 TIME–TEMPERATURE SUPERPOSITION

A property of viscoelastic materials is that they progressively stiffen as the temperature decreases or the rate of straining increases. However, the analysis of viscoelastic properties has been much simplified by the recognition of a general equivalence between strain rate and temperature. This equivalence can be expressed quantitatively by an expression which transforms data for any particular viscoelastic property, obtained at different strain rates and temperatures T, to a single function of rate at an arbitrary reference temperature T_s which is specific to each polymer. Such an expression was developed empirically by Williams, Landel, and Ferry (1955) and is generally known as the WLF equation:

$$\log_{10} a_T = \frac{-8\cdot86(T - T_s)}{101\cdot6 + T - T_s} \qquad (3.26)$$

$\log_{10} a_T$ is the shift along the log time or log rate axis required to super-impose isothermal viscoelastic data (e.g. of modulus or stress) on to a master curve at the reference temperature T_s. To take account of the temperature dependence of rubberlike elasticity (Ferry *et al.*, 1952), the viscoelastic data to be superimposed are multiplied by the factor $T_s\rho_s/T\rho$ where ρ_s and ρ are the densities at temperatures T_s and T respectively. The use for each polymer of an appropriate reference temperature T_s makes this transformation identical for a wide range of polymers; T_s is approximately $T_g + 50°C$, where T_g is the glass or 'second order' transition temperature. The transformation unifies apparently disconnected results into composite master curves and is applicable only to viscoelastic properties.

The methods and criteria for deciding the applicability of equation 3.26 to viscoelastic data are detailed by Ferry (1961).

3.3.5   THEORIES OF STRENGTH

It is reasonable to suppose that fracture of a rubber involves the rupture of the molecular chains crossing the rupture plane. Bueche (1959) attempted to estimate the strength of a rubber by summing the strengths of single chains. This involved estimating the number and nature of bonds to be broken and the forces required, but the strength so computed was at least a thousand times larger than that usually observed. This result implies that rupture occurs before the average chain bond in a vulcanisate is anywhere near the maximum breaking load. Bueche suggested that rupture is a process by which high local stresses are concentrated upon a few chains in the fracture cross-section. The increased load on neighbouring chains causes additional chain ruptures which, in turn, contribute to a self-accelerating process until rupture of the whole test-piece occurs.

This concept of a failure mechanism was obviously oversimplified and neglected inherent viscous forces which retard the response of real network structures. However, the idea whereby regions of high local stress occur is now well founded. Bueche's concept was that these regions were caused by a distribution of molecular weights between crosslinks, so short-network chains experienced abnormally high stresses because of finite extensibility. The points of rupture initiation can also be due to the presence of solid impurities or surface flaws, which can be caused by chemical attack or even defects in the mould used to shape the vulcanisates during curing. It is because these imperfections or flaws can alter local stress distributions that the observed strength is much less than the theoretical strength computed according to Bueche.

In general, the sort of flaws which generate high local stresses are of microscopic size. They give rise to macroscopic fracture by propagating through the material. The simple flaw theory states that fracture occurs when flaws become capable of propagation, and certain criteria have to be developed for this. For non-brittle materials, however, it is convenient to discuss separately the initiation of fracture from these flaws and their subsequent propagation.

3.3.5.1   *Energetic Approach*

Using Inglis' equation (Inglis, 1913) for the stress concentration at the tip of an elliptical flaw in an elastic material, Griffith (1921) derived an expression for the critical stress σ_{crit} at the tip of the flaw necessary for the material to fracture. He considered the elastic stored energy lost from a strained lamina by the introduction of an Inglis-type elliptical flaw. Assuming that the energy released is only available for the creation of a new surface, he found that

$$\sigma_{crit} = \left(\frac{2E\gamma}{\pi a}\right)^{1/2} \tag{3.27}$$

where E is Young's modulus, γ the surface energy, and a the half-length of the flaw.

Equation 3.27 indicates that a flaw or crack, once it has reached the critical length required to satisfy the equation, must always continue to satisfy it as it grows. Increase in a can only decrease the stress necessary to propagate the flaw. Thus the flaw must continue to grow. In fact, it will accelerate since the elastic energy released by growth increasingly exceeds that required in the form of surface energy. Thus, on the basis of the *simple* flaw theory, the attainment of the critical condition (equation 3.27) is a sufficient criterion for macroscopic fracture, but, because it uses classical elasticity theory in its derivation, it assumes Hookean behaviour, perfect elasticity, and infinitesimal strain. These conditions are not met in most real materials, especially rubbers. However, various approaches have been made to modify Griffith's equation to satisfy more realistic conditions.

Rivlin and Thomas (1953) proposed a theory which, whilst analogous to the modified Griffith treatment, avoids the use of classical elasticity theory. They considered the balance of energy between the strain body and the crack, and arrived at the result that fracture will occur at some critical value W_{crit} of the stored energy density in the bulk of the test-piece, given by

$$W_{crit} = \frac{\mathscr{T}}{KC} \qquad (3.28)$$

where $\mathscr{T}$ is a characteristic energy (per unit area of crack), C is the length of the crack in the sheet, and K is a constant for a given bulk strain. This is very similar to Griffith's expression (equation 3.27), which, after rewriting, gives

$$\frac{\sigma_{crit}^2}{2E} = \frac{\gamma}{\pi a} \qquad (3.29)$$

where the left-hand side is the energy density in the bulk of the specimen.

However, the modified flaw theories described above are essentially criteria for crack growth initiation and do not describe crack or flaw propagation. Thus, while the tensile fracture of brittle materials like PMMA and polystyrene can be regarded in this way, fracture involving non-brittle behaviour must be considered as 'propagation controlled'. To this end, Greensmith and Thomas (1955) measured $\mathscr{T}$, as a function of the rate of crack propagation and the temperature, for gum SBR. They found that $\mathscr{T}$ varied from 1 J/cm^2 at low temperatures to 10^{-2} J/cm^2 at high temperatures. No plastic deformation occurred at the tip of this material, but every volume element in the path of the propagating crack was subjected to a stress–strain cycle as the crack approached and passed its vicinity. An example of such a stress–strain cycle is shown in Fig. 3.2. The area under the strain-increasing curve represents the work W done per unit volume on stretching the rubber, whilst the area of the loop represents the hysteresis energy H lost during the cycle. Greensmith and Thomas found that the slower the cycle or the higher the temperature, the less energy was dissipated and this paralleled the time–temperature behaviour of $\mathscr{T}$. For a given temperature, they found that the rate of crack propagation was given by

$$\frac{dC}{dt} = B\mathscr{T}^n \qquad (3.30)$$

when n has a value of about 4 for SBR, and B is a function of temperature. This relationship is independent of the manner of loading or geometry of the specimen over several decades of rate of deformation. Dependence of the cut length C on time t for a test-piece *at constant extension* was obtained by substituting equation 3.28 into equation 3.30 and integrating, to obtain

$$t_b = \frac{1}{B(n-1)K^nW^n}\left(\frac{1}{C_0^{n-1}} - \frac{1}{C^{n-1}}\right) \tag{3.31}$$

where C_0 is the initial length of the cut, and t_b is the time to break. Constants B and n are determined from a logarithmic plot of dC/dt against $\mathcal{T}$.

It is seen that t_b becomes independent of $1/C^{n-1}$ when C is much greater than C_0. If n is large, the breaking time is determined almost entirely by the very early stages of propagation. Consequently, the size of the specimen undergoing rupture will have little effect on t_b, provided its dimensions are much greater than C_0.

For a test-piece extended at a *uniform rate of extension*, Greensmith (1964) calculated t_b to be given by

$$t_b = \frac{pn+1}{B(n-1)K^nW^nC_0^{n-1}} \tag{3.32}$$

where p is a fitting constant, and W refers to the point of fracture.

Greensmith showed that equations 3.31 and 3.32 correctly predicted the dependence of t_b on stored energy density at break, on strain at break, and on the nature of the test (constant strain and constant strain rate) for both notched and unnotched test-pieces. When notched test-pieces were used, the theory correctly predicted the dependence of t_b on C_0. Whilst there are difficulties in predicting absolute values, agreement is sufficient to justify the basis of the theory, at least over a limited range of time-scale. Results are therefore consistent with a mechanism of failure by crack growth from microscopic cracks or other stress raisers initially present in the elastomer.

The fundamental significance of the energy balance approach is that it relates conditions at the crack tip to externally applied forces or bulk deformations. However, further extension of the theory to predict the time dependence of strain and stress at break over larger time- and temperature-scales faces difficult problems: e.g. it requires an understanding of how the stored energy near the crack is affected by time.

3.3.5.2   Theory of Halpin (1964)

Halpin believes fracture to be a creep phenomenon. He considered the deformation and rupture of network chains at the tip of a growing crack, relating them to the behaviour of the bulk material. Halpin sets out to calculate from first principles what Greensmith takes as empirically given: the rate of propagation of a flaw as a function of stress and temperature in rubbers. The importance of viscous effects on rupture properties and rupture mechanism has been long recognised; it appears that the rate of propagation is determined by the dissipative behaviour of the rubber. In addition to viscoelasticity, energy dissipation, or hysteresis, can result from the breakdown of filler agglomerates and the breaking of weak crosslinks. Stress softening, a hysteresial phenomenon observed at moderate and high extensions, is probably also viscoelastic in origin (Section 7.10.1).

3.3.6   INFLUENCE OF HYSTERESIS ON STRENGTH

Recent studies have shown that the work done on stretching a rubber to break in a tensile experiment depends upon the amount of energy dissipated during the deformation (Harwood and Payne, 1968; Harwood, Payne, and Whittaker, 1969). Measurements were made at different temperatures of the hysteresis energy loss H_b at break, by performing a single stress–strain cycle whose maximum strain was very close to the predetermined breaking strain of the rubber. By plotting the logarithm of the work W_b done to break against the logarithm of H_b, a straight line was obtained of slope $\frac{2}{3}$ for the temperature range examined (Fig. 3.4). Values of W and H obtained at strains less than the breaking strain fell below the line in Fig. 3.4. It was found that W_b and

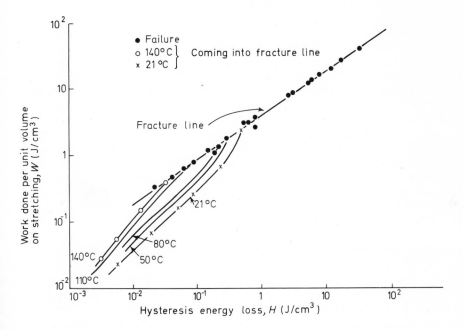

Fig. 3.4   Plot of work done per unit volume against hysteresis energy loss for SBR–gum mix

H_b for all the gum and filled amorphous rubbers examined could be related by the equation

$$W_b = k\left(\frac{H_b}{X}\right)^{2/3} \tag{3.33}$$

where k is a constant characteristic of the rubber, and $X = 1 + 2 \cdot 5c + 14 \cdot 1c^2$

where c is the volume fraction of the filler in the rubber. This equation thus gives quantitative expression to the effect of hysteresis on strength.

3.4   The Swelling of Rubbers by Liquids

3.4.1   INTRODUCTION

All rubbers can absorb liquids to a greater or lesser degree. The absorption of the liquid causes the rubber to increase in volume, and this is the phenomenon of swelling—a consequence of which is the deterioration in physical properties. Raw rubbers are completely soluble in certain liquids, but vulcanised rubbers are virtually insoluble. Strong bonds, such as chemical crosslinks between the rubber chains, prevent rubber molecules becoming completely surrounded by the liquid and restrict the deformation of the rubber.

The swelling of rubbers by liquids is a diffusion process. At the start of the process, the rubber at the surface of a component has a high liquid concentration while the liquid concentration in the bulk of the component is zero. Subsequently, the liquid molecules diffuse into the rubber just below the surface and eventually into the bulk of the rubbers. As the diffusion process proceeds, the dimensions of the rubber component increase until the concentration of the liquid is uniform throughout the component and equilibrium swelling is achieved. The amount of a given solvent that will diffuse into the rubber until it reaches equilibrium depends upon the number of crosslinks per unit volume of rubber. The greater the number of crosslinks per unit volume, the shorter the average length of rubber chains between crosslinks and the lower the degree of swelling. The degree of swelling can be expressed by the percentage increase in volume or by the volume fraction of rubber in the swollen gel. Apart from crosslinking restrictions on the equilibrium swelling, the degree of swelling depends upon the compatibility of the rubber and liquid on a molecular scale. It also depends on the amount and type of filler present in the rubber. The rate at which swelling proceeds depends upon the relative molecular size of the diffusing liquid molecule.

3.4.2   MOLECULAR COMPATIBILITY OF RUBBER AND LIQUID

If rubber and liquid molecules are compatible on a molecular scale, then the liquid will readily swell the rubber. In fact, if the rubber is unvulcanised, the liquid will dissolve the rubber. A general guide to the compatibility of rubbers and liquids is the relative values of their solubility parameters. The solubility parameter δ is given by

$$\delta = (\text{cohesive energy density})^{1/2} = \frac{\Delta H - RT}{M/\rho} \qquad (3.34)$$

where ΔH is the latent heat of vaporisation, R is the gas constant, T is the absolute temperature, M is the molecular weight, and ρ is the density.

In practice it is found that, unless there is specific interaction between dissimilar molecules, and in the absence of crystallisation, a rubber and liquid will be compatible if their values of solubility parameter are within unity of each other.

The degree of swelling is generally highest with chloroform ($\delta = 9\cdot3$) for all rubbers except butyl ($\delta = 7\cdot6$); in this case hexane ($\delta = 7\cdot3$) has been found to be a good swelling agent. Acetone ($\delta = 10$) is a poor swelling agent, except in strongly polar nitrile rubber (60 : 40) ($\delta = 9\cdot9$). Benzene ($\delta = 9\cdot2$) is also a good swelling agent for most rubbers (Section 9.2.3.1).

3.4.3   DIFFUSION OF LIQUIDS

Diffusion theory (Crank, 1956) predicts that, in the early stages of swelling, the mass of liquid absorbed per unit area of rubber is proportional to the square root of the time. An experiment by Southern and Thomas (1967) confirmed this prediction for flat sheets of rubber. They found that plots of the mass of liquid absorbed per unit area of rubber versus $\sqrt{t}$ were linear up to at least 50% of the equilibrium swelling.

With this knowledge, Southern and Thomas were able to calculate the penetration depth of the swelling liquid into the rubber. They considered that, because the boundary between the swollen and unswollen rubber is normally quite sharp, it was reasonable to assume that all the absorbed liquid is contained in a layer of swollen rubber of uniform concentration. It follows that the distance of liquid penetration is obtained by dividing the mass of liquid absorbed per unit area by the concentration of liquid in the surface. The depth of penetration is also proportional to $\sqrt{t}$. The penetration rate P (i.e. the rate of movement of the boundary between swollen and un-swollen rubber) is given by

$$P = \frac{M_t}{C_0 A \sqrt{t}} = \frac{l}{\sqrt{t}} \tag{3.35}$$

where M_t is the mass of liquid absorbed by a surface area A in time t; C_0 is the concentration (g/cm^3) of liquid in the surface of the rubber, and l is the depth of the swollen layer. The relation between P and the diffusion co-efficient D is given by

$$D = \tfrac{1}{4}\pi P^2 \tag{3.36}$$

This work on NR clearly demonstrates why thin components fail more rapidly than thicker components—it takes 100 times as long for liquid to penetrate 10 times the distance. When the liquid first contacts the rubber, the rate of penetration is a maximum; after this, the rate becomes pro-gressively slower until equilibrium is reached.

3.4.4   EFFECT OF LIQUID VISCOSITY

Southern (1967) measured the penetration rate P of a large number of hydrocarbon liquids into NR. He found that the rate decreased as the viscosity η of the swelling liquid increased (Fig. 3.5). It must be noted that the total amount of liquid absorbed does not depend on the viscosity but on the chemical nature of the swelling liquid. Thus, hexane and liquid paraffin

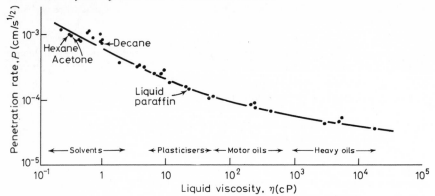

Fig. 3.5   *Effect of liquid viscosity on the penetration rate of liquids into natural rubber. From Allen, Lindley, and Payne (1967), courtesy Maclaren*

exhibit similar equilibrium swelling values, but the rate of penetration of the latter is one-ninth that of the former. On the other hand, acetone swells much faster than liquid paraffin but the equilibrium value is lower.

3.4.5   SWELLING OF FILLED RUBBERS

The influence of fillers on swelling is relatively small compared with the effect of rubber–liquid interaction. In fact, the effect of the filler sometimes does not amount to much above the average error of measurement. However, some effects of the filler on swelling can be detected without any doubt (Boonstra and Dannenberg, 1959). For instance, there is a very pronounced decrease in matrix swelling with some fillers when the filler loading is increased. This is the case with carbon black, where the decrease in matrix swelling is an almost linear function of the loading, and particularly with channel black and when strongly swelling liquids are used.

Non-black fillers cause a reduction in matrix swelling which, however, does not appear to depend so much on the filler loading as the reduction caused by carbon black fillers.

FOUR

Raw Polymeric Materials

Part A

BY G. F. BLOOMFIELD

4.1 Natural Rubber—NR

4.1.1 INTRODUCTION

Cis-polyisoprene, the hydrocarbon component of NR

$$\sim\!\!\text{CH}_2\!-\!\underset{\underset{\text{CH}_3}{|}}{\text{C}}\!=\!\text{CH}\!-\!\text{CH}_2\!\sim$$

is fairly widespread in nature, having been identified in about 2000 plant species (Bonner and Varner, 1965). Only *Hevea brasiliensis* is of any commercial significance; indigenous to the Amazon valley, it is the species on which all plantation rubber in Africa and the Far East is based.

Until the early 1960s the methods used in the preparation of NR had been virtually unchanged for more than 50 years. Processing followed the routine of tapping the trees for the latex, collection, coagulation, sheeting, drying, and baling—simple operations which have the merit of easy operation by both large and small producer, estate and small-holder, alike. Drying, whether by smokehouse or the more modern drying tunnels, has been a rather slow process taking four or five days—longer at ambient temperature. Dried sheets or crepes have been conventionally compressed into bales weighing up to 250 lb (112 kg) coated with a talc or chalk composition to prevent self-adhesion during shipment and storage.

4.1.2 MANUFACTURE

4.1.2.1 *Internationally Recognised Market Grades*

The internationally recognised market grades of NR are shown in Table 4.1. Each is defined by an approved verbal description, and most are covered

Table 4.1 INTERNATIONALLY RECOGNISED MARKET GRADES OF NR. FROM NAUNTON (1961), COURTESY EDWARD ARNOLD

'Green Book' section	Type	Grade
1	Ribbed Smoked Sheets	1XRSS, RSS Nos. 1–5
2	Thick Pale Crepes	1X, 1, 2, 3
	Thin Pale Crepes	1X, 1, 2, 3
3	Estate Brown Thick Crepes	1X, 2X, 3X
	Estate Brown Thin Crepes	1X, 2X, 3X
4	Compo Crepes	1, 2, 3
5	Thin Brown Crepes (Remills)	1, 2, 3, 4
6	Thick Blanket Crepes (Ambers)	2, 3, 4
7	Flat Bark Crepe	Standard, Hard
8	Pure Smoked Blanket Crepe	Standard

by reference samples (see *Green Book*, International Standards of Quality and Packing of Natural Rubber Grades, 1969).

For the manufacture of pale crepe and sheet rubber, incoming latex is strained and diluted, and then coagulated by the addition of formic or acetic acid.

White and Pale Crepes. The natural colour of crepe is pale yellow, mainly owing to its content of β-carotene. To produce the palest crepes, the β-carotene is removed by fractional (two-stage) coagulation of the latex and/or the latex is bleached by the addition of xylyl mercaptan (0·05% max. on rubber). In addition, sodium bisulphite suppresses any tendency towards darkening due to polyphenol oxidase. The coagulum is machined eight or nine times between grooved differential-speed rollers with liberal washing.

For sole crepe (for footwear), a number of plies of thin crepe are laminated to the required thickness, consolidated first by hand-rolling and finally through even-speed rolls. Market requirements of light colour and uniform thickness require the pale crepe to be specially selected.

Ribbed Smoked Sheet (RSS). Prior to coagulation (but after addition and distribution of coagulant) aluminium partitions are inserted vertically into the coagulating tanks so that sheets of coagulum are produced. These are floated into even-speed rolls provided with water sprays and adjusted to give progressively tighter nips and faster speeds. The last pair of rolls impresses a grooved pattern, and sometimes the producer's identification mark. Drying is in a smokehouse or tunnel at a temperature gradient of 43°–60°C.

Air-Dried Sheet (ADS) and Pale Amber Unsmoked Sheet (PAUS). The RSS procedure is followed, but the sheets are dried in hot air without smoking. The sheet surface is sometimes treated with *p*-nitrophenol solution to prevent mould.

Brown Crepes, Blanket and Compo Crepes, Flat Bark, etc. The source materials derive from pre-coagulated lump, cup lump (latex which has continued to flow after collection and has coagulated in the tapping cup), tree lace (latex which has dried on the tapping cut), earth scrap recovered from the ground, and undried sheet and coagulum from small-holder sources. These materials vary considerably in their cleanliness.

4.1.2.2   *Technically Classified Rubber (TCR)*

The rate of cure of NR in sensitive test mixes such as ACS1 (Table 5.2) (deficient in stearic acid) is affected significantly by the pH at which coagulation occurs and to a lesser degree by the dilution of latex prior to its coagulation. By control of these variables, estate production of RSS is maintained within relatively narrow bands of cure characteristics. TCR is supplied in three classes marked with a red (slow curing), yellow (medium), or blue (fast curing) circle, as detailed in NRPRA Tech. Info. Sheet 2, 1963. The differences in rate of cure are diminished by adjustment of curing system and by compounding with carbon black.

4.1.2.3   *'New Process' Rubbers*

The main emphasis in the visual grading scheme of Table 4.1 on appearance (especially freedom from bubbles) rather than on technical properties has permitted considerable variability. This can be minimised and NR's performance improved by preparing sheet rubber without the dilution of latex, otherwise necessary to achieve the visual perfection of RSS1 (Fleurot, 1965). However, visual appearance remained a major obstacle to such innovations until the early 1960s when radical new processes were introduced into the NR industry (Graham and Morris, 1968). Broadly, these comprise comminution or crumbling of the source material, drying in particulate form in deep-bed or apron driers within a few hours, and packaging in compressed small bales weighing 70–75 lb (32–34 kg) either wrapped in polyethylene sheet or lightly talced. A rather different approach utilises an extruder drier. Rubbers so produced are designated as 'block', 'comminuted', 'granulated', and 'pelletised'.

The main processes which have found market acceptance are given in the following paragraphs.

Heveacrumb. This is a mechano-chemical crumbling process developed and patented by the RRIM. A maximum of 0·7 p.p.h.r. of castor oil introduced into latex or sprayed on to wet rubber promotes crumbling on passing through ordinary creeping rolls (RRIM Tech. Bull., 1966). Any small amount of residual castor oil has no deleterious effect on the properties of the rubber; the amount of castor oil can be reduced by the inclusion of zinc stearate.

Comminution or Granulation Processes. Size reduction is by means of dicing machines, rotary cutters, granulators, hammermills, and allied equipment. These handle a variety of feedstocks. Latex coagulum from conventional coagulating tanks is sliced before passing to the comminutor; it is sometimes more convenient to use smaller tanks with partition plates to give pieces suitable for direct feeding. In the Decan system, the coagulum from a circular drum is floated in a bath of water and held against a band knife and an adjacent rotating roll to produce a veneer which is fed to a granulator, crumbler, etc. (Fleurot, 1965).

'Dricom' type rubber is dried (as thin brown crepe) prior to comminution.

Pelletisation. Developed by the Guthrie Corpn (Thompson, Howarth, and Smith, 1966), pelletisers operate on the same principle as the domestic meat mincer. Cutter knives work behind and in front of a dieplate drilled with approximately 2·5 mm holes.

Expellers. A screw carries the wet rubber from the inlet and squeezes it continuously while its water content is discharged through a vented casing. The rubber is then squeezed through a dieplate, and drying is completed on an apron drier or in an extruder drier.

4.1.3   SPECIFICATIONS

The technical classification scheme classified NR only according to rate of vulcanisation and has otherwise relied on visual grading to 'Green Book' standards. The introduction of new manufacturing processes for NR necessitated other technical specifications. Although several countries are producing NR to specification on similar lines, only the Standard Malaysian Rubber (SMR) scheme will be described here.

This scheme was evolved in 1965 for general purpose Malaysian NR in conventional and new forms presented in small bales, wrapped in polythene or other approved material, and free from other bale coatings. The current specifications are shown in Table 4.2.

The Plasticity Retention Index (PRI) (NRPRA Tech. Bull., 1967) is a measure of the resistance of raw rubber to breakdown during processing at temperatures above 100°C and to ageing, since both are oxidative processes. The PRI is defined as the Rapid Plasticity No. of a sample of the rubber after heating for 30 min at 140°C expressed as a percentage of the Rapid Plasticity No. before heating. (Plasticity measurements are with the Wallace Rapid Plastimeter—see Section 11.2.1.2). A high PRI denotes high resistance to breakdown, but its value is virtually independent of the initial plasticity of the rubber.

NR as normally prepared is well protected against oxidation, but the high level of natural protection may be reduced by excessive washing, excessive heat during drying, exposure to sunlight, or contamination with pro-oxidant substances. The PRI provides a sensitive indication of the presence of metallic impurities which can cause oxidative degradation of NR. Its introduction into the specifications eliminates the need for specifying maximum limits for copper and manganese (as in earlier 'Green Book' descriptions). High-grade NR of latex origin (e.g. pale crepe, RSS1, SMR5) normally has a PRI in excess of 70, but little technical significance attaches to differences in PRI when the values are above 60. The joint effect of the introduction of PRI specifications for each SMR grade, together with the elimination of the tedious analyses for copper and manganese, provides simplicity of approach to both producers and consumers.

Table 4.2 SPECIFICATIONS FOR SMR GRADES

Criterion	Manufactured from latex			Other grades		
	SMR EQ*	SMR 5L	SMR 5	SMR 10	SMR 20	SMR 50
Dirt (retained on 44 μm aperture sieve) (max., % wt)	0·02	0·05	0·05	0·10	0·20	0·50
Ash (max., % wt)	0·50	0·60	0·60	0·75	1·00	1·50
Nitrogen (max., % wt)	0·65	0·65	0·65	0·65	0·65	0·65
Volatile matter (max., % wt)	1·00	1·00	1·00	1·00	1·00	1·00
PRI (min., %)	60	60	60	50	40	30
Rapid Plasticity No. (min. initial value)	30	30	30	30	30	30
Cure indication	MOD 5, 6 or 7	†	†	—	—	—
Colour limit (Lovibond scale, max.)	3·5	6·0	—	—	—	—
Grade marker colour	Light blue	Light green	Light green	Brown	Red	Yellow
Plastic bale wrap colour	Transparent	Transparent	Transparent	Transparent	Transparent	Transparent
Plastic bale strip colour	Transparent	Transparent	Opaque white	Opaque white	Opaque white	Opaque white

*SMR EQ is to be initiated in 1971.
†MOD values may be given on a non-specification basis.

To ensure consumer confidence in the SMR scheme, the production of SMR is permitted only after registration by the Malayan Rubber Export Registration Board. Punitive action (including deregistration) can be taken against a producer who does not conform to the requirements of the scheme. To enable a conformance guarantee to be given with confidence, statistical methods are applied to the test results obtained from production lots.

4.1.4   SPECIAL TYPES

4.1.4.1   *Oil-Extended Natural Rubber (OENR)*

Early attempts to extend conventional forms of NR—sheets and crepes—with oil failed because the softening effect of the oil rendered extended sheet rubber unsuitable for smoke or hot-air drying. No such problem arises in the drying of OENR in crumb form, and the production of an oil-extended Heveacrumb or other new process granulated NR is already practicable (Sekhar, 1967). From the point of view of physical properties, however, there is little to choose between addition of oil in the factory and addition of oil at the plantations (NRPRA Tech. Bulls, 1966, 1968).

4.1.4.2   *Partially Purified Crepe (PP Crepe)*

Partially purified crepe contains less than half the amount of protein and mineral matter normally present in pale crepe and is prepared from latex which has been centrifuged to remove some of the naturally occurring non-rubber substances (NRPRA Tech. Info. Sheet 12, 1963).

4.1.4.3   *Rubber Powders*

There are several types of rubber powder made in different ways (NRPRA Tech. Info. Sheet 22, 1963). They may be lightly vulcanised and may contain appreciable quantities of anti-agglomerating agents to prevent massing on storage. They find applications in road surfacings.

4.1.4.4   *Skim Rubber*

When latex is concentrated by centrifuging, the by-product skim latex is coagulated and made into smoked sheet, thick crepe, or granulated rubber. It contains a higher proportion of non-rubbers than ordinary sheet or crepe and is rapid curing, but can sometimes be used with advantage where a clean low-priced rubber is needed (Baker, 1958; RAPRA Res. Mem. R410, 1958).

4.1.4.5   *Softened or Peptised Rubber*

Softened rubber is prepared by adding a small quantity of a softening agent or peptiser (Section 6.3.1) to latex prior to coagulation for sheet or crepe manufacture. Depending on the activity of the peptiser, the rubber may be softened during the drying process or it may be easily broken down to a suitable plasticity during mastication.

4.1.4.6 *Superior Processing (SP) Rubber*

Various types of superior processing rubbers are available: e.g. SP Smoked Sheet, SP Crepe, SP Air Dried Sheet, SP Heveacrumb, SP Brown Crepe, PA 80, and PA 57. The first four mentioned are made from a mixture of 20% vulcanised latex and 80% unvulcanised latex (by weight), coagulated and dried in the normal manner. To make SP Brown Crepe a wet crumb coagulum from an 80/20 mixture of vulcanised and unvulcanised latex is mixed with three parts of wet scrap on power mills and processed as Estate Thin Brown Crepe.

PA 80, a concentrated form of SP rubber, consists of 80% vulcanised and 20% unvulcanised rubber. PA 57, another concentrated form of SP rubber, consists essentially of 70 parts of PA 80 with 30 parts of a non-staining processing oil. The oil is added to give an easier-processing concentrated SP rubber.

The special value of SP rubbers lies in improved extrusion, moulding, and calendering, and they conform to technical specifications of swell on compound extrusion and Mooney viscosity (NRPRA Tech. Bull., 1965; NRPRA Tech. Info. Sheets 7, 34, 37, 38, 39, 67, 89, 1963–1966).

4.1.5 STORAGE HARDENING*

An early problem encountered in the introduction of NR to technical specifications was an apparently uncontrollable hardening during storage, the rate and magnitude being dependent on the initial plasticity of the freshly prepared rubber and the relative humidity of the surrounding atmosphere (Wood, 1952). At zero humidity the rate of change is greater at more elevated temperatures (up to 70°C) and, therefore, a convenient rapid laboratory assessment of the susceptibility to hardening of any given rubber is provided by storage at 60°C for 48 hours over phosphorus pentoxide (Sekhar, 1960).

The hardening under consideration here is not to be confused with 'freezing', i.e. crystallisation due to storing NR at moderately low temperatures. This occurs most rapidly at -26°C, but its rate between 0°C and 10°C is appreciable. Crystallisation, however, is a reversible physical change, and frozen rubber can be thawed without detriment to its quality. Storage hardening in the present context is irreversible, and evidence has been presented that a crosslinked structure is formed but oxygen is not directly involved (Wood, 1952; Sekhar, 1958).

The well-substantiated presence of carbonyl heterogroups in the NR

*Section 4.1.5 is based on an abridged version of text by Bloomfield (*J. IRI*, 2, 287, 1968).

molecule (Sekhar, 1962) provides not only a plausible explanation of spontaneous hardening through intermolecular condensation, but also suggests methods of bringing it under control or preventing it altogether.

4.1.5.1   *Viscosity-Stabilised Natural Rubber (CVNR and LVNR)*

When the specific aldehyde reagent Dimedone (5,5-dimethyl-1,3-cyclo-hexanediol) is added to latex prior to its conversion into dry rubber, a small amount becomes chemically bound to the rubber and storage hardening is inhibited. The aldehyde groups are effectively blocked. The minimum amount of reagent which is effective provides a numerical indication of the number of aldehyde groups per million molecular weight, and this figure ranges from 5 to 30 according to clonal type.

Of the more practical reagents, hydroxylamine gives the best compromise of cost, efficiency, and water solubility (Sekhar, 1967); 0·15% by weight (as hydrochloride) is required to ensure inhibition of hardening.

Hydroxylamine, being monofunctional, preserves the Mooney viscosity of NR at the level at which the rubber leaves the tree, and this can be as low as 35 with certain new clonal types. Experience is showing, however, that over a variety of clones, viscosity-stabilised NR (designated CV rubber) has a viscosity within the range 55–65 Mooney units, enabling premastication to be substantially reduced and sometimes eliminated altogether. A lower viscosity type (designated LV) contains 4 p.p.h.r. of a non-discolouring oil and requires no premastication; the amount of added oil is too low to affect properties detrimentally. If a high stabilised-viscosity level is required, a bifunctional amine can be used instead of hydroxylamine in order to introduce some crosslinking—it has long been known that benzidene can be used in this way to provide hard NR. Clearly there are objections to benzidene (discolouration and toxicity of residual reagent), but hydrazine has been successfully used and a hardened NR can thus be made available (provisionally designated HH).

4.1.6   MODIFIED NATURAL RUBBERS

4.1.6.1   *Cyclised Rubber Masterbatch*

Cyclised rubber masterbatch is prepared by heating specially stabilised latex with strong sulphuric acid, mixing with untreated latex containing an equal quantity of rubber, and coagulating. The coagulum is washed, machined, and dried in the usual way. Cyclised rubber masterbatch is useful in the preparation of stiff vulcanisates (NRPRA Tech. Info. Sheet 8, 1963).

4.1.6.2   *Heveaplus MG Rubber*

This modification of NR is made by polymerising methyl methacrylate monomer in latex so that polymer chains become attached to the rubber molecule. The resultant latex is coagulated, and the coagulum made into a crepe. Two products are available, MG30 containing 30% polymethyl-methacrylate, and MG49 containing 49%. Their special value is in adhesives and rigid mouldings (Allen, 1963; NRPRA Tech. Info. Sheets 9, 31, 71, 87, 1963–1966).

4.1.7 PROPERTIES OF RAW NATURAL RUBBER

4.1.7.1 *General*

The better types and grades of natural rubber contain at least 90% of the hydrocarbon *cis*-1,4-polyisoprene, in admixture with naturally occurring resins, proteins, sugars, etc. The composition of NR today varies somewhat between clones, some high-yielding types tending towards higher acetone-extract and protein content. Typical analyses reveal acetone-solubles 2·5–4·5%, nitrogen 0·4–0·6%, and ash 0·2%.

High-resolution nuclear magnetic resonance spectra demonstrate that the structure of the NR hydrocarbon is at least 99% *cis*-1,4. There is no evidence of any 3,4 units (Chen, 1966). A secondary, but technologically important, heterogeneity is present in oxygenated groups with carbonyl reactivity (Sekhar, 1962).

Estimates of the molecular weight of the hydrocarbon in the latex system of the tree reveal a magnitude of at least 1 000 000, together with some naturally occurring microgel. The raw material of commerce (sheet, crepe, block rubber) comprises a molecular weight mainly in the range of 500 000 to 1 000 000, with extensions into both higher and lower regions (Allen and Bloomfield, 1963; Nair and Sekhar, 1967; Ng and Schulz, 1969).

4.1.7.2 *Mastication**

It has long been known from practical experience that the breakdown of rubber on a mill or in an internal mixer occurs more rapidly at either high or moderately low temperatures than it does at temperatures around 100°C. It is now recognised that breakdown at the more elevated temperatures is due to oxidative scission and that, in the lower temperature range and under the stresses set up during mastication, mechanical rupture of primary bonds is possible, the free radicals thus produced becoming stabilised by addition of oxygen. When mastication is carried out in the absence of oxygen, e.g. in an inert atmosphere, either in an internal mixer or on a closed-in mill, little or no breakdown occurs. Under these conditions, radicals resulting from mechanical shear recombine.

The low-temperature breakdown of rubber is attributed to interaction of rubber radicals with oxygen to give oxygen-terminated fragments of lower molecular weight. Other radical acceptors can be expected to promote break-down of rubber in an inert atmosphere and, moreover, do so in a manner

*Section 4.1.7.2 is based on an abridged version of text by Naunton (*The Applied Science of Rubber*, p. 81, 1961), courtesy Edward Arnold.

capable of quantitative interpretation. This confirms the mechanical rupture hypothesis (Table 4.3) (Bristow and Watson, 1963).

Radical acceptors which promote breakdown of rubber are not to be confused with chemical plasticisers (peptisers). The former are only effective

Table 4.3   BREAKDOWN OF RUBBER IN NITROGEN ATMOSPHERE IN PRESENCE OF ADDITIVES (30 MIN MASTICATION AT 55°C). FROM NAUNTON (1961), COURTESY EDWARD ARNOLD

Additive	Amount (% on rubber)	Resulting Mooney viscosity [ML(1+4)100°C]*
None	—	75
Thiophenol	1·0	22
Iodine	2·35	45
o, o'-Dibenzamidodiphenyldisulphide	4·2	33
Diphenyl picryl hydrazyl	0·01	55
Oxygen	150 torr	19

* Mooney viscosity no., using large rotor, 1 min pre-heating and 4 min running at 100°C—see Section 11.2.1.1 and BS1673, Pt 3.3.6, 1969.

under conditions of cold mastication and do not cause breakdown above 100°C in absence of oxygen. Several of the common oxidation catalysts and peptisers (e.g. *o,o'*-dibenzamidodiphenyldisulphide) are also radical acceptors, and accordingly function as such in cold mastication; at higher temperatures they cause no breakdown in absence of oxygen but strongly catalyse oxidative

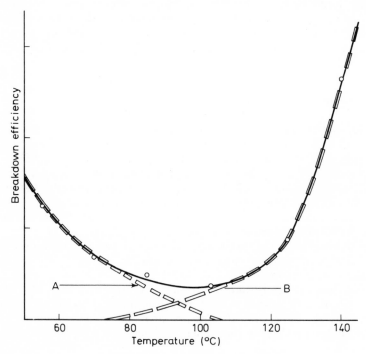

Fig. 4.1   Breakdown of NR after 30 min mastication at different temperatures. From Naunton (1961), courtesy Edward Arnold

breakdown, whereas some of the radical acceptors (e.g. pyrogallol, β-naphthol) retard it.

As the temperature reached in a mass of rubber during mastication is increased, the rubber becomes progressively softer and the rate of mechanical rupture is correspondingly reduced. Hence, the breakdown efficiency in air or with radical acceptors under nitrogen falls with increasing temperature (Fig. 4.1). A second type of breakdown due to conventional oxidative scission becomes apparent above 80°C, and this increases rapidly with temperature. The overall behaviour of rubber masticated in air is expressed by the continuous curve in Fig. 4.1, in which stabilisation by oxygen of radicals formed by mechanical rupture is responsible for the left-hand part (A) of the curve, and oxidative degradation is responsible for the right-hand part (B). Degradation is at a minimum at 105°C. The different reaction mechanisms responsible for the two portions of the breakdown curve explain why the right-hand portion responds to additions of common oxidation inhibitors (antioxidants) while the left-hand portion is insensitive to such additions, and, although requiring the presence of oxygen, is not dependent upon the amount of oxygen present (i.e. the oxygen pressure) over a wide range. The left-hand curve also explains why rubbers of widely different oxidisabilities undergo cold breakdown at comparable rates.

Part B General Purpose and Non-Oil-Resistant Synthetic Rubbers

BY G. J. VAN DER BIE, J. M. RELLAGE, AND C. VERVLOET

4.2 Styrene–Butadiene Rubbers—SBR

4.2.1 EMULSION SBR

Emulsion SBR, for long the only synthetic general purpose rubber, has blossomed from the few 'hot' types produced during World War II into the multiplicity of grades now available in 'cold' types: straight and oil-extended, black masterbatches, high-styrene resins, latices, etc.

In general, emulsion polymerisations give rise to non-stereospecific, and hence amorphous, polymers which will not crystallise, even on stretching their vulcanisates. Therefore, pure gum strengths are generally low, but high strength values and excellent abrasion resistance can be obtained by incorporation of reinforcing black or white fillers.

4.2.1.1 *Composition and Structure*

SBR is a copolymer of styrene (or vinyl benzene: $CH_2\!=\!CH\!-\!C_6H_5$) and butadiene ($CH_2\!=\!CH\!-\!CH\!=\!CH_2$). With the exception of some special grades, the styrene content is 23.5% wt, i.e. a molecular proportion in the chains of one styrene to about six or seven butadienes. The monomers are randomly arranged in the chain, whereas the butadiene fraction is for the greater part in the *trans* configuration. Table 4.4 gives a typical example.

Table 4.4 POLYMER PROPERTIES OF COLD STRAIGHT SBR (S-1500)

Property	Value
Arrangement of monomers	Random
cis-1, 4-Butadiene (% wt)	9
trans-1, 4-Butadiene (% wt)	76
1, 2-Butadiene (vinyl) (% wt)	15
Intrinsic viscosity (dl/g)	2·0
Gel content	Negligible
Mooney viscosity [ML(1 + 4) 100°C]	50
Specific gravity	0·94

Table 4.5 INGREDIENTS FOR EMULSION POLYMERISATION

Ingredient	Function during or after polymerisation	Types in current use	Effect on polymer
Emulsifier system*	To keep monomers and polymer in emulsion	1. Fatty acid 2. Fatty acid–rosin acid 3. Rosin acid	1. No tack, fast curing, non-staining 2. Some tack, slower curing, slightly staining 3. Sufficient tack, slow curing, staining
Catalyst system (redox type)	Initiator and activator	Organic hydroperoxide (oxidising agent) and sodium formaldehyde sulphoxylate (reducing agent); ferrous sulphate (reducing agent); EDTA† (sequestering agent)	
Modifier	Regulation of chain length; prevention of branching	Tertiary dodecyl mercaptan	Wider molecular weight distribution improves processing
Short-stop system	To stop the reaction at the required conversion	Soluble dithiocarbamates and sodium polysulphide or polyamine H (Cyanamid)	Short-stop remaining in the polymer will affect cure time
Stabiliser	Antioxidant	Staining, or non-staining	Keeps the polymer in good condition for at least one-year storage; can affect cure properties (e.g. speed, crosslinking) and end-use properties
Coagulant system	Flocculation of the latex	1. Salt–acid 2. Acid–alum 3. Glue–acid 4. Glue–alum	1. Polymer slightly hygroscopic 2–4. Polymer not hygroscopic; improved electrical properties

*In addition: a small amount of secondary emulsifier (e.g. a condensation product of formaldehyde and sulphonated naphthalene) and some electrolyte (e.g. potassium sulphate).
†Ethylene diamine tetraacetic acid.

Table 4.6 PRODUCERS OF GENERAL PURPOSE SYNTHETIC RUBBERS IN W. EUROPE

Rubber	Country	Producer	Trade names	Type*
Emulsion SBR	France	Firestone-France	FR-S	
		Elastomer	Ugitex S	
		Polymer Corpn	Polysar	
			Krylene	
			Krynol	
		Compagnie Française des Produits Chimiques Shell	Cariflex S	
	Italy	ANIC	Europrene	
	Netherlands	Chemische Industrie AKU-Goodrich	Hycar/Ciago	
		Shell Nederland Chemie	Cariflex S	
	W. Germany	Bunawerke Hüls	Buna Hüls	
			Duranit	
		Synthomer Chemie	Synthomer	
	UK	Doverstrand	Revinex	
			Butakon	
		International Synthetic Rubber	Intol	
			Intex	
Solution SBR	Belgium	Petrochim	Solprene	
	Spain	Calatrava, Empresa la Industria Petroquimica	Solprene	
	UK	International Synthetic Rubber	Unidene	
Solution IR	Netherlands	Shell Nederland Chemie	Cariflex IR	
Solution BR	France	Firestone-France	Diene	Low-*cis*
		Compagnie Française des Produits Chimiques Shell	Cariflex BR	High-*cis*
	Italy	ANIC	Europrene Cis	Medium-high-*cis*
	Spain	Calatrava, Empresa la Industria Petroquimica	Solprene	Low-*cis*
	W. Germany	Stereo-kautschuk-Werke	Buna CB	High-*cis* Medium-high-*cis*
	UK	International Synthetic Rubber	Intene	Low-*cis*
IIR	France	Compagnie Française Raffinage	Total Butyl	
		Société du Caoutchouc Butyl		
	Belgium	Polysar Belgium	Polysar	
	UK	Esso	Esso Butyl	
EPM and EPDM	Italy	Montesud Petrochimica	Dutral N, EPM, EPDM	
	Netherlands	Netherlandse Staatsmijnen	Keltan	
	UK	International Synthetic Rubber	Intolan	

*High-*cis* = 97%; medium-high-*cis* = 92%; low-*cis* = 35–45%.

4.2.1.2 *Manufacture*

The monomers, present in the weight ratio of 30 styrene to 70 butadiene, are emulsified in de-ionised water. The polymerisation reaction is carried out at either about 4°C ('cold' SBR grades) or about 42°C ('hot' SBR grades). The latter represents the older method.

The other ingredients, their function in the polymerisation process, and their influence on final properties are summarised in Table 4.5.

4.2.1.3 *Commercially Available Emulsion SBR*

The numerous grades available now are classified by a rather elaborate numbering system which covers the following variables: hot or cold polymerisation, styrene content, emulsifier, stabiliser, coagulant, the type and amount of oil (in the oil-extended grades), and the type and amount of black (in the black masterbatch and oil–black masterbatch grades). Table 4.6 lists the producers of emulsion SBR in W. Europe.

Information on available types and grades is given in the annual publication of the IISRP, *Description of Synthetic Rubbers and Latices*, and the IISRP numbering systems for emulsion SBR and other rubbers are given in Tables 4.7 and 4.8.

Table 4.7 THE IISRP NUMBERING SYSTEM FOR EMULSION SBR

Series	Rubber
1000	Hot-polymerised non-pigmented rubbers
1500	Cold-polymerised non-pigmented rubbers
1600	Cold-polymerised black masterbatch with 14 or less parts of oil per 100 parts of SBR
1700	Cold-polymerised oil masterbatch
1800	Cold-polymerised oil–black masterbatch with more than 14 parts of oil per 100 parts SBR
1900	Emulsion resin rubber masterbatches

Table 4.8 THE IISRP NUMBERING SYSTEM FOR STEREO AND RELATED RUBBERS

	Polybutadiene*	Polyisoprene*	Ethylene–propylene
Dry polymer	1200–1249	2200–2249	100–149
Oil-extended†	1250–1299	2250–2299	150–199
Black masterbatch	1300–1349	2300–2349	200–249
Oil–black masterbatch‡	1350–1399	2350–2399	250–299
Miscellaneous	1450–1499	2450–2499	350–399

*Includes copolymers.
†Dry polymer containing any quantity of oil.
‡Black masterbatch containing any quantity of oil.

4.2.1.4   *Properties*

The properties are influenced by the variables mentioned in Section 4.2.1.3. Some of these influences, namely those caused by ingredients for the polymerisation reaction, are listed in Table 4.5.

The influence of other variables on the properties of SBR is as given under the following headings.

Hot Polymerisation. Hot-polymerised grades are characterised by highly branched chains and contain substantial amounts of microgel. This branching generally makes the polymers more difficult to process, although in some respects (low mill shrinkage, good dimensional stability, and good extrusion characteristics) processing is better than that of 'cold' SBR. The vulcanisate properties of hot-polymerised grades are poorer than those of the cold grades.

Cold Polymerisation. In cold-polymerised SBR, the linearity of the chains is much improved and gel is absent. These grades do not require premastication and are easy to process.

Oil Extension. The successful development of the cold-polymerisation process made it possible to produce higher molecular weight polymers, which could be oil extended (OEP).

Generally, an easy-processing polymer with good vulcanisate properties should contain a high proportion of straight chains and have a wide molecular-weight distribution. Such a polymer contains 15–20% of low molecular weight material in the range up to $M = 30\,000$—which does not crosslink upon curing and acts merely as a processing aid—and it should also contain a high molecular weight fraction, e.g. M around 10^6, to obtain satisfactory final properties (Buckley, 1959). To a great extent, mineral oils can take over the role of the low molecular weight fraction in aiding processability and in realising economic-grade polymers. Three types of mineral oil, which are incorporated as an emulsion in the SBR latex prior to coagulation, are in general use: the *highly aromatic* which are general purpose, staining, and give excellent processability; the *aromatic*, which are general purpose, staining, and give good processability; and the *naphthenic*, which are for light-coloured products and non-staining applications; the processability of these oils is satisfactory, provided the paraffinic content is low, whereas the strength properties are generally lower than for the other types of mineral oil.

The affinity of the different types to SBR, and in general to diene elastomers, diminishes in the order aromatic–naphthenic–paraffinic. Paraffinic oils cannot be used with diene elastomers as they are incompatible; they 'sweat out', as also do the paraffinic vaselines and paraffinic waxes, forming a 'bloom'.

Black and Oil–Black Masterbatches. Although attempts to mix carbon black with NR in the latex stage were made soon after the establishment of blacks as reinforcing agents—i.e. around 1920—it was not until the development of emulsion SBR latex technology that manufacture on a commercial scale was realised. In the early processes, the dry black, dispersed by mechanical agitation in water containing an anionic surface-active agent, was mixed with the latex, whereupon the whole mass coagulated. These grades exhibited poorer final properties compared with those obtained from mill-mixed compounds, owing to the inactivation of part of the black by absorption of the anionic surfactant. The use of surfactant is avoided either by high-energy agitation processes (Scott and Eckert, 1967) or by blending the black with the wet coagulated crumb in an internal mixer. Better black dispersions than in conventionally mixed black compounds can be obtained in this way, and hence improved quality is realised. Other advantages of these master batches are: easier processing in the rubber-manufacturing plant; shorter mixing cycles; improved extrusion; uniformity of properties; cheaper and cleaner operation of the plant; reduction in number of weighing and mixing operations, and less power consumption (Janssen and Weinstock, 1961).

4.2.1.5   *High-Styrene Polymers*

Only a few grades contain smaller amounts of bound styrene than the 23·5% referred to above, the purpose mainly being to obtain improved low-temperature service properties. Grades, however, containing 30–50% wt of bound styrene find use in the treads of so-called high-hysteresis tyres, which exhibit a markedly improved road grip. Polymers with styrene contents up to 90% wt—known as high-styrene resins—are important compounding ingredients to impart increased hardness and stiffness, e.g. to obtain leather-like properties in shoe soling (Sections 6.2.4.5 and 6.3.9.3).

4.2.1.6   *Divinyl Benzene Crosslinked SBR*

Good processing properties, such as low die swell, dimensional stability of extrudates, smooth calendering, and intricate injection moulding, are obtained by giving the rubber a more thermoplastic character. This, however, is not to the benefit of good elastic properties in the vulcanisate. A good compromise is achieved by incorporating in the compound a polymer that has—at the emulsion polymerisation stage—been partially crosslinked with divinyl benzene; such an elastomer is available in the type 1009, for example.

4.2.2   SOLUTION SBR

About a decade ago the rubber industry made the acquaintance of the first members of a new family of rubbers, namely those which are prepared in solution. The rapid expansion of the solution rubbers was the direct consequence of the development of stereospecific catalysts. Ziegler and Natta developed systems based on a combination of a coordination catalyst (mostly chlorides of cobalt, nickel, vanadium, titanium, and other transition elements) with a reductive chemical. Of these, derivatives based on aluminium (mostly alkylated aluminium chlorides) are widely used (Horne *et al.*, 1956). Staveley and co-workers used catalysts based on lithium metal or a lithium

E

alkyl derivative (Staveley *et al.*, 1956). In addition, others, such as the oxides of nickel, cobalt, and chromium, were developed by Standard Oil of Indiana and by Phillips.

4.2.2.1   *Manufacture and Structure*

These grades of SBR are often manufactured in plants that can also produce low-*cis* polybutadiene. Polymerisation is usually carried out with an alkyl lithium type catalyst in a non-polar solvent, which results in a polymer with almost equal amounts of *cis* and *trans* configurations and a vinyl content below 10%.

Activation of the catalyst, e.g. by complexing with ethers, results in a more random microstructure. Such products have a *cis* content of about 30% and a vinyl content of about 25%. Increasing the polarity of the solvent gives a further increase in the vinyl configuration (Hsieh, 1965).

These differences in microstructure (i.e. in vinyl, *cis*, and *trans* configurations) are reflected in the hysteresis properties of the vulcanisate (Table 4.9) (Krol, 1968).

In various respects the low-vinyl-content solution SBR grades behave, therefore, like blends of butadiene rubber with the classical emulsion SBR, whereas their polymer purity, and hence water absorption, are superior, as is shown in Table 4.10.

The different possibilities of combining the two monomers styrene and butadiene, either randomly arranged or in blocks, yield a variety of products

Table 4.9   RESILIENCE OF SBR TYPES

	Solution SBR, random types		Emulsion SBR (1500)*
Microstructure (% wt)			
cis	40	30	10
Vinyl	8	25	12
Lupke resilience† of tread formulation (%)	62	52	48

*Random copolymer, cold-polymerised, staining type.
†See Section 11.3.12.

Table 4.10   COMPARISON OF FEATURES OF SOLUTION AND EMULSION SBR

	Solution SBR	Emulsion SBR
Styrene configuration	Block or random	Random
Ash (% wt)	0·1	0·75
Organic acid (% wt)	0–0·5	6·25
Rubber hydrocarbon (% wt)	98 +	92
Monomer incorporation	Controlled	Random
Molecular-weight distribution	Narrow	Broad
Long-chain branching	Slight	More extensive
Colour	White	Light to dark

with widely diverging characteristics, a survey of which is given in Table 4.11 (Krol, 1968). These SBR copolymers comprise not only rubbers in the conventional sense, but also those with a thermoplastic character (column 4 of Table 4.11), which do not need a curing system but contain a reversible vulcanisation mechanism (Section 4.7).

The rubbers of the conventional type can be divided into materials which are completely random (see columns 1 and 2 of Table 4.11), and those which

Table 4.11 CHARACTERISTIC PROPERTIES OF SBR MADE IN SOLUTION

Elastomer type	Classical			Thermoplastic
Vulcanisation	With sulphur, irreversible			Reversible via domain formation
Structure	Random		Certain degree of blockiness	Tailor-made blocks $S_xB_yS_x$
cis	40	30		
Vinyl	<10	25		
Average molecular weight	$(200-500)\times 10^3$			$(80-160)\times 10^3$
Glass transition point	$-40°C$	$-30°C$	No clear transition point	Two clearly separate glass transition points, $+90°C$ and $-70°C$
Amount of bound styrene (% wt)	10–25		25–50	30–45
Bound styrene present in blocks (%)	0		Approx. 60	100

have a certain degree of blockiness (column 3). In the latter case, approximately two-thirds of the bound styrene is present in the form of blocks with a size of more than five styrene monomer units. These blocks give a thermoplastic character to the elastomer, resulting in an improvement of certain aspects of processability (less nerve, improved extrusion characteristics, and stability of milled sheets) and also in harder but less elastic vulcanisates. To first approximation, therefore, the effect of this block formation appears to be identical with that of adding a high-styrene resin to a normal SBR.

These blocks result from the significant difference in the polymerisation rates of the monomers. When they are mixed in the final ratio required, mainly polybutadiene is initially generated, whereas towards the end of the reaction, mainly polystyrene with some built-in butadiene is formed. So 'plastic blockiness' is formed, particularly in the final stage of the reaction.

Random monomer distribution is obtained by polymerising a mixture of both monomers at a constant concentration of butadiene, i.e. by supplementing butadiene during the reaction.

Owing to their narrow molecular-weight distribution (Fig. 4.2), these rubbers process differently from emulsion SBR. For the random copolymers, compounding has often to be preceded by a premastication step. In the presence of small amounts of organic acids, the mechanical breakdown is

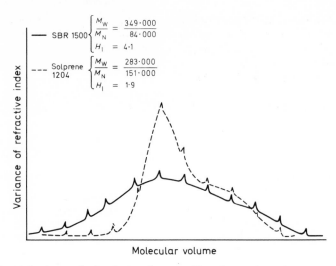

Fig. 4.2   *Molecular-weight distribution, determined by gel permeation chromatography (M_W = weight average molecular weight, M_N = number average molecular weight, H_1 = heterogeneity index): SBR 1500 is a random polymerised emulsion SBR, cold type; Solprene 1204 is a medium-cis high-vinyl random type solution SBR. Courtesy Phillips Petroleum Co.*

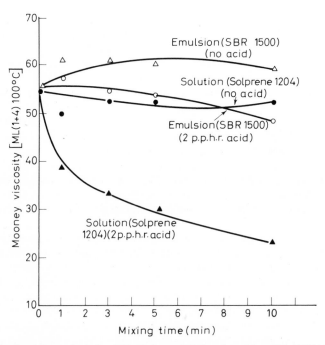

Fig. 4.3   *Breakdown of solution SBR (Solprene 1204) and emulsion SBR (1500) in the presence of stearic acid (all stocks mixed at 120°C, 60 rev/min, in midget internal mixer). Courtesy Phillips Petroleum Co.*

rapid (see Fig. 4.3). The block copolymers, on the other hand, are more comparable with emulsion SBR in that they do not show rapid breakdown.

4.2.2.2 *Commercially Available Solution SBR*

Table 4.6 lists details of suppliers of these rubbers.

4.3 Polyisoprene Rubbers—IR

4.3.1 MANUFACTURE

The lengthy search for a means of producing synthetic polyisoprene, which should be comparable with NR, culminated in the successful use of stereospecific catalysts.

Commercial production of such polyisoprene rubber became possible as

$$H_2C=\overset{\overset{\displaystyle CH_3}{|}}{C}-CH=CH_2$$

soon as the monomer isoprene could be produced cheaply.

4.3.2 COMMERCIALLY AVAILABLE POLYISOPRENE RUBBERS

The polymers are at present manufactured by several companies, in straight as well as oil-extended grades. Shell, who have manufacturing plants in the USA and in Europe, apply alkyl lithium as the catalyst (Diem, Tucker, and Gibbs, 1961; d'Ianni, 1961), whereas Goodyear and Goodrich produce IR by applying a Ziegler–Natta type catalyst, e.g. titanium tetrachloride with aluminium trialkyl (Table 4.6) (Lehr, 1963; Schoenberg, Chalfant, and Major, 1964).

4.3.3 STRUCTURE AND PROPERTIES

The analytical data on the different types of IR, in comparison with NR, are shown in Table 4.12.

The importance of a high-*cis*-1,4 content lies in the ability of the rubber to crystallise and yield high tensile strengths in pure gum formulations.

The linear structure of the 92% *cis* IR causes the cold-flow retention to be generally low. To compensate for this, such elastomers are polymerised until very high molecular weights are obtained ($M = 2\cdot5 \times 10^6$ for the straight types and $M = 4 \times 10^6$ for the oil-extended types, compared with 10^6 for NR).

The highly branched structure in 96% *cis* IR prevents cold flow at an LVN of 4–5 dl/g, but it often causes these rubbers to contain high amounts of gel, which can be either a disadvantage (when flow under low shear stress is important) or an advantage (dimensional stability, extrusion). Special grades are available from which the gel has been removed after polymerisation.

The very high molecular weight of the 92% *cis* IR, together with its narrow molecular-weight distribution, necessitates a small amount of pre-mastication, preferably in the presence of 0·1–0·25% of a peptiser such as

Table 4.12   ANALYTICAL DATA OF POLYISOPRENES

Property	Type of polymer		
	Ziegler IR	*Lithium IR*	*NR*
Cis content* (% wt)	96	92	98–100
LVN† (dl/g)	2·5–4·5	8–11	6–7
Gel content (% wt)	10–20	0	High level, depending on age
Macrostructure	Branched	Linear	Branched
Stabiliser content (p.p.h.r.)	1	0·5	2–3‡
Ash content (% wt)	0·15–0·30	0·05	Approx. 0·5
Total metal content (p.p.m.)	400–3000	70	Approx. 1000
Mooney viscosity [ML(1+4) 100°C]	60–90	—	Approx. 120
Colour	Yellow to amber	Waterwhite	Dark

*The remainder consists of 3, 4- and *trans*-1, 4-structures.
†Limiting viscosity no. is the viscosity no. at zero concentration (BS 1673, Pt 6, 1969).
‡Non-rubber constituents.
 (Where ranges are given in the table, they refer to the various grades available.)

Pepton 65 (Table 6.5). Fig. 4.4 illustrates the effect of mastication on the molecular-weight distribution of 92% *cis* IR and NR. Easy-processing grades, which do not require mastication, are also manufactured, but the *cis*-1,4 content of such grades is about 91%. The average molecular weights of 96% *cis* IR grades are medium high; premastication is not required in general.

The difference in macrostructure, in combination with the difference in microstructure, explains why the three types of polyisoprene have different

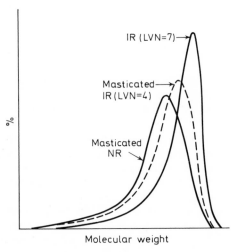

Fig. 4.4   Molecular-weight distribution of raw natural and synthetic (1R-305) isoprene polymers. From Rellage et al. (1969), courtesy Shell Research N.V.

rheological properties. Under high-shear-rate conditions, as for example occur during injection moulding (Krol, Verkerk, and Grijn, 1963), 92% *cis* IR exhibits excellent flow, whereas the flow of NR and 96% *cis* IR can be inhibited as a result of shear-induced crystallisation (Smit and Vegt, 1967, 1969). Under low-shear-rate conditions (milling, calendering, Mooney-viscosity determination), the flow is controlled by the macrostructure rather

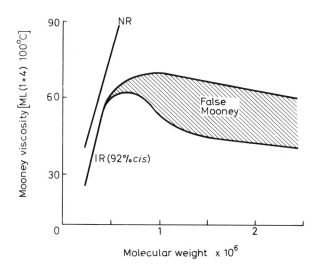

Fig. 4.5 *Variation of Mooney viscosity with molecular weight: 92% cis IR shows false Mooney values above molecular weights of 500 000; the coherence of the rubber is low, and it breaks up into lumps in the Mooney instrument. From Hannsgen (1961), courtesy Rubber and Technical Press Ltd*

than the *cis* content. As illustrated in Fig. 4.5 (Hannsgen, 1961), Mooney values of 92% *cis* IR at molecular weights of 5×10^5 and above are of no significance, and for the measurement of plasticity a static instrument such as the Hoekstra plastimeter should be used. Also as a result of difference in branching, for example, the green compound strength of 92% *cis* IR is lower than that of the other polyisoprenes (see Fig. 4.6). The incorporation of 'promoters' such as *N*-(2-methyl-2-nitropropyl)-4-nitrosoaniline ('Nitrol': Monsanto) or quinone dioxime (GMF) is often used to make stocks containing large amounts of 92% *cis* IR suitable for the building of carcasses.

The low viscosity of masticated 92% *cis* IR requires special precautions during compounding. The use of liquid or low-melting-point accelerators, addition of sulphur at an early stage and preferably as a masterbatch is often recommended. In this respect, 96% *cis* IR behaves more like NR (Todd, 1962).

For both synthetic polyisoprenes, calendering and extrusion are usually carried out at some 15°C lower when compared with NR. The good flow properties result in extrusion at lower pressure and in calendering to a more uniform thickness, while also higher speeds can be applied.

IR has proved to be the only synthetic rubber which, like NR, exhibits self-reinforcement at high elongations in combination with a very low hysteresis. The latter holds particularly for the 92% *cis*-1,4 elastomer, the heat build-up of which is even lower than that of NR. This explains why IR

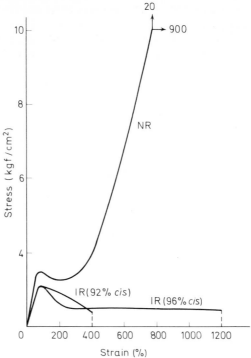

Fig. 4.6   Green strength of tread compounds based on NR and IR: strain rate, 10 min⁻¹; temperature, 25°C. From Rellage et al. (1969), courtesy Shell Research N.V.

has penetrated so deeply into the main application of NR, namely the truck tyre (Rellage *et al.*, 1969).

4.4   Polybutadiene Rubbers—BR

4.4.1   MANUFACTURE

In essence, three types of polybutadienes are manufactured (Bahary, Sapper, and Lane 1967; Sims, 1967):

1. The high-*cis*(97%)-1,4-polybutadienes, polymerised by a Ziegler–Natta type catalyst system, consisting of either a cobalt or a nickel salt, or organic compounds of these metals, with an alkylated aluminium halide.
2. The medium-high-*cis*(92%)-1,4-polybutadienes, also polymerised by a Ziegler–Natta type catalyst system, the transition metal being here titanium, instead of cobalt or nickel.
3. The low-*cis*(around 40%)-1,4-polybutadienes, polymerised in the presence of an alkyl lithium type catalyst.

4.4.2   COMMERCIALLY AVAILABLE POLYBUTADIENE RUBBERS

Today, solution polymerised BR is manufactured in straight as well as oil-extended grades by a large number of companies (Table 4.6).

Besides these solution types, there is some emulsion BR available which mainly finds its application in high-impact polystyrene. The manufacture of these rubbers is identical with that of emulsion SBR, and hence the structure equals that of the butadiene part of the latter copolymer (Duck and Locke, 1968).

4.4.3 STRUCTURE AND PROPERTIES

The microstructure of the various solution polybutadienes is shown in Table 4.13. For the sake of comparison and completeness, emulsion BR and Alfin-type BR—the latter polymerised with a catalyst consisting of the sodium salt of an alcohol—are also included.

With the exception of silicone, BR has the lowest glass transition temperature and highest resilience of known rubbers. The expected low heat

Table 4.13 MICROSTRUCTURE OF POLYBUTADIENES

Catalyst system	cis (%)	trans (%)	Vinyl (%)
Nickel-based	98	1	1
Cobalt-based	97	2	1
Titanium-based	92	4	4
Butyl lithium	36	54	10
Emulsion (Redox system)	9	72	18
Alfin	2	76	22

generation under dynamic condition, was not, however, found, owing to the hysteresis of BR being highly dependent on the degree of deformation (Table 4.14) (Krol, 1968).

This anomalous behaviour may be explained by three effects, which upset the structural uniformity of the chains. In the first place, the low molecular weight fractions of high-*cis* BR of poor structural uniformity and both *trans*-1,4 and vinyl configurations give rise to high heat build-up figures Secondly, high-*cis* BR is of a highly branched structure (Bahary, Sapper,

Table 4.14 HYSTERSIS PROPERTIES OF TREAD-TYPE VULCANISATES

Base polymer	Degree of deformation <2%	Degree of deformation 17·5%
	Lupke* resilience at room temperature (%)	Heat build-up over 38°C (°C)
BR	64	33
NR	60	20
IR (92% cis)	58	18
SBR (emulsion)	40	40

*See Section 11.3.12.

and Lane, 1967), which prevents the formation of an ideal elastomeric network upon vulcanisation. Finally, high-*cis* BR polymers are prone to *cis–trans* isomerisation when vulcanised with the usual sulphur-accelerator systems, and when peptised with the usual peptisers. This isomerisation can be inhibited by vulcanising the compounds in the presence of amines. Most of these, however, increase the scorchiness considerably and are therefore of little practical significance.

Still another peculiarity of BR, which is inherent in its structure, is the 'spontaneous crosslinking' which occurs during vulcanisation. BR therefore requires much less sulphur than other diene rubbers to obtain the optimum crosslink density, as can be seen from Fig. 4.7 for pure gum vulcanisates (Vervloet, 1969).

Microstructure also affects the crystallinity of BR polymer. Crystallinity has been observed in high-*cis*, high-*trans*, and high-vinyl polybutadienes. The melting point of 100% *cis* polybutadiene is 4°C, which decreases with

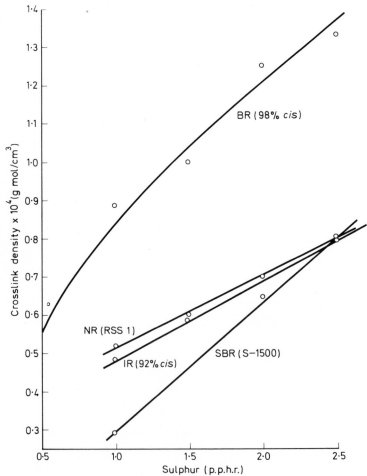

Fig. 4.7   *Crosslink densities in relation to sulphur content of pure gum vulcanisates. From Vervloet (1969), courtesy Shell Chemical Co. Ltd*

decreasing *cis* content. Hence, most high-*cis* polybutadienes do not crystallise at room temperature on stretching, and pure gum vulcanisates are not very strong; tensile-strength values fluctuate between 50 kgf/cm^2 and 80 kgf/cm^2.

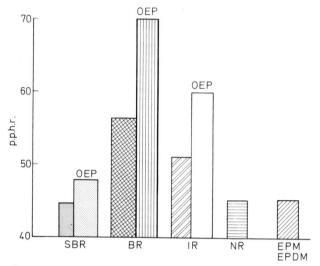

Fig. 4.8   Amount of HAF black required for a hardness of 60 IRHD

The high flexibility of the chains allows the incorporation of high amounts of black and oil, as is evident from Fig. 4.8, where the optimum situation for different general purpose elastomers is given.

Unlike NR and IR, BR and SBR resist breakdown under mechanical shear; BR polymers are stable in mill or internal mixer operations and present no processing difficulties in this respect. Owing to its low elastic recovery in the unvulcanised state, however, there is a strong tendency to bag when it is processed on a mill. The green strength is low; it can be improved by the use of high structure blacks, but at the cost of heat build-up.

4.5   Isobutene–Isoprene (Butyl) Rubbers—IIR

4.5.1   MANUFACTURE

Rubberlike copolymers of isobutene or 2-methylprop-1-ene

$$H_2C{=}\underset{\underset{CH_3}{|}}{\overset{\overset{CH_3}{|}}{C}}$$

and isoprene were first produced in the USA just before World War II.

$$H_2C{=}\underset{\underset{}{|}}{\overset{\overset{CH_3}{|}}{C}}{-}CH{=}CH_2$$

The monomers are polymerised in solvents such as methyl chloride, and at a low temperature ($-80°C$), using Friedel–Crafts catalysts such as $AlCl_3$ or BF_3.

4.5.2 COMMERCIALLY AVAILABLE BUTYL RUBBERS

A range (with isoprene contents between 0·5 and 3·0 mole %) is now commercially available (Table 4.6) (Buckley, 1959).

4.5.3 STRUCTURE AND PROPERTIES

Although the catalytic system is not essentially stereospecific, the polymeric chains contain a very regular structure, owing to the symmetrical nature of the isobutene monomer. Hence butyl elastomers are self-reinforcing, with a high pure gum tensile strength (250 kgf/cm^2). The abundance of methyl side-groups in the polymer chains brings about a considerable steric

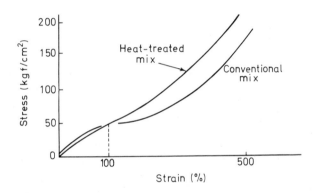

Fig. 4.9   Influence of heat treatment on stress–strain properties of IIR–MPC black elastomer. From Morton (1959), sponsored by the Rubber Division of American Chemical Society, Copyright 1959, Litton Educational Pub. Co., with permission

hindrance to elastic movements; although T_g values of around $-65°C$ have been measured, the resilience of vulcanisates at ambient temperatures is very low (about 14% rebound). On the other hand, the densely packed structure of these elastomers causes the gas permeability to be very low, and, because of this, for a long time the main application of butyl rubbers was for inner tubes.

Mainly as a result of the rather rigid, highly saturated chains, the polymer excels in ozone and weathering characteristics, heat resistance, chemical resistance, and abrasion resistance.

Perhaps the most important discovery in the compounding of butyl rubber is the effect of heating a masterbatch of butyl rubber and filler in an internal mixer at temperatures of $180°–230°C$, preferably activated by an aromatic nitroso compound such as *p*-dinitrosobenzene (Polyac), *p*-nitrosophenol (Butylac 33), or *N*-nitroso-*N*-methyl-*p*-nitrosoaniline (Elastopar).

The reaction between polymer and filler persists during subsequent processing and vulcanisation, and results in improved tensile and dynamic properties (Figs 4.9 and 4.10) (Section 9.4.1) (Gessler *et al.*, 1953).

The cure rate of polymers with low unsaturation is slow, but the more highly unsaturated grades will cure in the conventional manner with sulphur and accelerators. For the lower-unsaturated polymers, quinone dioxime (GMF) or dibenzoyl quinone dioxime (dibenzo GMF), together with red lead,

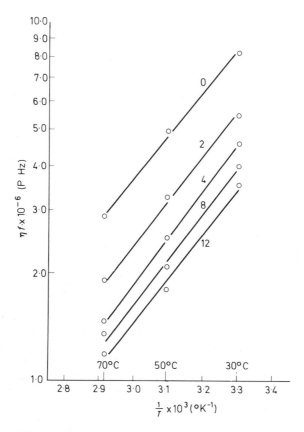

Fig. 4.10   Effect of heat treatment on damping properties of IIR; figures refer to the number of cyclic heat treatments. From Buckley (1959), courtesy American Chemical Society

is recommended. Curing with phenol formaldehyde resins results in vulcanisates with excellent ageing and heat-resistance properties.

A disadvantage is the incompatibility with diene-type rubbers: moreover, its low unsaturation prevents a good covulcanisation.

Butyl rubbers, in which a hydrogen of the isoprene units has been replaced by halogen, are available in the form of chlorobutyl (CLIIR) or bromobutyl. Because of increased polarity, these modified butyl rubbers are better able to be blended with general purpose rubbers. Moreover, because of higher activity of the adjacent double bond, covulcanisation is greatly improved (Brzenk and Booth, 1968).

4.6   Ethylene–Propylene Rubbers—EPM and EPDM

4.6.1   MANUFACTURE

These so-called polyolefin rubbers are produced in two main types: the saturated copolymers (EPM) and the unsaturated terpolymers (EPDM).

4.6.1.1   *Copolymers*

The monomers are copolymerised in solution with Ziegler–Natta type catalysts, such as vanadium oxychloride ($VOCl_3$) plus an alkyl aluminium. The ratio in which the monomers are polymerised does not depend on the ratio in which they are present, but on the nature of the catalyst. A fully saturated copolymer of ethylene and propylene is formed, having the following structure:

$$\left[- CH_2 - CH_2 - \right]_x \left[\begin{array}{c} CH_3 \\ | \\ -CH \end{array} \diagdown CH_2 - \right]_y$$

4.6.1.2   *Terpolymers*

The EPDM type, capable of sulphur vulcanisation, contains, in addition to these olefins, a non-conjugated diene as the third monomer; acetylene and conjugated dienes, like butadiene and isoprene, are not uniformly incorporated and, moreover, alter ethylene–propylene copolymerisation with the usual catalysts.

The choice of third monomers is, therefore, not wide, as their reaction speed in polymerisation should comply with that of ethylene and propylene, in order to obtain a random terpolymer chain. Until some years ago dicyclopentadiene (DCPD) was mostly used, but these rubbers are slow curing and, therefore, cannot be co-cured with diene rubbers. The recent trend is towards ever-faster curing grades, and most companies now incorporate ethylidene norbornene (ENB) as the third monomer:

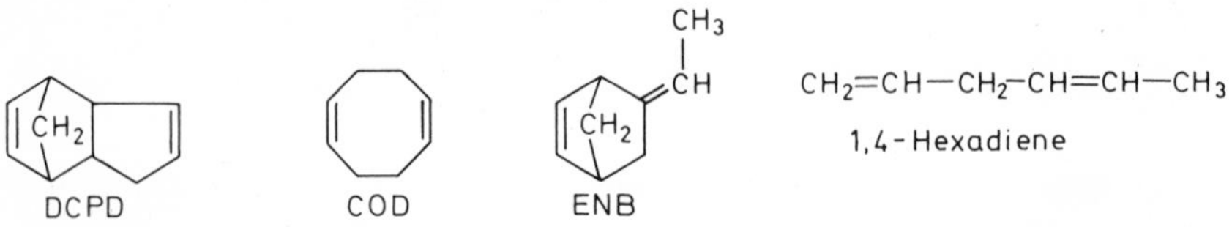

$$CH_2{=}CH{-}CH_2{-}CH{=}CH{-}CH_3$$

1,4-Hexadiene

DCPD   COD   ENB

Monomers used to provide unsaturation in EPDM rubbers

The effect of the vulcanisation speed of EPDM on the ability to co-cure with SBR is clearly illustrated in Fig. 4.11. It should be noted that the unsaturation in the EPDM terpolymers is located in a pendant side-group and not in the polymer backbone itself.

The high price of the third monomer for the terpolymers, sheds a somewhat sobering light on former statements that polyolefin rubber, by virtue

of the low cost of ethylene and propylene in comparison with styrene, butadiene, and particularly isoprene, would become the bulk rubber of the
future. Either the expensive peroxide for the vulcanisation of EPM or the

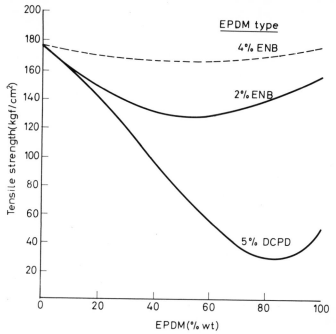

Fig. 4.11   Tensile strength of EPDM (Keltan)–SBR (1500) blends. Courtesy DSM

diene monomer in EPDM may well restrict the wide applications of polyolefin elastomers.

4.6.2   COMMERCIALLY AVAILABLE ETHYLENE–PROPYLENE RUBBERS

The saturated rubbers, which have to be cured with peroxides, initially
developed more rapidly. The development of the EPDM polymers took place
slowly after 1963 but has now overtaken that of the saturated type.
    The manufacture of these polymers was pioneered about 1955 by the
Montecatini Co. in Italy, which was immediately followed by other firms
(Table 4.6) (Natta *et al.*, 1963-1, 1963-2; German, Vaughan, and Hank, 1967).

4.6.3   STRUCTURE AND PROPERTIES

In order to obtain good elastomeric properties, the monomers have to be
arranged randomly in the polymer chains, while the ethylene:propylene
ratio fluctuates around 50:50 (although 65:35 is still possible).
    As a consequence, these polymers are predominantly amorphous and

non-stereoregular, and, therefore, the pure gum vulcanisates show low
tensile strengths.

Whereas the butyl elastomers are highly damping at ambient temperatures, the polyolefin elastomers are highly resilient and show a rebound–
temperature curve which approximates to that for NR (Fig. 4.12).

The saturated EPM grades are vulcanised with peroxides. Scission reactions can be controlled, crosslinking efficiency increased, and, consequently, vulcanisate properties considerably improved by the use of vulcanisation co-agents, such as sulphur and sulphur-donor agents, maleic

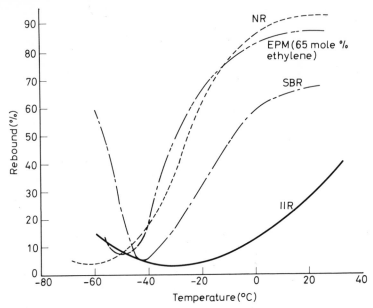

Fig. 4.12   *Rebound temperature curves of various elastomers. From Natta (1963-1), courtesy
American Chemical Society*

anhydride, quinone dioxime. The influence of sulphur on tensile properties
is illustrated in Fig. 4.13.

Sulphur provides polysulphide crosslinks, additional to carbon–carbon
crosslinks, so peroxide vulcanisates of EPM, in the presence of sulphur,
show better properties than those obtained with other co-agents.

The most striking feature amongst the properties of the vulcanisates are
the excellent resistance to oxygen, ozone, and ionisation effects. There is
only little difference between the ageing properties of vulcanisates of EPM
and EPDM, because the unsaturation in the latter is not contained in the
main hydrocarbon chain. An illustration of the excellent ageing properties
is given in Fig. 4.14.

Polyolefin rubbers have a high resistance to molecular breakdown on
milling. Consequently, grades with high Mooney viscosities are not suitable
for mill mixing. The ranges of Mooney viscosities found with currently available polyolefin rubbers are 25–35 (EPM) and 45–120 (EPDM).

Like butyl rubber, these polymers are essentially paraffinic and highly
saturated. Therefore, many of the processing principles developed for butyl

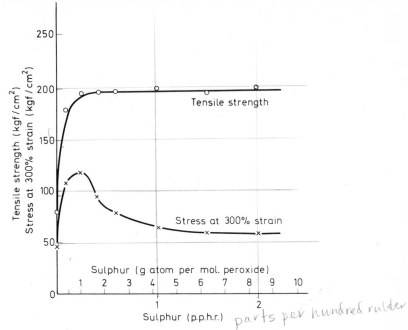

Fig. 4.13   Effect of quantity of sulphur on the tensile strength and the stress at 300% strain of EPM vulcanisates. From Natta (1963-1), courtesy American Chemical Society

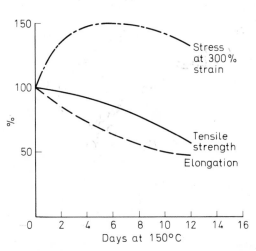

Fig. 4.14   Ageing properties of EPDM vulcanisate in air circulating oven at 150°C (properties of unaged vulcanisate: tensile strength, 195 kgf/cm²; elongation at break, 620%; stress at 300% strain, 87 kgf/cm²). Recipe: EPDM (COD terpolymer), 100; stearic acid, 1; zinc oxide, 5; Circosol 2xH, 5; HAF, 50; sulphur, 0·5; MBT, 0·75; TMTD, 1; vulcanisation at 150°C for 60 min. From Natta (1963-1), courtesy American Chemical Society

can be applied (e.g. heat treatment during compounding in the internal mixer, with or without promoters).

In order to obtain good properties, the use of reinforcing black or white filler is recommended. In general, EPM and EPDM are compatible with

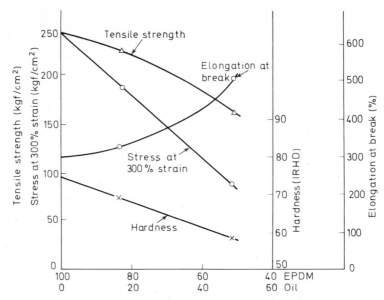

Fig. 4.15   *Effect of the amount of oil on the properties of EPDM vulcanisates. Recipe: EPDM [COD terpolymer, Mooney viscosity ML (1+4), COD 3·73%], 100–50; Circosol 2xH, 0–50; stearic acid, 1; zinc oxide, 5; HAF, 50; sulphur, 1·5; MBT, 0·5; TMTD, 1·5; vulcanisation at 150°C for 60 min. From Natta (1963-1), courtesy American Chemical Society*

oils of different types, although the higher affinities, and hence the better properties, are obtained with paraffinic and naphthenic oils. The oils can be added in large amounts, without too much loss in vulcanisate properties (Fig. 4.15).

EPM requires special extender oils because it is cured by peroxides, and hence residual acid and unsaturation should be absent (Capito, Innes, and Allen, 1968).

4.7   Thermoplastic Rubbers (Thermoelastomers)

In 1965 a new class of polymers, combining high elasticity, typical for rubber vulcanisates, with the processing properties of thermoplastics was announced. Reference to their manufacture has been made in Section 4.2.2.1.

4.7.1   COMMERCIALLY AVAILABLE THERMOPLASTIC RUBBERS

Two major categories are produced: those to be processed in the molten (plastic) state, and those used for solution processing.

At present, the sole commercial manufacturer of these rubbers is Shell: the trade names are Cariflex and Kraton.

4.7.2 STRUCTURE AND PROPERTIES

The polymers have an ordered structure of the general formula A–B–A, where A is a thermoplastic polymer block (polystyrene) with a glass transition point above room temperature, and B is an elastomeric polymer block (e.g. polybutadiene) with a T_g well below room temperature. A good balance

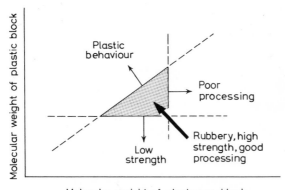

Fig. 4.16   Effect of the molecular weight of the blocks on the properties of thermoplastic elastomers. From Vlig (1968), courtesy Shell Research N.V.

between processing performance and elastomeric character can be obtained at a range of selected molecular weights for A and B. Fig. 4.16 shows this in a general way.

The structure–property relationships of such three-block copolymers have been discussed and published by Holden, Bishop, and Legge (1967),

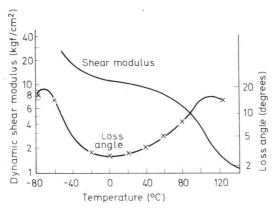

Fig. 4.17   Dynamic shear modulus and loss angle of thermoplastic rubber. From Bie et al. (1968), courtesy Shell Research N.V.

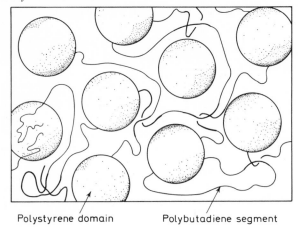

Polystyrene domain   Polybutadiene segment

Fig. 4.18   Model of an S B S block copolymer. From Holden, Bishop, and Legge (1967), courtesy Shell Chemical Co. (USA)

while network characteristics have been discussed by Bishop and Davison (1967) and by Meier (1967).

Owing to the inherent incompatibility of the two polymer blocks in S–B–S polymers, polystyrene and polybutadiene have the tendency to form separate phases as can be seen from the two distinct maxima in the viscous damping curve at temperatures corresponding to the maxima for the respective homopolymers (Fig. 4.17).

Hence at ambient temperatures the polymers contain a number of polystyrene aggregates referred to as domains, dispersed in a continuous matrix of polybutadiene. The domains serve both as multiple crosslinks and as filler particles. A diagrammatic presentation of the molecular arrangement is given in Fig. 4.18. These two effects provide a network with a high degree of phase stability and strength at low, ambient, and elevated temperatures up to the T_g of the polystyrene domain phase.

The behaviour of these polymers under end-use conditions at ambient temperatures is close to that of vulcanised conventional elastomers. A survey of the general characteristics of thermoplastic rubbers is given in Table 4.15.

Table 4.15   PROPERTIES OF THERMOPLASTIC RUBBERS

Elastomeric service	Thermoplastic processing
High elongation—low set	No scorch
Wide range of hardness	Re-use of scrap
Good low-temperature properties	Low viscosity
High grip	Good heat stability
Good abrasion resistance	Use of existing machinery

Fig. 4.19 illustrates the performance of thermoplastic rubbers at low temperature. These polymers behave as an elastic rubber down to temperatures of at least $-60°C$. The modulus of thermoplastic rubber is constant over a wide temperature range. In this respect it is even better than natural

rubber and, in fact, closely follows the pattern of butadiene rubber. Another illustration of the resemblance of S–B–S polymers to vulcanised rubber is given in Fig. 4.20. At room temperature, thermoplastic rubber occupies a position close to that of vulcanised isoprene rubber, as is shown in this stress–strain diagram. Thermoplastics such as plasticised PVC and EVA show a yield point in their curves, above which the deformation is no longer elastic but plastic. This is also reflected in the set at break values, which, notwithstanding the high elongation at break, are very low for thermoplastic rubber namely about 20%, whereas for PVC and EVA they are 150% and 550% respectively.

Owing to the dual chemical nature of the structure, interesting effects are moreover obtained in combination with solvents, resins, and the like (Vlig, 1968). If a solvent readily dissolves the polymer of one block, the other block can be solubilised. Beyond the extremes of the normal solubility range,

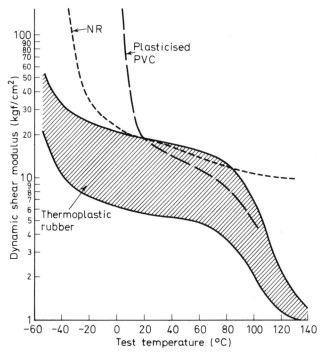

Fig. 4.19   *Dynamic shear modulus as a function of temperature. From Bie* et al. *(1968), courtesy Shell Research N.V.*

elastic gels, thermoplastic mouldable doughs, and high-viscosity solutions can be obtained. In most solution applications, e.g. several adhesives, true solutions are used. S–B–S block polymers generally form such solutions in solvents or solvent mixtures, with a solubility parameter in the range of 8–10.

The materials are easy to process in various types of internal and continuous mixers. A balance of properties can be obtained by addition of oils, fillers, and stabilisers. Conventional fillers act as diluents, but polymeric additives (e.g. polystyrene) have advantageous affects. Processing oils (naphthenic types) improve flow properties as well as reducing hardness and tensile strength.

Processing of these polymers in molten form becomes possible above a temperature of 150°C. At such temperatures the structure softens, and a

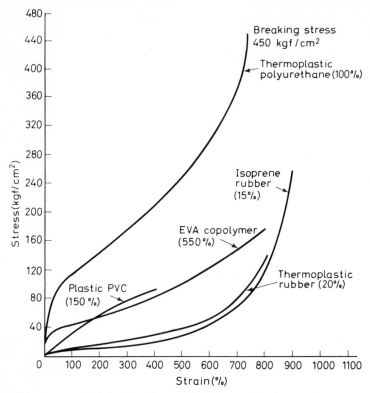

Fig. 4.20   Stress–strain behaviour of various flexible materials (figures in brackets represent set at break). From Breen et al. (1966), courtesy Rubber & Technical Press Ltd

further increase in melt temperature will increase the melt fluidity, so characteristic of thermoplastics. In this sense thermoplastic rubbers are distinctly different from conventional rubber which starts to crosslink under such conditions. Upon cooling the structure solidifies, and this process is reversible.

Part C   Special Purpose Synthetic Rubbers

4.8   Chloroprene Rubbers—CR

BY W. P. FLETCHER

4.8.1   INTRODUCTION

Polychloroprene, the world's first commercial synthetic rubber, became available to the rubber industry in 1935 and rapidly established a firm position as an important raw material, first in the USA, its country of origin, and very soon throughout the world. Soon after its introduction, chloroprene rubbers came to be known by the generic name Neoprene, a term still generally used. More recently manufacture has commenced in other countries, principally in the UK, Germany, Japan, and France.

4.8.2   MANUFACTURE

Chloroprene rubbers are manufactured by polymerising 2-chloro-1,3-butadiene in the presence of catalysts, emulsifying agents, modifiers, and protective agents. The polymer chain is built up through addition of monomer units, of which approximately 98% add in the 1,4 positions (Maynard and Mochel, 1954). About 1·5% add in the 1,2 positions, and these are utilised in the vulcanisation process since in this arrangement the chlorine atom is both tertiary and allylic. Accordingly, it is strongly activated and thus becomes a curing site on the polymer chain (Kovasic, 1955):

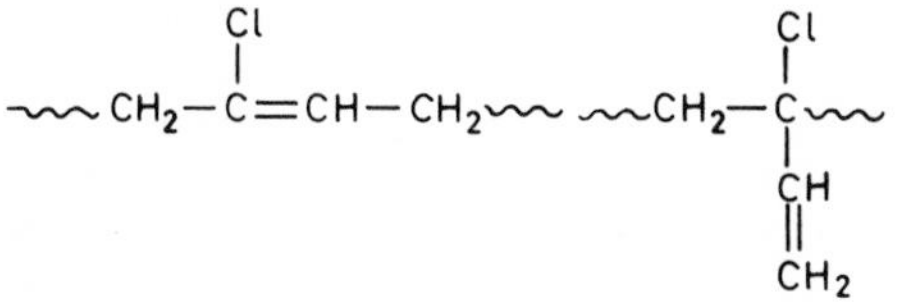

$$\sim\sim CH_2-\underset{\underset{Cl}{|}}{C}=CH-CH_2\sim\sim \qquad \sim\sim CH_2-\underset{\underset{\underset{CH_2}{\|}}{\underset{CH}{|}}}{\underset{\underset{Cl}{|}}{C}}\sim\sim$$

1,4 Addition   1,2 Addition

Crystallisation rate is reduced by modification of the polymer's molecular structure and/or incorporation of a second monomer in the polymerisation reaction.

4.8.3   COMMERCIALLY AVAILABLE CHLOROPRENE RUBBERS

The most commonly used types of CR fall into two classes: the sulphur-modified types, stabilised with thiuram disulphide; and those which contain

111

no sulphur, thiuram disulphide, or other sulphur donors. These differ in vulcanisation characteristics, processibility, and vulcanisate properties.

Within each of these two general classes, a number of polymers are available; these vary primarily in Mooney viscosity and in tendency to crystallise (Murray and Thompson, 1964). In Table 4.16 a short list is given of the most

Table 4.16   CHLOROPRENE RUBBERS PRODUCED IN W. EUROPE

Producer	Trade name	Modifier	Mooney viscosity*	Crystallisation rate†
Du Pont (UK)	Neoprene	Sulphur	M	M, L
		Non-sulphur	L	H
		Non-sulphur	M	H, M, VL
		Non-sulphur	H	H, VL
Bayer (Germany)	Baypren	Sulphur	M	M, L
		Mercaptan	L	H
		Mercaptan	M	H, VL
		Mercaptan	H	M, VL
Distugil (France)	Butaclor	Sulphur	M	M, L
		Mercaptan	L	H
		Mercaptan	M	H, M, VL
		Mercaptan	H	M, VL

*L = low (40 and below); M = medium (around 50); H = high (around 100).
†VL = very low; L = low; M = medium; H = high.

widely used chloroprene rubbers of W. European manufacture. In addition, a number of chloroprene rubbers have been developed, each exhibiting a special combination of raw polymer and/or vulcanisate properties, or with one characteristic greatly enhanced.

4.8.4   PROPERTIES

Pure gum vulcanisates of CR, like those from NR, show high levels of tensile strength. However, to provide optimum processing characteristics, hardness, and durability, the great majority of polychloroprene compounds contain fillers. Detailed information on compounding of CR is contained in a number of publications (Murray and Thompson, 1964; Penn, 1969-1). Within the range of normal compounding, CR vulcanisates have tensile strengths within the range 70–175 kgf/cm^2, elongation at break in the range 200–600%, and hardness in the range 40–95 IRHD. The effects upon basic physical properties of variations in type and quantity of filler generally match the effects in NR. The resilience of a pure gum polychloroprene vulcanisate is, however, lower than that of a similar NR compound. In both elastomers, as filler loading is increased, resilience decreases, though to a lesser extent with polychloroprene. Consequently, the resilience of most practical polychloroprene compounds is above that of NR of similar loading (Fig. 4.21).

CR vulcanisates are generally recognised to be highly resistant to oxidative ageing. When such ageing does occur at high temperatures, the vulcanisate becomes harder and less flexible. For service in air, CR is satisfactory for continuous exposure at 85°–90°C when protected by suitable antioxidants. One such combination which is very effective is a mixture of 4 parts octylated diphenyl amine and 1 part *p*-(*p*-tolylsulphonylamido)diphenylamine. The good long-term outdoor weathering properties of CR vulcanisates containing a minimum of 10% carbon black and antioxidants are well established. Ozone cracking is far less serious with CR vulcanisates than with many other rubbers, and for most applications their inherent ozone resistance is adequate. For very severe conditions, where the product is in service in areas

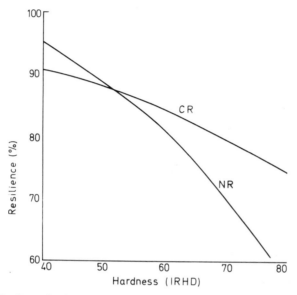

Fig. 4.21   Resilience–hardness curves for CR and NR. Courtesy Du Pont Co. Ltd

of high ozone concentration, or is subject to large surface strains, adequate additional protection is conferred by antiozonants of the *p*-phenylenediamine or other types.

CR vulcanisates show a high level of resistance to flex-cracking. For acute service conditions, this may be further enhanced by compounding with the available protective agents. Because of the chlorine in the molecule, CR has inherent flame resistance, and products made from it are normally self-extinguishing. In compounding, care is needed to avoid nullifying this property by incorporation of major quantities of flammable ingredients such as petroleum-based plasticisers.

Polychloroprene stiffens at low temperatures because of two factors: approach to a second order transition, and crystallisation. The former phenomenon is illustrated by a plot of torsional stiffness against temperature (Fig. 4.22). Stiffening temperatures of CR compounds can be lowered by incorporation of low-temperature plasticisers, such as butyl oleate. Fig. 4.23 indicates the substantial effect of this plasticiser on the stiffening temperature of a 60 p.p.h.r. carbon black loaded compound. Crystallisation of CR takes place under conditions of long exposure to low temperatures and is most rapid in the region of −10°C. The resulting gradual hardening, stiffening,

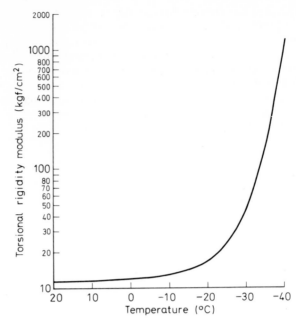

Fig. 4.22   Effect of temperature on stiffness of CR. Courtesy Du Pont Co. Ltd

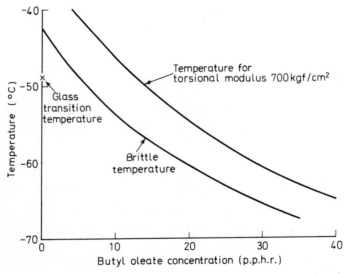

Fig. 4.23   Effect of plasticiser concentration on low-temperature properties of CR. Courtesy Du Pont Co. Ltd

and loss of flexibility are completely reversible on raising the temperature of the material. Crystallisation can occur in raw CR, in uncured compounds, and in vulcanisates, though the rate and extent of crystallisation is much reduced by vulcanisation; incorporation of certain resinous plasticisers and the use of sulphur in the vulcanising system also serve to control crystallisation.

CR vulcanisates are inherently resistant to swelling and to deterioration by a wide variety of animal and vegetable fats and oils, waxes, and greases, as well as most aliphatic hydrocarbons. Contact with aromatic hydrocarbons,

Table 4.17 · THE VOLUME INCREASE OF CR VULCANISATE AFTER 7 DAYS AT 25°C. COURTESY DU PONT CO. LTD

Liquid	Volume increase (%)	
	No carbon black	50 p.p.h.r. SRF carbon black
Acetone	55	40
Aniline	110	75
Benzaldehyde	365	140
Butter	5	5
Butyric acid	80	40
Carbon tetrachloride	350	170
Cotton-seed oil	5	5
Cyclohexane	119	75
Cyclohexanone	370	185
Dibutyl phthalate	200	110
Ethyl acetate	95	60
Ethyl alcohol	10	0
Ethylene glycol	0	0
Furfural	40	25
Isopropyl ether	80	35
Jet fuel JP-4	60	45
Kerosene	60	45
Lard	5	5
Linseed oil	20	10
Methylene chloride	335	150
Methyl ethyl ketone	170	85
Nitrobenzene	300	130
Oleic acid	80	40
Olive oil	5	5
Orthodichlorbenzene	500	375
Tetraethyl lead	75	35
Toluene	415	185
Trichlorethylene	450	190
Turpentine	215	100

certain ketones, esters, ethers, and chlorinated solvents, leads to greater swelling. Table 4.17, giving the volume increase of standard non-sulphur types of vulcanisate with and without carbon black reinforcement, provides comparative data on this type of rubber's suitability for use in contact with a wide variety of solvents and chemicals.

Mixing and processing of chloroprene rubbers are effected using conventional equipment and methods. In the raw state, chloroprene rubbers of the sulphur-modified and non-sulphur-modified types differ in their response to mastication. As is shown in Fig. 4.24, the former type softens considerably when masticated on a cold mill, whilst the latter are relatively unaffected. By addition of an agent such as piperidinium pentamethylene dithiocarbamate, the sulphur-modified types can be peptised very substantially by milling at ambient or elevated temperature; no effective peptiser has been found for other types.

In mixing, CR readily bands on a cool mill or masses in an internal mixer. To give maximum protection against scorch, it is customary in mill mixing to add magnesia first, followed by hard fillers and then soft fillers, followed by oils, waxes, or softeners, with zinc oxide and accelerators, if any, added in

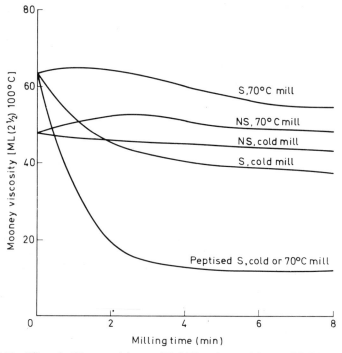

Fig. 4.24   *Effect of milling on sulphur-modified (S) and non-sulphur-modified (NS) chloroprene rubbers. Courtesy Du Pont Co. Ltd*

the final stage of mixing. With internal mixers, which provide more intensive shearing action, the mixing cycle and heat generation are often minimised by commencing oil addition after only a part of the filler has been incorporated.

Extrusion of polychloroprene compounds is carried out on standard rubber extruders, with either warm or cold feed facilities. In the case of warm feed, milling is required only to achieve uniform heating, and this working is generally less than is necessary with NR, for example. Best results are obtained using a cool barrel and screw, a warm head, and a hot (93°C) die. Accurate temperature control is necessary on calender bowls; high bowl temperatures reduce the nerve and roughness of calendered sheet, but increase the stickiness of compounds. In some cases, addition to the compound of a low molecular weight polyethylene is used to eliminate high-temperature sticking. Where soft sticky compounds are desirable, e.g. for frictioning, the ready peptisation of sulphur-modified types of polychloroprene makes them of particular interest.

4.9   Chlorosulphonated Polyethylene Rubbers—CSM

BY W. P. FLETCHER

4.9.1   MANUFACTURE

When polyethylene is reacted in solution with chlorine and sulphur dioxide it is transformed to a vulcanisable rubber, chlorosulphonated polyethylene. The basic structure of CSM may be represented as

where y is greater than x. Typical values for a general purpose CSM are $x = 12$, $y = 17$; variation of the amount and location of the chlorine and sulphonyl chloride groups gives products having a range of chemical and physical properties (Brooks, Strain, and McAlevy, 1953).

4.9.2   COMMERCIALLY AVAILABLE CHLOROSULPHONATED POLYETHYLENE RUBBERS

In Table 4.18 the commercially available types are listed. A more detailed discussion has been published (Straub, 1969).

Table 4.18   CHLOROSULPHONATED POLYETHYLENE RUBBERS MANUFACTURED IN THE USA BY DU PONT (TRADE NAME, HYPALON).

Type	Mooney viscosity	Special properties and applications
20	30	For flexible solution coatings and for blending to upgrade other elastomers
30	30	For stiffer solution coatings
40	45, 55, 115	Easy processing, for general moulding, extrusion, and calendering
45	40	Easy processing, higher modulus; can be used without curing
48	80	Higher chlorine content, greater solvent resistance

4.9.3   PROPERTIES

Properly compounded vulcanisates show levels of physical properties well within the range of values for good-quality vulcanisates from more general purpose rubbers. Tensile strengths of CSM vulcanisates range from 35 kgf/cm² to 200 kgf/cm², depending upon formulation. In resilience, the elastomer compares favourably with other synthetic elastomers though generally its resilience is lower than that of NR vulcanisates. Compression set of standard formulations is slightly higher than that of other elastomers, though this can be improved where necessary by special compounding. Flex and abrasion resistance are very good, and CSM will not support combustion. Electrical properties of a typical CSM vulcanisate intended for electrical insulation are: dielectric strength, 160–240 kV/cm; resistivity, 10^{14} Ω cm; dielectric constant at 1000 Hz, 5–7; power factor at 1000 Hz, 2–3%. Resistance to oxidative ageing is excellent. Fig. 4.25 shows a plot of the time taken at a given temperature for ultimate elongation to be reduced to 100% against the ageing temperature for a typical vulcanisate. At 100°C the 'life' so judged is greater than one year. For particularly severe heat-ageing service, special compounding is available to give two or three times longer life with some sacrifice in other properties such as colourability and water resistance.

CSM rubbers are resistant to ozone attack. In numerous tests involving exposure at up to 10 000 p.p.m. ozone concentration, at up to 250% elongation and 70°C, no specimen has shown any ill effects after exposure periods

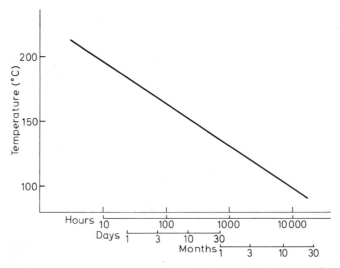

Fig. 4.25   Effect of exposure temperature of CSM on time for elongation at break to fall to 100%. Courtesy Du Pont Co. Ltd

up to 72 hours. The polymer's inherent resistance to oxidation allows the production of colour-stable weather-resistant products. The resistance of vulcanisates to chemicals is generally good, and products have found particular application in contact with strongly oxidising chemicals and acids, such as hydrogen peroxide, calcium hypochlorite, and sulphuric, chromic, and nitric acids. Table 4.19 gives the volume increases of a general purpose CSM vulcanisate in a representative range of oils and chemicals, and indicates the level of resistance.

Hypalon 48 produces vulcanisates which swell from 25% to 90% less than those from Hypalon 40, depending upon the fluid; in hydrocarbon oil, the swell is similar to that of a medium nitrile rubber (NBR) vulcanisate.

The types of CSM commonly used in dry compounds, namely Hypalons 40, 45, and 48, are, in the raw state, more thermoplastic than other commonly used rubbers. A comprehensive review of processing properties has been

Table 4.19 CHEMICAL RESISTANCE OF CSM. COURTESY DU PONT CO. LTD

Chemical	Temperature (°C)	Immersion time (days)	Volume increase (%)
ASTM* Reference Oil 3	100	3	39
ASTM* Reference Fuel B	21	3	33·5
Cotton-seed oil	70	7	2·8
Nitric acid (20%)	50	7	8·9
Sodium hypochlorite	70	14	1·4
Water	70	14	1·7

*ASTM D471.

published (Laan, 1969). CSM does not break down on milling but simply softens reversibly as it warms. Mixing is satisfactory on mills or in internal mixers, though the latter are preferred for speed and safer processing. Upside-down loading of internal mixers (fillers charged before CSM) gives the most efficient mix with most CSM compounds, with accelerators added on the sheet-off mill or later. Extrusion of CSM compounds is good, in keeping with their thermoplastic nature; rapid cooling of extrudates is recommended to avoid distortion of profiles. CSM compounds may be calendered with bowl temperatures in the range 60°–95°C for Hypalon 40, and somewhat higher for types 45 and 48. Incorporation into the compound of release agents, stearic acid, or wax, with the addition of low molecular weight polyethylene for higher-temperature calendering, provides satisfactory calender release. The thermoplasticity of uncured CSM makes building-up operations practical with warmed stock (50°–60°C).

4.10 Acrylonitrile–Butadiene (Nitrile) Rubbers—NBR

BY C. L. BRYANT

4.10.1 INTRODUCTION

In the course of work on the copolymerisation of butadiene with mono-olefines, Konrad and co-workers (1930) obtained a synthetic rubber based on butadiene and acrylonitrile, which, when vulcanised, had excellent resistance to oil and petrol. Pilot plant production of Buna N, as this rubber was first named, was started in Germany in 1934 by I.G. Farbenindustrie, but when the first full-scale production started operating in 1937 the trade mark 'Perbunan' was adopted.

The outbreak of World War II in 1939 stopped the import of nitrile rubber to the USA, where consumption had risen to about 150 Mg per annum. Local production of NBR was, therefore, started by the Standard Oil Co. of New Jersey, who had obtained control of Jasco Inc. to whom the American rights had been assigned by I.G. Farbenindustrie. Other rubber companies, notably Firestone Tire & Rubber Co., US Rubber Co. (now Uniroyal), Goodyear Tire & Rubber Co., and B. F. Goodrich also started production of nitrile rubber in the USA early in World War II, and were followed by Polymer Corpn in Canada in 1947.

Apart from German production which ceased in 1945 and was not resumed until 1952, commercial production of nitrile rubber in Europe did not commence until 1957 when two factories, one operated by British Geon Ltd (now BP Chemicals Ltd) and the other by Imperial Chemical Industries Ltd, came on stream in the UK. In 1969 the ICI production of 'Butakon' was taken over by Revertex Ltd, who through their subsidiary Doverstrand Ltd are associated with Standard Brand Chemicals Inc. of the USA. NBR production also started in many other countries.

A comprehensive review of nitrile rubbers has been published (Hofmann, 1963, 1964).

4.10.2   MANUFACTURE

Basically, nitrile rubbers are manufactured by emulsion copolymerisation of butadiene with acrylonitrile in processes similar to those used for other emulsion polymers, such as SBR. The main raw materials required are the monomers butadiene and acrylonitrile. At present, approximately 75% of the world production of the latter utilises the Sohio process (Veatch *et al.*, 1962; Schmidt, 1969), in which ammonia, propylene, steam, and air are passed through a fluidised bed of finely divided catalyst at about 450°–500°C. This and other similar processes are based on the reaction

$$CH_2{=}CH{-}CH_3 + NH_3 + \tfrac{3}{2}O_2 \longrightarrow CH_2{=}CH{-}CN + 3H_2O$$

The use of propylene as the basis for the production of acrylonitrile has now largely superseded older routes involving acetylene and hydrogen cyanide, for example, mainly on economic grounds.

In theory, the polymerisation reaction can be written

$$CH_2{=}CH{-}CH{=}CH_2 + CH_2{=}CH{\underset{\underset{CN}{|}}{}} \longrightarrow \left[(-CH_2{-}CH{=}CH{-}CH_2{-})_x \left(CH_2{-}CH{\underset{\underset{CN}{|}}{}} \right)_y \right]_n$$

The values of n and y depend on the precise polymerisation recipe and the temperature of polymerisation. It is important to bear in mind that, except under azeotropic conditions, the ratio of the two monomer units in the final polymer will not normally be the same as the ratio in which the monomers were charged to the reaction vessel. As the ratio of butadiene to acrylonitrile in the polymer largely controls its properties, as will be seen later, the design of the polymerisation recipe and the temperature at which this is carried out are important features of nitrile rubber production. The properties of the product may be influenced by other details of the recipe, such as the nature and amount of modifiers and emulsifiers.

The early nitrile rubbers were all polymerised at about 25°–50°C, and these hot polymers were characterised by their toughness resulting from the presence of a degree of branching of the polymer chains often referred to as 'gel'. By analogy with the developments in the emulsion polymerisation of SBR, since the early 1950s an increasing number of nitrile rubbers are being produced by 'cold' polymerisation at about 5°C; this results in more linear polymers containing little or no gel and which are easier to process than 'hot' polymers (Section 4.2.1.2).

4.10.3   COMMERCIALLY AVAILABLE NITRILE RUBBERS

According to the latest published information, there are 17 producers of dry nitrile rubber in the non-communist world who between them manufacture some 200 separately listed grades. Nitrile rubbers are also produced in certain communist countries, notably in the USSR and E. Germany. Table 4.20 lists some of the nitrile rubber producers in W. Europe, the trade names, and the range of grades offered.

4.10.4   PROPERTIES

The physical properties of nitrile rubbers are good when the rubbers are compounded with carbon black of suitable type; unfilled vulcanisates have very low tensile strength. NBR possesses generally better heat resistance than CR, but, like NR, is subject to ozone cracking. Products with low compression set properties can be made.

The commercially available nitrile rubbers differ from one another in three respects: acrylonitrile content, polymerisation temperature, and Mooney viscosity. The acrylonitrile content has by far the most profound effect on the properties of a vulcanised nitrile rubber, influencing its resistance to aliphatic hydrocarbons. Fig. 4.26 is a plot of acrylonitrile content against the volume of the 40% aromatic fuel SR6* absorbed for a range of rubbers. The mix consisted of:

Nitrile rubber	100	(parts by wt)
Zinc oxide	5	
Stearic acid	1	
SRF black	50	
Sulphur	1·5	
Dibenzthiazyl disulphide	1	

The curve is typical and shows that, in general, the higher the acrylonitrile content, the smaller is the volume change. In any given fuel or oil, swelling may be influenced by the filler and plasticiser content of the vulcanisate and

*60% diisobutylene, 5% benzene, 20% toluene, 15% xylene.

F¹

Table 4.20 NITRILE RUBBERS PRODUCED IN W. EUROPE

Country	Producer	Trade name	Range of acrylonitrile content	Polymerisation temperature	Special grades
UK	*BP Chemical	Breon	28–41	H, C	NBR232: liquid polymer
	*Revertex	Butakon A	27–40	H, C	XNR233: high gel, process aid
France	*Compagnie Francaise Goodyear	Chemigum	30–48	H, C	N8: crosslinked process aid
	*Plastugil	Butacril	20–40	H, C	Crosslinked process aid
	*Polymer Corpn	Polysar			
		Krynac	27–50	C	833: isoprene–acrylonitrile
Italy	ANIC	EuropreneN	20–40	H, C	Crosslinked process aid
	Montecatini	Elaprim	21–45	C	
Netherlands	Chemische Industrie AKU-Goodrich	Hycar	ML–H†	C	
W. Germany	*Bayer	PerbunanN	28–39	H, C	NS: antioxidant, approved for foodstuffs

*Those companies marked with an asterisk, and Ets Chevassus (France) and Rhein Chemie (Germany), supply NBR–PVC blends.
†ML–H = medium low to high.

by the immersion temperature. At any given acrylonitrile content, there are also differences in swelling behaviour according to the nature of the liquid, e.g. in benzene, toluene, or xylene, far larger volume changes occur, and much smaller volume changes occur in diisobutene or paraffin. For a fuller account

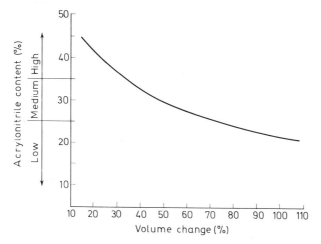

Fig. 4.26   Effect of acrylonitrile content of NBR on the volume of fuel absorbed after 24 hours immersion at room temperature. Courtesy Revertex Ltd

of the swelling behaviour of NBR, the literature should be consulted (Hofmann, 1963, 1964).

Polybutadiene has very good low-temperature properties, with a brittle point in the region of $-80°C$. As acrylonitrile units are incorporated in the polymer chain, the low-temperature flexibility of the resultant elastomer deteriorates progressively. Fig. 4.27 illustrates this graphically, and comparison with Fig. 4.26 makes it clear that, whenever nitrile rubbers are used, a

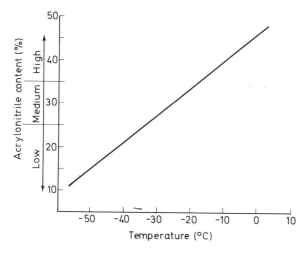

Fig. 4.27   Effect of the acrylonitrile content of NBR on the low-temperature flexibility. Courtesy Revertex Ltd

compromise between oil resistance and low-temperature properties is necessary.

Nitrile rubbers are not broken down by mastication in the same way as natural rubber and, therefore, the viscosity of the rubber as produced (measured usually in Mooney units) has an important effect on its processing properties, in addition to polymerisation temperature; 'hot' are usually more

Table 4.21   THE EFFECT OF ACRYLONITRILE CONTENT ON PROPERTIES OF NITRILE RUBBERS

High ⟵ ——————————— *Acrylonitrile* ——————————— ⟶ Low
⟵———————————— Resistance to fuels and oil improves
Low-temperature flexibility improves ————————————⟶
⟵———————————— Tensile strength improves
Resilience and elasticity increase ————————————⟶
⟵———————— Abrasion resistance and hardness increase
⟵———————————— Processing improves
⟵———————— Compatibility with plastics improves

difficult to process than 'cold' rubbers. Low Mooney rubbers absorb fillers and plasticisers more easily than high Mooney grades; they also cause less heat build-up during mixing, and faster extrusion rates can be obtained with them. On the other hand, when it is desired to produce soft compositions, it is preferable to use a high Mooney rubber as this will retain its strength more readily when only little filler and large amounts of plasticiser are used (BASRM, 1967). The effect of acrylonitrile content on swelling, low-temperature flexibility, and other properties of nitrile rubbers can be summarised as shown in Table 4.21.

4.10.5   COMPOUNDING

In general, NBR is compounded along lines similar to those practised with NR and SBR. There are a few noteworthy points. Dispersion of ground sulphur is often unsatisfactory, and the use of a magnesium carbonate coated sulphur is recommended to overcome the difficulty. Ester and polymeric plasticisers are widely used in NBR because they influence not only its processing but also such properties as hardness, low-temperature flexibility, and oil resistance. Small quantities of aromatic process oils can occasionally be used, particularly with those NBR types which do not contain more than about 30% acrylonitrile.

Whilst all types of carbon black and non-black fillers may be used with nitrile rubber, the most common practice is to use mainly the semi-reinforcing varieties because this is a means of obtaining suitable physical strength allied to low raw material costs. Nitrile rubbers may also be reinforced by the incorporation of phenolic resins (Ardley, 1963) and PVC. The interesting properties obtainable from blends of NBR and PVC have led to their introduction on a commercial scale by most producers of nitrile rubbers (Table 4.20).

The main advantages which can be obtained by the addition of PVC are resistance to ozone and weathering, better gloss of extrudates and mouldings, bright colours, higher abrasion and oil resistance, and, if suitable plasticiers are used, flame resistance (Abrams, 1962; Bryant, 1962).

4.11 Polyacrylic Rubbers—ACM

BY B. PICKUP

4.11.1 MANUFACTURE

Polyacrylic rubbers are usually copolymers of ethyl acrylate and a minor proportion of a second monomer containing chlorine, e.g. 2-chlorethyl-vinylether:

$$CH_2{=}CH{-}COOC_2H_5 \qquad\qquad CH_2{=}CH{-}O{-}CH_2{-}CH_2{-}Cl$$

    Ethyl acrylate   2-Chlorethylvinylether

Recently, monomers not containing chlorine have been utilised by some manufacturers.

4.11.2 COMMERCIALLY AVAILABLE POLYACRYLIC RUBBERS

Some polyacrylic rubbers are listed in Table 4.22, with details of viscosity, vulcanising systems, and special features.

4.11.3 PROPERTIES

Vulcanisates of polyacrylic rubbers are superior to those of nitrile rubbers in resistance to swelling and deterioration by hot hydrocarbon oils, extreme-pressure lubricants, and transmission and hydraulic fluids, and in resistance to heat, being suitable for intermittent use at temperatures up to 180°–200°C. Resistance to ozone, to sunlight, and to weathering is also good. Resistance to water and chemicals containing hydroxyl groups varies from poor to good depending on choice of polymer and curatives.

Most polymers give compounds suitable for use at temperatures down to $-20°C$. Special polymers have been developed to lower this limit to about $-30°C$. Where appropriate, special plasticisers are added to the compound to give flexibility down to $-40°C$, but these are liable to be lost by extraction by hot oils or by evaporation.

Reinforcing carbon blacks such as FEF are used. Synthetic graphite may be incorporated in compounds for rotating shaft seals, and reinforcing silicates are used in non-black mixes. Properties typical of acrylic rubber vulcanisates are given in Table 4.23 for a simple, 75 IRHD, compound containing 60 p.p.h.r. of FEF black. The precise values will vary with the polymer grade and the vulcanising system.

Table 4.22 SOME POLYACRYLIC RUBBERS

Country	Supplier	Trade name	Grade	Specific gravity	Mooney viscosity [ML(1+4) 100°C]	Vulcanising system	Special features
USA	Anchor Chemical	Cyanacryl	R	1·12	47	Ammonium salts, soap and sulphur, nitrile accelerator combinations	Easy processing; fast curing
			L	1·14	49	Ammonium salts, soap and sulphur, nitrile accelerator combinations	For special applications requiring flexibility at low temperatures
	B. F. Goodrich	Hycar	4021	1·09	55	Polyamines, HMDA carbamate, 2-mercaptoimidazoline	
			4031 4031X2	1·11 1·11	70 50	Polyamines, HMDA carbamate, soap and sulphur, dithiocarbamates	Low corrosion; chlorine free
			4032	1·12	40	Soap and sulphur, soap and sulphur donor, polyamines	For low-temperature applications
			4041	1·09	42	Soap and sulphur, soap and sulphur donor, polyamines	High temperature resistance
	Thiokol Chemical	Thiacril	44	1·08	50	Diamines, 2-mercaptoimidazoline	Improved processing
			76	1·09	60	Diamines, ammonium benzoate, dithio-carbamates, soap and sulphur	Low corrosion; chlorine free
Canada	Polysar (UK)	Krynac	882X2	1·08	30	Soap and sulphur, thioureas, organic ammonium salts, primary and second-ary amines	Suitable for low-temperature applications
Italy	Joseph Weil	Elaprim	AR153X	1·1	40	Guanidines, special curatives	Chlorine free

Table 4.23 TYPICAL PROPERTIES OF ACRYLIC ELASTOMERS (AFTER PRESS CURE AND POST OVEN CURE)

Property	Value
Hardness (IRHD)	75
Tensile strength (kgf/cm^2)	100–140
Elongation at break (%)	150–300
Compression set (70 h at 150°C) (%)	20–60
Heat resistance (70 h in air at 175°C):	
Change in elongation at break (%)	−30
Change in hardness (IRHD)	+5

Fluid resistance	Immersion		Volume change (%)	Hardness change (IRHD)
	Time (h)	Temperature (°C)		
ASTM* reference oil 1	70	150	+1	0
ASTM* reference oil 3	70	150	+15	−10
ASTM* service fluid 100	70	150	+2	+1
Extreme-pressure lubricant	168	125	+10	−10
Synthetic-ester-based lubricant	70	200	+50	−45
Distilled water	70	100	+15	−10

*ASTM D471.

4.12 Fluorocarbon Rubbers—FPM

BY W. P. FLETCHER

4.12.1 INTRODUCTION

During the past 15 years, considerable effort has been devoted by a few companies to the development of fluorine-containing elastomers (Radcliffe *et al.*, 1958; Tufts, 1959). The best-known polymers are those marketed by Du Pont under their trade name 'Viton'. The A grade is a copolymer of vinylidene fluoride and hexaflouropropylene and is vulcanisable to an

$$\sim\!\!\sim\!CF_2\!-\!CH_2\!-\!CF\!-\!CF_2\sim\!\!\sim$$
$$|$$
$$CF_3$$

elastomer which possesses a resistance to chemicals, oils, and solvents and to heat, which is outstanding in comparison with any other commercial rubber. Other polymers offered by Montecatini are based on vinylidene fluoride and 1-hydropentafluoropropylene. Variations in the polymerisation recipes and conditions and the use of undisclosed alternative or additional monomers have enabled the producers to supply a range of polymers varying in processing and vulcanisate properties but with the general qualities referred to above.

4.12.2   COMMERCIALLY AVAILABLE FLUOROCARBON RUBBERS

In Table 4.24, the polymers generally classed as fluorocarbon rubbers are listed, with brief details of their characteristics.

Table 4.24   COMMERCIALLY AVAILABLE FLUOROCARBON RUBBERS

Producer	Trade name	Type	Mooney viscosity (approx.)	Special features
Du Pont (USA)	Viton	A	65	General purpose
		AHV	160*	High-strength vulcanisate; better low-temperature properties
		A35	35	Good flow in extrusion and moulding
		B	74*	Superior resistance to heat, chemicals, and solvents
		B50	50	Safe processing; good flow
		LM	Semiliquid	Processing aid
		E60	60	Low compression set; easy processing
M M M (USA)	Fluorel	2140	115	General purpose
		2141	90	Safer processing
		2146	30	Higher filler acceptance; processing aid
		2160	60	Low compression set; easy processing
Montecatini (Italy)	Tecnoflon	SL	80	General purpose
		SH	110	General purpose
		T	90	Superior resistance to oils, chemicals, and solvents

*At 121°C.

4.12.3   PROPERTIES

Vulcanisates of the solid types with conventional compounding can range in hardness from 60 IRHD to 95 IRHD. To produce items at a hardness below 60 IRHD, which is the level for a pure gum vulcanisate, addition to the compound of a high-viscosity fluorosilicone fluid is necessary. Some consequent loss in strength may be partially offset by addition of a small quantity of an uncatalysed phenolic resin. FPM rubbers can be compounded to give tensile strengths up to 200 kgf/cm^2, with elongations at break in the range 150–300% according to hardness.

Of special interest is the high degree of resistance to heat ageing of fluorocarbon rubbers. A typical vulcanisate of hardness 75 IRHD has the ageing performance shown in Table 4.25. Continuous service limits for such vulcanisates are generally considered to be more than 3000 hours at 230°C, 1000 hours at 260°C, 240 hours at 288°C, and 48 hours at 315°C.

FPM rubbers are used successfully for long service at high temperatures in spite of the facts that actual tensile strengths at elevated temperatures are below room temperature values and hardness may also be reduced by up to 15 points.

Fluorocarbon rubbers, in general, show excellent resistance to oils, fuels, lubricants, and most mineral acids, and also resist many aliphatic and aromatic hydrocarbons such as carbon tetrachloride, and xylene which act as solvents for other rubbers. Resistance is not good to low molecular weight

Table 4.25 HEAT RESISTANCE OF A TYPICAL FLUOROCARBON RUBBER OF HARDNESS 75 IRHD. COURTESY DU PONT CO. LTD

Property	Unaged	Aged at 260°C		Aged at 288°C		Aged at 316°C	
		10 days	25 days	2 days	5 days	16 h	48 h
Tensile strength (kgf/cm^2)	172	67	49	70	46	59	39
Elongation at break (%)	330	400	290	340	210	350	200
Hardness (IRHD)	75	80	84	76	80	82	83

esters and ethers, ketones, certain amines, hot anhydrous hydrofluoric acid, and certain proprietary hydraulic fluids. Some examples of the performance of FPM rubbers in contact with fluids are shown in Table 4.26.

Fluorocarbon rubbers are flame resistant and do not support combustion. Parallel with their outstanding oxidation resistance, they are highly resistant to ozone attack. The performance of general purpose fluorocarbon rubbers at low temperature is moderately good as measured by brittle point, temperatures ranging from −40°C to −50°C being found. In the Clash–Berg stiffness test, the temperature at which a modulus of 700 kgf/cm^2 is reached is around −17°C. Low-temperature plasticisers have been found which effect some improvement in low-temperature flexibility, but in applications

Table 4.26 CHEMICAL RESISTANCE OF FLUOROCARBON RUBBERS. COURTESY DU PONT CO. LTD

Fluid	Immersion		Volume change (%)	Tensile strength (retained %)	Elongation at break (retained %)	Hardness change (IRHD)
	Time (days)	Temperature (°C)				
ASTM* reference oil 1	3	150	+0·2	103	100	+1
ASTM* reference oil 3	7	150	+4·3	95	100	−1
ASTM* reference fuel B	7	24	+2·5	90	100	+1
JP-4 fuel	3	205	+12	85	100	−3
Benzene	7	24	+20	73	92	−9
Carbon tetrachloride	7	24	+1·3	85	83	+2
Kerosene	28	150	+20	56	80	−4
Methylene chloride	28	38	+29	31	60	−18
Perchlorethylene	28	70	+7·5	87	100	−9
Toluene	730	38	+21	64	86	−23
Xylene	28	50	+18	66	85	−16
Phenol	28	150	+24	57	210	−19
Hydrochloric acid	7	70	+3·2	86	120	−7
Sodium hydroxide	183	38	−9·5	73	86	−5
Sulphuric acid	28	70	+4·8	88	90	+1

*ASTM D471.

where some part of the service is at very high temperatures, plasticiser is lost and the desired low-temperature flexibility is impaired. A special type of fluorocarbon rubber recently introduced shows an improvement of about 14°C in both brittle point and Clash–Berg stiffness temperatures.

The general levels of electrical properties of fluorocarbon rubbers are shown in Table 4.27. These are such that the elastomer is useful for low-voltage low-frequency insulation where heat and fluid resistance of a high order are necessary.

Fluorocarbon rubbers are mixed and processed on conventional equipment. In mixing, it is desirable to cool mills and internal mixers to avoid

Table 4.27 ELECTRICAL PROPERTIES OF FLUOROCARBON RUBBERS. COURTESY DU PONT CO. LTD

Property	Value
Dielectric strength (kV/cm)	160
Volume resistivity (Ω cm)	$2\cdot5 \times 10^{12}$
Power factor (%)	4·5

sticking and to retain maximum processing safety. Smooth extrusion of compounds is assisted by use of a cool die and by proper design of die shape and surface finish. Calendering is normally carried out by feeding mill-warmed stock to the calender controlled at around 40°C, avoiding handling. Higher calender temperatures are possible with the higher molecular weight polymers, owing to their higher green strength.

4.13 Silicone Rubbers—SI

BY C. M. BLOW

4.13.1 INTRODUCTION

The term 'silicone' was coined by Prof. Kipping (Kipping *et al.*, 1923), who carried out the pioneer work on organo-silicon compounds at Nottingham in the early part of the present century. It is now used as a generic term for materials based on compounds containing a chain of alternating silicon and oxygen atoms and available as fluids, greases, resins, and rubbers.

Silicone rubbers have been produced commercially for about 25 years, the first patent being applied for in 1944. Their high cost of manufacture, the initial difficulty of achieving good physical properties, and the need for special techniques were probably responsible for their technical and commercial development being, for the rubber industry, on unorthodox lines. The producers did not offer the raw polymers to the rubber processor to compound himself, but rather formulated and supplied ready-mixed materials requiring only forming, by one of the usual processes, and vulcanising. Subsequently, a few gums were made available and, over the years, some of the special technology required to be applied to them has been disclosed.

4.13.2 MANUFACTURE

Silicones, which chemically are polysiloxanes of the general formula

$$\left[\begin{array}{cc} R & R \\ | & | \\ -Si-O-Si-O- \\ | & | \\ R & R \end{array} \right]_n$$

where R, in commercially produced polymers, is methyl, phenyl, vinyl, or trifluoropropyl, are manufactured by hydrolysis of the appropriate dichloro-silane (R_2SiCl_2) to form cyclic tetrasiloxanes. These, in the presence of suitable catalysts, produce the long-chain siloxanes.

The first types available were the dimethyl siloxanes, followed shortly by methyl phenyl siloxanes in which the proportion of phenyl was small and which give an elastomer of lower stiffening temperature than the dimethyl polymer. Higher ratios of phenyl to methyl groups lead to a higher stiffening temperature, but to some degree of flame-retardant properties. The newer types of rubber contain an olefinic group—usually vinyl—to increase the reactivity of the polymer and provide much faster vulcanisation and more elastic vulcanisates; the mole % of vinyl groups is between 0·02 and 0·5 of the other organic groups—methyl and phenyl.

The incorporation of a proportion of fluorine- or cyano-containing groups in place of methyl confers excellent resistance to oils, fuels, and solvents, and the fluoro type in spite of its high price has found fairly wide application. The cyano or nitrile type, while offered on a small experimental scale some years ago, has not proved technically acceptable.

The introduction of boron atoms in the main siloxane chain at the rate of

$$\begin{array}{cccc}
 & \text{R} & & \text{R} \\
 & | & & | \\
\sim\!\!\sim\!\!\sim\text{Si} & -\text{O}-\text{B}-\text{O}- & \text{Si} & \sim\!\!\sim\!\!\sim \\
 & | & | & | \\
 & \text{R} & \text{R}' & \text{R}
\end{array}$$

one to every 200 to 500 silicon atoms confers a residual tack on the lightly crosslinked compound, and surface coalescence or welding of the surfaces takes place rapidly upon contact under slight pressure. Higher amounts of boron—one atom to every 3–100 silicon atoms—produce the now well-known bouncing putty with the unique properties of flow under even small loads and yet high resilience at high-frequency straining.

Passing reference should be made to the cold-cure or room-temperature vulcanising silicone rubbers, although their technology and use is outside the scope of this book. Catalysts such as organo-lead and organo-tin compounds promote condensation of silanols which are crosslinked through siloxane bridges, these reactions occurring at room temperature.

4.13.3 COMMERCIALLY AVAILABLE SILICONE RUBBERS

Table 4.28 gives a selected list of the most widely used silicone rubbers grouped under four headings—gums (i.e. raw uncompounded polymers), semi-compounded or base stocks, fully compounded stocks (which are usually available with or without the peroxide catalyst for vulcanisation), and non-solvent pastes.

4.13.4 PROPERTIES

The molecular-weight range of the heat vulcanising solid polymers is 300 000–1 000 000.

The outstanding characteristic of all types of silicone rubbers is the wide

Table 4.28 REPRESENTATIVE SILICONE RUBBERS COMMERCIALLY AVAILABLE IN THE UK

Supplier	Trade name and grade	Polymer type*	Description and applications
Gums—raw polymers without compounding or curing ingredients			
ICI	E301	SI	Devolatilised gum for general purpose stocks; benzoyl peroxide cured
	E303	VSI	Low-vinyl devolatilised gum for general purpose stocks; low compression set stocks
	E351	PVSI	Low-vinyl devolatilised gum, with extreme low-temperature flexibility
Base, semi-compounded stocks			
ICI	E367	VSI	Pyrogenic silica loaded stock to take filler addition for extrudable and low-cost rubbers
Midland Silicones	Polysil 2432	VSI	Low filler content for compounding to 30–80 IRHD rubbers
Dow Corning	Silastic 422	FVSI	For fuel- and oil-resistant aircraft seals, and electrical sleeving
Fully compounded stocks			
ICI	E313	VSI	General purpose range, 50–80 IRHD, for extrusions, mouldings, electrical sleevings, and cables
	E342	VSI	60–80 IRHD, with good oil resistance and compression set; for sealing applications
	E361	PVSI	High-strength 50 IRHD mix for sleevings and aircraft mouldings
	E323	VSI	Heat-resistant grade, 50 IRHD, for extrusions and mouldings; resists 300°C for short periods
	E343	VSI	80 IRHD grade for oil seals
	E330	VSI	50–80 IRHD range, vulcanised with vinyl specific peroxide; requires no post cure
Midland Silicones	Silastomer		
	2451–5	VSI	Easy processing 40–80 IRHD range with good heat stability
	2461–5	PVSI	Extreme-low-temperature 40–80 IRHD range, for mouldings
	2472–5	VSI	50–80 IRHD range with low compression set, good heat and oil resistance; low shrinkage
	2801U†	VSI	50 and 70 IRHD rubbers with high strength and resilience; also high flex resistance
	2438U	VSI	No-post-cure 70 IRHD grade; excellent compression set and oil resistance
	2457	VSI	Non-milling pelletised cable-insulation grade; steam or hot-air curable
	2811U	VSI	Translucent 50 IRHD grade for medical, pharmaceutical, and food applications
Dow Corning	Silastic		
	35, 55, 75U	VSI	High-performance tough abrasion-resistant grades for general purpose applications
	5503/5/7U	VSI	Translucent grades for medical, pharmaceutical, and food applications
	2351U	PVSI	High-strength flame-retardant grade for aircraft door and window seals
	745–8U	VSI	No-post-cure grade with low compression set, suitable for injection moulding
	LS53, 63U	FVSI	Fuel- and oil-resistant grades for seals
	LS2311U	FVSI	High-modulus fluorosilicone grade for O-rings
	LS2332U	FVSI	High strength and tear strength for shock mountings, sleeving, and seals

Table 4.28 *continued*

Supplier	Trade name and grade	Polymer type*	Description and applications
Cloth-coating grades			
ICI	E391	SI	Applied as solvent dispersion; coated cloth finds application in cable insulation and belting
Midland Silicones	Silastomer		
	125	SI	Non-solvent spreading paste; fabric coating for conveyor belting and ducting
	132	SI	Applied as solvent dispersion; for electrical tapes and glass braid sleevings

*SI = dimethyl siloxane; VSI = methyl vinyl siloxane; PVSI = phenyl methyl vinyl siloxane; FVSI = trifluoropropylvinyl-siloxane.
†U = peroxide addition required for vulcanisation.

range of temperature over which their properties are stable. No class of rubbers approaches them in this respect.

The T_g of dimethyl polysiloxane is $-123°C$ (Boyer, 1963). Elastomers based on this rubber, however, stiffen at about $-60°C$, at which temperature a first order, crystallisation, transition occurs (Weir, Leser, and Wood, 1951; Boyer, 1963). The introduction of a few phenyl groups suppresses this crystallisation so that, although the T_g is raised, the phenyl methyl polysiloxane elastomers have stiffening temperatures some $30°–40°C$ lower than the dimethyl polysiloxanes. As the proportion of phenyl groups is increased, the T_g is still further raised, and so also is the stiffening point.

At the high-temperature end, all silicone rubbers resist temperatures of 150°C for months without change of hardness or strength and withstand higher temperatures for lengthy periods as shown in Table 4.29.

Compression set of silicone rubbers, as a measure of their approach to perfect elasticity, at any temperature is equal or superior to that of any of the generally available natural and synthetic rubbers (Fig. 4.28).

Silicone rubbers are inert chemically, have no taste or smell, and are, with few exceptions, physiologically acceptable to animal tissue. They are unaffected by atmospheric exposure and do not show ozone cracking. Electrical

Table 4.29 USEFUL LIFE OF SILICONE RUBBERS AT VARIOUS TEMPERATURES. FROM DELGATTO (1965), COURTESY 'RUBBER WORLD'

Temperature (°C)	Useful life (retention of 50% of original elongation)
-50 to $+100$	Indefinitely long
120	10–20 years
150	5–10 years
205	2–5 years
260	3 months to 2 years
316	1 week to 2 months
370	6 hours to 1 week
420	10 min to 2 hours
480	2–10 min

properties are excellent (Table 4.30). Their permeability to gases is high (Table 2.1), this being used, medically, in the production of oxygen-permeable diaphragms (Folkman, Long, and Rosenbaum, 1966; Barnes, 1967).

The early silicone elastomers had very low strength, 20–30 kgf/cm². Later ones raised this figure to 35–50 kgf/cm² and for many years this was the

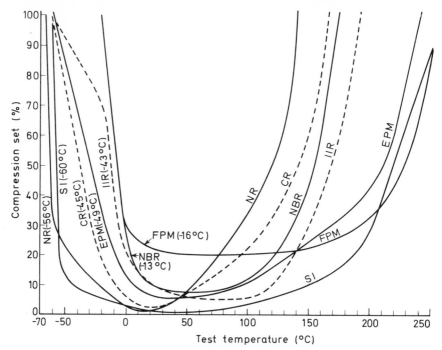

Fig. 4.28   *Dependence of compression set of elastomers on temperature (the temperature at which the torsional modulus is 700 kgf/cm² is stated on each curve). From Blow (1964), courtesy Bunhill Publications Ltd*

general level of this property that could be achieved, associated with correspondingly poor abrasion and tear strength. From time to time over the past 10 years, several high-strength materials have been offered by the manufacturers, but these have too often shown poor elastic properties and a considerably lowered heat resistance compared with the medium-strength products. Of recent years, however, the general level of strength has risen to around the 70 kgf/cm² mark and, furthermore, some new materials, with tensile strengths in the 90–120 kgf/cm² range, excellent tear strength, and good elastic properties have become available.

The lower strength of silicone rubbers has been offset by one feature, namely, a much lower loss of the value at elevated temperatures. Fig. 4.29 gives a plot of tensile strength against temperature for several rubbers, from which the advantageous position of silicone is obvious.

Although the dynamic properties of silicone rubbers were shown by Painter (1954) to be very good, their use in antivibration mountings and the like has been slow to develop, probably because of their poor tear strength and the rather low adhesion to metal that can be obtained.

What has been termed the 'confined heat stability' of some silicone rubber grades is not good. Softening or reversion takes place when the rubber is subjected to high temperatures in a confined space. At first, this phenomenon was associated with the restricted access of oxygen, but it has been shown by

Table 4.30 ELECTRICAL PROPERTIES OF TYPICAL SILICONE ELASTOMER

Property	Value
Dielectric strength (kV/mm)	16–20
Volume resistivity (Ω cm)	$(2\text{–}10) \times 10^{14}$
Permittivity at 1 MHz	3
Power factor at 1 MHz (%)	0·002

Turner and Lewis (1962) that the effect is largely due to moisture in the rubber, which, if not allowed to escape, promotes hydrolysis of the siloxane.

As indicated above, partly because of the manner in which the use of these rubbers developed and in contradistinction to other synthetic rubbers, the range of materials has been grouped into several grades, each of which has one particular property maximised: heat resistance, low-temperature flexibility, compression set, tensile strength, etc. This is not a true difference existing between silicone and other rubbers; in rubber technology all too often the product is a compromise with one or two essential properties optimised at the expense of others.

In considering the processing of silicone rubbers, a clear distinction must be drawn between the mixing and the forming operations. Silicone rubber compounded stocks can be calendered, extruded, and moulded, using conventional equipment and methods. Their low viscosity, stickiness, and non-black colour necessitate careful handling and precautions against contamination. Mention should be made of the so-called non-milling grades, sometimes

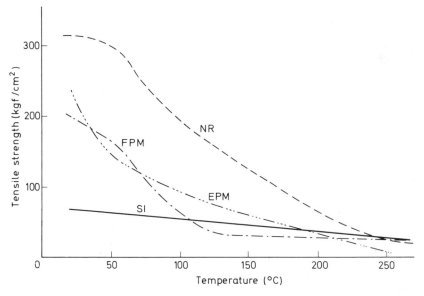

Fig. 4.29   Tensile strength of rubbers against temperature. From Blow (1968)

supplied as pellets, which can be fed direct to the extruder. Unlike the general run of rubbers, calendered and extruded products can be vulcanised in hot air by passing through a heated tunnel or direct to a well-ventilated oven at temperatures up to 250°C. Spreading can be carried out with conventional solvent dispersions and doughs or with the low-viscosity compounds available as pastes containing no solvent, which set and cure in hot air (Youngs and Konkle, 1964).

Peroxide crosslinking is rapid once the stock has reached the critical temperature for the peroxide, and scorching must be guarded against on the surface of hot moulds, for example (Baker and Riley, 1958). Transfer or injection moulding techniques appear very applicable, and very short cure times have been reported (DeSieno and Fuhrman, 1962).

4.13.5   COMPOUNDING

Of all rubbers, silicones show the lowest pure gum strength and, therefore, have to be reinforced in the same manner as other low-strength synthetic rubbers (Riley, 1962). Carbon black, in general, is not practicable for a number of reasons, and silica is the most widely used filler. Fume silicas with particle size in the 10–40 nm range show the greatest reinforcing effect. These, however, present a mixing problem in so far as an interaction occurs between them and the siloxane of such a magnitude that the stock either on the mill or on subsequent storage has the appearance of being partly vulcanised—termed 'crepe hardened'. A further feature of silicone rubber containing active silicas is the change of vulcanisate properties, in particular hardness, with the time of storage of the unvulcanised stock. Many compounded stocks therefore have to have a so-called 'bin ageing' of 1–3 weeks before being processed and vulcanised (Baker, Charlesby, and Morris, 1968; Southwart and Hunt, 1968, 1969, 1970).

In practice, to combat this structure and resultant behaviour of the material, chemicals, variously termed structure control additives and antistructure agents, are used. These are usually diols or substituted diols, such as diphenyl silanediol, or the reaction product of pinacol with dimethyl dichlorosilane:

Diphenyl silanediol   Pinacoxy dimethyl silane

Alternatively, silicas which have been pre-treated with chlorosilanes and washed free of the liberated hydrochloric acid, from their reaction with the hydroxyl groups, are used. These fillers do not display structure to the same extent and have the additional advantage of improving the water resistance, and hence the electrical properties.

Precipitated and ground silicas with larger particle sizes and correspondingly less reinforcing action are also widely used. Zinc oxide, which imparts tack, titanium dioxide for whiteness, inorganic and organo-metallic pigments

for colouring, and ferric oxide for heat stability are among the other compounding ingredients. Polytetrafluorethylene has been referred to as a
valuable additive at low loading to improve tear resistance (General Electric
Corpn, 1953; Safford and Bueche, 1953). The bonding of silicone rubber to
metals and fabrics presents no general problems provided the special primers
are used. An additional technique can be used; so-called integral bonding
aids added to the compound impart the property of spontaneously adhering
to metal and fabrics during vulcanisation (DeSieno, 1961; Korzun, 1962).

Crosslinking by organic peroxide takes place in the temperature range
110°–160°C. Post curing at 200°–250°C for approximately 24 hours is necessary to remove products of the reaction and impart the full properties, particularly heat resistance. The suppliers have, however, introduced grades
which do not require a post cure; these compounds use high-temperature-
acting peroxides (d'Adolf, 1963).

Part D Other Speciality Rubbers

BY W. COOPER

4.14 Polyurethane Rubbers—AU and EU

Although both ICI in the UK and Du Pont in the USA were pursuing similar work in the late 1930s, the first commercial exploitation of the diisocyanate polyaddition reaction, patented in 1942 by I.G. Farbenindustrie, was achieved by Farbenfabriken Bayer A.G., in the early 1950s.

4.14.1 MANUFACTURE AND COMPOUNDING

This group of elastomers is based on the reactions of polyols with diisocyanates. The main units from which the polyurethane chains are composed are polyethers and polyesters with terminal hydroxyl groups and average molecular weights in the region of 2000, endlinked by diisocyanates:

$$HO\left[(CH_2)_4 O\right]_n - H \qquad \text{Polytetrahydrofuran}$$

$$HO\left[CHMeCH_2O\right]_x - (CH_2)_2\left[OCH_2CHMe\right]_y OH \qquad \begin{array}{l}\text{Polypropylene oxide}\\ \text{(glycol initiated)}\end{array}$$

$$HO\left[(CH_2)_2O \cdot CO(CH_2)_4 CO \cdot O\right]_n CH_2CH_2OH \qquad \text{Polyethylene adipate}$$

$$HO\left[(CH_2)_5 CO \cdot O\right]_x - (CH_2)_2 - \left[O \cdot CO(CH_2)_5\right]_y OH \qquad \text{Polycaprolactone}$$

There are many other variants of polymer structure, e.g. tri- and tetra-functional polymers, and hydroxyl-terminated liquid polybutadienes. In elastomers, linear diols give the best properties, chain extension and branching being accomplished by a crosslinking agent. The diisocyanates used to link up the polyols are aromatic, e.g. toluene diisocyanate (TDI, a mixture of the 2,4 and 2,6 isomers), diphenylmethane-4,4'-diisocyanate (MDI), naphthalene-1,5-diisocyanate (NDI), and 3,3'-dimethyldiphenyl-4,4'-diisocyanate. Where good resistance to discolouration by light is required, aliphatic diisocyanates can be used, e.g. 1,6-hexamethylenediisocyanate and dicyclohexylmethane-4,4'-diisocyanate, but, as these are less reactive and more expensive, they are not of major importance at present.

The elastomers may be fabricated by casting processes, in which low molecular weight intermediate products containing terminal isocyanate groups (prepared *in situ* from the polyol and diisocyanate or as a prepolymer) are

138

chain-extended and crosslinked by the addition of low molecular weight polyols or diamines. Alternatively, conventional milling and moulding techniques can be applied to higher molecular weight polymers which are soft-millable gums. These are crosslinked either at the end-groups of the polymer or at reactive points in the chain; crosslinking agents are polyiso-cyanates, diamines, sulphur, and peroxides.

The dimensional stability of polyurethane elastomers is dependent to a considerable extent on hydrogen bonds between the appropriate groupings (allophanate, biuret, urea) in the polymer chains, and it is not essential for covalent crosslinkages to be present in the elastomer network. This fact

Table 4.31 POLYURETHANE FORMULATIONS AND PROPERTIES

*Thiokol formulation**		*Du Pont formulation†*	
Elastothane 455	100‡	Adiprene L100	100
MBTS	4	MOCA	12·5
MBT	2	Cure, 1–3 h at 100°C	
Activator ZC-456 (Thiokol Chemicals)	1		
Sulphur	1		
Cadmium stearate	0·5		
Press cure, 45 min at 140°C			
Tensile strength (kgf/cm^2)	286	Tensile strength (kgf/cm^2)	316
Elongation at break (%)	630	Elongation at break (%)	450
300% modulus (kgf/cm^2)	21	300% modulus (kgf/cm^2)	147
Hardness (IRHD)	42	Hardness (IRHD)	90
Compression set (22 h at 70°C) (%)	28	Compression set (22 h at 70°C) (%)	27

*From Thiokol Chemical Corpn Bulletin UE.1(a).
†From Du Pont Elastomer Chemicals Bulletin No. 7, October 1965.
‡30 p.p.h.r. of SAF black increases the modulus and hardness greatly with little effect on strength or elongation.

permits the synthesis of elastomers essentially composed of linear molecules which remain thermoplastic on moulding and which can be reprocessed under heat and pressure. Examples of all three types are available commercially (Table 4.32).

The compounding procedures and properties of the polymers are so diverse that only a brief outline of the processes can be given. In the casting process to make Vulkollan the prepolymer is made *in situ* from the polyester and naphthalene-1,5-diisocyanate, followed by addition of the crosslinking agent 1,4-butyleneglycol. The prepolymer Adiprene L contains reactive isocyanate groups and is received in sealed drums. This is converted to a tough rubber by reaction with diamines, typically MOCA (4,4′-methyl-bis-2-chloroaniline). In casting processes there are obvious limitations to the amounts and types of fillers that can be incorporated, but the gum rubbers are handled in much the same way as other solid rubbers.

The two formulations given in Table 4.31 may be taken as representative of a millable and a casting type urethane elastomer.

4.14.2   COMMERCIALLY AVAILABLE POLYURETHANE RUBBERS

The more important commercial types of polyurethane rubber are listed in Table 4.32.

Table 4.32 SOME COMMERCIALLY AVAILABLE POLYURETHANE RUBBERS

Supplier	Trade name	Grades	Characteristics	Vulcanisates
Du Pont (USA)	Adiprene	C	Millable gum: sulphur and diamine cured	Oil and ozone resistant
		L83, 100, 167, 200, 213, 315, 420	Viscous liquids: NCO content 2·8–9·5%; diamine cured	Hardness variable over wide range, 50–90 durometer
American Cyanamid Co. (USA)	Cyanaprene	4590 A8, 9 D5, 6, 7, 4060	Solids melting point 65°C; polyester based; cured with diamines; grade numbers indicate hardness of rubbers	Mouldings with wide range of properties: hardness, compression set, load bearing, etc.
Thiokol Chemical (USA)	Elastothane	455, ZR625	Sulphur or peroxide cure	Good processing characteristics
Ameripol (USA)	Estane	5701, 5740	Thermoplastic, polyester based	Suitable for injection moulding; process temperatures 120°–180°C
Witco Chemical (USA)	Formrez	MG2	Millable gum; sulphur or peroxide cured	
		Prepolymers	Liquid polymers of variable isocyanate content; amine or polyol curing	
General Tire & Rubber Co. (USA)	Genthane	S, SR	Millable gums; peroxide cured	
Anchor Chemical (UK)	Jectothane	A, D, E	Thermoplastic elastomers	Process temperatures 150°–200°C
Mobay Chemical (USA)	Multrathane	F60, 200, 242	Polyester based; cured with polyol	F242, casting type for high tear strength; F200 for soft elastomers
		F196	Polyether based; cured with MOCA	
Farbenfabriken Bayer (Germany)	Urepan	600, 601	Millable gums; isocyanate cure	Strong mouldings; Shore A hardness 80–90
		640	Peroxide cure	Softer elastomers than the other grades
Uniroyal (USA)	Vibrathane	6000 series	Liquids; diamine or polyol cure	Wide range of hardness; good mechanical properties
		5003, 5004, 5005	Millable gums; diamine cure	
Farbenfabriken Bayer (Germany)	Vulkollan	N	Polyester: naphthalene-1, 5-diisocyanate, diol crosslinked	Exceptional tear strength

4.15   Polyether Rubbers

Substituted ethylene oxides can be polymerised to high molecular weight polymers which may be crystalline, rubbery, or plastic, depending on the monomer, by a variety of initiators, of which the best known and most important are combinations of zinc diethyl–water and aluminium trialkyl–water–acetyl acetone. The polymerisations are conducted in solution in much the same way as the hydrocarbon 'stereo rubbers'. Commercially available polymers are those based on epichlorohydrin, and there is considerable interest in the copolymers of propylene oxide, although these are not yet fully established commercially.

4.15.1   EPICHLORHYDRIN RUBBERS

The two types of rubber currently in production are the homopolymer and a copolymer of epichlorhydrin containing about 40% of ethylene oxide (Vandenberg, 1965; Leach, 1966):

$$\left[-CH_2-CH-O-\atop\qquad\ \ |\atop\qquad\ CH_2Cl\right]_n \qquad \left[-CH_2-CH-O-CH_2-CH_2-O-\atop\qquad\ \ |\atop\qquad\ CH_2Cl\right]_n$$

4.15.1.1   *Homopolymer (CO)*

The homopolymer has a brittle point of $-15°C$ and is characterised by outstanding resistance to ozone, good resistance to swelling by oils (about 10% volume swell in ASTM reference oil 3 after 48 hours at 90°C), good heat resistance (satisfactory at 150°C for upwards of one week), and excellent weathering resistance. Of particular interest is its very low permeability to gases—about half to one-third that of butyl rubber. Like butyl, this elastomer has a low resilience.

4.15.1.2   *Copolymer (ECO)*

The copolymer is much more resilient and is more suitable for low-temperature applications (brittle point, $-40°C$). The air permeability is still low (about the same as a medium nitrile rubber), and there is little loss in resistance, compared with the homopolymer, to swelling by hydrocarbon solvents.

Neither the homopolymer nor the copolymer resists swelling by aromatic and chlorinated hydrocarbons, but both have good flame resistance.

4.15.1.3   *Compounding*

The polymers process readily, have good tack, and unvulcanised compounds can be extruded (usually at 50°–60°C) without difficulty. Medium fine black or non-black fillers can be used, and ester-type plasticisers are employed for softer products. A typical formulation with test results is given in Table 4.33.

Table 4.33 TYPICAL FORMULATION OF EPICHLORHYDRIN RUBBERS

	Homopolymer	*Copolymer*
Polymer	100	100
Zinc stearate	1	0·75
FEF black	50	50
Red lead	5	5
Nickel dibutyl dithiocarbamate	2	2
MBI	1·5	1·5
Press cure, 45 min at 155°C		
Tensile strength (kgf/cm^2)	148	155
Elongation at break (%)	260	240
Hardness (IRHD)	76	75

Other curing agents which may be used are diamines and polyamines or their derivatives (e.g. triethyl tetramine or hexamethylene diamine carbamate), lead oxides, and basic lead phosphite.

4.15.1.4 *Commercially Available Epichlorhydrin Rubbers*

These are listed in Table 4.34.

Table 4.34 COMMERCIALLY AVAILABLE EPICHLORHYDRIN RUBBERS

Producer	*Trade name*	*Grade*	*Type*
Ameripol (USA)	Hydrin	100	Homopolymer
		200	Copolymer
Hercules (USA)	Herclor	H	Homopolymer
		C	Copolymer

4.15.2 POLYPROPYLENE OXIDE RUBBERS (PO)

Polyproplene oxide rubber (Grüber, Briggs, and Meyer, 1963; Grüber *et al.*, 1964; Doyle and Vaughan, 1968) has been available on a semicommercial scale under the name Dynagen XP 139 (General Tire & Rubber Co.), but although its commercial future is not assured, a short description of the elastomer, which has many interesting features, may not be out of place. The polymers contain a proportion (up to 10%) of an unsaturated comonomer, such as allyl glycidyl ether, to permit sulphur vulcanisation:

$$\left[-\text{CH}-\text{CH}_2-\text{O} \right]_x \begin{array}{c} -\text{CH}-\text{CH}_2-\text{O}- \\ | \\ \text{CH}_3 \end{array} \quad \begin{array}{c} | \\ \text{O} \\ | \\ \text{CH}_2 \\ | \\ \text{CH}=\text{CH}_2 \end{array}$$

The polymer (specific gravity approx. 1·0) processes easily and is vulcanised with a butyl-type curing system. The normal types and amount of filler suited to hydrocarbon polymers are used. Vulcanisates are characterised by good physical properties, which depend to a significant extent on the micro-structure of the polymer, and have low-temperature and dynamic properties similar to those of NR. The heat-ageing resistance is moderate, being some-where between that of SBR and that of butyl, provided a suitable anti-oxidant is incorporated. The ozone and weathering resistance are good, and the polymer has oil resistance of the same order as that of polychloroprene rubber.

4.16 Polysulphide Rubbers (TR)

4.16.1 MANUFACTURE

Polysulphide rubbers are produced by the condensation of sodium poly-sulphide with dichloroalkanes:

$$RCl_2 + Na_2S_x \longrightarrow \sim\!\!\sim R\!-\!S_x\!\sim\!\!\sim$$

The polymer varies both in the character of R and x, and in the length of the polysulphide chain. Currently available polymers produced by the Thiokol Chemical Corpn are given in Table 4.35.

In addition, there is a series of liquid thiol-terminated polymers (Thiokol LP 2, 3, 4, 31, 32, and 33) based on the diethylene formal disulphide structure

Table 4.35 POLYSULPHIDE RUBBERS PRODUCED BY THIOKOL CHEMICAL CORPN

Polymer	R from	x	S(%)
Thiokol A	$(CH_2Cl)_2$	4	84
Thiokol FA	$(CH_2Cl)_2$	2	
	$CH_2(OCH_2CH_2Cl)_2$		47
Thiokol ST	$CH_2(OCH_2CH_2Cl)_2$	2·2	37
	2% trichloropropane		

but containing some branching, and Thiokol LP205 based on dibutylene formal disulphide. Low molecular weight polymers are made by the reductive cleavage of polysulphide bonds in solid rubbers by agents such as sodium sulphide. They have molecular weights in the range 600–7500 and viscosities 2·5–1400 P at 24°C.

4.16.2 SOLID POLYSULPHIDE RUBBERS

The solid polymers are used almost exclusively in applications where good resistance to solvents is required. This depends on the amount of sulphur in the molecule, and Thiokol A is resistant to every type of organic solvent. However, its processing characteristics, odour, and mechanical properties are very poor, and the other types which have moderate physical properties

Table 4.36 TYPICAL RECIPES FOR THIOKOL FA, ST, AND LP31

	Thiokol FA	*Thiokol ST*	*Thiokol LP31*
Thiokol FA	100	—	—
Thiokol ST	—	100	—
Thiokol LP31	—	—	100
Zinc oxide	10	0·5	—
Lead dioxide	—	—	5
Stearic acid	0·5	3	1
p-Quinonedioxime	—	1·5	—
Sulphur	—	—	0·15
MBTS	0·3	—	—
DPG	0·1	—	—
SRF black	60	60	50
Cure	50 min at 150°C	30 min at 145°C	1 day at 30°C
Tensile strength (kgf/cm^2)	82	88	73
Elongation at break (%)	380	310	440
Hardness (IRHD)	70	70	58
Low-temperature flexibility (°C)	−45	−53	—
Compression set (22 h at 70°C) (%)	100	37	—

and better all-round solvent resistance than chloroprene or nitrile rubbers are more widely employed.

Curing agents for Thiokols are diverse, but it is customary to use an organic accelerator (e.g. MBTS or TMTD with zinc oxide and stearic acid for type FA). The thiol-terminated polymers (Thiokol ST and the liquid Thiokols) can be crosslinked by metal oxides, metal peroxides, inorganic oxidising agents, peroxides, and p-quinonedioxime. Carbon black, usually SRF or FEF, in 40–60 loading is essential for adequate strength. Typical recipes for Thiokol FA and ST are given in Table 4.36.

4.16.3 LIQUID POLYSULPHIDE RUBBERS

The liquid polymers, which are used almost exclusively in sealant applications or as propellant binders, are crosslinked by lead or zinc peroxide or by epoxy resins, and it is customary to incorporate carbon black or white fillers and plasticisers, such as dibutyl phthalate. The products obtained have tensile strengths of 10–75 kgf/cm^2 (dependent on the type and amount of filler), with elongations of 250–550%; a recipe for Thiokol LP31 is given in Table 4.36.

Compression set of Thiokol-type elastomers, particularly at elevated temperatures, is poor, presumably because of polysulphide interchange reactions. Resistance to ozone and to weathering is good.

4.17 Polypropylenesulphide Rubbers

Reference should be made to polypropylenesulphide rubbers (Woodhams, Adameck, and Wood, 1965) in which a small proportion (3–10%) of an unsaturated comonomer (typically allyl glycidal thioether) permits crosslinking by sulphur. These polymers are not in commercial production but have physical properties and solvent resistance similar to that of Thiokol ST.

$$\left[-S-CH-CH_2-\right]_n S-CH-CH_2-$$

(structure)
$$\begin{array}{l} \left[-S-CH-CH_2-\right]_n \quad S-CH-CH_2- \\ \quad\ \ |CH_3 \qquad\qquad\ |CH_2 \\ \qquad\qquad\qquad\qquad\ |O-CH_2-CH=CH_2 \end{array}$$

4.18 Nitroso and Other Recently Developed Rubbers

This group of rubbers possess the structure:

$$\left[-N-O-CF_2-CF_2-\right]$$
$$\quad\ |R$$

4.18.1 COPOLYMER

The first of the nitroso series ($R = CF_3$) is an alternating copolymer of high molecular weight from trifluoronitrosomethane and tetrafluorethylene. It has a high specific gravity (1·9) (Barr and Hazeldine, 1955). The polymer is usually compounded with silica fillers (15–20 p.p.h.r.) and crosslinked by diamines or their derivatives. A typical formulation and physical properties are given in Table 4.37.

The resistance to ozone and exposure to light of this elastomer is excellent. It does not ignite even in pure oxygen and is almost completely resistant to swelling by hydrocarbons, alcohols, ketones, esters, and chlorinated hydrocarbons. Acid, bases, and concentrated hydrogen peroxide have little effect, and it is swollen only to the extent of 30% in 48 hours at 25°C by red fuming nitric acid, which destroys all other elastomers. Its main disadvantages are the rather poor mechanical properties and the limitations in its vulcanisation (Montermoso, Griffis, and Wilson, 1962).

Table 4.37 TYPICAL FORMULATION AND PROPERTIES OF A NITROSO RUBBER

Compound		Properties	
Nitroso rubber	100	Tensile strength (kgf/cm²)	49
HiSil 303-silica	15	Elongation at break (%)	600
TETD	1·25	300% modulus (kgf/cm²)	19
HMDA carbamate	2·5	Hardness (IRHD)	55
Press cure, 60 min at 120°C		Low-temperature stiffening	
Oven cure, 18 h at 100°C		Gehman T2	−31°C
		Gehman T10	−44°C

4.18.2 TERPOLYMER

Recently, terpolymers containing 0·5–1·5% nitrosoperfluorobutyric acid have been developed; these carboxy nitroso rubbers (Levine, 1969) can be cured by metal oxides, epoxides, amines, and organo-metallic compounds such as chromium trifluoroacetate. Metal oxides give the typical high tensile strength (above 200 kgf/cm^2) and high compression set associated with ionic crosslinkages, and, for good compression set, a diepoxide (e.g. dichloro-pentadiene diepoxide) is used as crosslinking agent. Within the hardness range 35–80 IRHD, tensile strengths of 40–130 kgf/cm^2 at 150–1200% elongation at break are achieved; compression set values are 10–15%. With fine silica or a thermal black filler (20 p.p.h.r.) tensile strengths approaching 140 kgf/cm^2 can be obtained. The high-temperature properties are good, and the elastomer is suitable for use in continuous service at up to 200°C.

Unlike the fluorocarbon rubbers, which are not readily plasticised, the nitroso rubbers can be blended with liquid fluorocarbon polyethers (up to 40 p.p.h.r.) to give soft but still relatively strong vulcanisates.

4.18.3   OTHER NEW RUBBERS

Two other rubbers, not commercially available, but of considerable interest, are the following: perfluorinated triazine rubber, developed by Hooker Chemical Corpn (Brochure 1184-R)

$$\left[-(CF_2)_6 - C \begin{array}{c} N \\ \diagup \ \diagdown \\ \\ N \\ \diagdown \end{array} C - \right]_n$$

and the perfluoronitroso polymer

$$\left[-N = P(OR)_2 - \right]_n$$

where $R = CH_2CF_3$ or $CH_2C_3F_7$, recently disclosed by Horizon Inc. (Rose, 1968).

The former gives an elastomer (specific gravity, 1·8) with good physical properties (tensile strength, 176 kgf/cm^2; elongation at break, 630%; glass transition temperature, 40°C; compression set 25% after 72 hours at 288°C) which has a reasonable service life at temperatures up to 400°C. The elastomer is also claimed to have good electrical properties and resistance to radiation, and good resistance to hot water and aviation fuels.

The other elastomer has only been briefly described but it is claimed to have excellent thermal, solvent, and chemical resistance (Anon, 1969-1).

FIVE

The Chemistry and Technology of Vulcanisation

BY S. H. MORRELL

5.1 Introduction

Previous to 1840 the uses of rubber had very largely been confined to those in which the rubber was supported by fabric, generally cotton in the case of garments, or silk in the case of surgical or medical apparatus. This was necessitated by the fact that raw natural rubber was relatively weak and it had to be masticated in order to dissolve it or process it in other ways, a procedure which further reduced its strength. Such rubber was sticky and it was customary to dust it with suitable powders, such as talc, china clay, whiting, litharge, or sulphur, to reduce the tackiness. It was perhaps a curious trick of fate which lead to the discovery that, when heated, a mixture of rubber and sulphur was converted into a non-tacky highly elastic tough material which was no longer soluble in solvents. Tradition holds that the first discovery was by Charles Goodyear in America, and that a sample of the vulcanised rubber was brought to Thomas Hancock in England by a friend. Hancock observed the sulphur bloom on the article and guessed correctly its identity. However it happened, Goodyear and Hancock announced the discovery of vulcanisation almost simultaneously in America and England.

The discovery resulted in a very sharp increase in the production of rubber goods, and also to a shortage of rubber. In fact, by the end of the first two decades, reclaim was being manufactured from rubber scrap to serve as a substitute (Section 6.4). Manufacturers were anxious to reduce the time of heating (about 8 hours), and it was not long before metallic oxides, such as those of zinc, calcium, magnesium, and lead, were being incorporated for this purpose. As an illustration of their use, some approximate figures are given in Table 5.1.

The use of these inorganic accelerators continued until after World War I, and in some cases they are even used today. For example, in the manufacture of large articles from NR, such as buffer springs, ships' fenders, and bridge bearings, a compound containing litharge as accelerator is used: this gives a long slow cure with some heat-resistant properties, enabling the inside of the article to be fully cured before the outside becomes overcured or oxidised.

In the early years of the present century, organic accelerators of vulcanisation came into use, and over the past 50 or 60 years progress has been continuous (Section 1.6). The discovery of mercaptobenzthiazole (MBT) in

Table 5.1   EARLY VULCANISING SYSTEMS FOR NATURAL RUBBER

Mix (parts by wt)	Cure (h at 141°C)	Tensile strength (kgf/cm²)	Elongation at break (%)
Rubber 100	Unvulcanised	25	1200
Rubber 100, sulphur 8	8	250	700
Rubber 100, sulphur 8, zinc oxide 5	4	250	700

1921 was undoubtedly one of the landmarks; at about the same time it was discovered that zinc oxide enhanced the action of organic accelerators and, from then onwards, zinc oxide came to be regarded as an activator whose addition to a compound was highly desirable. It is not clear when the use of stearic acid as a co-activator became customary, but it is certainly true that most recipes include it: the use of zinc stearate in place of zinc oxide and stearic acid does not produce the same activating effect.

Not only did organic accelerators enable the vulcanisation time to be reduced to minutes, but also their use gave products of much-improved properties, because less heating caused less degradative oxidation of the rubber and also the sulphur was used more efficiently (Section 5.2.2).

5.2   Sulphur Vulcanisation

5.2.1   PRACTICAL SYSTEMS FOR NATURAL AND SYNTHETIC OLEFIN RUBBERS

When synthetic rubbers were introduced, it was soon found that the sulphur vulcanising systems then in vogue for NR were of little use in SBR, NBR, or IIR. Such systems usually contained sulphur in amounts of 2·5–3 p.p.h.r. and accelerator in amounts of 0·5–1·0 p.p.h.r. The synthetic rubbers required a higher proportion of accelerator (say 1·5 p.p.h.r.), with a corresponding reduction in the amount of sulphur: often, equal parts of accelerator and sulphur gave a useful vulcanisate. Since synthetic rubbers are slower curing than NR, mixtures of accelerators are often used to bring curing times within the range customary for NR. The use of zinc oxide and stearic acid as co-activators is also usual, although the effect seems to be less marked than it is for NR. In all cases, approximately one crosslink per hundred monomer units is required to give a good 'technical' cure.

Curing curves for NR and SBR show distinct differences. Thus the tensile strength of NR reaches a peak and then declines. The modulus behaves very similarly, except that the peak is later than that for tensile strength. Rubbers which have passed the peak of the modulus curve are said to be 'reverted', since their properties are similar to those which would be expected from a reversal of the curing process (Fig. 5.1). SBR does not show any reduction of tensile strength on overcure; it remains approximately constant. The phenomenon known as 'marching cure' occurs, and the modulus gradually rises. Perhaps it is in elongation of break that the greatest difference is shown, for on overcure of SBR the elongation drops smartly, whereas in the case of NR little change is observed.

Nitrile rubbers show curing characteristics very similar to those of SBR, except that they exhibit a very pronounced improvement in compression

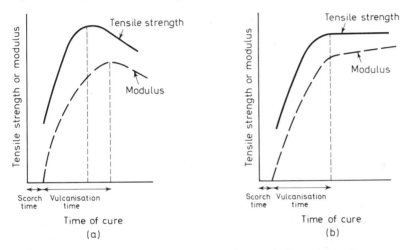

Fig. 5.1   Tensile strength and modulus against time of cure: (a) NR, and (b) SBR

set on prolonged cure. Since nitrile rubbers are often moulded on injection moulding machines, where shortness of cure is economically desirable, this is a serious disadvantage. Often the compression set can be improved by giving a further heating in an air oven after the initial cure has moulded the article to the correct shape, but care must be taken in precision mouldings to avoid undue shrinkage.

Butyl rubber possesses little unsaturation, and hence the more active accelerators are used, often at high temperatures. With butyl, however, reversion can occur, and so prolonged heating at curing temperatures must be avoided. Since sulphur is less soluble in butyl than in other rubbers, it is customary to reduce the amount used, often to as low as 0·5 p.p.h.r., and increase the amount of accelerator. Zinc oxide is necessary for activation, 3 p.p.h.r. being required, whilst higher amounts give improved resistance in severe ageing conditions. Stearic acid is not necessary for activation in butyl, but is sometimes added as an aid to processing. Since butyl is so saturated in nature, it is important to choose saturated softeners, otherwise the softener may react with the vulcanising ingredients and inhibit the vulcanisation of the rubber; pine tar, wood rosin, factice, and unsaturated acids are therefore unsuitable for this purpose.

The cure of EPDM is very similar to that of butyl; the level of unsaturation is low, and similar steps must be taken to avoid inhibition of vulcanisation. It is not certain whether or not reversion can occur with this polymer (Section 4.6.3) (Saagebarth, 1968).

Typical vulcanising systems for the above-mentioned rubber types are given in Table 5.2.

Table 5.2 TYPICAL SULPHUR CURING SYSTEMS FOR OLEFIN RUBBERS

	NR (p.p.h.r.)			SBR (p.p.h.r.)			NBR (p.p.h.r.)			IIR (p.p.h.r.)		EPDM (p.p.h.r.)	
Zinc oxide	6*	5	5	3	4	4	2	5	5	5	5	5	2·5
Stearic acid	0·5*	1·0	0·5	3	2	2	3	1	1	2	2	—	0·5
TMTD	—	—	0·04	—	—	—	—	—	—	1·5	0·75	—	—
MBT	0·5*	—	—	—	—	—	—	—	—	0·5	0·5	0·5	0·25
ZDC	—	—	—	—	—	—	—	—	—	—	0·75	—	—
MBTS	—	—	0·25	—	1·2	—	—	1·5	1·5	—	—	—	—
DPG	—	—	—	—	—	0·8	—	—	—	—	—	—	—
CBS	—	0·4	—	2	—	1·0	2	1·5	—	—	—	—	—
TMTM	—	—	—	—	0·15	—	—	—	0·4	—	—	1·5	0·75
Sulphur	3·5*	3·0	2·5	2	2	1·75	2	1·5	1·5	2	2	1·5	0·75
Cure time (min)	40*	30	20	30	25	25	30	25	20	40	30	25	55
Cure temperature (°C)	141*	141	141	153	153	153	153	153	153	153	165	165	165

*Standard compound of American Chemical Society (ACS 1).

5.2.2   THEORY OF SULPHUR VULCANISATION

In the previous section of this chapter, sulphur vulcanisation has been treated in an essentially empirical manner, and in fact the chemistry of sulphur vulcanisation is so complex that it is only within the last few years that a coherent theoretical treatment has been possible. Even today, only the main stages are proven and there is still much to be learned about the effect of additives of various types.

The unravelling of this complex reaction has not resulted from the efforts of one scientist, or indeed from the efforts of one school of scientists. Rather it is the result of a process which has continued for many years, an item of knowledge here added to an item of knowledge there, and so on. For a discussion of this earlier work reference may be made to Moore (1964) and Dibbo (1966), where it is given in detail, together with descriptions of later discoveries.

The unravelling process itself is an excellent example of the combined efforts of physical and organic chemists. It starts with measurement in 1910–1912, of the rate of the sulphur reaction, followed by the examination of the apparent loss of unsaturation by rubber in 1938–1945. The use of methyl iodide as a 'chemical probe' by Meyer and Hohenemser (1935) and by Selker and Kemp (1945) increased knowledge of the crosslinking reactions further. These latter workers, as well as Armstrong, Little, and Doak (1944), used 'model olefines' of low molecular weight to follow crosslinking reactions. These model substances were easier to handle than vulcanised rubbers and enabled intelligent guesses to be made as to the nature of the crosslinks themselves. The work was continued by Farmer and his school in this country, and by Craig, Juve, and others in the USA. Meanwhile, Scheele and co-workers in Germany (Scheele, Lorenz, and Dummer, 1956) had begun the study of the kinetics of vulcanisation, and the chemistry of the process began to be placed on a more quantitative basis, an advance which was further assisted by the discovery of the 'quantitative' crosslinking agents, the bis-azodicarboxylates and di-t-butylperoxide, by Flory, Rabjohn, and Shaffer, (1949) and Moore and Watson (1956) respectively.

Sulphur is combined in the vulcanisation network in a number of ways (Fig. 5.2), as enumerated by Porter (1969). As crosslinks, it may be present as monosulphide, disulphide, or polysulphide [Fig. 5.2(a)], but it may also be present as pendant sulphides [Fig. 5.2(b)], or cyclic monosulphides and disulphides [Fig. 5.2(c)]. An estimate of the number of sulphur atoms for

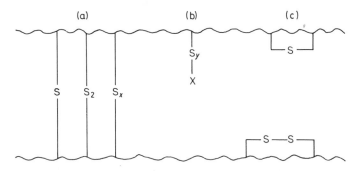

Fig. 5.2   Structural features of vulcanisate network

each crosslink formed has been made: an unaccelerated rubber–sulphur vulcanisate may give a figure of 40–45, whilst in conventional accelerated sulphur vulcanisates this 'inefficiency' figure may drop to 10–15. Special 'efficient' vulcanising systems can reduce it still further to 4 or 5, but for even lower values, so-called 'non-sulphur' vulcanising systems must be used (Section 5.3).

By the use of chemical probes, the relative amounts of mono-, di-, and poly-sulphide material can be assessed, and by measuring the degree of crosslinking as well, the pendant and intramolecular sulphur can be estimated. From this information and the amounts of nitrogen and sulphur combined with the rubber at various stages of vulcanisation, it is possible to deduce the general course of the vulcanisation reaction.

The initial step in vulcanisation seems to be the reaction of sulphur with the zinc salt of the accelerator to give a zinc perthio-salt XS_xZnS_xX, where X is a group derived from the accelerator (e.g. thiocarbamate or benzthiazyl groups). This salt reacts with the rubber hydrocarbon RH to give a rubber-bound intermediate

$$XS_xZnS_xX + RH \longrightarrow XS_xR + ZnS + HS_{x-1}X$$

and a perthio-accelerator group which, with further zinc oxide will form a zinc perthio-salt of lower sulphur content; this may, nevertheless again be an active sulphurating agent, forming intermediates $XS_{x-1}R$. In this way each molecule of accelerator gives rise to a series of intermediates of varying 'degrees of polysulphidity'. The hydrogen atom which is removed is likely to be attached to a methylene group in the α-position to the double bond, i.e. in natural rubber the hydrogen atoms at positions 4 and 5 are the most

$$\underset{(1)\quad(2)\ \ (3)\quad(4)}{\sim CH_2 - \underset{\underset{\underset{(5)}{CH_3}}{|}}{C} = CH - CH_2 \sim}$$

labile in this type of reaction.

The intermediate XS_xR then reacts with a molecule of rubber hydrocarbon RH to give a crosslink, and more accelerator is regenerated:

$$XS_x R + RH \longrightarrow RS_{x-1}R + XSH$$

Even this is not the whole story, for, on further heating, the degree of polysulphidity of the crosslinks declines. This process is catalysed by XS_xZnS_xX and can result in additional crosslinks. It is also evident that the crosslinks which were initially at positions 4 and 5 undergo an allylic shift, with the result that new configurations appear:

$$\sim CH_2 - \underset{\underset{\underset{\underset{R}{|}}{S_x}}{|}}{\underset{|}{C}}\!\!\!\overset{\overset{CH_3}{|}}{} = CH - \underset{|}{CH} \sim \longrightarrow \sim CH_2 - \underset{\underset{\underset{R}{|}}{S_x}}{\overset{\overset{CH_3}{|}}{C}} - CH = CH \sim$$

and

$$\sim CH_2 - \underset{\underset{\underset{\underset{R}{|}}{S_x}}{\underset{|}{CH_2}}}{\overset{}{C}} = CH - CH_2 \sim \longrightarrow \sim CH_2 - \underset{\underset{\underset{R}{|}}{S_x}}{\overset{\overset{CH_2}{||}}{C}} - CH - CH_2 \sim$$

At the same time, disappearance of crosslinks of the disulphide and polysulphide type occurs, with formation of conjugated trienes:

$$\sim CH_2 - \overset{\overset{CH_3}{|}}{C} = CH - \underset{\underset{\underset{R}{|}}{S_x}}{CH} - CH_2 - \overset{\overset{CH_3}{|}}{C} = CH - CH_2 \sim$$

$$\downarrow$$

$$\sim CH_2 - \overset{\overset{CH_3}{|}}{C} = CH - CH = CH - \overset{\overset{CH_3}{|}}{C} = CH - CH_2 \sim + RS_xH$$

This destruction of crosslinks is apparently associated with the formation of the cyclic sulphides [Fig. 5.2(c)], but this has not been investigated in detail.

A consideration of the above reactions leads to the conclusion that, if desulphuration proceeds rapidly as in the case of the mix depicted in Fig. 5.3, the final network will be highly crosslinked with mainly monosulphidic bonds, and there will be relatively few modifications of the cyclic sulphide or conjugated triene type: such a network is termed 'efficiently crosslinked'.

If on the other hand, desulphuration proceeds slowly as in the case of the compound depicted in Fig. 5.4, there will be opportunities for thermal decomposition, leading to reversion or loss of crosslinks and to networks

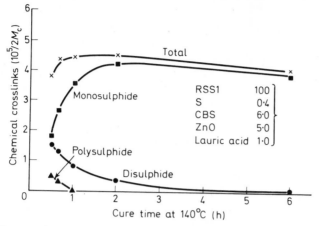

Fig. 5.3   Features of NR vulcanisate produced by an efficient crosslinking system (M_c = number average molecular weight between crosslinks). From Moore (1964), courtesy NRPRA

containing modifications: further, the crosslinks which do survive will be di- or poly-sulphidic and hence will be liable to further decomposition. These networks are inefficiently crosslinked.

Examination of a system containing HAF black shows that the remarks made above for a pure gum system generally hold for a natural rubber compound containing active black (Porter, 1969). The presence of HAF black increases the overall rate of reaction of the rubber and sulphur, and

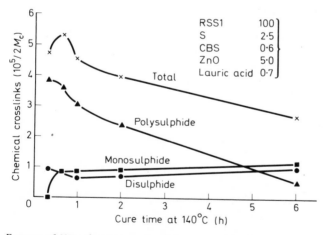

Fig. 5.4   Features of NR vulcanisate produced by a conventional crosslinking system (M_c = number average molecular weight between crosslinks). From Moore (1964), courtesy NRPRA

promotes the desulphuration reaction, thus leading to increased crosslink efficiency. A pictorial comparison of the properties of black compounds of various types is given in Fig. 5.8 (Section 5.6.1).

5.3   Non-Sulphur Vulcanising Systems for Olefin Rubbers

Almost as soon as the original discovery of vulcanisation had been made public, Parkes (1840) issued a patent in which he described the vulcanisation of thin sheets of rubber by exposure to the vapour of sulphur chloride. In later years, the rubber sheet was immersed in a solution of sulphur chloride in carbon disulphide or petrol. Tensile strengths were not very high, the vulcanisates were liable to smell of hydrogen chloride, ageing was relatively poor, and in any case the method was only suited to thin sheets. Nevertheless, the method, termed 'cold cure', was in use in this country until at least 1950 for the production of NR proofings. A monosulphide crosslink is formed, but chlorine is added to the molecule as well:

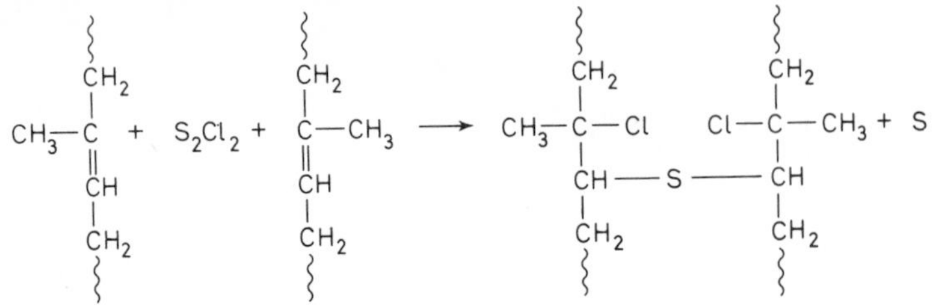

Presumably the method is applicable to all olefin rubbers, but there is little evidence of its being used for any polymer other than NR.

Tetramethyl thiuram disulphide (TMTD), besides being an accelerator for sulphur vulcanisation, can also be used in conjunction with zinc oxide as a non-sulphur vulcanising agent. Its use in this way dates back to 1921. Since one of the four sulphur atoms is readily removable to give the mono-sulphide, itself an accelerator, this was assumed to be the method of action, but prolonged and meticulous investigation by Scheele, Lorenz, and Dummer (1956), as well as by Moore (1964), has shown that the mechanism is highly complex, giving mainly monosulphide crosslinks. The TMTD reacts with the zinc oxide to give some tetramethyl thiuram polysulphide which reacts with the rubber and more zinc oxide to give a rubber-bound intermediate compound containing a polysulphide group. This then reacts with more rubber and zinc oxide to form a vulcanised rubber with polysulphide crosslinks which, as cure proceeds, give mainly monosulphide crosslinks. The remainder of the intermediate compound forms zinc diethyl dithio-carbamate, which is always found in the vulcanisate, although never in more than about 65% of the theoretical yield:

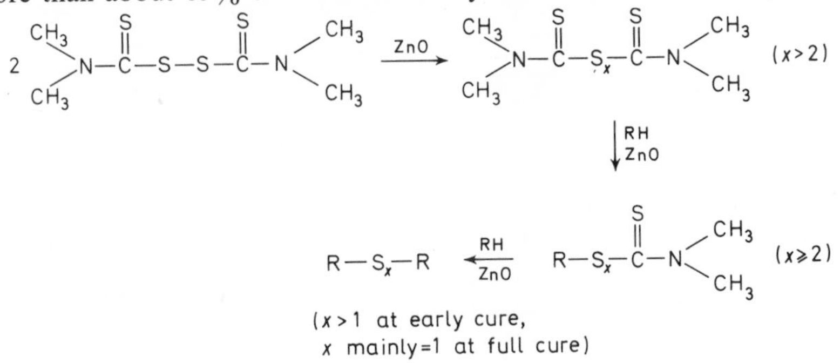

(x > 1 at early cure,
x mainly = 1 at full cure)

It is interesting to note that stearic acid or amines give an even more efficient crosslinking system, but the mechanism of this has not been investigated in full.

TMTD and TETD gives vulcanisates with NR which are lower in tensile strength than sulphur vulcanisates but have very good ageing properties. The cure may truly be described as a plateau cure, for an overcure of 12 times the optimum cure shows no sign of reversion. The thiuram disulphides may also be used to cure SBR and nitrile rubbers, but are too slow for use with butyl.

The use of bis-morpholinedisulphide as a vulcanising agent was reported by Sibley (1951) in an investigation which covered a number of organic sulphides. This agent gives very safe processing, coupled with good ageing properties, but may cause excessive cracking in tyre stocks. It is stated to give mainly monosulphide or disulphide crosslinks, and may be used as a curing agent for any of the olefin polymers.

Quinone dioxime can be used as a vulcanising agent for any olefin rubber, but its use in commercial practice appears to be limited to butyl rubber. An oxidising agent such as red lead, or even MBTS, is necessary, and the reaction has been explained as being due to the formation of dinitrosobenzene as an intermediate, followed by addition of this to two molecules of rubber

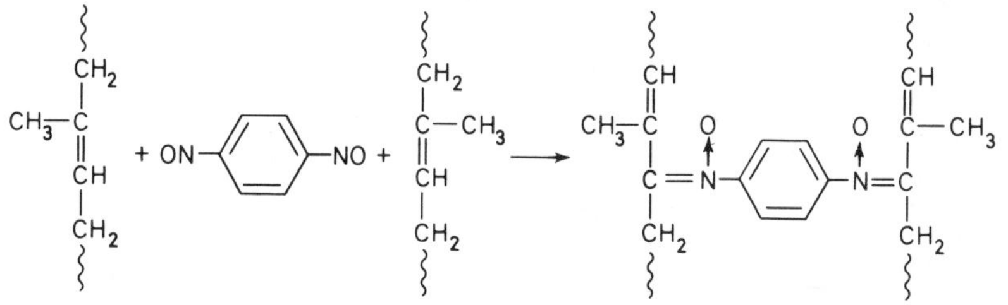

with removal of two hydrogen radicals, which react with the polymer or with more dinitrosobenzene, when quinone dioxime would be regenerated.

Phenol formaldehyde resins are also used to vulcanise olefin rubbers, but again practical use is confined to butyl, where stannous chloride can be used as an accelerator: alternatively, halogenated resins can be used. Possibly the action in the latter case is similar to that of a brominated phenol (Meer, 1944):

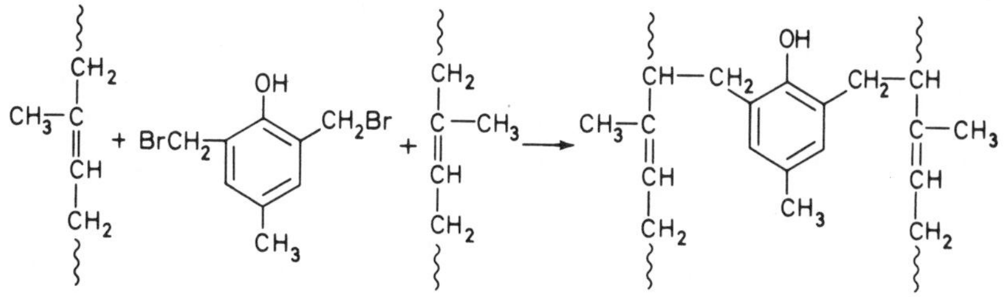

A vulcanising agent which shows great promise from the theoretical viewpoint is a bis-azodicarboxylate (Flory, Rabjohn, and Shaffer, 1949), which gives the theoretical amount of crosslinking required by the equation, and this can further be determined from the nitrogen content. The vulcanisates, further, do not contain inorganic particles which might lead to vacuoles.

$$CH_3-\underset{\underset{CH_2}{|}}{\overset{\overset{CH_2}{|}}{C}}\underset{\underset{CH_2}{|}}{\overset{\|}{\underset{CH}{|}}} + (CH_2)_{10} \begin{array}{l} O\cdot CO\cdot N{=}N\cdot CO\cdot OCH_3 \\ O\cdot CO\cdot N{=}N\cdot CO\cdot OCH_3 \end{array} \longrightarrow$$

$$CH_3-\underset{\underset{\underset{HN\cdot CO\cdot O\cdot(CH_2)_{10}\cdot O\cdot OC\cdot NH}{CH-N\cdot CO\cdot OCH_3\quad H_3CO\cdot OC\cdot N-CH}}{\overset{CH_2}{|}}}{\overset{\overset{CH_2}{|}}{C}}\quad C-CH$$

Another type of vulcanising agent for olefin rubbers is that which functions by thermal rupture into free radicals, which in turn remove active hydrogen atoms from the rubber molecules. The latter then crosslink by a combination of the free radicals one with another. The earliest use of a vulcanising agent of this type is due to Ostromislensky (1912), who vulcanised NR with benzoyl peroxide:

⟨C₆H₅⟩—CO·O·O·CO—⟨C₆H₅⟩ → 2 ⟨C₆H₅⟩—CO·O —

$$CH_3-\underset{\underset{CH_2}{|}}{\overset{\overset{CH_2}{|}}{C}} + ⟨C_6H_5⟩{-}CO\cdot O{-} \longrightarrow ⟨C_6H_5⟩{-}CO\cdot OH + CH_3-\underset{\underset{CH-}{|}}{\overset{\overset{CH_2}{|}}{C}}$$

$$CH_3-\underset{\underset{CH-}{|}}{\overset{\overset{CH_2}{|}}{C}} + \underset{\underset{-CH}{|}}{\overset{\overset{CH_2}{|}}{C}}-CH_3 \longrightarrow CH_3-\underset{\underset{CH}{|}}{\overset{\overset{CH_2}{|}}{C}}\ \underset{\underset{CH}{|}}{\overset{\overset{CH_2}{|}}{C}}-CH_3$$

The existence of the sparingly soluble benzoic acid in the vulcanisate is shown by the presence of a bloom on the surface. Vulcanisates produced by this method are relatively low in tensile strength, and some benzoylation of the molecule occurs at the same time.

Other peroxides have been used for vulcanisation, and one of the most common today is dicumyl peroxide, which gives a strong vulcanisate of good age resistance. NR, SBR, and nitrile rubber can all be cured by this means, but butyl is degraded by peroxide. Another peroxide which has been used in theoretical work is di-t-butylperoxide, but this is rather volatile for use in commercial practice. It is closely related to the compound 2,5-di-(t-butylperoxide)-2,5-dimethylhexane (Varox)

$$CH_3-\underset{\underset{CH_3}{|}}{\overset{\overset{CH_3}{|}}{C}}-O-O-\underset{\underset{CH_3}{|}}{\overset{\overset{CH_3}{|}}{C}}-CH_2-CH_2-\underset{\underset{CH_3}{|}}{\overset{\overset{CH_3}{|}}{C}}-O-O-\underset{\underset{CH_3}{|}}{\overset{\overset{CH_3}{|}}{C}}-CH_3$$

which is a commercial vulcanising agent. Peroxide vulcanisation can be modified by the presence of sulphur, and by the presence of amines or unsaturated substances, all of which serve to improve the level of tensile strength which is produced.

5.4   Non-Sulphur Vulcanising Agents for Non-Olefin Rubbers

5.4.1   METALLIC OXIDES

The various types of non-olefin rubbers must necessarily contain some active group by which they may be crosslinked.

5.4.1.1   *Polychloroprene Rubber*

In the case of polychloroprenes, the double bond is 'hindered' by the neighbouring chlorine atoms, and so vulcanisation with sulphur is no longer possible: this applies to both the sulphur-modified polymer and the homopolymer. The action of zinc oxide in the former case is said (Fisher, 1957) to be a crosslinking of the occasional 1,2 units, which react after an allylic shift of the chlorine atoms:

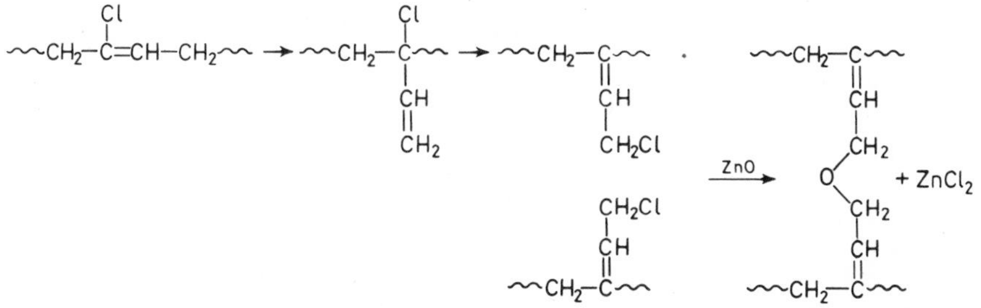

Magnesia is included in the formulation to act as a scavenger for the chlorine atoms also. The degree and rate at which the magnesia operates in this way is closely related to its activity as measured by iodine adsorption and profoundly affects the processing safety and rate of vulcanisation. More rapid vulcanisation is achieved by the use of organic accelerators, e.g. ethylene thiourea.

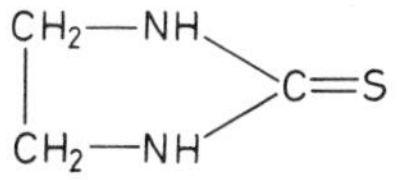

The crosslinking of the homopolymer, which requires an organic accelerator, is more complex (Section 5.4.2.1).

5.4.1.2   *Chlorosulphonated Polyethylene*

The vulcanisation of CSM is based on metal oxides, litharge, or magnesia, or a combination of magnesia and pentaerythritol, together with accelerators, of which dipentamethylene tetrasulphide (DPTS) is a suitable one. Thiazoles are satisfactory with litharge; dithiocarbamates give scorchy stocks which do not attain high states of cure. In all such curing systems it is believed that a mixture of ionic and covalent crosslinks is produced, the latter involving

Table 5.3   TYPICAL CURING SYSTEMS FOR HALOGEN-CONTAINING RUBBERS (P.P.H.R.)

	Polychloroprene		CSM			Fluorocarbon	
	S-types	Homopolymers					
Magnesia	4	4	5	20	20	15	15
Zinc oxide	5	5	—	—	—	—	—
Ethylene thiourea	—	0·5	—	0·5	—	—	—
Hydrogenated wood rosin	—	—	2·5	2·5	5	—	—
Litharge	—	—	10	—	—	—	—
MBTS	—	—	0·5	—	—	—	—
DPTS	—	—	0·75	1·0	1·0	—	—
HMDA carbamate	—	—	—	—	—	1·5	—
Dicinnamylidene hexamethylene diamine	—	—	—	—	—	—	3·0
Cure time (min)	20	20	30	30	30	45*	60*
Cure temperature (°C)	153	153	153	153	153	160	150

*Followed by oven cure of 24 h at 200°C.

sulphur crosslinks through chain unsaturation arising from complex degradative reactions of sulphonyl chloride groups.

CSM may also be vulcanised by exclusively organic chemicals, producing predominantly covalent crosslinks with low compression set vulcanisates (Table 5.3) (Maynard and Johnson, 1963).

5.4.2   POLYFUNCTIONAL AMINES

5.4.2.1   *Polychloroprene (Homopolymer)*

The crosslinking of the homopolymer usually requires an organic accelerator and is more complex. It has been proposed that the thiourea and the tertiary allylic chloride form a complex which reacts with zinc oxide and releases a molecule of urea. The residual zinc chloride mercaptide reacts with a second molecule of tertiary allylic chloride to form sulphur bridges between the two polymer chains (Table 5.3) (Murray and Thompson, 1964).

5.4.2.2 *Polyacrylic Rubbers*

Rubbers such as the polyacrylates often have reactive halogen groups included for the purpose of vulcanisation. Thus the copolymer of ethyl acrylate and 2-chlorethylacrylate, which contains a small proportion of the latter component, can crosslink by reaction with polyfunctional amines such as triethylene tetramine:

Hardening can be overcome by the addition of sulphur, MBTS, or TMT. Typical curing systems are:

$$
\begin{cases}
\text{Triethylene tetramine} & 1{\cdot}5 \\
\text{MBTS} & 2{\cdot}0
\end{cases}
$$

$$\text{HMDA carbamate} \qquad 1{\cdot}0$$

$$
\begin{cases}
\text{2-Mercaptoimidazoline} & 2{\cdot}0 \\
\text{Red lead} & 5{\cdot}0
\end{cases}
$$

Cure temperatures are in the range 150°–200°C, followed by a post cure in hot air of 16 hours at 175°C. Other more recent polyacrylate rubbers can be crosslinked by the use of ammonium salts, sulphur, and sodium and potassium soaps.

5.4.2.3 *Fluorocarbon Rubbers*

The basic ingredients for vulcanisation of fluorocarbon rubbers comprise a metal oxide (zinc, magnesium, or lead) with an amine such as hexamethylenediamine carbamate, ethylenediamine carbamate, bis-cinnamylidenehexanediamine or bis-*p*-aminocyclohexylmethane carbamate. In order to develop optimum mechanical properties, a two-stage cure is necessary: a press cure of 10–60 min at 120°–170°C, followed by an oven cure of 10–24 hours at up to 200°C.

The vulcanisation probably occurs in three stages; hydrogen fluoride is eliminated in the presence of basic materials to form regions of unsaturation; difunctional agents then react through additions to double bonds or through substitution of an allylic fluoride atom to form chemical crosslinkages; finally, during the high-temperature oven cure, conjugated double bonds are formed which undergo Diels–Alder condensation, and subsequent aromatisation leads to very stable aromatic crosslinks (Table 5.3).

5.4.3   PEROXIDES

5.4.3.1   *Ethylene–Propylene Copolymer*

Peroxides are useful for crosslinking saturated as well as unsaturated polymers. Ethylene–propylene may be crosslinked by this means: it is suspected that the point of attack is the occasional tertiary hydrogen atoms which occur where the chain branches. These are removed by the peroxide and a C—C crosslink results:

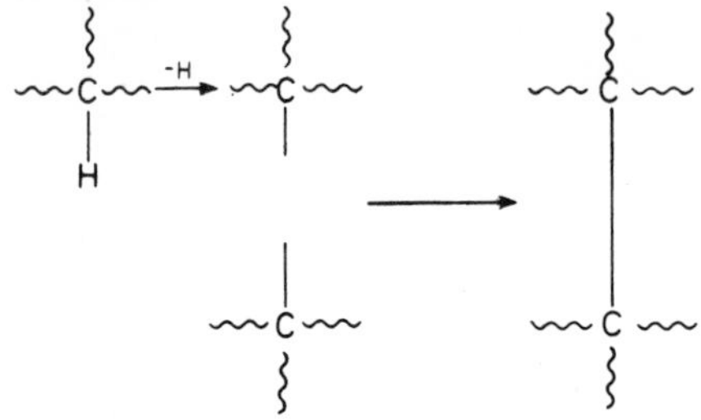

$$R\!-\!O\!-\!O\!-\!R \longrightarrow R\!-\!O\!- \xrightarrow{+H} R\!-\!O\!-\!H$$

Whilst crosslinking is the predominant reaction, there is some wastage of peroxide resulting from scission; this can be minimised by the inclusion of sulphur.

5.4.3.2   *Silicone Rubbers*

The crosslinking of dimethyl polysiloxane rubbers is due to attack by peroxides on the methyl group of the polymer:

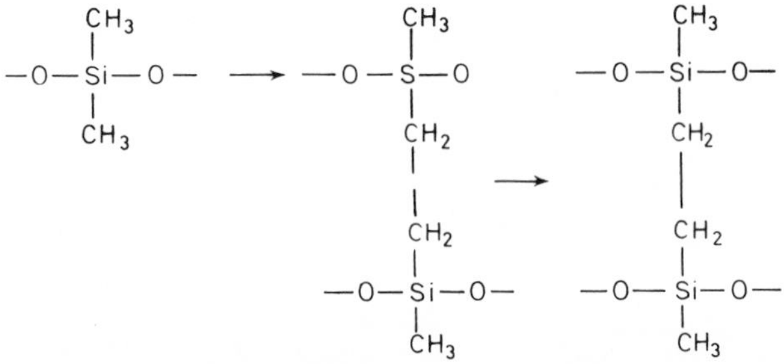

Where vinyl groups are also present in the polymer, the vulcanisation reaction is quicker and the main attack is undoubtedly at these reactive sites. Silicone rubbers with a high content of vinyl groups can be cured by sulphur and accelerators, but the vulcanisate is not so stable to heat as when peroxides are used and this has not been developed commercially (Section 4.13).

Table 5.4 lists some of the more widely used peroxides, with trade names and suppliers.

Table 5.4 ORGANIC PEROXIDES FOR VULCANISATION

Chemical name and formula	Trade name and producer	Form	Comments
Benzoyl peroxide $O=C-O-O-C=O$ (with two phenyl rings)	Lucidol (Novadel) Lucidol B-70 (Novadel) Lucidol S-50 (Novadel)	Damp powder, 80% active Paste in phthalate plasticiser, 70% active Paste in silicone oil, 50% active	Used for silicone rubbers; not suitable for olefin rubbers; dry powder explosive
2,4-Dichlorobenzoyl-peroxide $O=C-O-O-C=O$ (with two dichlorophenyl rings)	Perkadox SD (Novadel) Perkadox PDB-50 (Novadel) Perkadox PDS-50 (Novadel)	Damp powder, 50% active Paste in phthalate plasticiser, 50% active Paste in silicone oil, 50% active	Used for silicone rubbers; not suitable for olefin rubbers; dry powder explosive
Dicumyl peroxide $CH_3-\overset{\underset{\displaystyle }{CH_3}}{C}-O-O-\overset{\underset{\displaystyle }{CH_3}}{C}-CH_3$ (with two phenyl rings)	Perkadox SB (Novadel) Perkadox BC-40 (Novadel) Perkadox BM-50 (Novadel) Dicup (Hercules) Dicup 40 C (Hercules)	Powder, 95% active Powder, 40% active Solution in phthalate plasticiser, 50% active Powder, 96% active Powder, 40% active	Used for natural and synthetic olefin rubbers, and silicones
2,5-Di-(t-butyl peroxy)-2,5-dimethylhexane $CH_3-C-CH_2-CH_2-C-CH_3$ structure with CH_3, O, O, CH_3-C-CH_3 (CH_3) groups	Varox (Vanderbilt) Luperco 101-XL (Wallace & Tiernan)	Powder, 50% active Powder, 45% active	Used for natural and synthetic olefin rubbers, and silicone rubbers
Di-(t-butyl peroxy)-diisopropylbenzene structure with CH_3, C, benzene ring, $C-CH_3$, O, O, CH_3-C-CH_3 (CH_3) groups	Vulcup (Hercules)	Powder, 85% active	Used for natural and synthetic olefin rubbers, and silicone rubbers

5.5   The Assessment of State of Vulcanisation

5.5.1   THEORETICAL STUDY OF DEGREE OF CROSSLINKING

The degree of crosslinking is determined as $1/2M_c$, which equals the number of gram moles of crosslinks per gram of rubber in the network, where M_c is the number average molecular weight of the rubber chains between crosslinks; the number of network chains will be twice the number of crosslinks. The value of $1/2M_c$ can be arrived at from stress–strain measurements in simple strain using the expression

$$\sigma = \frac{\rho R T A_0}{M_c}\left(\lambda - \frac{1}{\lambda^2}\right)$$

where σ is the force to extend a sample of cross-sectional area A_0 to extension ratio λ, ρ is density, R is the gas constant, and T is the absolute temperature (Section 7.9).

Alternatively, the Flory–Rehner equation can be applied to the results of swelling the rubber in suitable solvents:

$$-\left[\ln(1-v_r)+v_r+\chi v_r^2\right] = \frac{\rho V_s v_r}{M_c}$$

where v_r is the equilibrium volume fraction of rubber in swollen gel, ρ the density of the rubber, χ an interaction constant characteristic of rubber and swelling liquid, and V_s the molecular volume of the swelling liquid (Saville and Watson, 1967).

5.5.2   PRACTICAL ASSESSMENT OF STATE OF CURE

The shape of the modulus and tensile curing curves for NR and for SBR have been described in Section 5.2.1, and it will be noted that there is an induction period before a noticeable modulus develops. This is known as the 'scorch time' (Section 5.6.2), and an appreciable level for this is a valuable asset to a compounding recipe, as it gives to the compound processing safety before 'set-up' or 'scorch' occurs. Once vulcanisation starts, the compound rapidly ceases to be thermoplastic and loses its ability to knit with other portions of stock: it also tends to crumb when milled.

Whilst the modulus curve is rising rapidly, vulcanisation is occurring. 'Cure time' can therefore be defined as the sum of the scorch time and the vulcanisation time, although the definition of 'optimum cure' is not so simple. In the first place, distinction must be made between a full cure, such as might be required for experimental purposes, and the 'technological optimum' which would be given under production conditions. Under laboratory conditions, it is possible to put the uncured stock into a hot mould, transfer to the press, heat for the required time at the temperature of cure, and immediately plunge the mould into water, thereby stopping the cure. Under factory conditions, it is customary to give the article a slight undercure as the cure will continue during the cooling process. This is certainly true of natural rubber vulcanisates, particularly where the compound and the conditions of cure (e.g. high temperature) tend to give a compound which readily reverts: the aim then is to obtain a material which will have the optimum properties when it has cooled down.

Formerly, assessment of cure necessitated making a series of cures of test-slabs, on each of which a number of physical tests were made. Generally, tensile strength, modulus at a convenient elongation, and hardness were measured, and from the values so obtained, the optimum cure was assessed. In the case of NR, a common method was to average the cures for optimum values of hardness (OV_H) and modulus (OV_M), and then take the mean of this value and the cure for optimum tensile strength (OV_T): i.e.

$$OV = \tfrac{1}{2}[OV_T + \tfrac{1}{2}(OV_M + OV_H)]$$

Since the tensile strength of NR is related to its tendency to crystallise, this optimum cure presents a compromise between the cure for optimum crystallisation and the cure for the optimum degree of crosslinking. It has, however, the disadvantage of taking several man-hours to arrive at the optimum cure for each compound.

In recent years, partly because of the growing use of synthetic rubbers, the former emphasis on tensile strength in the assessment of optimum cure has diminished and there is now a tendency to use modulus alone as the measure of cure. Whether one should look for the achievement of optimum modulus or a high proportion of that optimum varies with the polymer and the use to which the product is to be put. The result of this change in emphasis is that it has become of paramount importance to measure this quantity with the minimum of time and effort, preferably whilst the rubber is actually curing, by a machine which requires no special technique to operate and which automatically records the result in a convenient form. Descriptions of several machines for this purpose are given below; all are based on the principle that a reciprocating deformation is applied to the rubber whilst it is curing, and the increased resistance of the rubber to that deformation is measured (ASTM Special Technical Publn 383, 1965).

5.5.2.1 *The Agfa Vulcameter*

Peter and Heidemann (1957, 1958) described a machine in which a sinusoidal shear stress was applied to the rubber during cure: an optical system then projected the stress–strain ellipse on to a screen. The more recent Vulcameter imposes a sinusoidal shear strain of constant amplitude on the rubber and measures the stress by a proof ring and differential transformer. The makers claim that the optimum modulus is sharply defined and increases at the same time as the optimum value of the 300% modulus obtained in tension. Two test-specimens are required, each 20 mm × 30 mm × 5 mm in size, although another author has claimed that the use of specimens only 1 mm in thickness avoided the possible development of porosity with its resultant distortion of the trace. The instrument has also been used to measure scorch time, but this does not always coincide with the Mooney result (ASTM, 1965): this difference is thought to result from the severe shearing action in the latter instrument, which can break down the weak gel structure which exists at the start of cure.

5.5.2.2   *The Shawbury Curometer*

In the Shawbury Curometer (More, Morrell, and Payne, 1959), two small pellets of rubber (less than 1 cm³ each) are cured between electrically heated platens in a small hand-press on either side of a paddle to which a sinusoidal stress of constant amplitude is applied. The resultant strain, which diminishes as the rubber vulcanises, is amplified by a knife edge and lever system. A stylus on the other end of the lever executes a magnified copy of the oscillation of the paddle, and a chopper-bar causes it to mark a chart at the extremes of movement.

A typical trace for NR is shown in Fig. 5.5(a), and in this case the optimum cure is clearly visible as the start of the parallel portion of the curve. However, for most synthetic rubbers, the typical curve is different [Fig. 5.5(b)]

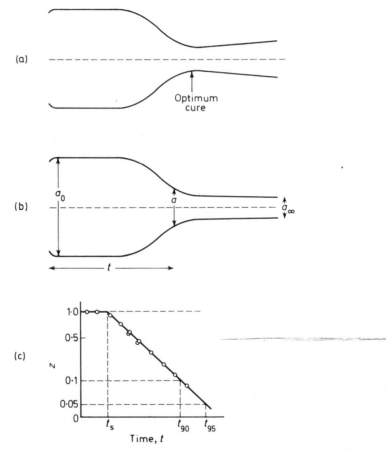

Fig. 5.5   *Shawbury Curometer traces: (a) NR, and (b) synthetic rubber; (c) plot of log z against t*

and here the estimation of optimum cure is not so easy by visual methods. Later models of the Curometer give single-sided curves, but the problem of estimation of optimum cure still remains. The time to 90% of optimum crosslinks has been arbitrarily chosen as the desirable optimum cure, and a relationship has been established (Payne, 1963) which enables this to be obtained from the reciprocal widths of the curve at various cures.

If z is the ratio of crosslinks still to be formed after time t to the total number which can be formed, and if a_∞, a_0, and a are the widths of the curves at the narrowest part, at the widest part, and at time t, then it can be shown that

$$z = \frac{a_\infty^{-1} - a^{-1}}{a_\infty^{-1} - a_0^{-1}}$$

A plot of log z against t is mainly linear, as shown in Fig. 5.5(c), and from this the time to 90% crosslinks (i.e. when $z = 0.1$) or the time to 95% crosslinks ($z = 0.05$) can be given. The value of z is not very dependent on the value of a_∞, so errors in this measurement do not greatly affect the 90% crosslinks figure. The value t_s where the plot meets the line $z = 1$ may be used as an estimate of the scorch time, whilst for routine work a direct correlation between a_0 and Mooney viscosity may be determined.

An assessment of variability was made on two batches of rubber of differing viscosity. The coefficient of variation on the measurement of a_0 was under 2%, whilst that relating to a_∞ was 2.5%. The measurement of t_s was subject to a coefficient of variation of 8%, and that of t_{95} to 7%.

The Curometer may also be used to investigate the kinetics of the vulcanisation reaction. The slope of the plot of log z against t may be used as a measure of rate of cure, and hence an Arrhenius plot of this slope against the reciprocal of the absolute temperature may be drawn. In such an exercise, a value was obtained for the temperature coefficient of the reaction of 1.8 per 10°C rise in temperature, in close agreement with the expected value.

5.5.2.3 The Firestone CEPAR Apparatus

The CEPAR apparatus was designed (Claxton, 1959) to measure cure, extrusion, plasticity, and recovery. The sample is heated in a cylindrical cavity into which a plunger is introduced periodically by the action of a spring, thus producing a deformation in the rubber which decreases with cure. The deflection of the plunger is mechanically magnified some 50 times and recorded on a drum of 15 mm diameter. The apparatus may be used as an oscillograph as well as for measurement of modulus in compression. By means of an orifice in the bottom of the test chamber, excess compound may be allowed to extrude out until a constant volume is left; this can be done whilst the compound is attaining the test temperature.

The chart is very similar to that of the Curometer, except that it is single sided, and the deflection equivalent to 95% crosslinks may be derived from the initial and final chart widths by means of a nomograph. Reference to the chart then gives the 95% cure time.

If required, the specimen can be further reduced in size. This is particularly of use when measuring state of cure at high temperatures. Another development is the provision of means for moulding a ring at the base of the CEPAR specimen. This ring can be cut off and used for the measurement of tensile properties, thus eliminating the need to prepare separate tensile slabs.

5.5.2.4   The Monsanto Rheometer

Superficially, the Monsanto rheometer (Decker, Wise, and Guerry, 1962) resembles the Mooney viscometer, but it will readily be appreciated that this is an oversimplification. The rubber specimen is contained in a rectangular cavity 50 mm × 50 mm × 10 mm and has embedded in it a biconical rotor of diameter 37 mm (or, in some cases, 40 mm) which is oscillated sinusoidally through an amplitude of between 2° and 10° by means of a motor-driven eccentric. The cavity and specimen are maintained at a temperature of $\pm 0.5°C$ by electric heaters regulated by thermistor controllers, whilst the dies which form the cavity are held together by a force of about 15 kgf/cm^2 exerted by means of a ram actuated by compressed air. On the arm of the eccentric are mounted strain gauges which measure the force required to oscillate the disc.

In use, the machine can be operated at 180 Hz or 15 Hz according to the measurement required. At the low frequencies, the stress attained at the moment of maximum strain in each alternative half-angle is recorded by means of a 'peak-picker', thus giving a plot against time of the low-frequency elastic modulus. At the high frequencies, a continuous record of the complex dynamic shear modulus is obtained, inertia in the pen of the recorder being eliminated by means of a counterbalance. Alternatively, the elastic and viscous components of the complex dynamic shear modulus can be fed to an oscillograph, and the loss factor at the higher frequency can be measured.

Plots of tensile modulus against rheometer torque show in some cases a linear relationship for each compound, but the slope of this line is different for each polymer, and the linear nature is dependent on the filler content. Measurements of dynamic shear modulus show a correlation with rheometer torque when the experiment is carried out under strictly analogous conditions, and the Yerzley effective dynamic modulus is directly proportional to the rheometer torque when a low ($\pm 1°$) amplitude is used. Agreement has also been shown between the Mooney plot at curing temperatures and the rheometer plot, at least until a five unit rise in value has occurred in either case. This applies to the low frequency of oscillation and the low amplitude, as more vigorous conditions cause heat build-up in the specimen.

5.5.2.5   The Goodrich Viscurometer

This instrument (Juve *et al.*, 1963) also resembles the Mooney viscometer at first sight. It consists of a cylindrical die cavity in which a cylindrical rotor with bevelled edges is oscillated through a fixed arc of rotation at a rate of two oscillations per minute, the torque required being measured by means of a strain gauge which forms part of the rotating shaft. The die cavities are heated electrically and give a control of temperature to $\pm 1°C$. The rotor is slightly different from the conventional Mooney rotor and has been designed to give a lower maximum torque, thereby preventing any tendency for the stock to slip or the rubber to tear at the edge of the rotor. The peak torque envelope is plotted, and this gives a double-sided curve which is the inverse of the early Curometer curves, being narrow when the stock is uncured and wide when it is cured.

The minimum distance apart of the curves has been shown to be loosely correlated with the Mooney viscosity of the compound, whilst the scorch time (the time corresponding to an increase of two units above the minimum) corresponds very closely with the Mooney scorch time. Similarly, the time to 90% cure is claimed to be highly correlated with the optimum cure from the 100% modulus of press-cured sheets when measured in the same way.

A statistically planned programme investigated the effect of a number of compounding variables on the values given by the viscurometer. It was shown that the amplitude of the oscillation had no effect on the optimum cure. The temperature of test affected the cure time in a manner which agreed with Arrhenius' law and suggested an activation energy of 21·5 kcal and a temperature coefficient of approximately 2 per 10°C rise of temperature. It was found that the presence of black in the compound affected the stiffness of the uncured stock much more than it did that of the fully cured vulcanisate, whereas increasing oil affected the stiffness of the cured compound, relative to that of the uncured, less at high crosslink densities than at low values, perhaps because it increased the elastic response. The room-temperature modulus could be predicted from the width of the trace providing a correction was made for the varying black and oil in the compounds used, although the author appreciated that this was purely empirical and did not take into account the theoretical factors involved.

5.5.3 DISCUSSION OF METHODS OF MEASURING CURE

The machines listed above all attempt to give the time of cure which would be expected to produce 90%, 95%, or 100% optimum crosslinks in a sheet of rubber 2·5 mm in thickness. However, certain design factors must be considered before accuracy can be expected.

In the first place, the temperature of the specimen must be considered. Rubber is a relatively poor conductor of heat, and the length of time required to raise the temperature of the inside of a thick sheet of rubber may be relatively long. Furthermore, if there is an unheated rotor of substantial size in the centre of the rubber, this again may serve to cool down the layers of rubber near to it, particularly if it is connected by a metal shaft to relatively cool parts of the machine. The author has found that when running the conventional Mooney viscometer at curing temperatures, the inside of the rubber may take 8 min to reach the temperature of dies, even when the rotor was heated to that temperature prior to the introduction of the specimen. On the other hand, when the slab of rubber is placed in a hot mould and compressed to 2·5 mm thickness, the inside of the rubber reaches the mould temperature in about 15 s. This would seem to suggest that results with a machine which requires a large specimen thickness should be viewed with suspicion.

The second factor which should be considered is the amount of work done on the rubber in the course of measurement. Vulcanised rubbers, and particularly commercial compounds with relatively large amounts of filler, have high heat build-up values, and any movement of large amplitude or at high frequency must result in a substantial rise in the interior temperature of the material. For that reason, measurements at the low rate of the viscuro-meter and the lowest rate of the rheometer are to be preferred, although care must be taken to see that the frequency is sufficient to give an adequate curve for quick-curing compounds at high temperatures.

Consideration should also be given to the nature of the cured product. If there is insufficient pressure on the mould, it will be porous and the resultant trace will not be indicative of the modulus of a press-cured speci-men. If the trace is being used merely to indicate cure, this will not matter as long as the shape of the curve is not altered by the porosity.

Lastly, when the compound contains a substantial amount of plasticiser, this may be brought to the metal–rubber interface in the course of cure and cause slipping. With the Curometer, this has given rise to the suggestion that nitrile vulcanisates can exhibit reversion, but it is strongly suspected that the other forms of curemeter are not entirely above suspicion in this respect.

5.5.4   CALCULATION OF CURE IN THICK ARTICLES

Having arrived at a suitable time of cure for a thin article, how does one pro-ceed to estimate the cure for a thicker one? Hitherto, this would have been an area where an inspired guess by an experienced rubber technologist would be required, and in the case of irregularly shaped sections this is still true to a large extent. However, theoretical calculations on heat transfer have been made by Cuthbert (1954) and by Atkin and Nye (1969), and enable the equivalent cure for a thick article to be determined. Unfortunately, the mathematics are complex and time-consuming, but can be avoided in

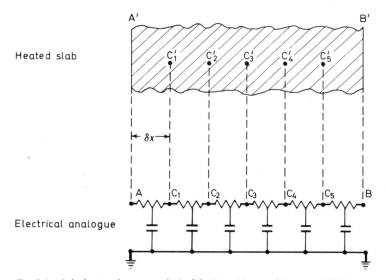

Fig. 5.6   Calculation of cure in a thick slab. From Heap and Norman (1966), courtesy RAPRA

individual cases by building an electrical analogue consisting of a network of resistors and capacitors (Heap and Norman, 1966).

If A′ B′ is a heated slab of rubber (Fig. 5.6), then its electrical analogue can be represented by the model shown, where voltages are applied at A and B equivalent to the temperatures of the surfaces A′ and B′. The voltages at

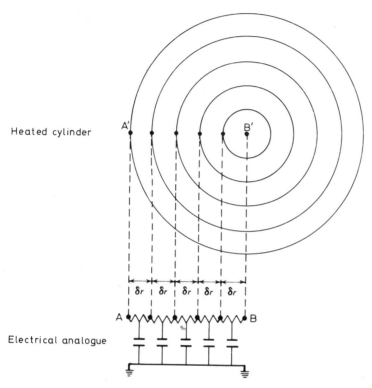

Fig. 5.7   Calculation of cure in a large-diameter cylinder. From Heap and Norman (1966), courtesy RAPRA

points C_1, C_2, C_3, . . . then represent the temperatures at the various corresponding points C'_1, C'_2, C'_3, . . . through the slab, where the thickness δx of each layer is small.

Similarly, for a cylinder of infinite length, the analogue is as shown in Fig. 5.7, where the resistors are proportional to $1/r$ and the capacitances to r, the mean distance of the corresponding sector of the cylinder from the cylindrical axis.

5.6   The Relation between Curing System Type and Properties (Sulphur and Polyolefin Rubbers)

In the discussion in Section 5.2.2, it was made clear that the particular sulphur–accelerator combination used for the vulcanisation of polyolefin rubbers affects not only the speed of cure but also the number and type of crosslinks introduced—in particular, the degree of polysulphidity of the crosslinks themselves.

5.6.1   VULCANISATE PROPERTIES

Consideration must now be given to the effect of the different vulcanisate structures on the properties of the elastomer. Generally speaking, such properties as tensile strength, modulus, resistance to swelling in liquids, resistance to fatigue, and resilience are increased as the degree of crosslinking is increased, i.e. as the number of crosslinks is increased. If, however, operation

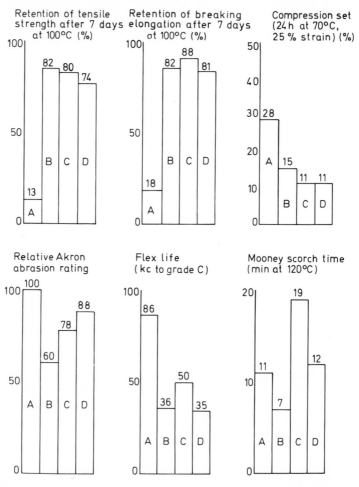

Fig. 5.8   Effect of vulcanising system on the properties of natural rubber vulcanisates [HAF black 45 p.p.h.r. and antioxidant 1 p.p.h.r.; vulcanisates cured for 40 min at 140°C to an equivalent hardness of (59+1) IRHD]: A, conventional 'tread' compound (CBS, 0·4 p.p.h.r.; sulphur, 2·5 p.p.h.r.); B, TMTD sulphurless compound; C, efficient vulcanisation system (CBS 5·0 p.p.h.r.; sulphur, 0·33 p.p.h.r.); D, efficient vulcanisation (sulphur donor) system (MBTS, 2·08 p.p.h.r.; bis-morpholinedisulphide, 1·47 p.p.h.r.). From Moore (1964), courtesy NRPRA

at elevated temperatures is required, the type of crosslink will be important, as will be apparent from what has been stated earlier in this chapter. For example, labile polysulphidic crosslinks will break down at elevated temperatures and the material will show high permanent set if under strain. For good set performance, therefore, the curing system and cure time must be selected to produce stable crosslinks.

Ageing, or the change of physical properties with time, is generally confused with the change in physical properties after the rubber has been exposed for prolonged periods at elevated temperature (accelerated ageing), since the two processes may have very similar results. There are also two quite different effects, the effect of oxygen in breaking the molecules and reacting with the crosslinks, and the non-oxidative scission of polysulphidic crosslinks. Again, the effect of the curing system may be complex: e.g. a curing system such as a TMTD sulphurless one gives stable crosslinks, and the byproduct of its action, ZDMDC, is also an efficient antioxidant.

Fatigue is another property which is usually associated with ageing. It may be defined as the ability to withstand repeated flexing without the development of severe cracking. Whereas a mainly monosulphidic crosslink system, such as that produced by a TMTD or other sulphur-donor system, gives good ageing even at elevated temperatures, it does not give good fatigue resistance. On the other hand, a polysulphide crosslinking system, such as given by MBT or CBS with a high sulphur content, has very good fatigue resistance. Accordingly, 'semi-efficient' vulcanising systems have been introduced which effect a compromise between the two extremes, giving both good ageing and good fatigue properties. Fig. 5.8. compares the properties for four vulcanising systems.

Most practical compounds contain an antioxidant (Section 6.5) as well, and hence their ageing is improved to some extent. However, it should be remembered that, although a good antioxidant improves a compound, it is a poor substitute for a good curing system or an inherently more resistant polymer.

5.6.2   SCORCH TIME

In practical compounding, another factor enters, that of the necessity of avoiding 'scorch', or premature vulcanisation, during the fabrication stage: this is particularly true of NR compounds.

During compounding and fabricating, a rubber compound is continually subject to heat. This is particularly evident during the mixing process and is the reason why accelerators are usually added towards the end of the mixing cycle, when the temperature of the mill or internal mixer has fallen. In the case of internal mixing, the sulphur is usually added on the sheeting mill for the same reason. This heat history is continued through the extrusion or calendering of blanks, and during the initial moulding of the final product. Ideally, it is only after the shaping processes have been completed that a substantial amount of cure should begin to take place, and then for economic reasons the compound should cure rapidly.

During the development of the use of accelerators in rubber compounding, it became evident that some were 'safer' than others. The term 'delayed action accelerator' was employed to describe an accelerator such as MBTS which was relatively inactive below 120°C, in contrast to MBT which has a measurable rate of cure at 100°C. In this category, the sulphenamide accelerators have proved a valuable addition.

For the same reason, mixtures of accelerators were introduced, in which the degree of crosslinking or the speed of cure of the primary accelerator was boosted by the presence of small amounts (usually under 0·4 p.p.h.r.) of the secondary accelerator without greatly affecting the scorch tendencies of the compound. A useful series of graphs has been given by Thompson and Watts (1966) which shows the relation between scorch time and cure time for a number of such mixtures.

Another way of retarding the scorch of compounds is to add a 'retarder'. Most acid substances act as retarders, and phthalic anhydride (0·5–2·0 p.p.h.r.) is one of the most common. Another material used as a retarder is *N*-nitrosodiphenylamine (1–5 p.p.h.r.). Both retarders are marketed under a number of trade names.

5.7   Accelerator Types

Table 5.5 lists some of the principal accelerators in each of the main types, with notes on their properties. The abbreviations, in some cases, originated in trade names and have now passed into the language of the industry.

Table 5.5   ACCELERATORS

Type and formula	Chemical name	Abbreviation	Property
Aldehyde amines			
	Hexamethylylene tetramine (hexamine)	HMT	Occasionally used as secondary accelerator on Continent; not used in UK because of high toxicity
	Ethylidene aniline	EA	Medium accelerator; activates thiuram sulphides and dithiocarbamates
Guanidines			
	Diphenyl guanidine	DPG	Medium accelerator (must be 200 mesh or will not disperse); now used mainly in conjunction with other accelerators
	Triphenyl guanidine	TPG	Slow accelerator
	Di-*o*-tolylguanidine	DOTG	Medium accelerator; also used as plasticiser for Neoprene; disperses more readily than DPG

Table 5.5 *continued*

Type and formula	Chemical name	Abbreviation	Property
Thiazoles			
(benzothiazole) C—SH	Mercaptobenz-thiazole	MBT	Semi-ultra accelerator; apt to be scorchy
(benzothiazole) C—S—S—C (benzothiazole)	Dibenzthiazyl disulphide	MBTS	Delayed-action semi-ultra accelerator
(benzothiazole) C—S⁻Na⁺	Sodium salt of MBT	—	Water-soluble; used in latex compounding
Sulphenamides			
(benzothiazole) C—S—NH—⟨cyclohexyl⟩	N-Cyclohexylbenz-thiazylsulphenamide	CBS	Semi-ultra accelerator; delayed action
(benzothiazole) C—S—N⟨morpholine⟩O	N-Oxydiethylbenz-thiazylsulphenamide	NOBS	Semi-ultra accelerator; delayed action
Dithiocarbamates			
⟨piperidine⟩N—C(=S)—S⁻ ⁺NH⟨piperidine⟩	Piperidine pentamethylene dithiocarbamate	PPD	Ultra-accelerator
((C₂H₅)₂N—C(=S)—S⁻)₂ Zn⁺⁺	Zinc diethyl dithiocarbamate	ZDC, ZDEC	Ultra-accelerator
(C₂H₅)₂N—C(=S)—S⁻ Na⁺	Sodium diethyl dithiocarbamate	SDC	Water-soluble ultra-accelerator; used for latex work
((C₂H₅)(C₆H₅)N—C(=S)—S⁻)₂ Zn⁺⁺	Zinc ethyl phenyl dithiocarbamate	—	Scorch-resistant ultra-accelerator
Thiuram sulphides			
(CH₃)₂N—C(=S)—S—S—C(=S)—N(CH₃)₂	Tetramethyl thiuram disulphide	TMT, TMTD	Ultra-accelerator and vulcanising agent
(C₂H₅)₂N—C(=S)—S—S—C(=S)—N(C₂H₅)₂	Tetraethyl thiuram disulphide	TET, TETD	Ultra-accelerator and vulcanising agent
(CH₃)₂N—C(=S)—S—C(=S)—N(CH₃)₂	Tetramethyl thiuram monosulphide	TMTM	Ultra-accelerator
⟨piperidine⟩N—C(=S)—S₄—C(=S)—N⟨piperidine⟩	Dipentamethylene thiuram tetrasulphide	DPTS	Ultra-accelerator and vulcanising agent; tendency to scorch
Xanthates			
((CH₃)₂CH—O—C(=S)—S⁻)₂ Zn⁺⁺	Zinc isopropyl xanthate	ZIX	Ultra-accelerator
(CH₃)₂CH—O—C(=S)—S⁻ Na⁺	Sodium isopropyl xanthate	SIX	Water-soluble ultra-accelerator for latex work
(C₄H₉—O—C(=S)—S⁻)₂ Zn⁺⁺	Zinc butyl xanthate	ZBX	Low-temperature ultra-accelerator
Morpholine disulphide			
O⟨morpholine⟩N—S—S—N⟨morpholine⟩O	Bis-morpholine-disulphide	—	Used principally as vulcanising agent

SIX

Materials for Compounding and Reinforcement

6.1 Carbon Blacks

BY J. B. HORN

6.1.1 INTRODUCTION

Carbon blacks are essentially elemental carbon and are composed of particles which are partly graphitic in structure. The carbon atoms in the particle are in layer planes which, by parallel alignment and overlapping, give the particles their semi-graphitic nature. The outer layers are more graphitic than those in the centre (Fig. 6.1). The particles range in size from 10 nm to 400 nm in diameter, the smaller ones being less graphitic.

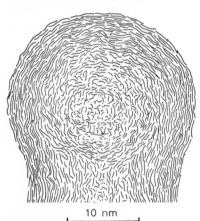

Fig. 6.1   *Concentric layer orientation of carbon black particle*

10 nm

6.1.2 MANUFACTURE

Carbon blacks are produced by converting either liquid or gaseous hydrocarbons to elemental carbon and hydrogen by partial combustion or thermal decomposition (Figs. 6.2, 6.3, and 6.4).

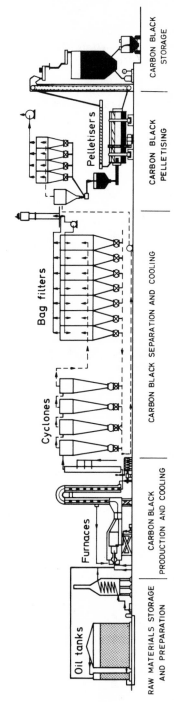

Fig. 6.2 Diagrammatic representation of carbon black manufacture by the furnace process

Oil tanks

Furnaces

Cyclones

Bag filters

Pelletisers

RAW MATERIALS STORAGE AND PREPARATION | CARBON BLACK PRODUCTION AND COOLING | CARBON BLACK SEPARATION AND COOLING | CARBON BLACK PELLETISING | CARBON BLACK STORAGE

6.1.2.1   *Furnace Process*

Most of the carbon black used today is made by the furnace process (Fig. 6.2). This process consists of the incomplete combustion of natural gas or heavy aromatic residue oils from the petroleum or coal industries in refractory-lined steel furnaces, ranging in length from 1·5 m to 6 m and in diameter from 0·15 m to 0·75 m. Blacks produced from natural gas are a very small proportion of the total furnace black output.

The black is separated from the combustion gases by means of cyclones and filter bags. It is then pelletised by either a dry or a wet procedure. In the former the black is rotated in large horizontal drums. In the latter, it is mixed with water in pin-type mixers and then dried in horizontal rotating drums. (This process was invented by Wiegand in 1927, improved by Cabot in 1934 and Huber in 1938, and was a big step forward in processing carbon blacks—Section 1.5.)

The yield from the furnace process varies from 25% to 70% of the available carbon, depending upon the particle size. Particle diameter ranges from 20 nm to 80 nm.

6.1.2.2   *Thermal Process*

The thermal process (Fig. 6.3) consists of the thermal decomposition at 1300°C of natural gas in the absence of air in 4 m diameter, 10 m high cylindrical furnaces filled with an open checkerwork of silica bricks. Two furnaces are used, the process being cyclic. Natural gas is admitted to one which has already been heated by firing a mixture of hydrogen and air; carbon and hydrogen are formed. The carbon is collected as carbon black, and the hydrogen is used to heat the other furnace. The recovery is about 40–50% of the available carbon, and the blacks obtained range in particle diameter from 120 nm to 500 nm.

Thermal black is also made from oil by the Jones process. This is

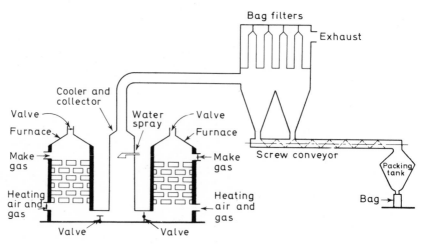

Fig. 6.3   The thermal process

cyclical in operation. Oil is burnt in a current of air to heat the checkerwork of refractory bricks in the reaction vessel, followed by a period when a mixture of steam and oil is admitted to the vessel. The oil cracks to give medium thermal carbon black and a mixture of gases, chiefly hydrogen, methane, and carbon monoxide.

Acetylene black is a thermal-type black but differs from the standard thermals in manufacture. The decomposition of acetylene gas is exothermic, so heat is supplied only to start the reaction.

6.1.2.3   *Channel or Impingement Process*

The impingement process (Fig. 6.4) also uses natural gas or natural gas enriched with oil. The feedstock is piped to thousands of small burner tips. The small flames impinge either on to a large rolling drum or on to slowly reciprocating channel irons. The deposited black is taken off by scrapers and

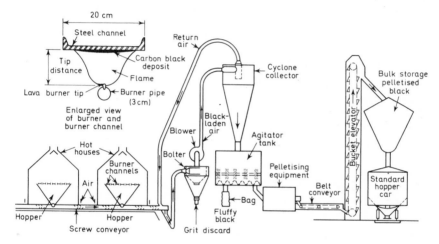

Fig. 6.4   The channel process

collected. The yield from this very inefficient process is 5% or less, but very fine blacks can be made, the particle diameters ranging from 9 nm to 30 nm.

6.1.2.4   *Lampblack Process*

A further process is the lampblack one, in which the black is made by burning petroleum or coal tar residues in shallow pans. The 'smoke' is then conducted to settling chambers in which the carbon flocculates.

6.1.3   PROPERTIES

The five most important properties of carbon blacks are: (a) particle size,

(b) structure, (c) physical nature of the surface, (d) chemical nature of the surface, and (e) particle porosity.

6.1.3.1   *Particle Size*

The particles of carbon black are not discrete but are fused. 'clusters' of individual particles. The fusion is especially pronounced with very fine blacks. These aggregates can be seen in vulcanised rubber, so they appear to be the working unit. However, the reinforcement conferred by the black is not influenced to any extent by the size of the unit but greatly by the size of the particles within the unit.

The electron microscope is used to determine particle size, and, as it has shown the ultimate particle to be basically spheroidal, the measurement of surface area by adsorption methods is also used to find the size of the particles. The surface area is measured, and, assuming each particle to be a true sphere, the average particle diameter is calculated. The most reliable adsorption procedure is that of Brunauer, Emmett, and Teller, using the nitrogen isotherm (BET area). The adsorption of iodine, however, from an aqueous iodine–potassium iodide solution can be related to nitrogen adsorption, and therefore iodine adsorption is more generally quoted because the determination is rapid and simple to perform. Fig. 6.5 shows the particle sizes of various

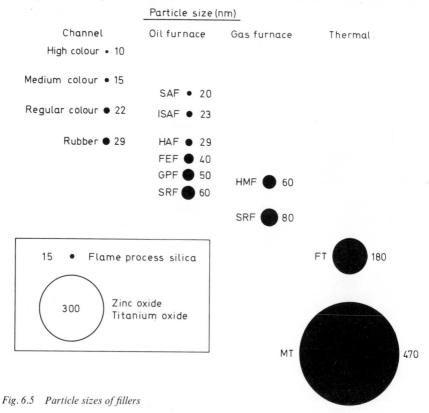

Fig. 6.5   Particle sizes of fillers

types of carbon blacks compared with those of other fillers used as rubber-compounding ingredients. The particle diameters of the various types and grades of black are given in Table 6.2.

6.1.3.2 Structure

The term 'structure' refers to the joining together of carbon particles into long chains and tangled three-dimensional aggregates. It should not be confused with the intraparticle semi-crystalline structure. This aggregation of particles takes place in the flame during the manufacture of carbon black. Neighbouring particles at the same stage of growth have more layers of carbon deposited on them until they finally fuse together. These primary aggregates can also form loose associations or 'secondary structure'. This latter structure is weak and much of it is lost during pelletisation and the remainder when the black is mixed into rubber.

The thermal process produces blacks with little or no structure. There is some particle aggregation in blacks produced by the impingement process. On the other hand, the oil furnace process, using highly aromatic raw materials, gives blacks of high structure (Fig. 6.6). The range of structure

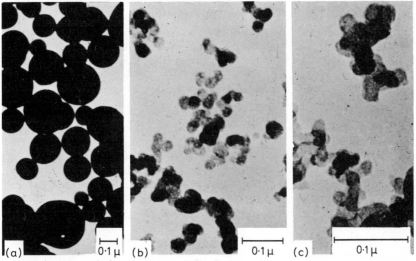

Fig. 6.6 Electron micrographs of different types of blacks, showing different structure levels: (a) thermal, (b) channel, and (c) furnace

within blacks from the furnace process is large. The use of flame additives to modify the flame ionisation potential considerably lowers structure. High structure blacks are produced by modification of furnace design and choice of raw materials.

The higher the structure of a carbon black, the more irregular the shapes of the aggregates, hence the less these aggregates are capable of packing together. Structure is normally measured by determining the total volume of the air spaces between aggregates per unit weight of black. The test is done by measuring the volume of a liquid (dibutyl phthalate, DBP), required to fill the voids. The instrument developed for this, the absorptometer, consists of a small chamber in which a weighed amount of the black is stirred by rotors as the liquid is added at a constant rate. When all the voids are filled, the whole mass stiffens up abruptly with an increase of the torque required to turn the rotors. The machine is set to shut off at this point and the DBP adsorption is read off and recorded as cm^3 per 100 g of black (Eaton and Middleton, 1965; Horn, 1969).

6.1.3.3   *Physico-Chemical Nature of Particle Surface*

As stated earlier, the carbon atoms in a carbon black particle are present in layer planes. Diffracted-beam electron micrographs have shown that, in blacks with low reinforcing potential (thermal blacks), the layers are highly orientated. They are mainly parallel to the surface, have regular spacing, and are quite large with very few defects in their network structure. Blacks with a high reinforcing potential, on the contrary, show less crystallite orientation. Micrographs show their particles to be more irregular in shape, and, as they are much smaller, it follows that the layers are smaller, less frequently parallel, and have more defects. This may indicate the presence of significant amounts of 'non-graphitic' carbon (i.e. interstitial carbon between the layers, aliphatic chains at layer edges, and incomplete networks at lattice defects) (Harling and Heckman, 1969).

It is possible that the nature of the carbon atoms at the surface of a particle may affect rubber reinforcement. The carbon atoms can be relatively un-reactive if they are an integral part of the layer plane, more reactive if attached to a hydrogen atom, and very reactive if present as a resonance-stabilised free radical.

6.1.3.4   *Chemical Nature of Particle Surface*

Carbon blacks consist of 90–99% elemental carbon. The other major constituents are (combined) hydrogen and oxygen. The hydrogen comes from the original hydrocarbon and is distributed throughout the carbon black particle. As the particles are formed in the reducing atmosphere of the flame, the oxygen appears subsequently and is, therefore, confined to the surface.

The principal groups present are phenolic, ketonic, and carboxylic, together with lactones. As heating in the range 400°–1600°C is necessary to devolatilise carbon black, it appears that the surface groups are not physically absorbed but are chemically combined. In addition to the oxygen and hydro-gen groups, carbon blacks may contain very small amounts of sulphur depending upon the nature of the hydrocarbons used in their manufacture. Most of this sulphur is chemically combined and appears to be inert and does not cause crosslinking of rubber.

6.1.3.5   *Particle Porosity*

The surfaces of carbon black particles are not smooth owing to the attack on them by high-temperature oxidising gases immediately after their formation.

Oxidation takes place at the 'non-graphitic' atoms and can progress into the particle to give pores.

A measure of porosity can be obtained by comparing the surface area (a) obtained by calculation using the particle diameter as measured by the electron microscope and assuming that the particle is a true sphere, and (b) as measured by adsorption (e.g. by using nitrogen). The latter method measures not only the external surface but any pores and cracks in the particle which are available to the nitrogen molecule.

6.1.4   COMMERCIALLY AVAILABLE CARBON BLACKS

Table 6.1 gives the trade names of the various carbon blacks grouped according to type. The names of the manufacturers are also shown.

All producers try to make blacks of each type completely equivalent to those made by other producers. Because of differences in such factors as raw materials and equipment, they do not always succeed completely and slight differences can sometimes be found between products of different manufacturers.

6.1.5   CLASSIFICATION

The naming of the different types of carbon blacks has developed over many years with no logical basis; manufacturers have arbitrarily given new products new type names as well as trade names.

The system of type names as given in column 2 of Table 6.1 has always been cumbersome, but the continuing development of new types has now made it impossible, e.g. there is now a new black which reinforces rubber more than HAF (High Abrasion Furnace) but not as well as ISAF (Intermediate Super Abrasion Furnace). What should it be called? One suggestion has been 'Intermediate Intermediate Super Abrasion Furnace'!

Recently several proposals have been made to change the system, three of which have been published: ASTM D1765-67 (1967), Heal and Beddoe, and Wolf and Westlinning (Anon, 1968-2).

6.1.5.1   *ASTM D1765-67*

This ASTM system consists of one letter followed by three numerals.

The letter indicates rate of cure, N and S being used for normal and slow respectively.

The first numeral gives the surface area of the carbon black as measured by iodine adsorption. Here, some attempt has been made to relate the proposed system to the type name system, e.g. 1 = SAF, 2 = ISAF, 3 = HAF, but it has meant that the IISAF black, already mentioned, cannot be fitted into this proposed system either. Another disadvantage is that the iodine number ranges are too wide in the ASTM system, and different types of black overlap.

The remaining two numerals are selected arbitrarily, but one rule applies. In the case of a black with a standard level of structure, the second numeral is

Table 6.1 MAJOR TRADE NAMES FOR CARBON BLACKS

Type name	Type code	Cabot	Cofrablack	Columbian
Super abrasion furnace	SAF	Vulcan 9	Cofrablack SAF	Statex 160
Intermediate super abrasion furnace, high structure	ISAF-HS	Vulcan 6H	Cofrablack ISAF-HS	Statex 125H
Intermediate super abrasion furnace	ISAF	Vulcan 6	Cofrablack ISAF	Statex 125
Intermediate super abrasion furnace, low modulus	ISAF-LM	Vulcan 6 (LM)	—	—
Intermediate super abrasion furnace, low structure	ISAF-LS	Regal 600	—	Neotex 150, Neotex 130
—	(N285)	Vulcan 5H	—	Statex 120H
High abrasion furnace, high structure	HAF-HS	Vulcan 3H	Cofrablack HAF-HS	Statex RH
High abrasion furnace	HAF	Vulcan 3	Cofrablack HAF	Statex R
High abrasion furnace, low structure	HAF-LS	Regal 300	Cofrablack HAF-LS	Neotex 100, Neotex 100H
Fine furnace	FF	Regal 99	—	Statex B
Fast extrusion furnace	FEF	Sterling SO	Cofrablack FEF	Statex M
All purpose furnace	APF	Sterling 105	Cofrablack APF	—
General purpose furnace	GPF	Sterling V	Cofrablack GPF	Statex G
High modulus furnace	HMF	Sterling L	Cofrablack HMF	Statex 93
Semi-reinforcing furnace	SRF	Regal SRF, Sterling S	Cofrablack SRF	Furnex SRF
Lampblack	Lampblack	—	—	Magecol 888
Extra conductive furnace	XCF	Vulcan XC-72	—	—
Super conductive furnace	SCF	Vulcan XXX	—	Conductex SC
Conductive furnace	CF	Vulcan C	—	—
Acetylene	Acetylene black	—	—	—
Channel	Impingement types	—	—	Micronex Std, Micronex W6
Fine thermal	FT	Sterling FT	—	—
Medium thermal	MT	Sterling MT	—	—

Continental Witco	Degussa	Ketjen	Huber	Phillips Philblack	United	Miscellaneous
Continex SAF	Corax 9	Ketjenblack SAF	Huber SAF	Philblack E	United 85	—
Continex ISAF-HS	—	—	Aro 200, Aro 250	Philblack I-H	United ISAF-HS	—
Continex ISAF	Corax 6	Ketjenblack ISAF	Huber ISAF	Philblack I	United 70	—
Continex ISAF-LM	—	—	—	—	—	—
—	Corax 600	Ketjenblack LHI	—	—	United 70LM	—
Continex N285	—	Ketjenblack N285	Aro 180	Philblack N285	United 67H	—
Continex HAF-HS	—	Ketjenblack HAF-H	Aro 100, Aro 150	Philblack OH	United 60H, United 65SPF	—
Continex HAF	Corax 3	Ketjenblack HAF	Huber HAF	Philblack O	United 60	C31 (SPA)
Continex HAF-LS	Corax 300	Ketjenblack CR	Huber CRF	Philblack 55, Philblack CRF	United HAF-LS	—
—	—	Ketjenblack FF	—	—	—	—
Continex FEF	Corax A, Corax B	Ketjenblack FEF	Huber FEF	Philblack A	United 50	—
—	—	Ketjenblack APF	—	Philblack APF	—	C27 (SPA)
Continex GPF	Corax G	Ketjenblack GPF	Huber GPF	Philblack G	Ukarb 327, Dixie 45, Kosmos 45, United 35	—
Continex HMF	—	Ketjenblack HMF	—	—	United 40	—
Continex SRF	—	—	Essex SRF, Aro 30	Philblack SRF	United 20	—
—	Durex O, Durex 101	—	—	—	—	Conservo (Cledca), VGC (Lummerz-hiem)
—	—	—	Aromex 115	—	—	—
—	Corax L	—	—	—	—	—
Continex CF	—	Ketkenblack CF	Huber CF	—	United CF	—
—	—	—	—	—	—	Shawinigan P1250 (VSP)
—	CK3/CK4	—	Aro MPC, Wyex EPC	—	—	Texas E, Texas M (S. Richardson)
—	—	—	—	—	United FT	P33 (Vanderbilt)
—	—	—	—	Sevacarb MT	United MT	Thermax (Vanderbilt)

always a repeat of the first numeral and the last is zero, e.g. SAF = N-110 and ISAF = N-220.

6.1.5.2 *Heal and Beddoe*

This system is similar to that proposed by ASTM and consists of one letter followed by three numbers. The letter is again either N or S. The first number is based on surface area as measured by iodine number but with different ranges so that no overlap occurs. The IISAF black, for instance, can be included in this scheme, which has the disadvantage of having no digit for a black having the surface area of the present thermal types (MT and FT). The second and third numerals indicate the structure from a rating based on measurements of oil (dibutyl phthalate) absorption and pour density. Thus, this method describes each black more completely than does the system proposed by ASTM.

6.1.5.3 *Wolf and Westlinning*

This system is a four-numeral one and bases the classification of carbon blacks on the effects they have in rubber. The first numeral represents the surface area as measured by nitrogen adsorption. This system does not have overlapping surface area ranges, and differs from the other systems in using the highest numeral to represent the highest surface area. The second numeral represents the filler activity coefficient (Section 7.9) and is a measure of the effect of the carbon black on the deformation of rubber no matter the network density. The third numeral indicates the effect on time to scorch, and the fourth numeral indicates the effect on the constant of the crosslinking reaction, i.e. on rate of cure. The data used to determine the last three numerals are obtained on a curemeter.

Table 6.2 shows the relationship of the various systems.

6.1.6   EFFECTS ON RUBBER PROPERTIES

Carbon black is the most important filler for carbon-backboned rubbers.

However, each of the five properties discussed in Section 6.1.3 has its effect on the properties of the rubbers in which carbon black is incorporated, and careful selection of grade is necessary to ensure a correct balance of processing and vulcanisate properties (Horn, 1968).

In general terms, the smaller the *particle size*, the poorer the processability and the higher the reinforcement. By reinforcement is meant the enhancement of tensile strength, abrasion resistance, and tear resistance (Section 7.2). Reference has been made in Section 2.2.3 to the importance of the finer particle size blacks in making high-strength rubbers from non-crystallising polymers such as SBR, IIR, and NBR.

The effect of *structure* is more noticeable on processing properties than on the properties of the vulcanisate. In general, the higher the structure, the stiffer and less 'nervy' the unvulcanised compound and the harder the

Table 6.2 PROPOSED NOMENCLATURE FOR CARBON BLACKS

Type code*	ASTM D1765	Heal and Beddoe	Wolf and Westliming	Iodine adsorption (ASTM D1510) (mg/g black, approx.)	Mean particle diameter (electron microscope surface mean averages) (nm, approx.)	Structure (ASTM D2414) (cm³ DBP/100 g, approx.)
SAF-LS	—	N-115	8355	140	22	90
SAF	N-110	N-130	8555	140	22	115
SAF-HS	—	N-135	8655	140	22	135
ISAF-LS	N-219	N-215	7355	120	28	80
ISAF-LM	N-231	N-225	7455	120	28	90
ISAF	N-220	N-230	7555	120	28	115
ISAF-HS	N-242	N-235	7655	120	28	130
CF	N-293	N-230	7555	140	—	115
SCF	N-294	—	—	190	—	110
—	N-285	N-335	6655	100	30	125
EPC	S-300	S-315	5253	†	32	95
MPC	S-301	S-315	6263	†	32	95
HAF-LS-SC	S-315	S-320	5253	80	32	70
HAF-LS	N-326	N-415	5355	80	32	70
HAF	N-330	N-425	5555	80	32	105
HAF-HS	N-347	N-435	5655	80	32	125
SPF	N-358	N-445	5755	80	32	140
XCF	N-472	—	—	230	—	190
FF	N-440	N-515	4355	50	42	60
FEF-LS	N-539	N-615	3355	42	47	110
FEF	N-550	N-630	3555	42	47	120
FEF-HS	N-568	N-635	365$	42	47	130
HMF	N-601	N-720	3455	38	61	85
GPF	N-660	N-720 American N-825 British	2555	35	70	90
APF	N-683	N-730	2655	35	70	135
MPF	N-785	N-830	2655	26	83	140
SRF-LM	N-761	N-815	2365	26	83	65
SRF-LM-NS	N-762	N-815	2365	26	83	65
SRF-HM	N-770	N-815	2365	26	83	70
SRF-HM-NS	N-774	N-815	2365	26	83	70
Lampblack	—	N-930	1655‡	18	—	120
FT	N-880	—	1165	†	190	33
MT-NS	N-907	—	0075	7·5	300	33
MT	N-990	—	0075	†·	300	33

* Abbreviations included additional to those in Table 6.1 are: EPC, easy pocessing channel; SC, slow curing; MPC, medium processing channel; SPF, super processing furnace; MPF, medium processing furnace; NS, non-staining.
† Iodine adsorption method ASTM D1510 is not applicable to these grades owing to interference with iodine adsorption.
‡ There are various grades of lampblack which may have different classifications from that indicated.

H'

vulcanised material. Table 6.3 shows in detail the effects of decreasing particle size and increasing structure on both the vulcanisate properties and the rheological behaviour of unvulcanised compounds (Horn, 1969).

Table 6.3   EFFECTS OF CHANGES IN PARTICLE SIZE AND STRUCTURE ON RUBBER PROPERTIES

Processing properties	Decreasing particle size	Increasing structure
Loading capacity	Decreases	Decreases
Incorporation time	Increases	Increases
Oil extension potential	Little	Increases
Dispersability	Decreases	Increases
Mill bagging	Increases	Increases
Viscosity	Increases	Increases
Scorch time	Decreases	Decreases
Extrusion shrinkage	Decreases	Decreases
Extrusion smoothness	Increases	Increases
Extrusion rate	Decreases	Little
Vulcanisate properties	*Decreasing particle size*	*Increasing structure*
Rate of cure	Decreases	Little
Tensile strength	Increases	Decreases
Modulus	Increases to maximum then decreases	Increases
Hardness	Increases	Increases
Elongation	Decreases to minimum then increases	Decreases
Abrasion resistance	Increases	Increases
Tear resistance	Increases	Little
Cut growth resistance	Increases	Decreases
Flex resistance	Increases	Decreases
Resilience	Decreases	Little
Heat build-up	Increases	Increases slightly
Compression set	Little	Little
Electrical conductivity	Increases	Little

The role of the *physico-chemical nature of the surface* in rubber reinforcement is still not fully understood. It has been postulated that a black with high structure gives a high-modulus rubber, not because the carbon black agglomerates restrict the crosslinked network but because the high shear forces during mixing break these agglomerates down to give active free radicals capable of reacting with the rubber. High structure, however, does not increase either tensile strength or tear resistance—the two properties usually associated with reinforcement.

On the contrary, it has been shown that, as a carbon black is progressively graphitised by heat treatment, tensile strength and tear resistance progressively decrease, indicating that the physico-chemical nature of the surface is important.

There appears to be no direct connection between the *chemical nature of the surface* (i.e. the individual chemical groups on the particle surface) and the properties which the black confers on rubber. Electrical conductivity and rate of cure, however, are affected to a considerable degree by the surface chemistry of the carbon black. Conductivity is basically dependent on particle size, but the chemical groups on the surface must be below a certain proportion to allow the black to conduct at all (e.g. channel black which has a high proportion of chemical groups is non-conductive, whereas a furnace black of the same particle size is very conductive).

The crosslinking rate of a rubber is affected by phenolic and carboxylic groups; the black slows the rate of cure in proportion to its total acidity. Channel black, for example, is slower curing than furnace black.

Carbon black particles with *pores and cracks* have higher surface areas than blacks of similar particle size without such features. This can result in cure retardation owing to the increased adsorption and inactivation of rubber curatives. If a carbon black is porous, there is an increased number of particles per unit weight. Thus, porous blacks give decreased resilience and increased electrical conductivity when compared with equal weight loadings of non-porous blacks.

In summary, particle size is the most important feature of carbon black. Structure is also of some importance, and porosity has a direct effect on certain rubber properties. The effect of the carbon black surface may be a minor one, but the present knowledge of the relationship between surface chemistry and rubber properties is not complete.

6.1.7 APPLICATIONS IN RUBBER COMPOUNDING

Between 90% and 95% of the total carbon black produced is used in the rubber industry, and approximately 80% of this is used in the manufacture of tyres and tyre products, such as inner tubes and retreading compounds.

There is little, if any, lampblack used in tyres, but all other blacks find application in tyre components. The thermal blacks are used in inner linings and inner tubes, the low reinforcing furnace grades in carcasses, the high reinforcing ones in treads, and the channel blacks in treads and also in the breaker areas of radial tyres. The high structure, high reinforcing furnace grades are used in synthetic rubber treads to give increased tread life and reduced groove cracking. The low structure, highly reinforcing grades are used in natural rubber off-the-road treads and, to a limited extent, in truck treads to increase their resistance to cutting and chipping.

Thermal blacks provide a low degree of reinforcement and can be used at high loadings. They are used in V-belts because of their low heat build-up, and other applications are mats, sealing compounds, and mechanical goods.

Lampblacks have coarse particles but have high structure. They are used in high-hardness compounds which are required to be resilient and have low heat build-up when used dynamically. Examples of applications are tank treads and engine mounts.

Acetylene black is used mainly to confer electrical conductivity on rubber compounds but, like lampblack, it can be used to produce high-hardness compounds which will process easily and be fairly resilient.

Channel blacks are expensive and their use is on the decline. Their retarding effect on cure is advantageous for bonding purposes and for compounds which require high reinforcement but must process easily. Their lack of electrical conductivity makes them useful for certain cable applications.

By far the largest amount of carbon black used today is of the furnace type. Furnace blacks are found in practically every type of black-filled rubber article. The fine particle blacks are used where high strength and resistance to abrasion are required, e.g. in conveyor belt covers and certain types of footwear. The coarse particle blacks are used in such articles as hose, cables, footwear uppers, mechanical goods, and extrusions. As indicated in Table 6.3, high structure improves processing characteristics and often the choice of grade will be determined by this factor (e.g. the high structure FEF and APF blacks for extrusion). Another instance of the selection of a grade for other than its reinforcing properties is the use of SCF black in conductive and antistatic goods because of the porous and chemically clean particle surface.

6.2   Non-Black Fillers and Colouring Materials

BY D. N. SIMMONS

6.2.1   INTRODUCTION

In this section, general information is given regarding non-black fillers which are used to reduce cost, to improve processing by reducing 'nerve', and to reinforce the polymer by increasing hardness, tensile strength, tear resistance, abrasion resistance, and other properties in the production of white or coloured compounds.

They are usually classified as:
1. Fillers used mainly to reduce cost.
2. Semi-reinforcing fillers.
3. Reinforcing fillers used to achieve high performance in non-black products.

This classification is not rigid as materials derived from similar sources may vary in function according to their methods of preparation and treatment. It has been considered expedient, therefore, to group them here according to their source. The specific gravities of the most common fillers referred to in this section are given in Table 6.4.

6.2.2   MINERAL FILLERS

6.2.2.1   *Aluminium Hydroxide (Hydrated)*

Hydrated aluminium hydroxide is a fine white powder, obtained by precipitation processes, used as a reinforcing filler in shoe soling and in mechanical goods. The retarding effect on cure is usually counteracted by the addition of glycols or amines or by increased accelerator dosage.

6.2.2.2   *Aluminium Silicate (Hydrated)*

Hydrated aluminium silicate is also a fine white powder, obtained by precipitation processes, used as a reinforcing filler in shoe soling and in mechanical goods. It can be used in compounds requiring high electrical resistivity and has little effect on cure except in very large amounts.

Table 6.4 SPECIFIC GRAVITIES OF RUBBER-COMPOUNDING INGREDIENTS

Ingredient	Specific gravity	Ingredient	Specific gravity
Mineral and fibrous ingredients			
Aluminium silicate	2·10	Lead chromate	5·70
Ammonium carbonate	1·59	Lignin	1·30
Antimony sulphide	3·30	Litharge	9·35
Antimony trioxide	5·40	Lithopone (30 % ZnS)	4·15
Asbestos	2·70	Lithopone (50 % ZnS)	4·06
Barium sulphate (barytes)	4·50	Magnesium carbonate	2·21
Barium sulphate (blancfixe)	4·25	Magnesium oxide	3·60
Cadmium sulphide	4·40	Mica	2·95
Calcium oxide (lime)	2·20	Pumice	2·35
Calcium silicate	2·10	Silica	1·95
Carbon black	1·81	Sulphur	2·05
China clay	2·50	Titanium dioxide (anatase)	3·90
Chromium oxide	5·21	Titanium dioxide (rutile)	4·20
Cotton fibre	1·05	Ultramarine blue	2·35
French chalk (talc)	2·80	Whiting (ground)	2·70
Glue	1·27	Whiting (precipitated)	2·62
Graphite	2·25	Woodflour	1·25
Ground glass	2·60	Zinc carbonate	3·30
Iron oxide (red)	5·14	Zinc oxide	5·57
Kieselguhr	2·20	Zinc stearate	1·06
Organic ingredients			
Bitumen	1·04	Mineral rubber	1·02
Ceresine wax	0·92	Paraffin wax	0·90
Coumarone–indene resin	1·11	Petroleum jelly	0·90
Dibutyl phthalate	1·04	Phenol–formaldehyde resin	1·27
Dibutyl sebacate	0·94	Pine oil	0·93
Dioctyl phthalate	0·98	Pine tar	1·08
Factice	1·04	Rosin oil	0·99
Mineral oil (aromatic)	1·02	Tricresyl phosphate	1·18
Mineral oil (naphthenic)	0·93		

6.2.2.3 *Antimony Trioxide*

This fine white semi-reinforcing filler is used in flame-resistant compounds and sometimes as a colouring agent.

6.2.2.4 *Barium Ferrite*

This filler used at high loadings gives flexible magnetic materials.

6.2.2.5 *Barytes (Barium Sulphate)*

The coarser-ground natural barytes is used as an inert filler to reduce cost and in chemically resistant compounds for tank linings, etc. Purer materials, produced by precipitation from solutions of barium salts, are employed as inert fillers in pharmaceutical products.

6.2.2.6   *Calcium Carbonates and Oxide*

Ground Limestone. Ground limestone is an off-white powder with particle size below 100 mesh. It is used where low cost is a main consideration; high loadings can be used to give only moderate hardness.

Ground Chalk, or Whiting. Ground chalk, or whiting, is a white powder of particle size varying to below 30 nm. It is used in low-cost compounds, giving moderate hardness and fairly high resilience at high loadings but with poor tensile strength and tear resistance. Moderate loadings can be used in moulded products and to improve the surface appearance of extrusions.

Precipitated Whitings. Precipitated whitings may be byproducts of water-softening processes or produced by precipitation from solutions of calcium salts, with particle sizes from about 20 µm to 50 nm. As semi-reinforcing fillers, high loadings can be used in mechanical goods and proofings to give low-cost products of good appearance with moderate hardness and better physical properties than the ground material.

Treated Whitings. These are precipitated whitings, the particles of which have been coated with up to 3 % of a stearate. They are more easily dispersed and readily give highly loaded compounds free from aggregates. As semi-reinforcing fillers, giving products of moderate cost, good appearance with fairly high resilience, and tensile strength, they are used in the higher grades of coloured mechanical products.

Calcium Oxide, Calcium Hydroxide, Lime. Hydrated lime was used as an early accelerator of vulcanisation. Quicklime, dispersed preferably in an oil or ester, has of recent years become important as an ingredient of mixes to be cured in salt baths or fluidised beds to avoid porosity arising from moisture (Section 8.9.2.4).

6.2.2.7   *Calcium Silicate (Hydrated)*

Hydrated calcium silicate is a fine powder, of particle size down to 1 µm, produced by precipitation from solution, and it gives compounds of fair tensile strength, high hardness, abrasion resistance, and high electrical resistivity. It is a reinforcing filler of medium cost and low tinctorial power, but has slight retarding effect on cure, which is usually compensated by increased acceleration. It is used, often with high-styrene resins, in shoe soling and mechanical goods.

6.2.2.8   *Calcium Sulphate (Hydrated)*

As ground gypsum or hydrated Plaster of Paris, calcium sulphate is an inert filler of lower specific gravity and whiter colour than barytes.

6.2.2.9   *China Clays*

China clays are essentially hydrated aluminium silicates derived from natural deposits.

Soft Clays. Soft clays are off-white powders, of particle size greater than 2 μm, which are used as semi-reinforcing fillers often at high loadings for low-cost compounds. Their hardness and tensile strength are greater but resilience is less than those obtained with calcium carbonates. They are used mainly in mechanical goods, but cannot be used for bright-coloured products.

Hard Clays. Hard clays, which are whiter than soft clays, of particle sizes less than 2 μm, are used as reinforcing fillers of moderate cost. They give compounds with higher tensile strength, tear resistance, and abrasion resistance than soft clays. Compounds may have high electrical resistivity. The main usage is in mechanical goods, hose, flooring, and shoe soling. A slight retarding effect on cure may be observed at high loadings.

Calcined Clays. Hard clays, calcined to remove combined water, behave as reinforcing fillers with good white colour. Hardness, tensile strength, and electrical resistivity are higher than with ordinary hard clays, and calcined clays are used especially where colour or electrical properties are important.

Treated Clays. Hard clays, treated with amines, often off-white, give greater reinforcement with similar accelerator systems than untreated clays. They are used in high-grade mechanical goods.

6.2.2.10   *Graphite*

The powdered graphitic form of carbon is used in heat-resistant materials and to reduce surface friction of rubber products—especially by application to surfaces before cure.

6.2.2.11   *Lead*

Finely powdered metallic lead is used at high loadings to give compounds impermeable to radiation. When mixing, great care is required to avoid aggregation in the mill nip.

6.2.2.12   *Litharge*

Litharge, or lead monoxide, is mainly used as an accelerator activator, and also to make compounds of high specific gravity or low permeability to radiation. One application is in balancing patches for tyres.

6.2.2.13   *Lithopone*

Lithopone is a mixture of zinc sulphide and barium sulphate, usually co-precipitated in equimolecular proportions; it is used mainly in cheaper white or coloured compounds as a whitening agent.

6.2.2.14   *Magnesium Carbonate*

The light grade of magnesium carbonate is a finely ground powder, acting as a reinforcing filler and giving high tensile strength and tear and abrasion resistance. It also stiffens uncured extruded products and reduces collapse in open steam cures. Uses are in cycle covers and in light-coloured extrusions. Compounds containing it can be translucent, if suitably formulated.

The heavy grade is less finely ground powder and again is a semi-reinforcing filler of low colouring power.

6.2.2.15   *Magnesium Oxide*

The light grade, of fine particle size, is used mainly as an acid acceptor in halogen-containing polymers, its activity being dependent on particle size, among other factors.

The heavy grade has larger particle size, and is less active than light grades as acid acceptors. Heat-resistant compounds used in seals, gaskets, and the like often contain magnesium oxide.

6.2.2.16   *Mica Powder*

Natural mica, washed and ground to pass 200–300 mesh, provides a filler of laminar type, imparting good resistance to heat and lower permeability to gases.

6.2.2.17   *Molybdenum Disulphide*

Molybdenum disulphide is used to reduce surface friction of products. It is more expensive and only marginally better than graphite for this purpose.

6.2.2.18   *Silicas**

Ground Mineral Silica. Mineral silica or sand (silicon dioxide) ground below 200 mesh is a cheap filler for heat-resistant compounds. It has no effect on cure.

---

* **Note:** Precautions should be taken when handling all silicas to avoid inhaling them owing to the risk of silicosis.

Kieselguhr. Kieselguhr (diatomaceous earth) is a finely ground siliceous material also used for heat-resistant compounds.

Hydrated Precipitated Silicas. These silicas are silicon dioxide containing 10–14% water, with particle sizes in the range 10–40 nm. They are reinforcing fillers giving compounds of high tensile strength, tear resistance, abrasion resistance, and hardness, and are used in translucent and coloured products, mechanical goods, and shoe soling. In combination with reinforcing blacks they improve tear resistance and adhesion to fabrics. Hydrated silicas retard cure and require increased dosage of accelerator or the addition of materials such as glycols or amines to promote curing. They respond to hot mixing techniques in a manner similar to reinforcing blacks giving increased modulus or resilience (Sections 4.5.3 and 9.4.1).

Fume Silica or Pyrogenic Silica. Fume or pyrogenic silica is silicon dioxide, containing less than 2% combined water, usually prepared by burning volatile silicon compounds. These silicas are highly reinforcing fillers of very small particle size, giving high tensile strength, tear resistance, and abrasion resistance, particularly to silicone rubbers. The retarding effect on cure in organic rubbers requires increased amounts of accelerators or the use of glycols or amines. The use of wax, polyethylene, or other lubricating materials is recommended to avoid sticking to rolls in processing and to moulds after cure.

6.2.2.19   *Slate Powder*

This finely ground natural material, consisting mainly of silica and silicates, is a very cheap inert filler used in heat-resistant gaskets, etc.

6.2.2.20   *Talc, French Chalk*

This finely ground natural mineral, consisting mainly of silicates of magnesium and aluminium, is used as an inert filler in heat-resisting compounds for gaskets, autoclave jointing, etc. Specially ground grades of great fineness are reinforcing fillers for higher-grade light-coloured goods. 'Platy' talc, a laminar form of the mineral, helps to reduce the permeability of rubbers to gases in hose linings, inflating bags, etc. (see also Mica). Talc is also widely employed as a lubricant to prevent uncured stocks sticking to themselves and other surfaces.

6.2.2.21   *Titanium Dioxide*

The pure material is extracted from natural minerals, precipitated, calcined, and ground. It is a reinforcing filler comparable on a volume basis with zinc oxide, but is mainly used for its whitening power in tyre sidewalls, hospital accessories, floor tiles, etc., and as an excellent heat-resisting filler for silicone rubbers. The anatase form is preferred where extreme whiteness is required.

The rutile form gives a rather creamier colour but is more stable at high temperatures.

6.2.2.22   *Zinc Carbonate*

This fine white powder, prepared by extraction and precipitation, is cheaper than titanium dioxide and is used when maximum whiteness is not required. It also functions as a neutral acid acceptor in place of zinc oxide in halogen-containing polymers.

6.2.2.23   *Zinc Oxide*

A white powder is prepared either (a) by the 'direct' process which involves the calcining of ores in reducing conditions and subsequent burning of the zinc in air, or (b) by the 'indirect' process in which purified zinc is sublimed and burned in air; the products of either process are subsequently ground to the fineness required. An accelerator activator and reinforcing filler, which gives compounds with high tensile strength and resilience but only moderate hardness, it is used at high loadings to produce plastic and easily moulded compounds; these tend to stick to mills, calenders, etc. Low-lead grades are preferred for coloured materials.

6.2.2.24   *Zinc Stearate*

Zinc stearate is a source of soluble zinc for accelerator activation for translucent compounds or in cases where maximum activity with minimum quantity is desired. It is used as a lubricant for uncured stocks, having the advantage over talc of dissolving in the rubber during cure.

6.2.3   FIBROUS FILLERS

6.2.3.1   *Asbestos**

Asbestos is a naturally occurring siliceous fibre. Both the shorter fibres and the ground material are used in flame-resistant or heat-resistant compounds for gaskets, brake shoes, etc.

6.2.3.2   *Flocks*

Short fibres of cotton, rayon, or nylon stiffen uncured stocks and increase tensile modulus, tear resistance, and abrasion resistance of vulcanisates. They are used particularly in shoe soling and similar products.

* **Note**: Precautions should be taken to avoid inhalation when mixing any form of asbestos.

6.2.3.3 *Woodflour*

The finer grades of woodflour (finely ground wood) are obtained from harder woods. Cheap hard compounds for flooring tiles are the main application for woodflour.

6.2.4 ORGANIC FILLERS

The processing aids, softeners, plasticisers, and extenders, are dealt with in Section 6.3.

6.2.4.1 *Cork*

Natural cork, ground to various degrees of fineness from about 3 mm particles to powder, gives compounds with a high degree of resilience and compressibility and is used in flooring, tiles, gaskets, etc.

6.2.4.2 *Glue*

Animal or fish glue added as a powder or in hydrated form imparts a degree of oil and fuel resistance to natural rubber for such items as hose linings and gaskets, where cheapness and very moderate fluid resistance are required.

6.2.4.3 *Cyclised Natural Rubber*

Cyclised natural rubber (Section 4.1.6.1), usually supplied blended with natural rubber, gives compounds of high modulus and hardness with low specific gravity, and also imparts to extrusions some degree of resistance to collapse in open steam or air cures. A moderate amount assists in giving a good surface finish to moulded goods, particularly when bright colours are required.

6.2.4.4 *Heveaplus*

Heveaplus (Section 4.1.6.2), which is natural rubber modified, by graft polymerisation, with 30–50% polymethylmethacrylate, is similar to cyclised rubber in giving compounds of high tensile modulus, hardness, and resilience, with low specific gravity. Properties, however, are maintained at higher temperatures than other thermoplastic fillers. The electrical properties are good, and products have a fine finish.

6.2.4.5 *High-Styrene Resins*

Copolymers of butadiene and styrene, with 50–80% bound styrene, are alternatives to cyclised rubber for high modulus, tear-resistant, and abrasion-resistant compounds. In combination with silica or silicates, their main use is in hard-wearing non-marking shoe soling (Sections 4.2.1.5 and 6.3.9.3).

6.2.4.6   *Phenolic Resins*

Reinforcing phenolic resins, usually of the novolak type, are supplied either ground with hexamethylene tetramine or in lump form requiring the addition of a methylene donor to the compound. They are reinforcing agents giving compounds with high tensile strength, hardness, and resilience. These properties may be enhanced by interaction with other fillers such as blacks or silicas, more particularly with the highly reinforcing varieties. Such compounds are much more easily processed than those of equivalent hardness obtained with highly reinforcing fillers only. They are used in moulded or open steam cured mechanical goods where maximum combination of hardness and resilience is required. In nitrile or chloroprene rubbers, these resins can react with the polymer and hexamethylene tetramine if the temperature rises above about 120°C during mixing or processing, giving an effect resembling scorch (Section 6.3.9.4).

6.2.5   COLOURING MATERIALS FOR NON-BLACK RUBBER COMPOUNDS

A number of inorganic materials are used for colouring rubbers. They are usually chosen for their stability to curing conditions and their complete freedom from staining or 'bleeding'.

6.2.5.1   *Antimony Sulphides*

Crimson Antimony. Crimson antimony is antimony trisulphide which may contain some free sulphur and antimony tetrasulphide. It gives strong colours with a crimson base, and imparts good resistance to ageing.

Golden Antimony. Golden antimony is antimony trisulphide which may contain free sulphur and sometimes calcium sulphate. It gives colours of a red-orange type.

6.2.5.2   *Cadmium Sulphide*

According to the method of preparation, cadmium sulphide gives rise to colours ranging from deep red, through orange to yellow.

6.2.5.3   *Cadmium Sulphoselenide*

Cadmium sulphoselenide, which may contain a proportion of lithopone, gives a similar range to the plain sulphides.

6.2.5.4   *Chromium Oxide*

This gives a rather dull green, particularly stable at high temperatures.

6.2.5.5   *Iron Oxide*

Iron oxide (mainly ferric oxide), according to method of preparation, gives a range of colours from deep red, through orange to yellow.

6.2.5.6   *Mercuric Sulphide (Vermilion)*

Mercuric sulphide gives a strong bright red.

6.2.5.7   *Nickel Titanate*

Nickel titanate gives yellow colours of extremely high stability to heat, light, and chemicals.

6.2.5.8   *Ultramarine Blue*

Ultramarine blue is a sulphur-containing complex of silicates, originally extracted from lapis lazuli but more usually obtained by calcining and extracting suitable mixtures. The colours obtained vary from a deep blue to greenish shades. The colour of rubber compounds is subject to appreciable change in cure.

The inorganic materials usually give rather dull colours in rubber compounds, so for brightly coloured materials it is desirable to use more expensive synthetic organic pigments which are available in a very large range of colours and shades. Pastel shades are generally obtained by combining such materials with light-coloured inorganic fillers, notably titanium dioxide.

6.2.5.9   *Organic Pigments*

Research by the manufacturers of organic pigments has produced a variety of such materials which are stable to curing conditions and to light, and are non-bleeding either to adjacent rubber compounds or to other finishes. Many of these pigments are available as pastes or as masterbatches in rubber, which greatly assists in dispersion and leads to appreciable economies in the amount needed to produce the colour required in the final compound. Certain dyestuffs soluble in rubber are also used to produce delicate shades in translucent materials.

Colour matching in rubber compounds is somewhat of an art, and, in case of difficulty, valuable advice can be obtained by consulting the manufacturers of the colours.

Details of suppliers and grades available of the materials described above may be found by reference to *British Compounding Ingredients for Rubber* published by Heffer & Sons Ltd for RAPRA, *Compounding Ingredients for*

Rubber published by the *Rubber World*, New York, and from suppliers' advertising media.

6.3   Plasticisers, Softeners, and Extenders

BY B. PICKUP

The materials described in this section, unless otherwise stated, are added to rubber compounds primarily to aid the processing operations of mixing, calendering, extruding, and moulding.

6.3.1   PEPTISERS

Peptisers are used to increase the efficiency of mastication of rubbers, i.e. to increase the rate of molecular breakdown, particularly in natural rubber, where usage is normally less than 0·5 p.p.h.r. Higher loadings are usually required in synthetic rubbers. Peptisers are added to rubber at the start of mastication, and other compounding ingredients only when mastication has been taken to the required stage (Sections 4.1.7.2 and 9.3.1). Sulphur inhibits their action.

The effectiveness of a peptiser increases as the mastication temperature is increased. Some are particularly suitable for mastication in internal mixers,

Table 6.5   SOME COMMERCIALLY AVAILABLE PEPTISERS

Producer	Trade name	Chemical name	Minimum operative temperature (°C)
Anchor Chemical	Pepton 22 Pepton 65	Di(*o*-benzamidophenyl)disulphide Zinc-2-benzamidothiophenate	115 65
Bayer	Renacit IV Renacit VII	Zinc salt of pentachlorothiophenol Pentachlorothiophenol with acti- vating and dispersing additives	70 70
ICI	Vulcamel TBN	33% thio-*β*-naphthol, 67% wax	70
Robinsons Bros.	Robac TBZ	Zinc thiobenzoate	70

and others for open mill mastication. They normally have no effect on the properties of the vulcanisate.

In Table 6.5 some commercially available peptisers are listed, together with their minimum operative temperatures.

Plasticising oils which contain chemically active petroleum sulphonate compounds can also be classed as peptisers; Ancoplas OB and ER (Anchor Chemical) are of this type.

6.3.2 PROCESS OILS AND EXTENDER OILS

In contrast to peptisers, petroleum oils and petroleum jelly function in a physical rather than a chemical manner; their effect is not dependent on the temperature of mixing. From 5 to 10 p.p.h.r. act as a plasticiser during processing, causing a reduction in viscosity and easing filler incorporation.

Petroleum oils are also used as extenders to reduce the cost of rubber compounds. They may be incorporated during the manufacture of certain polymers, e.g. oil-extended SBR and EPDM (Section 4.2.1.4), or may be added during compounding, together with substantial quantities of fillers, to offset their softening effect on the vulcanisate.

The types of oils used as processing aids and extenders are broadly classified under the headings paraffinic, naphthenic, and aromatic, according to the value of the viscosity gravity constant (VGC):

$$VGC = \frac{G - 0.24 - 0.022 \log_{10}(V_2 - 35.5)}{0.755}$$

where G = specific gravity at 15.5°C, and V_2 = Saybolt viscosity at 99°C. Oils with VGC around 0.8 are classified as paraffinic, and those with VGC above 1.0 as extremely aromatic.

As a generalisation, aromatic oils give best processability but are likely to have detrimental effects on staining, colour stability, and ageing resistance. Paraffinic oils are usually less effective as process aids, but have little effect on ageing performance, contact staining, or colour stability. Performance at low temperatures is also better than that of aromatic oils. Naphthenic oils fall between aromatic and paraffinic types in their effects on the performance of rubber.

Chlorinated paraffinic hydrocarbons are sometimes used as flame-retardant plasticisers, e.g. Cereclor (ICI) (Section 6.6.2).

6.3.3 ESTER PLASTICISERS

Certain esters of organic acids or phosphoric acid and high molecular weight alcohols are used as plasticisers in circumstances where petroleum oils may be unsuitable, e.g. because of incompatibility with the polymer. They are used particularly in NBR and CR polymers to give suitable viscosity for processing, to aid the incorporation of fillers, to give the required vulcanisate stiffness, and to provide flexibility at lower temperatures. Examples are

Table 6.6 POLYMERISABLE ESTER PLASTICISERS

Supplier	Trade name	Chemical name
Anchor Chemical	SR-206 SR-297 SR-350	Ethylene glycol dimethacrylate 1,3-Butylene glycol dimethacrylate Trimethylol propane trimethacrylate
Lennig Chemicals	Monomer X-970 Monomer X-980	1,3-Butylene glycol dimethacrylate Trimethylol propane trimethacrylate

dibutyl sebacate, diisooctyl phthalate, and trixylyl phosphate. The higher molecular weight polymeric esters show lower losses by volatility or extraction.

Polymerisable plasticisers are also used under certain circumstances with peroxide vulcanising systems. They act as plasticisers and tackifiers during processing and are caused to polymerise by the peroxide during cure. They contribute to high hardness and high stiffness in the vulcanisate, and increase the state of cure by participating in the crosslinking reaction. Typical materials are given in Table 6.6.

6.3.4   FACTICE

Brown (dark) factice is manufactured by reacting special vegetable or marine oils with sulphur at 140°–160°C. For the production of golden factices, the reaction is conducted in the presence of an accelerator, which enables lower temperatures to be used and better colour to be obtained. The product has a friable nature and springy consistency. It is sold either as blocks or ground.

There are many types of dark and golden factices, normally classified as first, second, or third (commercial) grades, depending on the acetone extract. There is, however, no clear dependence of performance characteristics on acetone extract.

The main application for dark and golden factices is as a processing aid in natural and synthetic rubber extrusions, where 5–30 p.p.h.r. control die swell, improve surface quality, and prevent distortion of shape, particularly during vulcanisation. Factice also controls 'nerve' during mixing, causing improved incorporation and dispersion of powders and reduced power consumption. Control of the thickness of calendered sheet is also improved. Large loadings of factice, e.g. 100 p.p.h.r., are used in very soft compounds, for such applications as printers' rollers, both to aid the incorporation of liquid plasticisers and to reduce extractability.

Dark and golden factices normally accelerate the cure of sulphur-vulcanised compounds. White factice, used in the manufacture of erasers and in rubber compounds cured at room temperature with sulphur chloride, is manufactured by vulcanising selected oils with sulphur chloride, and is sold in granulated form.

Factice is manufactured in the UK by J. Allcock & Sons, Anchor Chemical, Croda International, and Hubron Rubber Chemicals.

6.3.5   FATTY ACIDS AND THEIR SALTS

Small quantities of stearic acid are normally present as part of sulphur vulcanising systems. The stearic acid also acts as a plasticiser, aids dispersion of black and other fillers, and minimises any tendency for sticking to the mill rolls. Under some circumstances zinc stearate may be used in place of stearic acid and zinc oxide.

Zinc laurate and zinc salts of high molecular weight unsaturated fatty acids are sometimes used as processing aids, e.g. Aktiplast (Rhein-Chemie).

Other fatty materials recommended as process aids include Manosperse (Hardman & Holden) and Skliro (Croda).

6.3.6 PINE TAR

Pine tar, a dark viscous liquid, is manufactured in Scandinavia and the USA by the distillation of pine wood. It is used at 3–7 p.p.h.r. as an aid to the incorporation of carbon blacks into rubber and also to improve tackiness. Some former applications for this product have been replaced by synthetic materials.

6.3.7 WAX

Paraffin waxes of melting point approximately 55°C are used as processing aids. They bloom to the surface and protect ozone-sensitive elastomers against cracking under static stressing. Microcrystalline waxes have less tendency to bloom.

6.3.8 BITUMEN

Petroleum bitumens, sold as 'mineral rubber', are used as low-cost low-gravity extenders and process aids. They are lustrous black thermoplastic materials, supplied either as irregular lumps or as powder, and are used at 5–20 p.p.h.r. to aid the incorporation of fillers, to lower processing temperatures, and to decrease any tendency to scorch during processing. Grades are available with melting points of 120°C, 130°C, and 160°C.

6.3.9 RESINS

6.3.9.1 *Coumarone Resins*

Coumarone resins are manufactured by polymerisation of styrene, coumarone, indene, and related materials occurring in certain fractions of coal tar. By controlling reaction conditions, a range of resins can be produced which vary in appearance from thick viscous liquids to hard clear resinous solids, and in colour from dark brown to pale straw.

Both solid grades, with melting points in the range 65°–110°C, and liquid grades are used as tackifiers and plasticisers, particularly in synthetic rubbers. The liquid grades usually impart greater tack. Both grades also help to restrain bloom from uncured and cured rubbers. They are essential ingredients in non-black SBR compounds if optimum physical properties are to be developed; for this purpose the solid grades are the more effective.

6.3.9.2 *Petroleum Resins*

Petroleum resins are similar in general characteristics and applications to solid coumarone–indene resins. They are manufactured by polymerisation of olefins in steam-cracked heavy-hydrocarbon petroleum fractions. Trade names include Escorez (Esso Petroleum), Piccopale (R. H. Cole), and Celanese PR (Hubron Rubber Chemicals).

6.3.9.3   *High-Styrene Resins*

Styrene–butadiene copolymers, containing up to 85% styrene, are used as thermoplastic processing aids in addition to their reinforcing action, particularly in compounds required to have high vulcanisate hardness. To ensure

Table 6.7   SOME COMMERCIALLY AVAILABLE HIGH-STYRENE RESINS

Supplier	Trade name	Grade	Comments
Anchor Chemical	Marbon	8000A	High-styrene resin with 85% styrene
Hubron Rubber Chemicals	Pliolite	S-6F	
Polysar (UK)	Polysar	SS250	55% bound styrene
		SS260	63% bound styrene
Revertex	Butakon	S8551	
Shell	Cariflex	SP145	45% high-styrene resin, 55% SBR

good dispersion, the softening point of the resin (95°C) must be exceeded during incorporation into the rubber mix (Sections 4.2.1.5 and 6.2.4.5). Some grades are listed in Table 6.7.

6.3.9.4   *Phenolic Resins*

Thermosetting alkyl phenol–formaldehyde resins act as processing aids in similar circumstances to those in which high-styrene resins are used (Section 6.2.4.6). They give excellent flow characteristics in moulding, calendering, and extruding. Their effect on the vulcanisate properties is discussed in

Table 6.8   SOME MISCELLANEOUS PROCESSING AIDS

Supplier	Trade name	Use
Anchor Chemical	Interlube A	Improved mould flow and release
	NE Plasticiser	Non-extractable plasticiser
British Solvent Oils	Rubitac, Rubolene	Tackifying plasticisers
Du Pont	IML-1	Internal lubricant
Hubron Rubber Chemicals	Biltac	Tackifying plasticiser
	Kenflex	Aromatic hydrocarbon polymer
	Softack	Tackifying plasticiser
	Wingtack 95	Tackifying resin
Shell	Dutrex RT	Tackifying plasticiser

Section 9.4.1. Typical materials of this type are Cellobond H831 (BP Chemicals UK), Synphorm R3000 (Anchor Chemical), and Cataplast (Wilfrid Smith).

Non-reactive phenolic resins are used as tackifiers in synthetic rubbers. They give a higher level of tack than can be achieved with coumarone resins, although at higher cost. Typical products are Synphorm T2000 (Anchor Chemical), Amberol ST137X (Lennig Chemicals), and Arrcorez 17 (Rubber Regenerating).

6.3.10 MISCELLANEOUS

There are numerous compounded processing aids which do not fall under any of the above headings. Some are listed in Table 6.8.

6.4 Reclaimed Rubber *

BY R. SINGLETON

6.4.1 INTRODUCTION

By the application of heat and chemical agents to ground vulcanised waste rubber, a substantial regeneration of the rubber compound to its original plastic state is effected, yielding a product known as 'reclaim' or reclaimed rubber, capable of being processed, compounded, and revulcanised. The process is essentially one of depolymerisation. Reclaimed rubber has become widely accepted as a raw material which possesses processing and economic characteristics that are of great value in the compounding of natural and synthetic rubber stocks.

There are four principal reclaiming processes in use today, of which the digester and reclaimator processes are responsible for the major portion of the annual output in the UK.

6.4.2 MANUFACTURING PROCESSES

The raw material for reclaiming is scrap rubber in a wide variety of forms, but tyres, as is to be expected, form the major quantity.

The first stages, in all processes, are the cracking and grinding of the scrap rubber to reduce it to a crumb passing through a 20–30 mesh screen.

6.4.2.1 *Digester Process*

The ground waste is loaded into a digester along with water, reclaiming oils, and other additives, such as activated black (for minimum staining grades). The digester is a cylindrical jacketed pressure vessel fitted with a horizontal agitator, and steam can be supplied to both interior and jacket,

* Section 6.4 is based on text by Singleton (*Rubb. J.*, **146**, 46, 1964), courtesy Maclaren.

thus enabling a uniform temperature to be maintained throughout the mass. The contents of the digester are then heated to about 190°C and maintained at this temperature for some 4–10 hours with continuous agitation. The digester is then 'blown down', and the contents deposited on to a conveyor. Any necessary adjustments to the specific gravity and plasticity by addition of plasticiser, carbon black, or fillers are carried out in a ribbon blender, and the stock is then automatically conveyed to extruders for straining, refining, and leafing on to a drum from which it is removed in slabs.

6.4.2.2 Reclaimator Process

It can be shown that ground vulcanised rubber heated in a temperature range of 120°–200°C undergoes a rapid initial increase in plasticity, and, on continued heating, passes through an inversion point and rehardens until, after prolonged heating, a further but slower increase in plasticity is attained. It follows, therefore, that three points of equal plasticity occur in this cycle (Fig. 6.7). Although this behaviour is characteristic for certain vulcanised rubbers, the resultant plasticity–time relationship will vary with the type of

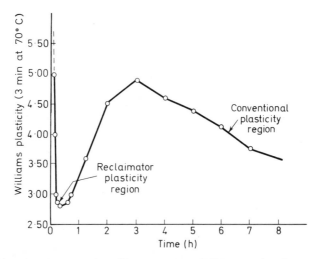

Fig. 6.7   Plasticity–time curve for rubber in steam at 170°C; note that the more plastic the material, the lower the Williams plasticity value because it is the pellet thickness after compression (Section 2.11.2)

rubber, reclaiming agents, and physical conditions used. Synthetic rubbers behave in a similar manner, but the rehardening process tends to predominate, so the initial increase in plasticity is less marked and can only be achieved under special conditions.

Whereas some earlier processes completed the cycle mentioned above, the reclaimator process depends on arresting the reaction at the point where the initial rapid increase in plasticity is achieved in the presence of plasticisers, oxygen, and catalytic materials. The rapid heating of the rubber while it is being worked mechanically gives the desired plasticity in a matter of 3–6 min. The total cycle time is only 30 min, and the process is entirely dry.

The reclaimator machine, from which the process takes its name, is of the screw extrusion type with a hopper at one end into which the crumb previously mixed with oil and chemicals is automatically fed at a predetermined rate. It generates its own heat for depolymerisation by mechanically working the finely divided rubber crumb under pressure, and then discharges it as reclaimed rubber. Temperatures in the machine are controlled by alternating oil and water jackets.

The raw material for this process is whole tyres, and, in the preparation of the crumb, fabric is removed mechanically by a series of operations in which

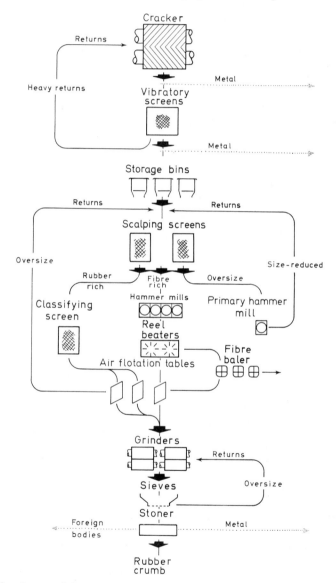

Fig. 6.8   *Flow diagram of the separation process*

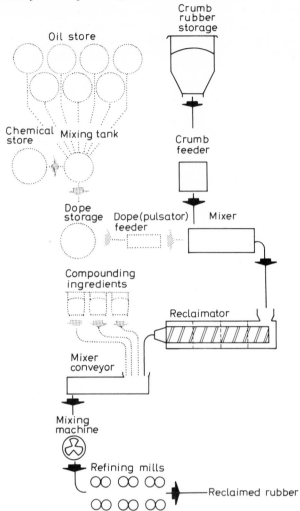

Fig. 6.9　*Flow diagram of the reclaimator process*

fractionating and sizing screens, hammer mills and reel beaters, air flotation, and gravity tables are used. The reclaimed rubber is compounded as necessary and refined and sheeted using conventional machines. The full sequence of the above operations is shown in the flow charts (Figs 6.8 and 6.9).

6.4.2.3　*Pan Process*

The raw material, fabric free, is cracked on the same type of equipment as for whole tyre. The ground waste intimately mixed with the required oils is loaded into steam vulcanisers and subjected to steam heating at pressures of the order of 10 kgf/cm^2. This is followed by straining and refining in the conventional manner. Not only NR and SBR scraps but also IIR, CR, and NBR compounds are reclaimed by this method.

6.4.2.4   *Engelke Process*

This process is extremely simple as it requires no wetting of the stock or elimination of the fibre. Vulcanised scrap rubber containing fabric is subjected to very high temperatures for short periods of time, 10–20 min and upwards, in small autoclaves. Any fibre is completely carbonised *in situ* throughout the mass. The cracked stock, when necessary, can be premixed with plasticising oils and peptisers. The reclaim is strained and refined in the usual manner.

6.4.3   TYPES OF RECLAIM

A variety of grades of reclaimed rubber is offered today, but mention is made here of only the important ones.

6.4.3.1   *Whole Tyre Reclaim*

Whole tyre reclaim is the one produced in the largest quantity. First-quality reclaim made from whole tyres contains about 45% rubber hydrocarbon by weight. The remaining 55% consists of valuable carbon black, a little mineral filler, and softeners, all of which are substantially unchanged by the reclaim manufacturing operation, and may be considered to function as virgin materials.

6.4.3.2   *Minimum Staining Reclaim*

Minimum staining reclaim can replace the conventional whole tyre material when occasion demands. As implied, it has a much lower tendency to stain, by either migration or contact, than conventional reclaim. The reduction in staining characteristics is achieved by the use of activated-carbon non-staining oils and by selecting tyres containing a higher proportion of natural to synthetic rubber.

6.4.3.3   *Drab and Coloured Reclaims*

As the names imply, drab and coloured reclaims are made from non-black scrap. The digester process is usually employed and, when fabric is present, a small addition of caustic is made in order to destroy it. The period of heat treatment is usually several hours at 195°C.

6.4.3.4   *Butyl Reclaim*

Whereas reclaimed rubbers have been successfully produced from scrap CR, NBR, SI (d'Adolf, 1962), and other speciality rubbers, the only one of substantial commercial importance is butyl reclaim. The starting material for this is butyl inner tubes. A modified digester process is adopted, every precaution being taken to avoid contamination by NR or SBR, because of their adverse effect on the curing characteristics of the butyl. Extensive control tests are necessary to ensure that the curing properties are satisfactory. The nerve of butyl reclaim is much reduced compared with that of the original polymer. Because of this, compounds containing butyl reclaim will mix, calender, and extrude faster and more smoothly than similar compounds based on virgin rubber.

6.4.4   TESTING AND EVALUATION OF RECLAIMED RUBBER

The testing and evaluation of reclaimed rubber includes both chemical and physical testing. Although the list of tests given in Table 6.9 is very comprehensive, the decision as to which tests must be applied to a particular

Table 6.9   TESTS APPLIED TO RECLAIMED RUBBER

Chemical	*Physical*
Acetone extract	Visual inspection for smoothness and freedom from metal contamination
Carbon black content	
Ash content and composition	Plasticity
Total polymer content and composition	Specific gravity
Chloroform extract	Rate of vulcanisation
Alcoholic potash extract	Physical tests on the vulcanised reclaim (i.e. tensile, modulus, elongation, tear strength, and hardness)
Cellulose content	
Acidity or alkalinity	Staining characteristics
	Colour and odour
	Extrusion rate
	Filler incorporation characteristics

reclaim will depend upon the end product and the specification agreed between the consumer and manufacturer. In general the test methods are those laid down by the BSI for the testing of vulcanised rubber.

6.4.5   PROPERTIES OF RECLAIMED RUBBER AS A COMPOUNDING INGREDIENT

Because reclaimed rubber has been given a large amount of mechanical working during manufacture and contains all of the fillers which were incorporated in the original compound, mixing time and power consumption are reduced by its use. Furthermore, less heat is developed during mixing and subsequent processing compared with compounds based on new rubber. This is an important property especially in compounds containing high carbon black loadings, which are sometimes scorchy if too high a temperature is allowed to develop.

Mixings containing reclaimed rubber are generally faster processing during extruding and calendering, less affected by continued milling, and less thermoplastic; they hold their shape during processing and open steam cure, a desirable property particularly in extruded articles such as tubing, hose, and weatherstripping. Extruder die swell and calender shrinkage are minimised by the proper use of reclaimed rubber.

On vulcanisation, reclaimed rubber stocks have less tendency to revert, and they age well. Ball and Randall (1949) stated that:

'Reclaim has excellent ageing properties, when used in either the cured or uncured condition. The severe treatment to which it has already been subjected includes oxidation, heating, digestion, washing and mastication in air, and these appear to have stabilised the hydrocarbon against further changes.'

On the other hand, the tensile strength, modulus, resilience, tear resistance, and abrasion are all reduced as the proportion of whole tyre reclaim is increased.

Reclaims made from mechanical waste find applications in a wide variety of moulded goods where smooth processing is not essential and reduced cost is the more important consideration. It should be appreciated that reclaim itself can be compounded and reinforced with carbon black in the same manner as other polymers.

6.5   Antioxidants and Antiozonants

BY T. J. MEYRICK

6.5.1   MECHANISM OF DEGRADATION

The basic principles of the mechanism of the oxidative degradation of rubbers are now well understood and documented. Based on the study of model substances, it is now established that attack commences on the α-methylenic carbon atom in the chain. A hydrogen atom is abstracted and, in the presence of oxygen, an oxidative reaction chain is initiated which, if unchecked, propagates autocatalytically:

1. $RH \rightarrow R^{\cdot}$
2. $R^{\cdot} + O_2 \rightarrow ROO^{\cdot}$
3. $ROO^{\cdot} + RH \rightarrow ROOH + R^{\cdot}$
4. $ROOH \rightarrow$ chain scission with free radical formation.

In order to prevent the autocatalytic development of the reaction, an inhibitor must break the reaction chain either (a) by capturing the free radicals formed, and/or (b) by ensuring that the peroxides and hydroperoxides produced decompose into harmless fragments without degrading the polymer and without initiating new free radicals capable of propagating the reaction chain. Possible inhibition reactions are:

1. $R^{\cdot} + AH \rightarrow RH + A^{\cdot}$
2. $ROO^{\cdot} + AH \rightarrow ROOH + A^{\cdot}$
3. $ROOH + AH \rightarrow$ harmless fragments.

These theoretical considerations assume that there is sufficient antioxidant present and mobile in the rubber compound to reach the points of oxidative attack. Recent work (Cain *et al.*, 1968) has shown that mobility of the antioxidant is not essential; workers at NRPRA at Welwyn have demonstrated that certain chemical structures, in particular nitrosoamines, which can be made to react with the rubber hydrocarbon chain during vulcanisation, are efficient antioxidants and are not extracted by powerful solvents or water.

Under certain circumstances of use (e.g. with latex thread), water leaching could pose a problem for the rubber compounder. It is clear that water leaching of antidegradants could have far-reaching effects upon a wide variety of rubber articles including tyres. Evidence so far available based on work on thin sections of rubber (Latos and Sparks, 1969) confirm this supposition. Work at ICI Dyestuffs Division, as yet unreported, under more realistic conditions including actual tyre tests, throws doubt upon the significance of the effect on fairly thick articles under practical conditions.

6.5.2   EXTERNAL CAUSES OF DEGRADATION

The oxidation reaction may be initiated by different external factors, each giving rise to an essentially different form of degradation. These can be classified as follows: oxygen, ozone, heat, light, flexing (fatigue), and metal catalysis.

Under service conditions, it is rarely possible to isolate any single factor as being solely responsible for degradation; there is invariably a random interplay of factors and conditions which can never be properly reproduced by a single laboratory ageing test.

Although it is reasonable to regard the oxidative ageing of rubber (except ozone attack) as derived essentially from one fundamental reaction, this implies that the relative efficiencies of inhibition should not vary according to test conditions. In fact, there is a discernible pattern in the effectiveness of different inhibitor structures depending upon the test conditions used. The concept is clearly an oversimplification, and, in reality, no single inhibitor structure is equally effective against all forms of degradation. The selection of an inhibitor, therefore, represents a compromise of the best balance of technical properties and price for a particular end use.

6.5.3   MECHANISM OF OZONE ATTACK

The mechanism of ozone attack and its inhibition are not so well understood. It is probable that ozone attacks rubber by the following mechanism:

$$\text{>C=C<} + O_3 \longrightarrow \text{>C}\overset{O_3}{\diagup\diagdown}\text{C<}$$

$$\text{>C}^+OO^- + \text{>C=O} \longrightarrow \text{>C}\underset{O}{\overset{O-O}{\diagup\diagdown}}\text{C<}$$

(1)   (2)

It is believed that an antiozonant reacts with either the zwitterion (1) or the ozonide (2) to form an inert protective film which, every time it is broken, is repaired by the formation of fresh film produced from the three reactants (rubber, ozone, antiozonant) *in situ*.

Useful review articles covering the mechanism of oxidation and ozone attack have been published by Shelton (1957), Biggs (1958), Ambelang *et al.* (1963), Kuzminskii (1966), Cunneen (1968), Loan, Murray, and Story (1968), and Hill (1969).

6.5.4 COMMERCIALLY AVAILABLE ANTIDEGRADANTS FOR RUBBER

The majority of the commercially available inhibitors belong to two main chemical classes: amines and phenolics, which represent respectively staining and non-staining types, judged by whether or not they cause significant discolouration of a rubber compound on exposure to light. The main characteristics of the principal structures are summarised in Tables 6.10 and 6.11. It should be emphasised that each table is self-contained, and the ratings are, intercomparable and relative only within each table. Intertable comparisons of ratings are not justified. The tabulated data lay no claim to being comprehensive; they list products typical of each class which are reasonably widely used in the UK.

6.6 Special Purpose Additives

BY T. J. MEYRICK

6.6.1 CHEMICAL BLOWING AGENTS

In the manufacture of cellular rubbers from solid rubber, many types of chemicals are used. The most commonly used is sodium bicarbonate, though ammonium carbonate and bicarbonate are still in use: these liberate essentially carbon dioxide, and yield an open pore structure.

More sophisticated chemicals, such as those based on dinitrosopentamethylene tetramine, benzene sulphonyl hydrazide, and azodicarbonamide, liberate nitrogen and find applications either alone or in combinations with the inorganic agents for applications where controlled cell structure is important (Section 10.6).

For the manufacture of microcellular materials, the organic blowing

Table 6.10 CHARACTERISTICS OF THE PRINCIPAL STAINING INHIBITOR STRUCTURES

Class	Typical commercially available products (in the UK)	
	Supplier	Trade name
Phenyl-(α, β)-naphthylamines 	Anchor Chemical ICI ICI Monsanto	AgeRite Powder Nonox AN Nonox D PBN
Ketone–amine condensates (a) Acetone–anilines R may be H, alkyl, or alkoxy	Anchor Chemical Monsanto Monsanto Monsanto Monsanto	AgeRite AK Flectol B, H, and Flakes Santoflex AW Santoflex DD Santoflex R
(b) Acetone–diphenylamine complex mixtures of products	Rubber Regenerating Anchor Chemical Rubber Regenerating Rubber Regenerating Rubber Regenerating ICI Monsanto	Aminox Antioxidant B Betanox Special BLE-25 BXA Nonox B, BL Santoflex DPA
Substituted diphenylamines (a) R = alkyl, usually C_7—C_9	ICI Rubber Regenerating Rubber Regenerating	Nonox OD Octamine Polylite
(b) R = alkoxy (e.g. CH_3O—)	—	—
Substituted p-phenylenediamines (a) R, R′ = phenyl	Anchor Chemical Monsanto ICI Rubber Regenerating	Antioxidant DPPD DPPD Nonox DPPD JZF
(b) R, R′ = β-naphthyl	Anchor Chemical ICI Monsanto	AgeRite White Nonox CI Santowhite CI

(c) R = phenyl
   R′ = sec-alkyl (C_3—C_8)
   or cycloalkyl (C_6)
	Anchor Chemical
	ICI
	Monsanto
	Monsanto
	Monsanto
		Antiozonant IP
		Nonox ZA
		Santoflex CP
		Santoflex IP
		Santoflex 13

(d) R, R′ sec-alkyl (C_7—C_8) or cycloalkyl (C_6)	Monsanto Monsanto Universal-Matthey	Santoflex 77 Santoflex 217 UOP 88

Efficiency in inhibiting degradation initiated by					Comments
Oxygen	Heat	Flexing	Metal catalysis	Ozone	
Good	Good	Good	Moderate	None	α-Isomer has a higher compatibility with rubbers
Good	Very good	Generally poor	Poor	None	Alkylation (e.g. C_{12}) in the 6-position improves flexing; alkoxylation (e.g. C_2H_5O) in the 6-position improves flexing and imparts some measure of antiozone activity
Good	Very good	Moderate	Poor	None	Liquid products have very good flex-cracking properties
Moderate	Moderate	Moderate	Poor	None	Causes comparatively little discolouration; especially useful in synthetic rubbers, particularly polychloroprene types
Good	Moderate	Very good	Poor	None	Rarely used alone; usually found as a constituent of proprietary antiflex-cracking antioxidant blends
Good	Very good	Very good	Good	Some	Limited by its low compatibility with rubbers, it is usually found as a constituent of antiflex-cracking heat-resistant antioxidant blends: e.g. AgeRite HPX (Anchor), Nonox HFN (ICI), Santoflex 75 (Monsanto)
Good	Good	Poor	Very good	None	Comparatively little discolouration; much used, especially in latex foams, for its copper-inhibiting properties; poor light resistance in latex foam
Good	Very good	Excellent	Excellent	Excellent	Broadly speaking, antiozonant and flexing properties tend to fall off with increasing size of alkyl substituent; on the other hand, scorchiness tends to improve; compounds in this group synergise strongly with waxes
Good	Good	Very good	Very good	Very good	Excellent antiozonants under static conditions; inferior to the previous group under dynamic; scorch properties vary

Table 6.11 CHARACTERISTICS OF THE PRINCIPAL NON-STAINING INHIBITOR STRUCTURES

Class	Typically commercially available products (in the UK)	
	Supplier	Trade name
Substituted phenols [structure: phenol ring with R, R', R" substituents] R, R', R" may be the same or different; usually alkyl, cycloalkyl, or aralkyl	Anchor Chemical Anchor Chemical Monsanto Anchor Chemical Monsanto ICI ICI ICI ICI Goodyear	Antioxidant MP Antioxidant SP BHT Cyanox LF Montaclere Nonox HO Nonox SP Nonox TBC Nonox WSL Wingstay S
Phenyl alkanes (a) *o*-bridged [structure: two phenol rings bridged by C with R, R', R", R'''] R, R', R", R''' may be the same or different alkyl or cycloalkyl (b) *p*-bridged [structure: two phenol rings bridged by C, with R, R' substituents]	Anchor Chemical Anchor Chemical Rubber Regenerating Rubber Regenerating ICI ICI ICI Monsanto	Antioxidant 425 Antioxidant 2246 Naugawhite Naugawhite Pdr Nonox EX, EXP, EXN Nonox WSO Nonox WSP Santowhite Pdr
Phenyl sulphides *p*-bridged [structure: two phenol rings bridged by S, with R, R' substituents] R, R' = alkyl	Monsanto	Santowhite crystals
Mercaptobenzimidazole [structure: benzimidazole with SH]	Bayer	Antioxidant MB

* None of the chemical structures in this table possesses antiozonant properties.

Efficiency in inhibiting degradation initiated by*					Comments
Oxygen	Heat	Flexing	Metal catalysis	Staining	
Poor to good	Poor to good	Moderate to good	Poor	Good to excellent	Broadly speaking, this class shows least discolouration, but migration staining can occur
Good to very good	Good	Moderate	Good	Moderate to good	This class contains the most efficient phenolic antioxidants so far developed, which, however, discolour rather more than the substituted phenols
Moderate to good	Moderate	Fair	Moderate	Moderate to good	Generally speaking, these two classes occupy a position midway between the first two
Moderate to good	Moderate	Poor	Moderate	Moderate to good	
Fair	Fair	Poor	Poor	Very good	The performance of this antioxidant is extremely sensitive to choice of curing system; it is of special interest as a synergist for other antioxidant types especially in the presence of metal contamination, or for use with modern EV curing systems; Nonox CNS (ICI) is a blended system based on this principle

Table 6.12   NITROGEN-GENERATING BLOWING AGENTS

Producer	Trade name	Abbrev-iation	Chemical composition
Anchor Chemical	Ancablo A	—	Zinc amine complex
Bayer	Porofor	BSH DNO	Benzene sulphonyl hydrazide Dinitrosopentamethylene tetramine
ICI	Vulcacel BN94	DNPT	Dinitrosopentamethylene tetramine
Whiffens	Genitron	AC AZDN OB BSH	Azodicarbonamide (azobisformamide) Azoisobutyronitrile pp'-Oxybisbenzene- sulphonylhydrazide Benzene sulphonyl hydrazide

agents are definitely favoured, the preferred type being dinitrosopenta-methylene tetramine. The main ones available in the UK and W. Europe are given in Table 6.12.

6.6.2   FLAME RETARDANTS

With the increasing use of polymeric materials in civil engineering, including domestic and public building, greater emphasis is being placed upon reducing the fire hazards normally attendant upon the use of rubbers and plastics.

Flame proofing, or at least reduction of the fire hazard, is achieved by the use in rubber compounds of mixtures of inorganic and organic materials. Antimony trioxide (Section 6.2.2.3) and chlorinated derivatives of paraffin hydrocarbons provide the most-used combination, though zinc borate is also frequently included (Table 6.13). The inherent flame resistance of poly-chloroprene rubbers is improved by the addition of antimony trioxide. The oxide is used as the bulk filler to reduce the fire hazard, the borate acts as a crust-forming agent, while it is believed that the halogenated body decom-poses to yield chlorine which immediately combines with the antimony oxide to form antimony trichloride, which acts as a flame suppressant. If

Table 6.13   FLAME RETARDANTS

Trade name	Supplier	Chemical composition
Cereclor range Timonox —	ICI Mond Division Associated Lead Manufacturers Joseph Storey	Chlorinated paraffin hydrocarbons Antimony trioxide Zinc borate

softening agents or plasticisers are necessary, non-flammable materials such as tritolyl phosphate are recommended (Section 9.4.6.2).

6.6.3 ANTISTATIC AGENTS

In certain rubber articles, e.g. conveyor belting, textile rollers and spinning cots, and trolley wheels for use in hospitals, the build-up of static electricity is undesirable. As an alternative to rubber compounded with conductive

Table 6.14 ANTISTATIC AGENTS

Trade name	Producer	Chemical composition
Arquad 2HT–75%	Armour Hess Chemicals	Quaternary salts
Negomel AL 5	ICI Dyestuffs Division	Ethylene oxide condensate
Negomel GM	ICI Dyestuffs Division	Ethylene oxide condensate

blacks, antistatic agents, which are frequently based on quaternary ammonium salts or ethylene oxide condensates, have been designed to assist in producing vulcanisates with low electrical resistance (Table 6.14).

6.6.4 ABRASIVES

The types of abrasive used in the compounding of rubber vary quite widely, depending upon the application. In the manufacture of erasers, the abrasives used must be cheap fine powders and thus pumice and glass powder find frequent use, though such a unique material as powdered brick is not unknown in some of the less sophisticated countries. In the manufacture of polishing and grinding wheels, carborundum powder (silicon carbides) is favoured.

Pumice, glass, and carborundum powders are available from Wilfred Smith Ltd, British Drug Houses, and The Carborundum Co. respectively.

6.6.5 INTEGRAL BONDING ADDITIVES

The introduction of cobalt salts, e.g. cobalt naphthenate, into rubber compounds to improve the adhesion to metals, especially steel, has been known and practised for many years and is of major importance in the manufacture of steel-reinforced tyres. Ionisable cobalt is, like copper and manganese, a pro-oxidant for sulphur-cured rubber and thus should only be used in rubber compounds which are likely to have only limited contact with air or oxygen.

More recently, great interest has been aroused in adding resorcinol and p-formaldehyde or some other formaldehyde donor to rubbers to improve their adhesion to textiles, e.g. rayon and nylon, and to metals. It is suggested, too, that the addition of silica along with the other additives results in some further increase in adhesion. The patent position is a little obscure at the

I

present time and agencies in the UK for various products are changing, but it would appear that Bayer and ICI will be offering suitable products.

Integral bonding additives such as alkyl hydrogen polysiloxane are occasionally used for improving the adhesion between silicone rubber compositions and many substrates, especially metals. Midland Silicones offer ESP 2437 for this purpose (Sections 8.5.2.4 and 8.7.4).

6.6.6 STIFFENING AGENTS

At one time benzidine and *p*-aminophenol were favoured as stiffening agents for uncured stocks but, owing to their toxic nature, their use has ceased. Nowadays several methods are used to impart stiffness and increase the viscosity of 'green' stocks so as to reduce flattening, sagging, or distortion in storage and during open steam cure, and to improve the moulding of very soft compounds:

1. The introduction of crosslinked rubbers, e.g. PA 80 (Section 4.1.4.6) and SBR 1009 (Section 4.2.1.6).
2. The use of certain types of factice (Section 6.3.4).
3. The addition to the rubber compound of dihydrazine sulphate and similar substances, e.g. Anchor Chemical Stiffener DSC.

6.7   Textiles

BY J. O. WOOD

6.7.1   PRODUCTION OF TEXTILES

Most of the textile used to reinforce rubber products is accounted for by tyre cord, belting fabric, and hose reinforcement. Forty years ago natural fibres only were available for the purpose, cotton then being outstanding in its combination of physical properties and economic suitability. Nowadays, cotton has been largely superseded by rayon, nylon and polyester, steel, and glass.

Textile in the form of continuous filament yarn is obtained from solution or from melt by extrusion through multi-hole dies or spinnerets. The resultant bundle of filaments is either regenerated in the case of rayon to give almost pure cellulose, or, in the case of nylon, it solidifies by cooling. After solidification, the fibre has to be attenuated or drawn by some 400% in order to obtain sufficient strength and stability. Typical examples of industrial yarns are shown in Table 6.15.

In this table, twist in turns per metre is defined as the number of turns by which one end of a straight length of yarn must be rotated to remove all twist, the number obtained being divided by the twisted length. Denier (den.) is the weight in grams of 9 km; tex is the weight in grams of 1 km.

Textile strength measurements pose a problem. Cross-sectional area is ill-defined, making direct measurement of ultimate tensile strength subject to inaccuracy. Instead it is customary to quote specific strength or tenacity, e.g.

$$\text{Tenacity in grams per denier} = \frac{\text{strength in grams}}{\text{denier}}$$

At present there is a growing preference for tex units, and tenacity may equally or perhaps preferably be expressed in grams per tex.

It is important to note that specific strength is frequently of greater direct physical relevance than ultimate tensile strength, hence its use in aerospace technology where maximum strength per unit weight is a prime objective.

Table 6.15   TYPICAL EXAMPLES OF INDUSTRIAL YARNS

Fibre	*Diameter* (mm)	*Denier*	*No. of filaments in yarn*	*Weight per unit length*		*Twist* (turns/m)
				den.	tex	
Rayon	0·0125 (av.)	1½	1100	1650	180	60
Nylon	0·025	6	140	840	90	20

Similar considerations apply to most textiles used for rubber reinforcement. Rayons of higher tenacity than those required for apparel are needed, and nylons of highest tenacity are also sought, nylon 66 being preferred, though nylon 6 is used to a lesser extent.

Cord is made from yarn by taking two or more yarns and inserting twist into them separately, after which the assembly of yarns is twisted in the same or opposite direction. Further similar assemblages may then be twisted together to yield a larger corded structure, and so on until a rope is obtained. As an example, a typical V-belt cord is written:

> 3/4/1000 denier polyester
> Singles twist   turns per metre   40 s
> First folding twist   turns per metre   80 s
> Second folding twist   turns per metre   80 z

i.e. 4 yarns each of 1000 denier having twist of 40 turns per metre are twisted together to give a first folding with 80 turns per metre. Three such first foldings are then twisted together in the opposite direction at 80 turns per metre. The direction of twist is conveyed by use of the letters s and z. If these are placed on top of a twisted yarn running from top to bottom of the page, the filaments will lie in the direction of the central part of one or other of the letters; if the direction corresponds to the direction of the central portion of the letter s then the yarn is said to contain s twist; z twist is of course the opposite, corresponding to a right-hand screw thread.

In general, tyre cords are as simple in construction as possible, consistent with performance requirements, e.g.

> 2/1650 denier rayon
> Yarn twist   turns per metre   470 z
> Folding twist   turns per metre   470 s

Tyre cord is usually woven into a fabric for convenience of processing. Cords are spaced at some 6–14 per centimetre, say, a total of some 2000 cords lying in parallel-sheet array, this array being known as the warp. Lightweight weft threads are interlaced alternately over one warp cord and under the next, at right angles to the warp direction, to form a woven fabric. The number of weft or pick threads is minimised at about 1 per centimetre. Because the number of warp and weft is nearly equal in conventional fabrics, tyre fabrics are frequently referred to as weftless fabrics.

Non-woven fabrics have so far found only limited use in the rubber industry. Non-wovens consist of fibrous webs, the fibres being more or less randomly arranged. Their increasing general use for disposable and near-disposable articles, e.g. dish cloths and undergarments, could encourage further incursion into rubber reinforcement applications.

6.7.2   REQUIREMENTS OF TEXTILE FOR REINFORCEMENT OF RUBBER PRODUCTS

6.7.2.1   *Strength, Dimensional Stability, and Flexibility*

Strength, dimensional stability, and flexibility are the basic requirements needed, to which must be added others arising from particular end uses. The densities, strengths, and moduli of industrial fibres are given in Table 6.16,

Table 6.16   MECHANICAL PROPERTIES OF INDUSTRIAL FIBRES

Material	Density (g/cm³)	Strength (kgf/mm²)	Specific strength		Modulus (kgf/mm²)	Specific modulus	
			(cm × 10⁶)	(g/den.)		(cm × 10⁶)	(g/den.)
Rubber (tread)	1·1	2	0·19	0·2	0·7	0·062	0·1
Silk	1·25	60	4·6	5	350	27·5	30
Cotton	1·54	70	4·6	5	1 000	62·5	70
Rayon (dry)	1·52	70	4·6	5	1 900	12·5	140
Nylon	1·14	90	8·1	9	500	38·0	40
Polyester	1·40	110	8·1	9	1 100	76·0	90
Polyvinylalcohol	1·28	100	8·1	9	1 600	125	120
Polypropylene	0·92	70	8·1	9	700	76·0	85
Glass	2·56	350	13·7	15	7 500	280	320
Steel	7·8	280	3·8	4	21 000	280	300
Carbon fibre	1·95	210	11·2	12	42 000	2 150	2 400

showing the wide variety of properties available. In general, modulus increases with strength. There are, however, exceptions, notably the low modulus of nylon. To minimise this disadvantage, a hot stretching process is now applied, usually after cord manufacture or after weaving.

6.7.2.2   *Heat Resistance*

Strength loss at elevated temperature is of two types. First, there is reversible loss in strength, which is that reduction in strength at elevated temperature which is totally recovered on return to the lower reference temperature.

Fig. 6.10 shows the temperature-dependent nature of strength in rayon and nylon. Other polymeric fibres are very similarly influenced by temperature (Wood, Goy, and Daruwalla, 1959). Secondly, degradative loss is that part

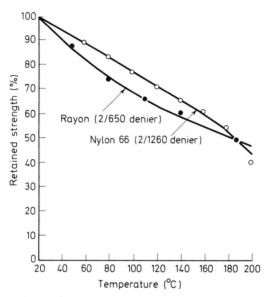

Fig. 6.10   Variation of strength with temperature for rayon and nylon

of the loss which is not recovered on return to the reference temperature, and is due to either chemical or physical change in the material under test.

Agents known to influence degradation are oxygen, water, and compounding ingredients. Embedment in rubber provides protection from the first. Both nylon and polyester, however, are susceptible to moisture at elevated temperatures. Polyester is probably attacked chemically by water, amine groupings in the compounding ingredients increasing the rate of degradation. Significant loss of strength during cure at 150°C has been observed when conditions have prevented moisture from escaping (Chapman, 1967; Gardner, 1969). At higher temperatures still, around 170°–180°C, nylon fibre is liable to attack by small amounts of moisture, disorientation of the fibre structure taking place.

6.7.2.3   Fatigue Resistance and Durability

If simple twistless yarns were used, tyres, for example, would have negligible durability. It is necessary to employ a particular twisted structure in which two or three yarns are twisted separately in one direction, the yarns then being twisted as a bunch in the opposite direction to form the cord. In general, the higher the twist the higher the fatigue resistance, provided excessive twist levels are not employed. Typical tyre cord specifications are given in Table 6.17.

It is generally accepted that, in conventional cross ply tyres, fatigue ratings place nylon as superior to either polyester or rayon, in that order. Fatigue failure in tyres occurs in those zones where cord undergoes longitudinal compression. Fatigue testers are therefore designed to produce cyclic longitudinal compressive stress in laboratory-made specimens, fatigue resistance assessments then being based upon strength loss over a period of

Table 6.17　TYPICAL TYRE CORD SPECIFICATIONS

| Fibre | Yarn denier | Cord construction | Twist (turns/m) | | Application |
			Yarn z	Cord s	
Rayon	1 650	2/1650	470	470	Tyre casing
Nylon	840	2/840	470	470	Tyre casing
Rayon	2 200	2/2200	430	430	Tyre casing
Rayon	2 200	2/2200	320	320	Tyre breaker
Polyester	1 000	2/1000	470	470	Tyre casing

time of flexure. Studies of cord behaviour during compression in transparent rubber have explained the need for a plied twisted structure (Wood and Redmond, 1965). Fig. 6.11 shows a specimen containing a sequence of 2/1650 denier cords, the twist of which increases from right to left of the photograph. Low-twist cords buckle, but those above about 300 turns/m yield by pushing the rubber radially outwards, displacement being uniform and without buckling. It is reasonable to conclude that such cords should be twisted to at least 300 turns/m to avoid fatigue damage.

　　Twist levels around 470 turns/m are, in fact, conventionally employed, the

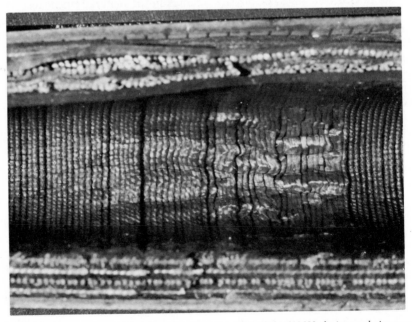

Fig. 6.11　Cord behaviour during compression; the twist of the 2/1650 denier cords increases from right to left. From Wood and Redmond (1965), courtesy The Textile Institute

improvement in fatigue resistance apparently being brought about by the reduction in lateral displacement of constituent yarns in each cord, with increasing twist and consequent reduction in self-abrasion. Nylon is noted for its high abrasion resistance.

Compression fatigue is rarely a factor in the durability of belting and hose. In conveyor belting, fabric damage is caused by impact of rocks, coal, etc. Durability is, therefore, best obtained by use of strong resilient materials. The latter requirement conflicts somewhat with the need for high modulus lengthwise and so, for example, a resilient nylon weft is adopted as a compromise; this meets both requirements of durability and troughability. Transmission V-belts contain low-twist heavy denier cords, e.g. 3/3/1000 denier polyester. In this case, fatigue occurs as the result of cyclic variation in tensile stress.

Steel cords resist compression rather than yield to it. Instead of high twist, low twist is employed, one of the most typical steel tyre cord constructions consisting of a centre strand of three wires, each of 0·15 mm diameter, surrounded by six identical strands, strand twist amounting to 120 turns/m, the twist in the whole cord also being 120 turns/m.

Impregnated glass fibre yarn also finds application in the breakers of belted bias tyres in the USA, a measure of fatigue resistance being obtained through the effect of the impregnant in insulating individual fibres and preventing interfibre abrasion. At low twist, these cords function along similar lines to steel.

Steel is used in certain conveyor belts (Section 10.2.1) and in some types of hose (Section 10.3) where high strength is needed. Glass is widely used as the reinforcement in positive drive tooth belts, which must not stretch appreciably under load if true meshing of the rubber teeth and sprocket is to be maintained. In these applications the textile reinforcement must have high tensile strength, high modulus, and a measure of flexibility, but the ability to withstand compressive stresses is not essential.

6.7.2.4 *Growth*

Hose, carrying fluids under pressure, conveyor belts, V-belts, and similar articles operating under protracted load are all subject to growth. Growth may be defined as the total extension which occurs over a period of time, being made up of two components, initial extension and creep.

Conventional textiles, rayon, nylon, and polyester grow appreciably during service; rayon in particular is prone to high creep, more especially at elevated temperatures, while nylon exhibits high initial extension. Polyester shows some improvement over rayon and nylon, but none of them is as dimensionally stable as glass or steel (Table 6.18).

6.7.2.5 *Force–Temperature Effects*

When nylon fibre is heated, it tends to shrink or, if prevented from shrinking, it exerts a retractive force. Polyester behaves similarly, but rayon remains

Table 6.18 PERCENTAGE EXTENSION OF CORDS UNDER 2 KGF LOAD

Test temperature (°C)	2/2650 denier rayon		2/840 denier nylon		2/1000 denier polyester	
	Under load for ½ min	Under load for 1000 min	Under load for ½ min	Under load for 1000 min	Under load for ½ min	Under load for 1000 min
20	1·8	3·5	3·2	4·2	1·4	2·3
60	1·7	2·9	3·2	4·0	1·6	2·3
100	2·3	3·9	2·8	3·5	3·5	4·0
140	2·5	4·9	3·3	4·3	3·8	4·0

Table 6.19 EFFECTS OF TREATMENT* ON MECHANICAL PROPERTIES OF CORDS

Tyre cord	Extension at 4·5 kgf (%)		Free shrinkage at 180°C (%)		Force to hold to length at 180°C (gf)	
	Before treatment	After treatment	Before treatment	After treatment	Before treatment	After treatment
Nylon 2/840 denier cord						
Specimen A	11·7	8·9	8·3	6·4	413	535
Specimen B	11·4	9·2	9·5	7·4	431	530
Polyester 2/1000 cord						
Specimen A	7·1	5·9	7·2	6·0	209	157
Specimen B	7·9	4·3	18·8	8·5	490	480

V-belt cord	Extension at 45 kgf (%)		Free shrinkage at 150°C (%)		Force to hold to length at 150°C (gf)	
	Before treatment	After treatment	Before treatment	After treatment	Before treatment	After treatment
Polyester 4/3/1000 denier	6·6	5·5	8·3	3·0	3 300	1 500

* The treatments were:

Nylon tyre cord	6% stretch, 60 seconds in an oven at 235°C.
Polyester tyre cord A	0% stretch, 60 seconds in an oven at 235°C.
Polyester tyre cord B	0% stretch, 60 seconds in an oven at 280°C.
Polyester V-belt cord	
First stage	2% stretch, 8 seconds at 240°C.
Second stage	0% stretch, 12 seconds at 210°C in fluidised bed.

relatively unaffected. These tendencies are apt to cause distortion in products during vulcanisation, thus leading to defects. These defects may be reduced, but not eliminated, by appropriate hot stretching. Table 6.19 shows how the magnitudes of shrinkage and developed tension are modified by treatment. It should be noted that modulus (indicated by percentage extension at a fixed load) is also changed by treatment (Goy and Möring, 1964).

6.7.2.6   *Energy Loss*

Measurement of dynamic elastic modulus (E_1) and loss modulus (E_2) has been carried out on a variety of rubber compounds and tyre cords. Underlying theory and techniques of measurement have been described by Meredith (1956). The percentage contributions from each major component of a tyre to the total energy loss have been calculated from these values so obtained and are given in Table 6.20 for two sizes chosen to be representative of the extremes.

From the analysis of these tyres it would appear that, within the range, the cord component will be at the high level of 30–40% of the total drag (Collins, Jackson, and Oubridge, 1965). Clearly, since the cord system component is

Table 6.20   PERCENTAGE CONTRIBUTION OF TYRE COMPONENTS TO TOTAL ENERGY LOSS

Component	*9·00-20 tyre*	*2·25-8 tyre*
Tread compressive component $\propto E_2/(E^*)^2$	32	22
Tread bending component $\propto E_2$	27	29
Ply rubber component $\propto E_2$	12	4
Sidewall component $\propto E_2$	—	5
Cord system component $\propto {}_T E_2$	29	40

E^* is the complex modulus.
${}_T E_2$ is the loss modulus of the cord.

directly proportional to the loss modulus, it is desirable that it should be minimised in order to reduce the power required to roll the tyre and the heat build-up. Nylon has the lowest modulus amongst standard textile materials, which suggests one reason why it finds particular application in giant tyres where heat generation has always been a problem.

6.7.3   TEXTILE REINFORCEMENT IN FOOTWEAR AND OTHER PRODUCTS

Wellington boots, canvas shoes, and the like employ fabrics calling for strength, modulus, abrasion resistance, and resistance to bacterial attack. By using cotton or staple rayon fabrics, woven or knitted, good mechanical adhesion to rubber is obtained (Wake, 1954).

Tennis balls are covered with a dense fabric, which may be of the Melton type, i.e. a fabric woven, with floating woollen weft and a cotton-base weave, felted and raised. Nowadays, a proportion of synthetic fibre is blended with wool to impart improved abrasion resistance. Dense non-woven fabrics are also finding favour as alternatives to Melton.

6.7.4   NEW DEVELOPMENTS

It is worth noting that until recently no really new industrial fibre had appeared for 15–20 years. Instead, development has concentrated on improvements to those already existing: the introduction of nylons of increasing molecular weight, and higher tenacity polyester; limited uses in Japan of new higher tenacity PVA; and limited use of glass fibre in the USA. Perhaps the time is now ripe for the introduction of new fibres. If so, an industrial-quality fibre along the lines of Du Pont's Qiana may be the first of a new series.

Fibres with better heat resistance, e.g. Nomex (Du Pont), are expected to develop and find application in rubber manufacture.

SEVEN

Reinforcement by Fillers

BY B. B. BOONSTRA

7.1 Introduction

The use of fillers in rubber is almost as old as the use of rubber itself. As soon as rubber-mixing machinery was developed, fillers, such as ground whiting, barytes, or clay, were added to cheapen the products and were found, in natural rubber, not to detract too much from the final vulcanisate properties.

Zinc oxide, originally used for whiteness, was the first 'active' filler, a fact recognised after the work of Ditmar in 1905. The discovery, in 1904, and development of carbon black to become the most important powder used in rubber has been discussed in Chapter 1.

In 1939, the first reinforcing siliceous filler was introduced (Sellers and Toonder, 1965). Developments over the next 10 years led to the emergence of the two types: the precipitated silicas, containing about 85–90% SiO_2, with ignition losses of 10–14%; and the pyrogenic silicas, which contain 99·8% silica. Because of its much higher price, the latter is mainly used as a filler for high-cost compounds, such as silicone rubber.

Organic fillers have made their appearance at various times in the literature and on the market without ever obtaining a significant place on it. In most cases they were incorporated in the latex phase rather than in dry rubber. Reinforcing resins were introduced when LeBras and Piccini (1950, 1951-1, 1951-2) prepared a resorcinol–formaldehyde condensate to the intermediate resol stage, and added the aqueous solution to rubber latex, which also contained dispersed curing agents. After gelation, drying, and vulcanising at low temperature, considerable reinforcement was observed in tensile strength, modulus, and elongation. Even before that, Meer and Wildschut (1942) had found that cresol–formaldehyde condensation products in the early stage of condensation (paracresol–dialcohol) could be added to dry natural rubber and that this combination would cure at normal vulcanising temperatures.

In 1947, a lignin, a byproduct from paper manufacture, was introduced as a reinforcing filler (Keilen and Pollak, 1947). This material had to be added to natural or synthetic latex in the form of an aqueous solution to observe the reinforcing effect in the dried and vulcanised final product. More recently,

227

starch derivatives have been reported as having certain reinforcing effects (Buchanan *et al.*, 1968).

Alphen (1954) and Houwink and Alphen (1955) published reports on the use of aminoplasts (aniline–formaldehyde, melamine–formaldehyde, and urea–formaldehyde condensation products) formed in stabilised acidified latex as reinforcing fillers. This method was different from that of Piccini; the resin was also formed *in situ* in the latex, but the composite was coagulated, leached out with water, dried, and then mixed on the mill with the curing ingredients. Higher tensile strength and elongation than for 50 p.p.h.r. of EPC black were observed in some cases with natural rubber. All of these organic fillers, formed *in situ* in the latex or in the rubber, have until now not developed farther than technological curiosities.

In a somewhat different category fall fillers such as high-styrene resins, copolymers with butadiene in general purpose rubbers, and phenol–formaldehyde resins in nitrile rubbers (Sections 4.2.1.5, 4.10.5, 6.2.4, and 6.3.9) (Burke, 1965). These resins are milled into the dry rubber at a temperature above their softening point and are mechanically dispersed and mixed as much as possible. They impart some of their stiffness at room temperature to the rubber and should not be considered as real reinforcing agents but rather as polyblends similar to blends of PVC with nitrile rubber as a softener (Section 4.10.5).

The latest development in resin fillers is represented by the particulate thermoset melamine–formaldehyde and urea–formaldehyde resins produced by CIBA in Switzerland (Renner, Boonstra, and Walker, 1969). These powders of very fine particle size and high surface areas (up to 200 m^2/g) are incorporated in rubber by dry mixing. They give higher modulus than inorganic white pigments at equal volume loading and have the advantage of lower specific gravity (about 1·5, compared with an average of 2·2 for silica).

7.2   The Reinforcement Concept

The concept of reinforcement relates basically to composites built from two or more structural elements or components of different mechanical characteristics whereby the strength of one of these elements is imparted to the composite and combined with the set of favourable properties of the other component: e.g. concrete with embedded steel rods or cable, where the high strength of the steel is imparted to the concrete to give it increased flexural and impact strength. The composite material should be readily shapeable into the required form and be capable of solidifying and hardening to a solid mass within a practical time limit. Similarly, in a glass fibre reinforced polyester, the enormous strength of the glass fibres is combined with the easy processing of the polyester. In these cases, dimensionally strongly anisometric members, fibres or rods, with length–diameter ratios of many thousands are usually overlapping each other over large sections and are bonded together by the matrix so that their strength is transmitted from one region to another as is the case with felt or leather. Basically what has been done is to bond a loose network of strong, long members together by a high viscosity or solid matrix so that the members support each other.

A totally different mechanism must be responsible for the reinforcement imparted to elastomers by particulate solids which, although not actually spherical in shape, still are not so strongly anisometric that they can be said to overlap each other over large proportions of their length. Nor are they

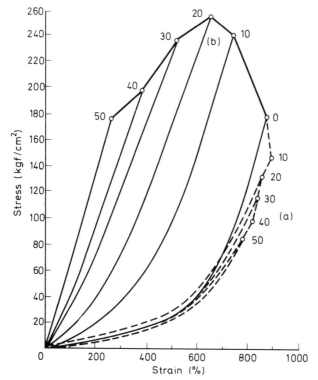

Fig. 7.1 Stress–strain curves of basic compound with increasing volume percentage of (a) barytes, and (b) EPC black. From Rossem (1958), courtesy Prof. A. van Rossem

actually so strong that they would be expected to impart additional strength to the composite. Furthermore, practically the same flexibility as the matrix is maintained. A clear definition is required of reinforcement of rubbers by fillers.

A pure gum vulcanisate of general purpose SBR has a tensile strength of no more than 22 kgf/cm^2 but, by mixing in 50 p.p.h.r of a reinforcing black, this value can easily be raised to 250 kgf/cm^2; the reinforcement in strength is obvious. However, a natural rubber pure gum vulcanisate made from latex with proper precautions can have a tensile strength of 450 kgf/cm^2 at 700% strain, and the best natural rubber SAF black compound will only give about 350 kgf/cm^2 at 550% strain. Even without considering the effect of elongation, it is obvious that there is in this case no improvement in tensile strength, so tensile strength is not a criterion for reinforcement.

Modulus at 300% strain has been considered as a measure, but many relatively inert fillers will raise this modulus without improving any failure properties. Therefore, the best definition is probably: *a reinforcing filler is a filler that improves the modulus and failure properties (tensile strength, tear resistance, and abrasion resistance) of the final vulcanisate.* The best single criterion for this reinforcement is the energy at rupture, introduced by Wiegand (1920) as 'resilient energy'. The energy at rupture can be obtained from the stress–strain curve as the area between the curve and the elongation axis.

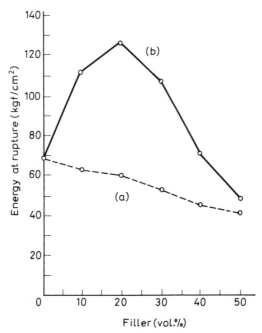

Fig. 7.2   Change of energy at rupture with increasing volume percentage of (a) barytes, and (b) EPC black. From Rossem (1958), courtesy Prof. A. van Rossem

The effect of increasing loadings of carbon black and of an inert filler (barytes) is demonstrated in Figs 7.1 and 7.2, taken from Rossem (1958).

7.3   Factors Influencing Elastomer Reinforcement

Figs 7.1 and 7.2 show the increase of strength and energy at rupture obtained with carbon black as a function of loading. The fact that there is an optimum loading indicates that there are two opposing factors in action when a reinforcing filler such as carbon black is added.

First, there is the increase of modulus and tensile strength, which is very much dependent on the particle size of the filler; small particles having a much greater effect than coarse ones. Since particle size is directly related to the reciprocal of surface area per gram of filler, the effect of small particles actually reflects their greater extent of interface between polymer and solid material.

Secondly, the reduction in properties at higher loading is a simple dilution effect, general to all fillers, merely due to a diminishing volume fraction of polymer in the composite. If there is not enough rubber matrix to hold the filler particles together, strength rapidly approaches zero. Actually, before this stage is reached, the stiffness of the compound has attained a point where it could be called brittle and at the normal rate of testing (50 cm/min) such a

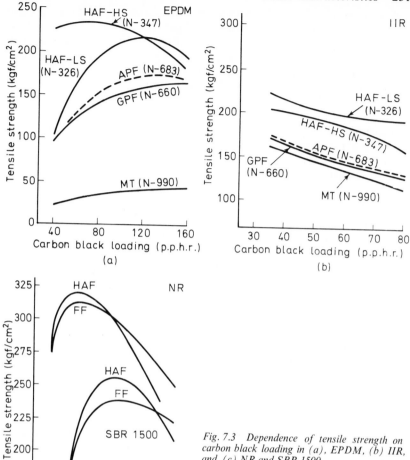

Fig. 7.3 *Dependence of tensile strength on carbon black loading in (a), EPDM, (b) IIR, and (c) NR and SBR 1500*

brittle compound would show poor strength. At much lower rates of stretching the decline in strength with higher loading would be less; in other words, for one particular filler the height and place of the maximum in the strength—loading curve is rate dependent; it will occur at a higher loading when testing is at slower rates. The place of the maximum is also dependent on the particle size of the filler, and this is illustrated in Fig. 7.3.

7.4   Filler Characteristics

The interaction of particulate fillers with an elastomer is dependent on a number of factors than can be classified as extensity, intensity, and geometrical factors.

The *extensity factor* is the total amount of surface area of filler per cubic centimetre of compound, in contact with the elastomer.

The *intensity factor* is the specific activity of this solid surface per square centimetre of interface, determined by the physical and chemical nature of the filler surface and, to some extent, of that of the elastomer.

Geometrical factors are: (a) the 'structure' of the filler, which can be determined by its void volume under standardised packing conditions; and (b) the porosity of the filler, usually a minor factor, which can be varied over a wide range with carbon blacks. Because the weight of individual spongy particles is lower than that of solid particles, the number of particles per cubic centimetre of compound at a given weight loading is greater for spongy than solid particles.

7.4.1   THE EXTENSITY FACTOR—THE TOTAL INTERFACE

The total surface area of a particulate solid is directly related to its particle size. If all particles are spheres of the same size, the surface area A_s per gram of filler is given by the equation

$$A_s = \frac{\pi d^2}{\frac{1}{6}\pi d^3 \rho} = \frac{6}{d\rho}$$

where d is the diameter and ρ is the density. In the preferred units for surface area (m^2/g) and diameter (nm or 10^{-9} m), the formula becomes

$$A_s = \frac{6000}{d\rho}$$

So, for a pigment with a density of 2 g/cm^3, the particle size of 30 nm corresponds to a surface area of 100 m^2/g. This is a rule of thumb that gives a good idea of the order of magnitude of the relation between surface area and particle size.

In reality there is always a distribution of sizes that can be averaged in various ways (Dannenberg and Boonstra, 1955), and particles are usually far from round (Heckman and Medalia, 1969). Particle size or surface area is a factor of the greatest importance in reinforcement because it can vary over such a wide range. Coarse inorganic fillers may have surface areas of about 1 m^2/g, fine silicas up to 400 m^2/g, and carbon blacks to 1000 m^2/g. In the range of rubber grade carbon blacks, there is a variation from 6 m^2/g for medium thermal blacks (N-991) (Section 6.1.3.1) to 250 m^2/g for conductive blacks (N-472), so there is a factor of 40 between the highest and the lowest surface area. No other factor in reinforcement varies over such a large range.

What counts is the area of interface between solid and elastomer per cubic centimetre of compound, and this is dependent on the surface area per gram of the filler, and also on the amount of filler in the compound, which brings in another large factor to the already wide range of variation. A tread compound containing 50 p.p.h.r. of an ISAF (N-220) black has about 35 m^2 of interface per cubic centimetre of compound, and an HAF (N-330) about 25 m^2/cm^3; these values give a practical idea of the reinforcement obtained. Below about 6 m^2/cm^3, not much reinforcement as defined in Section 7.2

Table 7.1 EFFECT OF PARTICLE SIZE OF CARBON BLACK AT 50 P.P.H.R. ON MAIN MECHANICAL PROPERTIES OF SBR

Type	ASTM designation	Average particle size (nm)	Tensile strength (kgf/cm²)	Relative laboratory abrasion	Relative roadwear
SAF	N-110	20– 25	250	1·35	1·20
ISAF	N-220	24– 33	230	1·25	1·15
HAF	N-330	28– 36	225	1·00	1·00 (standard)
EPC	S-300	30– 35	220	0·80	0·90
FEF	N-550	39– 55	185	0·64	0·72
HMF	N-683	49– 73	160	0·56	0·66
SRF	N-770	70– 96	150	0·48	0·60
FT	N-880	180–200	125	0·22	—
MT	N-990	250–350	100	0·18	—

is obtained. A schematic picture of the particle sizes in carbon black and the effect of particle size on mechanical properties is given in Fig. 6.5 and in Table 7.1.

It is interesting to consider that 36 m² of interface in a cube of 10 mm side can be obtained by dividing the cube into 216×10^{12} small cubes of 160 nm side by three perpendicular sets of 60 000 parallel planes, 160 nm apart.

7.4.2 THE INTENSITY FACTOR—THE SPECIFIC SURFACE ACTIVITY

The nature of the solid surface may be varying in a chemical sense, having different chemical groups: hydroxyl or metaloxide in white fillers, organic carboxyl, quinone, or lactone groups in carbon black, etc. In a physical sense they may be different in adsorptive capacity and in energy of adsorption.

Although with rubbers of a polar nature, such as CR and NBR, there will be a stronger interaction with filler surfaces showing dipoles such as OH groups or chlorine atoms, with the general purpose hydrocarbon rubbers no dramatic influences on reinforcement are noticeable when carbon black surfaces are chemically modified.

Chemical surface groups play an important role in their effect on the rate of cure with many vulcanising systems. In specific cases (heat treatment of butyl compounds and antioxidant action in polyethylene), properties are strongly affected by the concentration of hydroxyl or other oxygen-containing groups on the surface of carbon black. However, on the whole, the physical adsorption activity of the filler surface is of much greater importance than its chemical nature for the mechanical properties of the general purpose rubbers.

Fillers of sufficiently small particle size (sufficient surface area) will all give about the same order of magnitude in reinforcement if there are no great differences in particle shape.

Heat of adsorption measurements have shown that the adsorptive activity of reinforcing carbon blacks is not homogeneously distributed but concentrated at a number of sites of much greater activity than most of the surface. These active sites only represent a small percentage, less than 5%, of the total surface. This effect is illustrated in Fig. 7.4 (Taylor and Atkins, 1966), where the differential heat of adsorption Q_s is plotted as a function of coverage. The first few percent of the surface to be covered is the most active, as shown by the much higher heat of adsorption.

The importance of these active sites can be demonstrated by heat treatment (graphitisation) of the carbon black to a temperature of 1600°–3000°C.

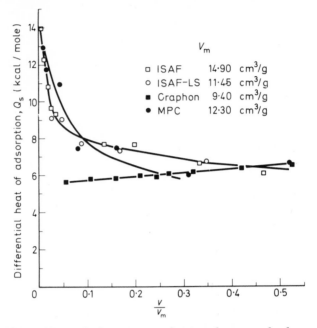

Fig. 7.4   *Differential heat of adsorption as a function of coverage for four carbon blacks (V = volume adsorbed, V_m = volume adsorbed to form monolayer). From Taylor and Atkins (1966), courtesy American Chemical Society*

Through this treatment the activity of the sites is lost, recrystallisation occurs, and the surface flaws in the lattice, which (probably) constitute these more active spots, disappear. The surface becomes homogeneous in adsorptive activity, as is illustrated in Fig. 7.4 for Graphon* (Schaeffer and Smith, 1955; Dannenberg, 1966). This loss of active sites has a most profound effect on the mechanical properties of the vulcanisate made with this black (Section 7.4.3).

Another aspect that needs discussion is the formation of chemical bonds between filler surface and polymer. There is one instance in which one can be reasonably sure that chemical bonds actually form, and that is when brass powder is used as a filler in a vulcanisate containing sulphur. Although very small particles cannot be obtained and rather large amounts of sulphur must be used to effect curing, the result is evident. One obtains a very high modulus but poor tensile strength and elongation (Rigbi, 1956, 1957). Low swelling of this vulcanisate in good rubber solvents indicates high crosslink density either in the rubber matrix directly or via the brass particles.

* Cabot's trade name for graphitised channel black, introduced in 1934 and made available for research purposes.

Experiments with model fillers consisting of both polystyrene particles and styrene–butadiene copolymer particles as fillers in SBR vulcanisates have been carried out by Morton, Healy, and Denecour (1967). In carefully controlled experiments they could show that, in the case of the polystyrene particles, no chemical bonds were formed during vulcanisation; this was to be expected since the polystyrene chain molecule has no olefinic double bonds and hence is not reactive. By copolymerisation with 11 % butadiene, chemical unsaturation is introduced into the molecular chain, and particles made from this copolymer showed evidence of being chemically bound to the rubber. Remarkably, the vulcanisate having polystyrene particles as a filler had a higher tensile strength than the one containing the corresponding volume of styrene–butadiene copolymer particles as a filler. The latter, however, had a considerably higher modulus. Obviously then, strong chemical bonding between filler and polymer does not lead to desirable vulcanisate strength properties but causes high moduli (load at 300 % strain).

In some cases (e.g. butyl rubber) there seems to be insufficient interaction between filler and rubber, but vulcanisate strength and rebound properties can be improved by heat treatment with mild bonding agents (Leeper *et al.*, 1956) and introduction of oxygen-containing groups at the filler surface (Gessler, 1964). It seems, therefore, that both total surface area and its specific activity are important factors in reinforcement. Perhaps,

$$\text{Total area} \times \text{specific surface activity} = \text{reinforcement factor}$$

If either one of the two factors is zero, the whole product is zero; in other words, neither is of value without the other.

Unfortunately, no precise and rapid method exists to measure surface activity. The measurement of the differential heat of adsorption is too cumbersome to be of practical use. Water adsorption (Dannenberg and Opie, 1958) at various relative humidities has been used but is dependent on the polarity of surface groups. More appropriate is propane adsorption, but no routine test method has been developed. Also indicative of surface activity is the amount of bound rubber formed during mixing of filler and rubber (Section 7.7).

7.4.3   GEOMETRICAL CHARACTERISTICS

7.4.3.1   *Structure and Aggregation—Void Volume—Anisometry*

'Structure' used in the sense of aggregation is basically nothing but deviation of the primary aggregates from the spherical shape and their size. Non-spherical particles have a packing which is less dense than that of spheres (74·01 %), leaving a greater volume of voids in between the particles (i.e. greater than the 25·99 % between spheres). Void volume is used as a measure of structure, by measuring compressibility of the dry filler or by determining oil absorption according to a standard procedure (Section 6.1.3.2).

Another way of looking at structure has been indicated by Medalia (1967, 1970) analysing electron micrographs. The method is illustrated in Fig. 7.5.

A typical particle aggregate silhouette is subjected to computerised integration, and its maximum and minimum moments of inertia are determined. Then an ellipse is constructed that has the same two moments of inertia around two perpendicular axes. The ratio of lengths of these two axes

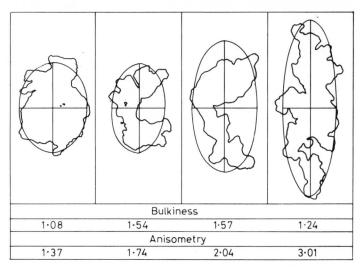

Bulkiness			
1·08	1·54	1·57	1·24
Anisometry			
1·37	1·74	2·04	3·01

Fig. 7.5　Analysis of black structure. From Medalia (1967), courtesy Cabot Corpn

K_A/K_B is the anisometry Q; the ratio of silhouette area over ellipse area is called bulkiness B; and $(QB - 1) = T$, the structure factor. The trouble with this method is that many aggregates or particles have to be reviewed before representative average values for anisometry and bulkiness can be reached, so the measurement is very time consuming.

In the inorganic or mineral fillers there is a great deal of difference in geometry of particles, depending on the crystal form of the mineral. The minimum anisometry is found with materials that form crystals with approximately equal dimensions in the three directions, i.e. isometric particles. More anisometric are particles in which one dimension is much smaller than the two others, i.e. platelets. The most anisometric are particles which have two dimensions much smaller than the third, so that they are rod-shaped. When compounds are compared with fillers of equivalent surface area and chemical

Table 7.2　EFFECT OF ANISOMETRY ON PROPERTIES (VISCOSITY AND MODULUS DATA AT 20% VOLUME LOADING OF FILLER)

Property	Natural ground calcium carbonate	Kaolin clay	Acicular clay (Attacote)
Particle shape	Round	Platelets	Needles
Mooney viscosity in SBR [ML(1+4)100°C]	40	43	66
Mooney viscosity in NR [ML(1+4)100°C]	42	46	Too high
Extrusion shrinkage (%)	58	56	46
Stress at 300% strain in NR (kgf/cm²)	35	49	74

nature but different anisometry, the modulus increases with increasing anisometry (Table 7.2).

In rubber technology it is customary to associate high structure of a filler with high modulus of the vulcanisate, and this is to a large extent true of a range of blacks.

As already discussed (Section 7.4.2), graphitisation of black causes re-orientation of the crystallite structure at the surface and thereby removes the sites of high adsorptive activity. The analytical data of Table 7.3 show that no great reduction in surface area has resulted through the heat treatment (2700°C), and the structure, if anything, seems to have increased. If one looks

Table 7.3   EFFECT OF BLACK GRAPHITISATION ON MAJOR PROPERTIES

Property	High structure ISAF		ISAF	
	Original	*Graphitised*	*Original*	*Graphitised*
Surface area (N_2 adsorption*) (m^2/g)	116	86	108	88
Oil absorption (cm^3/g)	1·72	1·78	1·33	1·54
Water adsorption at 55 r.h. (%)	2·4	0	1·85	0
Density (dry black) (g/cm^3)	0·283	0·295	0·353	0·392
Propane adsorption at				
$P/P_0 = 0·001$ (cm^3/g)	1·03	0·25	0·93	0·31
Bound SBR (%)	25·1	0–2	18	0·4–2
Extrusion shrinkage (%)	30	37·5	39·6	43·5
Mooney viscosity [ML(1+4)100°C]	83	87	73	76
Scorch at 135°C (min.)	10·5	17	18	20
Dispersion (%)	99	99	99	98·2
Tensile strength (kgf/cm^2)	270	240	280	230
Stress at 300% strain (kgf/cm^2)	150	36	105	30
Abrasion loss ($cm^3/10^6$ rev)	62	181	67	142
Elongation (%)	450	730	630	750
Hardness (IRHD)	73	68	68	65
Hysteresis	0·238	0·315	0·204	0·297

* See Section 6.1.3.1.

at the vulcanisate properties of these fillers in SBR, dramatic changes become apparent. Through graphitisation of the black, the optimum vulcanisate modulus has dropped to about one-quarter of the value of the untreated black compound, while tensile strength has only lost some 10–15% of its value. The most important failure property, abrasion resistance, has declined parallel with the modulus to about one-quarter of that obtained with un-treated black. What has changed is the adsorptive activity of the black sur-face, as illustrated by the drop in bound rubber, and in propane and water adsorption.

This emphasises again the relation between specific surface activity and structure, and how both are necessary to provide reinforcement. High struc-ture without surface activity does not result in (high) reinforcement, as the data of Table 7.3 show. High surface activity without structure should not result in any higher modulus increase than dictated by the hydrodynamic factor of Einstein, Guth, and Gold (Section 7.8). Although the graphitised high structure black still gives somewhat higher modulus than the corre-sponding regular structure material, the effect of surface activity on reinforce-ment completely dominates this small effect of structure as such.

The interpretation of these results may be that the main effect of aniso-metric (i.e. highly structured) particles on modulus is the resistance they offer against orientation of the matrix molecules in the direction of the flow lines during extension or other deformation. This resistance is stronger, the intenser the interaction between rubber and filler surface. When the sliding of polymer molecules along this surface is unhampered by strongly adsorbing sites, as is the case with graphitised black, the resistance against orientation of the particle is minimal—the matrix flows freely around it and a low modulus is the result.

It is of interest to point out here that this high modulus is apparently a viscous phenomenon, an observation that will be used again in the theory of reinforcement.

In Table 7.4 the oil absorption values of a number of different blacks and white fillers are listed (Bachman *et al.*, 1959). It is remarkable that silicas and

Table 7.4   OIL ABSORPTION (VOID VOLUME) OF VARIOUS REINFORCING FILLERS, AND MECHANICAL PROPERTIES OF SBR VULCANISATES WITH 20% VOLUME LOADING

Carbon black		*Oil absorption, manual* (cm^3/g)	*DBP absorption, automatic* (cm^3/g)	*BET* area* (m^2/g)	*Stress at 300% strain* (kgf/cm^2)	*Tensile strength* (kgf/cm^2)	*Elongation at break* $(\%)$	*Hardness* (IRHD)
ASTM	Type							
N-220	ISAF	1·35	1·15	115	105	225	520	59
N-330	HAF	1·30	1·05	75	110	210	500	59
N-770	SRF	0·70	0·70	20	70	140	540	55
S-300	EPC	—	0·95	100	85	255	560	63
White fillers								
Pyrogenic silica (Cab-o-sil)		1·50	1·20	190	46	325	660	69
Precipitated silica (Hi-Sil233)		1·95	1·90	150	33	250	690	59
Precipitated calcium silicate (Silene EF)		1·20	1·10	80	32	160	590	58
Clay (Suprex)		—	0·25	5–10	45	180	590	62

* See Section 6.1.3.1.

silicates or other white inorganic fillers have their lower moduli at inter-mediate strains (100–300%) than carbon blacks of corresponding particle size and structure.

As a consequence, the white filler vulcanisates have a low energy at rupture (integrated area under the stress–strain curve) than the corresponding black vulcanisates, and, in most cases abrasion resistance is lower than that of corresponding black vulcanisates.

Persistent and Transient Structures. Reinforcing fillers, certainly carbon black in particular, undergo a number of mechanical treatments to make them easily processable in rubbers. Carbon black immediately after formation is in a very fluffy state and may have a tap density of 0·08–0·16 g/cm³. In this state it is difficult to incorporate in rubber or plastic and therefore it is 'pelletised' (Section 6.1.2.1). The density increases to about 0·4 g/cm³ whereas the specific gravity of carbon black is about 1·85, i.e. four-fifths of the bulk is air. During this relatively mild process, the most fragile structures are broken and, when the pellets are subjected to the operations necessary to determine the structure or void volume by oil absorption, some more of the 'structure' is broken down. In this respect, the hand operation is milder than the automated method with DBP in the absorptometer; the latter always giving somewhat lower values for the void volume. Fig. 7.6 gives a schematic picture

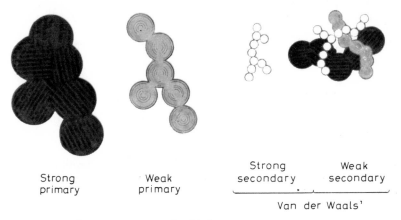

Strong	Weak	Strong	Weak
primary	primary	secondary	secondary

Van der Waals'

Fig. 7.6   Variety of structure types in carbon black

of some types of structure that may be found. The distinction between primary and secondary structure is rather arbitrary. However, the most rigorous test that the structure has to pass is the final mixing with high-viscosity rubber on a roll mill or in an internal mixer.

It is possible to get an idea of this breakdown of structure during rubber mixing by recovering the black from the compound (rubber is distilled off in a stream of nitrogen) and measuring its oil absorption again.

As Fig. 7.7 shows, a considerable reduction in structure occurs with HAF black of high structure after mixing into rubber for increasing periods of time.

Similar phenomena were observed by Gessler (1967), who interpreted the breakdown of carbon black structure as a main source of active sites at the fracture surface. Voet, Aboytes and Marsh (1969), on the other hand, failed to observe structure breakdown in their experiments using a colloid mill to homogenise the black rubber solution before the rubber-milling process.

It seems logical to assume that the more rigorous the mechanical treatment of a reinforcing filler, such as carbon black, the less of its structure will be left. What is important to the practical compounder is the structure as it exists in the rubber compound, because it is this structure that influences the rheological behaviour of the compound—in particular, its extrusion through a narrow opening. Electron microscopic evidence for this breakdown has been supplied by Heckman and Medalia (1969).

To judge this rheological behaviour, an extrusion shrinkage test is run in which a length of compound is extruded under standardised conditions

(ASTM D2230-63T). After a fixed time of recovery, usually 1 hour, one meter of extrudate is measured off, cut, and weighed (weight W). From the specific gravity ρ of the stock and the cross-sectional area A_d of the extrusion die

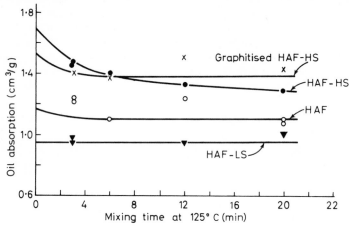

Fig. 7.7   *Breakdown of structure on mixing*

(A_d is usually 0·178 cm^2 since the diameter is usually 4·8 mm), the length L_t that the extrudate should have had if it had exactly the same diameter as the die can be calculated:

$$L_t = \frac{W}{\rho A_d}$$

This theoretical length L_t (in centimetres) is then used to calculate the percentage extrusion shrinkage E:

$$E = \left(\frac{L_t - 100}{L_t}\right)100$$

A related entity, the percentage die swell D, is given by

$$D = \left(\frac{A_e - A_d}{A_d}\right)100$$

where A_e denotes the actual cross-sectional area of the extrudate. Therefore

$$D = \left[\frac{(W/100\rho) - (W/L_t\rho)}{W/L_t\rho}\right]100$$

$$= L_t - 100$$

and the relation between the two entities is

$$E = \frac{100D}{D + 100}$$

Die swell or extrusion shrinkages are the best practical measures of filler structure in the final compound. High structure fillers incorporate more slowly in most rubbers but are more easily dispersed than corresponding medium or low structure fillers (Section 7.7).

Breakdown of filler structure also happens on deformation of the final vulcanisate, as is shown by measuring dynamic modulus and electrical conductivity at increasing amplitudes of shear. (Payne and Watson, 1963; Payne, 1966). The picture of a network of carbon particles penetrating the entire rubber sample is probably not a correct one, but there is undoubtedly some structure (aggregates or conglomerates) that is broken down as the amplitude of the cyclic shear is increased. This is illustrated in Fig. 7.8 (Payne, 1966).

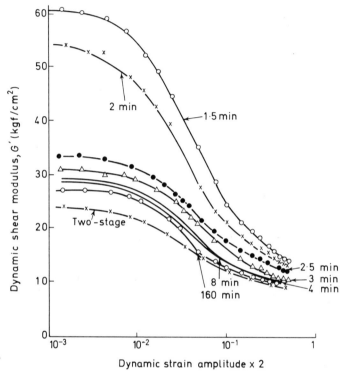

Fig. 7.8 *Variation of dynamic shear modulus with dynamic amplitude of straining (double amplitude) for a series of rubber–black mixes. From Payne (1966), courtesy American Chemical Society*

The curves relate to a series of rubber–black mixes of increasing mixing time and therefore of increasing degree of dispersion. The shortest mixing times, i.e. the poorest dispersions, show the highest dynamic elastic modulus but also the largest decrease in modulus with increase of amplitude; the lowest results appear for the compound obtained by a two-stage mixing process. This behaviour is due to breakdown of the agglomerates formed as the primary product of the mixing (Section 7.7).

Fig. 7.9 (Payne, 1966) shows that a similar behaviour is found for the electrical conductivity and that conductivity and dynamic modulus run closely parallel. Both are dependent on the presence of structural entities that are broken down by the larger strain amplitudes. Voet and Cook (1967, 1968) found that at extremely small amplitudes the dynamic shear modulus

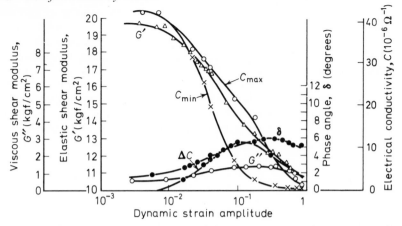

Fig. 7.9   Variation in electrical and mechanical dynamic properties with increasing amplitude (C_{min} and C_{max} are minimum and maximum conductivity during flexing; ΔC is the difference between maximum and minimum conductivity); the mix contained 50 p.p.h.r. HAF black. From Payne (1966), courtesy American Chemical Society

diminishes again, probably because the amplitude is too small to influence existing agglomerates; the electrical conductivity follows a similar course.

7.4.3.2   *Porosity*

Although a special type of porosity can be found with many particulate fillers, porosity is a characteristic property of carbon black that can be relatively easily controlled. The nature of the carbon black particles (Harling and Heckman, 1969) is such that crystallite regions at the surface alternate with short disordered sections, and the interior is definitely much less ordered. By controlled oxidation it is possible to remove part of the less orientated material from the particle so that more or less porous particles remain. In the extreme case, just an empty shell remains (Fig. 7.10).

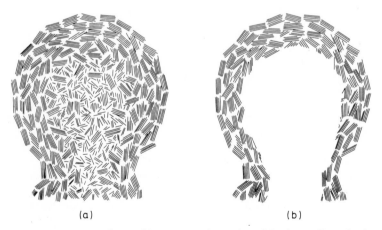

(a)   (b)

Fig. 7.10   Concentric crystallite models: (a) normal particle, and (b) drastically oxidised particle

In most cases the pores are too small for polymer chains to enter, although some smaller molecules in the compound may do so; the specific gravity of a porous black compound is usually not significantly lower than that of the corresponding compound with solid particles. A direct consequence of the porosity is that the diameter–surface area relation of Section 7.4.1 no longer holds. To determine the effective outside surface area, the so-called 't' method of de Boer, Linsen, and Osinga (1965) can be used. Evidently, only the outside surface area is effective in reinforcement. A number of values of total and outside surface area are given in Table 7.5 (Dannenberg and Boonstra, 1955; Smith, 1970).

Large internal surface areas may be detrimental in so far as a certain proportion of accelerator may become immobilised and inactivated.

Another consequence already mentioned (Section 7.4.1) is the number of particles per gram weight. If the internal pore volume per particle is on the

Table 7.5 SUMMARY OF BET AND 't' SURFACE AREAS FOR CARBON BLACKS AND MINERAL FILLERS

Carbon black		BET area (total) (m^2/g)	't' area† (outside) (m^2/g)
ASTM	*Type*		
N-472	XCF (Vulcan XC-72)*	243	143
S-301	MPC (Spheron 6)*	113	92
N-110	SAF (Vulcan 9)	138	138
N-220	ISAF (Vulcan 6)	114	115
N-242	ISAF-HS (Vulcan 6H)	116	116
N-285	ISAF-HS (Vulcan 5H)	101	101
N-330	HAF (Vulcan 3)	82	82
N-326	HAF-LS (Regal 300)	82	85
N-550	FEF (Sterling SO)	44	43
Channel-type colour (Monarch 81)*		130	116
Channel-type colour (Monarch 71)*		525	264
Pyrogenic silica (high surface area)		380	387
Pyrogenic silica (medium surface area)		207	215
Pyrogenic alumina		94	99
Precipitated silica		125	129

* In essence, all the rubber fillers are non-porous except the conductive furnace black XC-72 and the channel blacks; colour blacks are much more porous.
† Using the fine thermal black master-curve as standard.

average 15%, then each particle is 15% lighter than a solid particle of that size and the number of such porous particles in one gram will be 15% higher than for solid particles. Properties dependent on volume loading and interparticle distance are therefore influenced by porosity, so porous black will give higher viscosities and higher electrical conductivity than solid blacks. Polley and Boonstra (1957) have shown that, if the chemical nature of types of carbon black of vastly different particle size is the same, then the electrical conductivity is governed by the average particle distance or, rather, the average distance of nearest neighbours.

Since porosity in carbon blacks is obtained by oxidation, these blacks often contain many oxygen groups on the surface. This would counteract the easy passage of conductance electrons and increase the resistance of the rubber compound made with such a black, so one cannot state that porous carbon blacks will also give highly conductive compounds. This is only true if the surface chemistry is approximately the same as for the solid particle type carbon black.

Summarising, it can be stated that, although porosity is a factor that cannot be overlooked in assessing a filler's influence on vulcanisate properties, its effect on reinforcement is a secondary one.

7.5 Main Effects of Filler Characteristics on Vulcanisate Properties

Although rubber properties are interconnected and relate to the combination of all filler properties, a brief summary of the main influence of each of the four filler characteristics is given below:

1. *Smaller particle size* (larger external surface area) results in higher tensile strength, higher hysteresis, higher abrasion resistance, higher electrical conductivity, and higher Mooney viscosity, with minor effects on extrusion shrinkage and modulus.
2. *An increase in surface activity* (physical adsorption) results in higher modulus at the higher strains (300% upwards), higher abrasion resistance, higher adsorptive properties, higher 'bound rubber', and lower hysteresis.
3. *An increase in persistent structure* (anisometry, bulkiness) results in lower extrusion shrinkage, higher modulus at low and medium strains (up to 300%), higher Mooney viscosity, higher hysteresis, and longer incorporation time. Higher electrical conductivity and heat conductivity are found for higher structure blacks.

   This property is interrelated with surface activity, structure changes on fillers without surface activity (graphitised black) showing the effects indicated above only rather faintly. At constant high activity, the structure effects are most pronounced.
4. *Porosity* results in higher viscosity and higher electrical conductivity in the case of carbon blacks.

7.6 Influence of Fillers on the Crosslinking Process

Fillers are known to influence the crosslinking reaction during vulcanisation; there is, for instance, the retardation of cure by channel blacks as compared to furnace blacks, of hard clay as compared to whiting, of some silicas as compared to silicates, all corresponding in particle size. In most cases the cause of this retardation can be traced to the greater or lesser acidity of the filler (indicated to some extent by the pH of its aqueous slurry) which influences the kinetics of the crosslinking reaction. The slurry pH of a channel black is 4–4·5, that of a furnace black 7–9; for clay and whiting the values are approximately 4·5–5·5 and 8–10; for silica it varies from 3·5 to 7, whereas the silicates approach 10.

Manufacturers of certain types of silicas suggest that better properties are obtained in an accelerator–sulphur cured rubber–silica compound when the zinc oxide is omitted and that this ingredient, which is actually an activator for the curing reaction, reduces the viscosity of the unvulcanised compound. The effect has been explained by assuming that the zinc oxide reacts with the most active spots on the filler surface to form a zinc silicate.

Work by Cotten (1969) has shown that the chemistry of the carbon black surface plays an important part in the initial steps preceding the actual cross-linking reaction as well as in the crosslinking rate itself. A convenient technique to study effects on rate of cure and crosslinking is by means of cure-meters such as the oscillating disc rheometer, which is used more and more to evaluate fillers for their reinforcing potential (Section 5.5.2) (Wolff, 1969). Significant for filler reinforcement is the value of ΔL_{max}, the maximum change in torque during vulcanisation. The effect of the loading of carbon black on this value is shown in Fig. 7.11.

The relative increase due to the addition of X p.p.h.r. of the filler (in this case carbon black) is expressed by:

$$\frac{\Delta L_{max(X)} - \Delta L_{max(gum)}}{\Delta L_{max(gum)}}$$

When this entity is plotted against the loading X, a straight line is obtained, the slope of which has been called α_F by Wolff (1970).

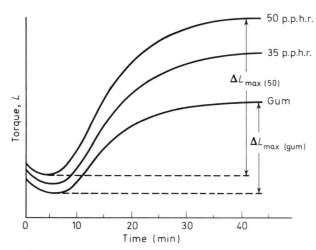

Fig. 7.11   Effect of black loading as shown on the oscillating disc rheometer trace

This parameter α_F is used to characterise the structure of fillers, in particular of carbon blacks, and it is of interest to note that this is one of the properties that changes very little after deactivation of the surface by graphi-tisation of the carbon black. Values of ΔL_{max} on original and deactivated carbon black illustrating this fact are shown in Table 7.6. Other vulcanisate properties used to measure crosslinking, such as modulus and swelling, undergo large changes. As mentioned in the section on structure, the modulus imparted by graphitised black is only 25 % of that imparted by the untreated carbon black. The restrictive effect of carbon black on the swelling of rubber vulcanisates disappears completely when the black is deactivated by heat treatment at a temperature of 1500°–2500°C (Boonstra and Taylor, 1965).

Table 7.6 VALUES OF ΔL_{max} ON ORIGINAL AND GRAPHITISED BLACKS (AT 3° ARC)

Original carbon black	ΔL_{max}	Graphitised carbon black	ΔL_{max}
SBR gum	56·4		
SBR + 50 p.p.h.r. MT black	66	MT	66
SBR + 50 p.p.h.r. HAF black	65	HAF	69
SBR + 50 p.p.h.r. ISAF black	85	ISAF	88
NR + 50 p.p.h.r. HAF black	66	HAF	62
NR + 50 p.p.h.r. EPC black	64	EPC	64

In principle, ΔL_{max} is closely related to the modulus at rather low strain (3° angular movement of the rotor is equivalent to about 20% strain). Since the relationship between moduli at 20% and at 300% strain depends on the shape of the stress–strain curve, there is no general rule relating the torque at 3° angle and 300% 'modulus' for different compounds.

7.7   The Mixing Process

The mixing stage in the course of the production of a vulcanisate introduces more variance in its mechanical properties than any other step; it actually introduces more variance than all of the other steps (curing, sample preparation, and testing) together. Notwithstanding the fact that this is known, relatively few studies have been made on the subject.

Microscopic evaluations of early stages of mixing (Boonstra and Medalia, 1963-1, 1963-2) show that one must visualise the primary process in mixing of carbon black and rubber as a penetration of the voids between the aggregates by rubber; the primary products are concentrated agglomerates held together by the rubber vehicle. When all voids are filled with rubber, the black is considered 'incorporated' but not yet dispersed. Immediately after, and even during, their formation these concentrated agglomerates are subjected to high shear forces that tend to break them down again into smaller and smaller units until the final dispersion is reached. This picture can be considered as representing the general process of mixing of a particulate filler with a high-viscosity vehicle, although the effect will not be so evident with coarse fillers.

It is of interest to look at the mechanical properties of compounds and vulcanisates obtained from very short to more than adequate mixing procedures (Table 7.7).

It is to be noted that after only 1·5 min of mixing, a tensile strength of about 175 kgf/cm² is reached (compared with 20 kgf/cm² for the pure gum vulcanisate). Even though the microscopic rating gives only 23·6% black dispersed in aggregates smaller than 6 μm, the tensile strength has attained already two-thirds of its maximum value. The abrasion resistance is about 40% of that of the best dispersion (two-stage). Another point of interest is the change in 100% modulus which drops from 34 kgf/cm² for the short-time mixed vulcanisate to less than half that value for the vulcanisate from the fully mixed compound. The major part of this reduction in modulus is due to further breakdown of the carbon black aggregates in the early stages of mixing. This idea is supported by the very low extrusion shrinkage of the compounds mixed for the shortest times and its increase on continued mixing. Torsional hysteresis also shows a steady decrease as mixing progresses.

These data show how profoundly reinforcing abilities are influenced by variations in mixing procedure, apart from the degradation of the elastomer.

Table 7.7 EFFECT OF MIXING TIME ON PROPERTIES OF SINGLE-STAGE MIX OF ISAF BLACK IN OIL-EXTENDED SBR 1712*

Property	Mixing time							
	1·5 min	*2 min*	*2·5 min*	*3 min*	*4 min*	*8 min*	*16 min*	*Two-stage†*
Tensile strength (kgf/cm^2)	173	220	245	265	260	265	255	265
Stress at 100% strain (kgf/cm^2)	34	31	27	20	17	15	12	15
Stress at 200% strain (kgf/cm^2)	86	87	86	69	63	57	56	54
Stress at 300% strain (kgf/cm^2)	130	146	142	127	128	122	119	123
Elongation (%)	380	460	490	540	530	540	530	540
Hardness (IRHD)	65	65	64	62	61	59	59	57
Tear strength (kgf/cm)	40	40	40	43	41	41	39	41
Abrasion (Akron, volume loss, cm^3/10^6 rev)	289	194	142	133	136	—	—	122
Cut growth (De Mattia, kc to 25 mm)	5	6·5	9	8·5	11	—	—	27
Torsional hysteresis at 100°C	0·273	0·278	0·266	0·238	0·226	0·228	0·213	—
Goodrich flexometer								
Static compression (%)	24·4	24·8	26·6	28·6	28·5	28·5	28·6	—
Permanent set (%)	4·6	4·2	4·2	3·8	3·4	2·9	2·8	—
Heat build-up (°C)	51	49·5	48·5	46	44·5	41	38·5	—
Mooney viscosity [ML(1+4)100°C]	133	122	114	97	83	68	63	35
Extrusion shrinkage (%)	29·1	39·7	44·2	46·8	45·7	41·7	36·1	43·2
DC resistivity (Ω cm)	124	88	108	175	300	440	760	728
Specific gravity	1·148	1·152	1·152	1·152	1·153	1·153	1·152	—
Dispersion rating (%)‡	23·6	71·4	86·4	96·9	99·3	100	100	100

* Recipe:
SBR 1712	137·5
ISAF black	69
Stearic acid	1·5
Zinc oxide	3
Antioxidant	1
Sulphur	2
CBS	1·1

Vulcanisation time: 60 min at 144°C for sheets,
                   70 min at 144°C for thicker specimens.
† Stock prepared by two-stage high-viscosity mix with 69 p.p.h.r. black in first stage.
‡ Rating calculated by revised procedure assuming that $A = s$ and $v = 0·4$ (Medalia, 1961).

The heat degradation is relatively small in SBR, much larger in NR; and in BR, at least at temperatures below 148°C, it is far less than SBR and all other elastomers.

The composition of the primary formed agglomerate can be determined from the oil (or DBP) absorption test since, in this test, the voids between particles and aggregates are filled with oil; during mixing, they are filled with rubber. For a filler with an oil absorption of 1·25 cm^3/g, the primary agglomorate has a composition of 125 cm^3 (118 g) of rubber and 100 g of filler, or 100 g of rubber and 85 g of filler. This, therefore, is the maximum loading of filler that is still dispersible. It is possible to make higher loadings, but these have such intensive particle-to-particle contacts and interparticle interaction that they are no longer dispersible within a practical time. Master-batches intended for dilution, to arrive at better dispersions than by direct mixing, should not exceed this critical concentration, otherwise poor disper-sion with large lumps will result. A micrograph of such a poorly dispersed masterbatch is shown in Fig. 7.12 in comparison with a well-dispersed black.

The lower the void volume (oil absorption) of a filler, the higher is the critical loading that can be tolerated in a masterbatch before the masterbatch

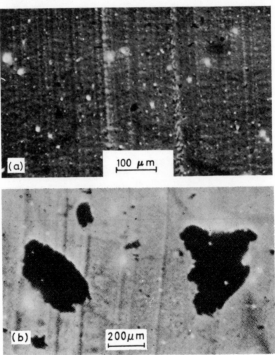

Fig. 7.12   Examples of (a) a good, and (b) a very bad dispersion of ISAF black in SBR; the magnification of (a) is twice that of (b)

becomes indispersible. As a consequence of the mixing mechanism, therefore, low-viscosity rubbers will penetrate faster (incorporation times are shorter) than highly viscous ones, but subsequent dilution to good dispersions is slow, owing to the shearing forces being small. High structure fillers incorporate more slowly than low structure fillers (more void volume to be filled) but, once incorporated, the former disperse more easily and rapidly than their low structure counterparts.

7.8   The Role of Bound Rubber

During the milling process, rubber chain molecules become attached to reinforcing fillers, so they are no longer soluble in the usual rubber solvents;

this process is the basis for the formation of 'bound rubber'; it continues after mixing and eventually a system of interconnecting chains and particles results, which appears as an insoluble fragile black gel containing all the black and part of the rubber, the bound rubber (Section 2.6.2.1). The process may continue for weeks at room temperature and is accelerated by increasing the temperature. Bound rubber is usually expressed quantitatively as the percentage of the rubber originally present, i.e. a compound of 50 parts of black and 100 parts of rubber, with a bound rubber of 35 %, has 35 parts of rubber bound to 50 of black.

Highly reinforcing fillers acquire high percentages of bound rubber, coarse fillers form practically none. An ISAF black, compounded at 50 p.p.h.r. in SBR or NR, will have about 35 % bound rubber immediately after mixing, and this percentage will rise somewhat when the stock is heat-treated near vulcanising temperature (of course, without any vulcanising ingredients). The gel remaining after rubber extraction (usually with benzene) is extremely rich in solvent and highly fragile; it usually contains 20–30 times more solvent than rubber. Bound rubber can be considered as a measure for the surface activity of the black or white filler (Brennan and Jermyn, 1965).

The high swelling ratio—i.e. low v_r, where v_r is the volume fraction of rubber in the swollen gel (Section 7.9)—would indicate a very low crosslink density in the bound rubber; but work by Endter (1952, 1954) has shown the bound rubber to consist of an open network that might well occlude a considerable amount of solvent, so the actual swelling may be much less than the apparent swelling and the crosslink density may be higher than calculated from the apparent swelling. Cotten (1966), however, showed that, after destruction of the gel by mechanical milling, the bound rubber is still present in the form of small flocs that can be concentrated by ultracentrifuging. After this procedure, the total bound rubber volume is about the same as before the mechanical disruption of the carbon gel. The conclusion then is that bound rubber as such does not contribute appreciably to the crosslink density of the final vulcanisate (Boonstra and Dannenberg, 1958).

However, as mentioned in Section 7.4.3.1, bound rubber can be considered as a measure of surface activity and, because of this, the percentage of bound rubber may run parallel to rubber properties related to surface activity such as modulus, abrasion resistance, and hystersis (Section 7.5).

The formation of bound rubber is usually explained by assuming that mechanical breakdown of polymer chain molecules results in the appearance of free radicals at the newly formed chain ends. Reactive sites on the filler surface then combine with these free radicals to form the bound rubber. Since there are many sites on a filler particle, it can act as a giant crosslink. The amount of bound rubber first increases with milling and then goes down as the polymer (natural rubber) breakdown becomes the dominating factor. This effect for both carbon black and white fillers was shown by Watson (1954, 1955), and is illustrated in Figs. 7.13 and 7.14.

A somewhat different concept is proposed by Gessler (1967), who suggests that the reaction between polymer and carbon black occurs when carbon black aggregates are mechanically broken during mixing. The newly exposed fracture surfaces are so active that they react with either normal chains or chain ends activated by mechanical breakdown and form the insoluble carbon black gel. Although there are a number of arguments that speak for

K

this concept, it does not alter the conclusions on the influence of bound rubber on reinforcement. The amount of bound rubber on a filler with 100 m²/g at 50 p.p.h.r. loading may be about 30%. This would amount to a layer of about 6 nm surrounding each particle contributing to the surface area.

It has been suggested (Brennan and Jermyn, 1965) that this bound rubber should be considered as part of the volume occupied by the filler. This would

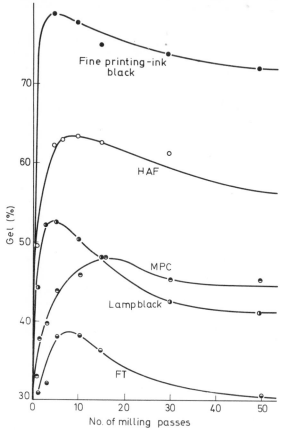

Fig. 7.13   Gel production on milling with different grades of carbon black. From Watson (1954)

influence the hydrodynamic properties, such as viscosity, according to the Einstein, Guth, and Gold equation:

$$\eta_f = \eta_u(1 + 2 \cdot 5c + 14 \cdot 1c^2) \tag{7.1}$$

In this equation η_f and η_u are the viscosities of the filled and unfilled compound, and c is the volume fraction of filler; it is obvious that the viscosity would be more than doubled in the case cited above, since 50 p.p.h.r. of black would constitute a volume fraction of 0·2, which makes the factor in parentheses equal to 2. If the 30% bound rubber is added to the filler fraction, c rises to 0·44 and the factor $1 + 2 \cdot 5c + 14 \cdot 1c^2$ rises to 4·9, so the viscosity becomes about $2\frac{1}{2}$ times as high as without the bound rubber.

It is well known that reinforcing blacks give a much greater increase of viscosity than expressed by the hydrodynamic formula (equation 7.1) using for c only the volume fraction of the filler. In some cases this increase is just about double that expected from the computation, and approximately of

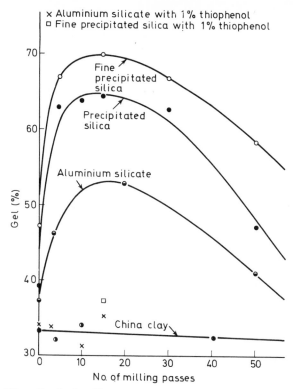

Fig. 7.14   Filler gel with white fillers. From Watson (1954)

the same order of magnitude as given by including bound rubber as part of the filler. However, this agreement does not exist, for instance, in the case of graphitisation of carbon black. This process reduces the bound rubber to almost nothing and has negligible effect on viscosity; actually, a higher viscosity may be found for the graphitised black compound owing to less breakdown of the rubber matrix during milling.

According to Guth (1948) and to Cohan (1948), a similar relationship (equation 7.2) is valid for Young's modulus of vulcanisates crosslinked to the same degree with various amounts of filler:

$$E_f = E_u(1 + 2 \cdot 5c + 14 \cdot 1c^2) \tag{7.2}$$

This has been found to hold even for the 300% modulus for spheres, but, for vermiculate particles, the formula

$$E_f = E_u(1 + 0 \cdot 67 fc + 1 \cdot 62 f^2 c^2)$$

gave better correlation with experiment. However, the exceptions are many, and the theoretical basis is not strong. The shape factor f is the ratio of length over diameter of the rod-shaped particles. For some reinforcing carbon blacks $f = 6$ gives good agreement with the experimental values (Bolt *et al.*, 1960).

7.9   Reinforcement and Crosslink Density

Comparison of filled vulcanisates with pure gum vulcanisates, of otherwise identical formulations and cured to their optimum, shows two important characteristic differences: the modulus at 300% strain is greatly increased, and the swelling of the elastomer in solvents is reduced. Both modulus and swelling are used to determine crosslink density. The modulus is related to crosslink density by the well-known formula of the kinetic theory of elasticity, in its simplest form:

$$\sigma = RTv\left(\lambda - \frac{1}{\lambda^2}\right) \tag{7.3}$$

in which v is the number of crosslinks per cubic centimetre (equal to half the number of active chains per cubic centimetre, since there are, to a first approximation, twice as many chains as there are crosslinks); λ is the extension ratio (at 100% elongation the extension ratio is 2, at 200% elongation $\lambda = 3$, etc.). According to equation 7.3, the modulus at a certain extension ratio and at a given temperature can only be increased by increasing v, the crosslink density. The addition of a reinforcing filler increases modulus, so its effect is the same as an increase in crosslink density.

An uncrosslinked rubber dissolves in a suitable solvent, but, if the rubber is held together by crosslinks between the molecular chains, it cannot dissolve; instead it swells to an extent determined by the solvent power of the liquid, which tends to extend the rubber gel, on the one hand, and the crosslinks, which hold the molecular chains in the gel together, on the other hand. Evidently, for a given solvent, the higher the crosslink density of the rubber the lower the swelling, and conversely, for a given degree of crosslink density, a more powerful solvent will give a higher degree of swelling. This relationship is quantitatively expressed by the Flory–Rehner equation:

$$v = \frac{1}{V_s}\left[\frac{\ln(1-v_r)+v_r+\chi v_r^2}{v_r^{1/3}-\frac{1}{2}v_r}\right] \tag{7.4}$$

which is used frequently to calculate v, the crosslink density, from swelling measurements.

In this equation, V_s is the molecular volume of the solvent, and v_r is the volume fraction of rubber in the swollen gel; χ is the interaction constant—for natural rubber usually of the order of 0·4 in good solvents, and determined by the cohesive energy density of solvent and polymer and the swollen gel.

If this equation is applied to rubbers containing reinforcing fillers, one finds that v_{rf} of the rubber phase in the swollen gel (corrected for the volume of filler, since the filler is assumed not to swell) is always much higher than for the pure gum v_{r0}, so the ratio v_{r0}/v_{rf} decreases with increase of filler loading. This ratio represents the degree of restriction of the swelling of the rubber matrix due to the presence of filler.

Fig. 7.15 shows that, as the volume fraction c of filler in the vulcanisate increases, the restriction of swelling increases; and, as the solvent power of the swelling medium (characterised by its v_{ro}) increases, the amount of increase is greater. Non-adhering fillers (glass beads) show an increase in v_{ro}/v_{rf} because of the pockets of solvent forming around the particles. Graphitised blacks constitute a completely unique class of fillers in that they neither

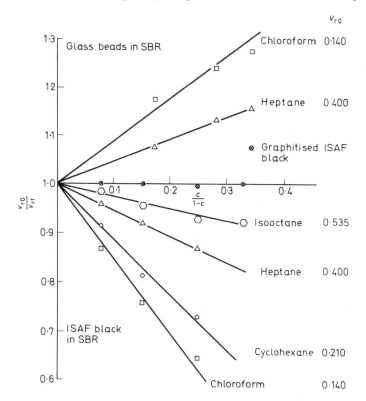

Fig. 7.15   *The relation between the volume fraction c of filler in a vulcanisate and the swelling, for glass beads and ISAF black in SBR*

restrict nor enhance but leave the swelling unaffected at any loading in any solvent.

The question now arises as to whether or not this restriction of swelling is due to additional crosslinks formed either with the filler surface or in the matrix, owing to a catalytic effect of the filler on the vulcanisation reaction. Only in the case of peroxide cures of natural rubber (Parks and Lorenz, 1961), has it been shown that the crosslink reaction and density in the rubber matrix are independent of the presence of fillers, such as carbon black. However, even in that case, considerable restriction of swelling, due to the presence of reinforcing fillers, occurs. Therefore the influence of reinforcing fillers on swelling points in the same direction as their influence on modulus, namely, they appear to increase crosslink density.

Additional crosslinks apparently introduced into the matrix by reinforcing fillers are different from chemical crosslinks which consist of sulphur bridges or covalent carbon–carbon bonds (Table 7.8).

It can therefore be concluded that the 'crosslinks' introduced by reinforcing fillers are of a mobile nature, allowing more creep and involving frictional energy dissipation.

This conclusion fits well with the concept of mobile adsorption (de Boer, 1953; Ross and Olivier, 1964), which is well established in adsorption

Table 7.8 EFFECTS OF CHEMICAL CROSSLINKS AND REINFORCING FILLERS

Property	Chemical crosslinks	Reinforcing fillers
Creep and cold flow	Reduced	Increased
Hysteresis and heat build-up	Reduced	Increased
Modulus at high temperature	Increased	Reduced
Abrasion resistance	Reduced	Increased

physics. The principle can best be demonstrated by a graph showing the energy of adsorption at various points along the surface of the solid material (Fig. 7.16).

Fig. 7.16(a) shows typical mobile adsorption: a molecule sitting at location A needs an energy E to remove it from the surface, but only the amount x of the small energy barrier to move it to location B. This barrier is of the same order as the thermal energy of the molecule. The adsorbed molecule can be a rubber segment. It implies that the adsorbed rubber segment can freely move over the surface by its own thermal motion and certainly when helped by stresses from the outside.

Fig. 7.16(b) the opposite of mobile adsorption (i.e. local adsorption on active sites) is depicted. A molecule located at A_1 needs an energy E_1 to remove it from the surface and, to move it from A_1 to B_1, that same high energy is needed since the molecule or rubber segment has to be removed completely before it can be readsorbed.

Fig. 7.16(c) illustrates a combination of the other two types; a more realistic picture of the situation that prevails in carbon black is shown in Fig. 7.17. In Fig. 7.17(a) the heterogeneity of the surface of a reinforcing carbon black is depicted in the form of energy troughs of varying depth (see also Fig. 7.4). In general, these troughs are caused by defects in the crystal lattice or, in the case of carbon black, also by end effects at the edges of the crystallographic plane layers.

After heat treatment, defects have been eliminated and the plane layers have grown together to a large extent; the result is illustrated in Fig. 7.17(b). No great energy is involved in moving a molecule or rubber segment from one place on the surface to another, but considerable energy is necessary to completely remove it from the surface. The swelling behaviour reflects these situations. Whereas the regular black restricts the swelling because the molecular chains are held at these sites of high energy, the graphitised black does not affect swelling at all. This is a very unusual behaviour since many white fillers, so-called non-adhering fillers, will form pockets filled with fluid around the particles; these pockets make the swelling seem to increase with loading, as was shown by the glass beads of Fig. 7.15. The graphitised black neither lets go of the polymer chains adsorbed on its surface, nor puts any restriction on its swelling. It can be shown that this must mean that the adsorbed rubber segments move freely along the surface with hardly any

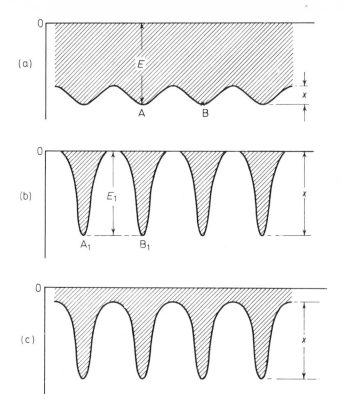

Fig. 7.16 Schematic representation of energy barriers to the translation of adsorbed molecules along the surface. From Ross and Olivier (1964), courtesy Wiley-Interscience

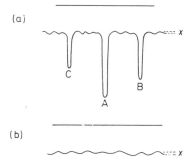

Fig. 7.17 Potential energy profiles of (a) a normal carbon black, and (b) a graphitised carbon black surface; points A, B, and C represent sites of high adsorptive energy in decreasing order. From Boonstra and Taylor (1965), courtesy Cabot Corpn

restriction on their mobility except that they (i.e. a definite average number of segments) stay adsorbed at the surface.

It is highly significant that the restriction of swelling by reinforcing blacks and some reinforcing white fillers occurs only in solvents of sufficiently high swelling power ($v_{r0} < 0.63$) (Boonstra and Taylor, 1965). In solvents with low swelling power ($v_{r0} > 0.63$), no restriction of swelling by these reinforcing fillers is found and they behave much the same as graphitised blacks in the more powerful swelling media. This would indicate that, with untreated reinforcing fillers, some limited lateral mobility of adsorbed segments is allowed, but larger movements are restricted probably by attachments at the active sites.

7.10   The Mechanism of Reinforcement and its Implications

The previous sections have been leading up to a picture of reinforcement that can now be presented in full. The strength of non-crystallising rubbers, such as SBR, NBR, and EPDM, is poor since in the test-piece there is unequal distribution of stress. The crosslinks are introduced at random at a temperature at which, owing to the movement of the molecular chains, the involved chain sections may not be in their most probable positions. The result is not only a random distribution of the chain lengths between crosslinks but also the presence of local microstresses in some of these.

As a consequence, when the sample is subjected to the tensile (or other) test, a number of chains, the shortest or most strained ones, will break early in the test, followed by the next most strained ones, until at the moment just before break only a few chains effectively carry the load. This is not only true for pure gum rubber vulcanisates but for all solid materials. All give experimentally a tensile strength lower, by a factor of 20–1000, than that computed by adding the strengths of all bonds running through a cross-section. If the chains which are most strained are given a chance of slipping to relieve the tensions caused by stretching, added to those built in already, they will not break prematurely in the earlier part of the strain experiment but will survive to the very moment before rupture. More chains then effectively carry the load, and a higher strength results.

A schematic picture of such a slippage process is drawn in Fig. 7.18, which shows three chains of different lengths between two carbon black particles in the direction of stress. As the stretching process proceeds from stage 1, the first chain slips at the points of connection A and A' until the second chain is also taut between B and B' (stage 2) and, on continuing elongation, starts to slip there so that finally a stage 3 is reached in which all three chains are stretched to their maximum and share the imposed load equally. The homogeneous stress distribution gives the high improvement in strength. In stage 4 the tension is relieved and the test-piece has retracted. The situation now is not the same as the original (stage 1) since, owing to the slippage, the three chains now have about equal lengths.

A repeated elongation will now give a lower modulus than the first cycle since the energy of slippage does not have to be furnished, as was the case originally. This explains the Mullins effect (stress softening). It also identifies part of the modulus value at the initial cycle to be of a viscous nature and the slippage process as one of energy dissipation; this energy would otherwise be used to break bonds in or between molecular chains and particles. From this picture it also becomes evident that high modulus is due to the sites of high (adsorptive) energy, and high strength is mainly due to the energy necessary for slippage. The last is also important for abrasion resistance, but Table 7.3 has shown that one needs points of attachment since the graphitised (inactivated) black had lost three-quarters of its abrasion resistance.

Support for this point of view is given by calculations made by Wake (1959), who computed the entropy changes which hydrocarbon gas molecules undergo when they become adsorbed on surfaces of carbon black or other fillers.

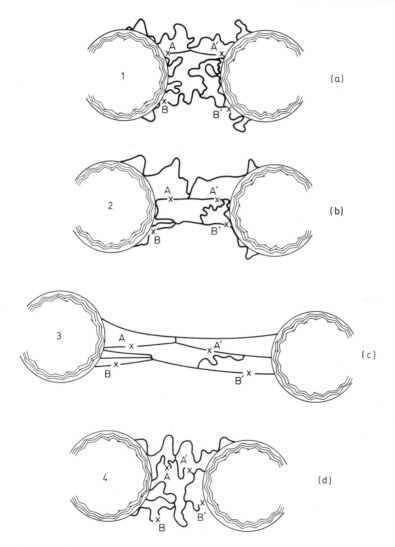

Fig. 7.18 Molecular slippage model of reinforcement mechanism: (a) original state; (b) intermediate elongation—shortest chain slipped beyond A and A'; (c) all chains fully stretched; (d) after retraction, all chains have equal lengths between particles

Wake found smaller changes in entropy for adsorption on carbon black than for ionic white fillers, indicating residual two-dimensional mobility of these molecules in the adsorbed state on the carbon black surface.

7.10.1   STRESS SOFTENING

Criticism of the above mechanism to explain stress softening (Mullins effect) comes from Mullins, Harwood, and Payne (1965, 1966), who showed that, if elongated to the same stress, pure gum natural rubber shows the same phenomenon, so the effect is not due to the filler but to the polymer. However, to obtain the same stress as a vulcanisate containing carbon black, natural rubber gum has to be subjected to such a high elongation that crystallisation occurs. The rubber crystallites act in the same way as reinforcing fillers, i.e. as stress homogenisers, so one can expect a similar stress softening as with reinforcing fillers (Dannenberg and Brennan, 1966). Of course, the picture is schematic and breakage of chains on deformation is not completely avoided. This is shown by work of Brennan, Dannenberg, and Rigbi (1967) and of Peremsky (1963). These authors found that the amount of stress softening expressed as the percentage of strain energy retained at repeated cycles is, generally speaking, a function of the energy input—which function is nearly the same for all rubber–filler combinations. The higher the energy input, the higher is the percentage energy lost in softening. Only in very precise experiments, made under highly standardised conditions, do differences between fillers, rubbers, and types of crosslinks become apparent.

Stress softening is a temporary effect: in the long run the stress-softened sample will recover and its modulus will approach the original value except for a percentage which is permanently lost. The value of the permanent energy loss depends again on the energy input and evidently represents the chains or bonds that are actually broken. As the level of the energy at rupture is approached, the permanently lost proportion becomes larger. This is shown in Fig. 7.19, where even after 24 hours recovery only 20% of the input energy is recovered.

Equally important for the effect is the amount and type of filler, as is

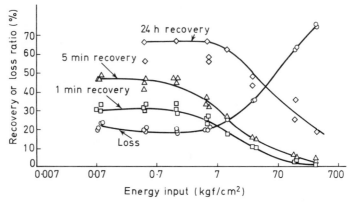

Fig. 7.19   Recovery ratio compared to loss ratio as a function of energy input at different intervals. From Brennan, Dannenberg, and Rigbi (1967)

illustrated in Fig. 7.20. This figure shows that higher energy losses are experienced at 50 p.p.h.r. loading of black than at 20 p.p.h.r., and that a peroxide cure gives lower losses than a sulphur–accelerator one. The last is understandable since peroxide cures give more permanent crosslink bonds

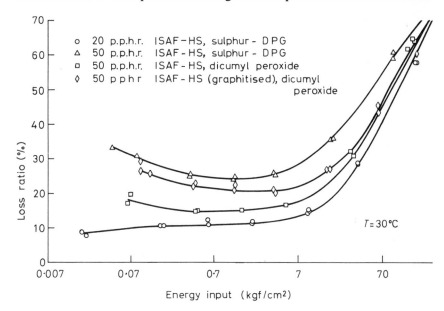

Fig. 7.20   *Loss ratio of vulcanisates at low and medium levels of energy input*

than the polysulphide links obtained by sulphur–accelerator combinations, and the stability of the crosslinks plays a part in the stress softening effect.

Fig. 7.20 also shows that the graphitised black causes more stress softening than its non-graphitised counterpart. Again this is expected since graphitised black no longer has the sites of high adsorptive capacity on its surface which restrain the adsorbent molecules from easy slipping. All these and other similar factors play a part in stress softening. Its importance for reinforcement, particularly for resistance to wear of actual tyres, is based on the following consideration.

7.10.2   EFFECT OF ABRASION

The abrasion of an automobile tyre is a cyclic process. After each contact with the road, the tread segment is at rest for the remainder of its travel, tracing a cycloid curve until it hits the road surface again. During the contact with the road, abrasion and stressing occurred and part of the energy was dissipated in the form of the otherwise unwanted heat. After this process, recovery must take place to such a degree that, at the next contact, energy dissipation can occur again. The rate of recovery is dependent on the (number of) persistent bonds in the elastomer network and between network and filler surface; evidently the presence of many strong bonds runs against the large energy dissipation concept which is at the basis of high abrasion resistance.

It is impossible to have both energy dissipation and recovery at their maximum since one impairs the other. Therefore, a compromise must be found between these two opposing factors, and it depends on the conditions of the road test which is the more important. It could be stated that, for use at higher speeds, which leave a shorter time cycle for recovery, more strong bonds between filler surface and polymer are necessary at the expense of energy dissipation. At higher speeds, tyre wear increases rapidly. Under these conditions, a filler with a more active surface is wanted. In many cases, more active surfaces are present in carbon blacks having higher structure. The reason for the higher activity may be the breakdown of carbon black structure, as proposed by Gessler, which would generate fresh and active surfaces *in situ*. Alternatively, it may be caused by incidental manufacturing processes. The fact remains that, under high-severity conditions, treads containing high structure black are vastly superior to those with normal structure blacks, whereas, under mild conditions, differences are hardly noticeable.

7.10.3   FILLER REINFORCEMENT IN POLYMER BLENDS

All previous general considerations for reinforcement of single polymers hold as well for blends of rubbers. However, there are some special effects that are characteristic of these blends. Basically, these effects are caused by the fact that most polymers are not miscible and blends usually consist of microdispersions of one in the other or of intermingled microregions often having dimensions around $0.1-1.5$ μm. When fillers are mixed into such a blend, a situation may develop in which the filler is unevenly distributed between the two phases, and this distribution affects the compound properties.

An example of such behaviour is found in the blend of natural rubber and polybutadiene rubber as described by Hess, Scott, and Callan (1967). These authors showed that the best mechanical properties were obtained when the polybutadiene contained the larger proportion of reinforcing carbon black as determined by electron microscopic techniques. It also appeared that, if black was mixed in to a 50–50 blend of natural rubber and polybutadiene rubber, the black was preferentially incorporated in the natural rubber.

Both phenomena can be easily understood; during the mixing, the soft polymer will first penetrate the voids between black particles, leading to a higher concentration of black in the softer rubber. As the mixing continues, however the viscosity of the softer rubber increases because of its higher filler content, after which the second polymer shares in the black uptake until the viscosity of the two phases is about the same. The other phenomenon, i.e. better properties being given by a product with more black in the butadiene phase, is due to the fact that polybutadiene as a gum vulcanisate is weak and needs carbon black for reinforcement, whereas natural rubber by itself is already strong and has built-in reinforcement because it crystallises and its crystallites fulfil a function similar to a reinforcing filler.

In most blends, the effects are not so pronounced as cited here (Marsh and Voet, 1968). In addition to filler distribution, the degree of crosslinking must be considered. Since one component will not cure at the same rate as the other in identical formulations, one phase may be stiffer than the other one.

Theoretical concepts of reinforcement are well developed and are in agreement with many of the experimental results. There are still a number of unexplained phenomena and unanswered questions which will occupy the thoughts of scientists in the years to come. New and unexpected developments in the field of reinforcement may well appear.

EIGHT

Processing Technology

BY B. G. CROWTHER AND H. M. EDMONDSON

8.1 Introduction

Processing is a general term which includes all the operations which are carried out on the rubber and which alter its physical shape or chemical composition.

The raw polymer, natural or synthetic, can be *softened* either by mechanical work, termed *mastication*, by heat, or by chemicals known as *peptisers*. The increase of plasticity or decrease of viscosity brought about by mastication and peptisers is permanent; by heat it may be permanent or temporary, depending on the nature of the polymer.

Details of the wide variety of materials which are added to the polymer to give it the desired processing and vulcanisate properties are given in Chapter 6. When the rubber contains all the ingredients needed it is known as a *compound* or, preferably, a *mix* (BS 3558). If some ingredients have been withheld deliberately, the partially completed compound becomes a *masterbatch*. The masterbatch is converted to the compound by the addition of the withheld ingredients, which are usually the *vulcanising* or *curing* ingredients.

8.2 Mastication and Mixing

8.2.1 MACHINERY

A useful survey of mixing machinery was published in 1963 (Anon, 1963).

8.2.1.1 *Mills*

Just over a century ago, when the processing of rubber was first being undertaken commercially, two-roll mills were used. The roll axes were horizontal and parallel, and the distance between the rolls was made adjustable by having the bearing blocks of the front roll resting against adjusting screws. These mills remain substantially unchanged to the present day, the main differences being only refinements of the original design.

Double Geared Mills. Where both front and back roll are driven from a common backshaft, the mills are known as double geared. Early mills were nearly always double geared, and the backshaft was originally driven from a steam engine. The mills were turned on or off by engaging or disengaging a dog clutch connecting to the backshaft.

In later years the steam engine was replaced by an electric motor and reduction gear, and the number of mills driven from a single motor has gradually reduced until single mills, with their independent motor and drive, are quite common. The more stringent safety regulations together with advances in engineering design have led to a reduction in the number of mills operating from a single drive.

Single Geared Mills. Single geared mills have developed from the double geared mills as they are cheaper to manufacture. In the single geared mill, the back roll is driven from the gearbox or backshaft, and the front roll from the back roll through roll-end gears. The gears will come out of mesh on a single geared mill if the nip, i.e. the distance between the rolls, is set too wide; furthermore, there is always a danger of stripping the teeth if the nip is opened so that the gears are working on the tips of the teeth. The maximum nip on a double geared mill is greater than that for a single geared mill; the latter is about 12 mm and is usually adequate.

The present tendency is to mount a single geared mill, complete with motor and drive, on a sub-frame. The sub-frame can then be stood on a firm level concrete floor without the need for the elaborate foundations which are otherwise necessary. The imperfections in the floors, etc., are usually taken up by fitting rubber–metal bonded insulators between the sub-frame and the floor.

Friction Ratio. The speeds of the two rolls are often different, the friction ratio depending upon the mill's use. For natural rubber mixing a ratio of $1:1.25$ for the front-to-back roll is common. High friction ratios are used for refining compounds, and even-speed rolls on feed mills to calenders, i.e. when mills are being used other than for mixing. For mixing some of the synthetic rubbers, a near-even speed is best, or even an inverse friction ratio, i.e. less than 1.0.

Other synthetic rubbers are very difficult to mix on mills, so internal mixers are often used for these.

Cooling. Two alternative types of cooling are used: the principal one employs *cored rolls*, i.e. water is sprayed on to the outside of an axially drilled central core; with *peripherally drilled rolls*, the water is circulated through a labyrinth of passages about 50 mm under the roll surface.

Safety. Mills are usually fitted with various safety devices, some to protect the operator, the others to protect the mill.

The operative is protected by the Lunn safety bar, now fitted to all mills in the UK. This is a breast-high bar which will stop the mill if pressure is applied. The Lunn bar is fitted in such a position that an operative cannot get near the mill nip, but if, by some mischance, he should be caught up and dragged towards the nip, then pressure on the Lunn bar will stop the mill.

Maximum stopping distances of mills are established, and braking of the motor is normal practice, either electrically to the motor or by means of spring-loaded brake shoes which operate as soon as power is cut from the mill.

The mill is protected by fitting an overload cut-out on the drive motor and by fitting cast-iron 'breaker plates' between the roll nip adjusting screws and the roll-end bearing housing. A recent development is to replace the cast-iron breaker plates by a hydraulic relief system which allows the nip to open in the event of overload, whether a 'sustained' or a 'shock' load.

Ancillary Attachments. Mills are fitted with a metal tray under the rolls to collect droppings from the mill, and with guides or cheeks, which are plates fitted to the ends of rolls to prevent rubber being contaminated with grease, etc.

Other attachments which may be added, depending upon the use of the mill, are:

1. A mill apron, which replaces the mill tray and consists of a belt running under the nip rolls up to the top of the back roll; the belt returns to the mill any powders which fall from the nip.
2. A stockblender, which is a device for blending the rubber mechanically instead of depending upon the operator.
3. Cutting knives mounted against the roll to produce strips for feeding other equipment or for removing compound from the mill.
4. Scraper blades to remove soft stocks.

8.2.1.2 *Internal Mixers*

The internal mixer, because of its versatility, rapid mixing, and large through-put, is tending to replace the mill as a batch mixer (Edmondson, 1969).

An internal mixer consists of two horizontal rotors with nogs or protrusions, encased by a jacket. The Bridge 'Banbury mixer' has a friction ratio between the rotors, and the work required to incorporate the powders is carried out between the rotors and the jacket. In the Francis Shaw 'Intermix', however, the rotors run at even speed, and the nogs are designed to produce a friction ratio between the rotors. The work in this machine is done between the rotors, rather than between rotors and jacket.

Both machines are fitted with a pneumatically operated ram to ensure that the rubbers and powders are in contact. Fig. 8.1 shows a line drawing of the Francis Shaw Intermix, and Fig. 8.2 a line drawing of a Banbury mixer.

Machine Sizes. A range of sizes of machine is made (Table 8.1). The data given in the following discussion relate to the most popular size of machine in the UK, namely one with a batch load of about 200 kg of compound.

Rotor Speeds. Internal mixers of the 200 kg size can be obtained with rotor speeds in the range of about 20–66 rev/min as manufacturers' standard. Table 8.1 gives the details of rotor speeds for each machine size. By the use of two-speed electric motors, rotors speeds can be obtained of either 20 or

40 rev/min, or alternatively 30 or 60 rev/min, i.e. a 2:1 speed difference. It is possible, but more expensive, to obtain other speed combinations by the use of two drive motors instead of one, both feeding into a common gearbox; in some instances, variable-speed motors are used.

If it is considered that, to carry out a mixing, a certain number of rotor revolutions are needed, then the mixing time in the machine is directly proportional to rotor speed. The loading and discharge time will normally be approximately a constant for any given machine, so it can be deduced that:

Output of mixer = (loading plus discharge time) + (constant × rotor speed)

Ram Thrust. The thrust applied to the ram affects the output of an internal mixer in three ways.

The normal range of ram thrusts on a 200 litre machine is from 1·5 Mg to 30 Mg; increases in batch weights can be obtained by increasing the ram thrust up to a certain level to reduce the voids in the machine. Once the

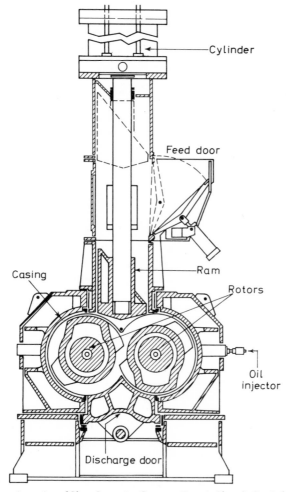

Fig. 8.1 Diagrammatic section of Shaw Intermix. Courtesy Francis Shaw & Co. Ltd

voids are full, increasing ram pressure will not enable more material to be forced into the machine. Secondly, increasing the ram thrust increases the speed of engagement of materials in the mixer by the rotors, and hence reduces the mixing cycle; and thirdly, it increases the contact between the materials in the machine and so gives more rapid ingredient absorption.

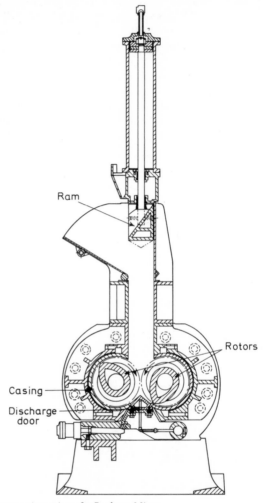

Ram

Rotors

Casing

Discharge door

Fig. 8.2   Diagrammatic section of a Banbury Mixer

This better contact of materials reduces the tendency of the batch to 'crumb' and gives greater reproducibility in mixing.

The effect of a ram thrust on output is not as simple to define as the effect of rotor speeds, but it has been noted that a seven- or eight-fold increase in thrust, from, say, 2 Mg to 15 Mg has produced a 30% increase in output. There must, however, be a maximum thrust beyond which any further increase results in no significant increase in output since the voids in the machine have been filled and maximum contact has been achieved between the rotors and the materials, and between the materials themselves.

Power Input. Checks have indicated very little difference in energy absorption per unit weight of compound mixed, whether mixing is done in a high-speed or low-speed internal mixer, in a small or large mixer, or even on a mill. It follows, therefore, that the greater the compound output arising from increases in rotor speed and ram thrust, the greater must be the energy input into the machine in the same unit of time. Whereas, with low ram thrust machines, of about 200 litre capacity, power inputs as low as 7·5 kW per

Table 8.1 CAPACITIES OF INTERNAL MIXERS

	Bridge Banbury				Shaw Intermix				
	3	*9*	*11*	*27*	*K4*	*K5*	*K6*	*K7*	*K10*
Capacity of compound at specific gravity 1 (kg)	47	130	170	420	45	70	90	165	460
Standard rotor speeds (rev/min)	35–100	21–60	20–60	32	22–66	22–66	22–66	22–66	44–66
Motor power (kW)	150–450	190–450	300–1100	1100–1900	110–330	150–450	240–700	375–1100	1900–3000
Machine weight without motor (Mg, approx.)	17	30	48	110	23	34	57	70	90

revolution per minute of rotor speed are possible, with the higher ram thrusts, a power demand of 15–22 kW is normal. Even higher power inputs per revolution per minute of rotor speed are not unknown.

Cooling Arrangements. The principal limitations to power input are the strength of the internal mixer and the rate of cooling which can be achieved. The energy put into the machine has to be dissipated into the compound, the surroundings, or the cooling water, mainly in the form of heat. The greater the energy input, the more important it becomes to remove excess heat, and the more difficult are the design problems to achieve this. It is usually the problem of heat removal which provides the limit to energy input.

At the outgoing end of the internal mixer, other handling equipment is needed, such as mills, a tuber slitter, or roller die, for converting the compound to sheet form.

The sheeted compound must then be cooled as quickly as possible by, for example, forced air, water, or antitack solution, to prevent the compound from scorching or sticking to itself.

8.2.2 MASTICATION

The mastication stage in processing is, in general, only applied to natural rubber, and the internal mixer is an efficient machine for this. The quicker energy can be put into the rubber, the shorter the masticating cycle; the limitations to shortening the mastication cycle are the time to load the machine, and the handling of the rubber after discharge from the mixer.

The shape of the rubber being loaded and the bulk density of the rubber affect the loading of the machine. The modern 32 kg bale of natural rubber is ideal for loading whole into the type of mixer referred to, or alternatively rubber cut from larger bales to the same shape. Wedge-shaped pieces of rubber as obtained from a star-bladed bale cutter are totally unsuitable, as these become fast in the throat of the machine.

The low bulk density of certain crepes means that, to obtain similar batch weights, larger volumes must be loaded into the machine than for other higher-density rubbers. These low-density rubbers can take longer to load than to masticate, and hence are at an economic disadvantage in processing.

The handling and storage of masticated rubber in sheet form is not very satisfactory unless the viscosity and dimensional stability of the rubbers can be assured. Mastication costs money—it reduces the viscosity, but at the same time results in a reduction in many physical properties of the finished vulcanisate. If the viscosity reduction is obtained by using softeners, then a saving in compound volume cost results. The reduction in physical properties of the vulcanisate is probably no greater from using softeners than from extra or separate mastication of the polymer.

With the availability of constant-viscosity and low-viscosity natural rubber (Section 4.1.5.1) and of synthetic rubbers of suitable viscosity for direct mixing, the need for mastication may well be eliminated in due course.

When the bulk of mixing is done on open mills, rubber is masticated to improve the rate of addition of powders. Owing to the increased softening obtained by cooling and remilling, a separate mastication process is advantageous, especially if this is done in an internal mixer or Gordon plasticator. Incorporation of large quantities of process oils can be time consuming on a mill, so mill mixing cycles are generally shorter if small quantities of oil are used and if the rubber is adequately masticated. Modern internal mixers enable oil to be incorporated into rubber quickly so that the reverse applies and mastication is less important.

8.2.3   MILL MIXING

In operation, the ingoing side of the nip is at the top of the rolls and rubber is added to the nip. A band of rubber comes through the nip and is formed around the front roll. Varying degrees of difficulty arise in forming these bands, depending upon the polymer. Natural rubber, if not premasticated, is difficult to band initially, but, after a time, depending upon mill nip and temperatures, a band forms. This starts with many holes and, after continued working, the band becomes smooth. Powders, process oils, etc., are now added into the nip, and any which fall into the tray underneath the rolls are returned to the nip until they are absorbed by the rubber.

The compound is blended to homogenise the additives and give adequate distribution of the ingredients. Mills generally give good ingredient dispersion—this word, usually, but not always, referring to the microdistribution of ingredients.

Mills have a fairly high power input. A 1·5 m mill, i.e. with rolls 1·5 m long, will handle about 50 kg of rubber at specific gravity 1·0–1·2 and will require 45–75 kW. A 2·1 m mill will handle 100–150 kg of compound and needs a motor of 115–225 kW. The majority of electric power is converted to heat so, to prevent compounds vulcanising on the mill with the heat produced, the rolls are water cooled, as referred to above (Section 8.2.1.1).

8.2.4 INTERNAL MACHINE MIXING

With internal mixers having low rotor speeds and low ram thrust, a batch of compound can take from, say, 5 min to 10 min to mix. An error of $\frac{1}{2}$ min in dumping the batch on a 10 min cycle does not often prove overcritical, and an extra $\frac{1}{2}$ min does not result in a marked temperature increase in the compound when dumped. However, as mixing cycles are reduced, it becomes increasingly important to control the time because this same time error on a short cycle of, say, 2 min is far more important than on a longer cycle of 10 min.

Since one of the main problems with internal mixers is to achieve enough cooling, it is not considered necessary to pre-heat the machines for most rubbers. The full operating temperature, and consequently minimum mixing cycle, is achieved after about five consecutive batches starting from a cold machine under full cooling conditions.

On the short mixing cycles it becomes impractical to work to a fixed mixing time, because of the large change in rate of mixing with increasing mixing temperature of the mixer. Two methods other than time are used to determine the end of a mixing cycle: the integrated power input into the machine, and the temperature of the compound. It is found that the completeness of mixing correlates reasonably well with both the amount of power consumed and the batch temperature.

Experience has shown that time cycles established as correct when using a cold internal mixer result in compounds scorching as the mixer warms up, after perhaps three or four batches; on the other hand, if the mixing cycle is based on time fixed for the mixer when it is hot, then the first batch or two may be found to be incompletely mixed (Evans, 1969-4).

8.2.5 MASTERBATCHING

It is generally accepted that masterbatching improves the physical properties of those compounds where a high degree of carbon black dispersion must be achieved. Where a masterbatch is used, the proportion of ingredients in the masterbatch can be varied from that used in the compound, e.g. the masterbatch recipe may (a) be the same as the compound recipe, but without curatives; (b) contain only rubber, filler, and some softeners; or (c) be richer in filler than the final compound.

Each of these types of masterbatch is capable of improving black and filler dispersion in a tyre tread or other reinforced compound, but type (c) has been found less satisfactory than the other two. The problem is the dispersion in the hard filler masterbatch of the relatively soft polymers added in the next stage. If a high-speed high-pressure internal mixer is being used for masterbatching, it is inadvisable to incorporate accelerators in the master-batch, as the temperature of the rubber is likely to exceed that of their activation temperature.

For many years carbon black has been mixed into synthetic rubbers, such as SBR, at the latex stage, to form a black masterbatch. The masterbatches have poor black dispersion, and a milling operation is essential not only to add the remaining ingredients to obtain a compound, but to achieve adequate dispersion of the black. Since the power consumed in this operation is high, there is little economic or technical advantage to be achieved from these latex–black masterbatches; plant contamination with black dust is, however, avoided.

It has been found that an aqueous dispersion of carbon black, if mixed with droplets in water of a polymer solution under conditions of high agitation, can give good black dispersion (Burgess, Hirshfield, and Stokes, 1965; Scott and Eckert, 1966, 1967). The high degree of agitation is needed so that the polymer solution sweeps out the carbon black particles to form a polymer–black mixture without reagglomeration of the carbon black. This is known as HSMB (hydrosolution masterbatch). Curatives can be added to the masterbatch in the conventional manner, and the resulting compound has physical properties at least as good as, and generally technically superior to, those which can be obtained by mixing in an internal mixer. Whilst the economics of the process are not yet known, the originators claim that the HSMB is a viable proposition and plant is now being laid down for commercial manufacture. There is every possibility that this type of masterbatch can succeed where the latex black masterbatch has failed, owing to the definite technical advantage to be obtained with HSMB.

8.2.6   CONVERSION OF MASTERBATCHES

Masterbatches can be converted to compounds by the addition of curatives using either a mill, an internal mixer, or a continuous mixer. A 200 litre internal mixer can use 20 rev/min, 30 rev/min, or 40 rev/min for this operation, depending upon the cooling and associated handling facilities. The curing ingredients themselves may be masterbatched; there is considerable divergence of opinion in the industry as to which system gives the best distribution of ingredients.

8.2.7   CONTINUOUS MIXERS

Many rubbers are in the wrong physical form for metering; they need pelletising or dicing, then must be chalked to prevent sticking. For every 100 kg of rubber, weights of fillers up to 200 kg are frequently used, and even 400 kg may be needed on occasions, and these must be accurately mixed with perhaps grams of accelerators. The number of ingredients is seldom less than 10, and can rise to 20, and 30 ingredients in 'one mix' are not unknown.

The capital cost of such equipment is very high, and the savings in labour costs which can be made are relatively low. The problems of compound quality and the frequency of compound changes are detrimental to success in this area at the present time. Considerable progress is only likely to be made in those production units requiring a very large output of a very few compounds, e.g. tyres. Compare the data given in Table 8.2 for the Transfermix with those of Table 8.1 for internal mixers.

All continuous mixing operations face the same fundamental problem of how to weigh continuously the multiplicity of very variable weights of rubbers and their compounding ingredients, with the accuracy required.

Mention, however, must be made of the attempts since World War II to develop a continuous mixing machine to replace the batch mixing process

Table 8.2 CAPACITIES OF TRANSFERMIX (FROM COLD MASTERBATCH)

	Transfermix						
	R3·25	*R4·5*	*R6*	*R8*	*R10*	*R15*	*R21*
Output of compound (kg/h)	340	655	1 200	2 130	3 350	6 800	15 600
Screw speeds (rev/min)	74	53	40	30	24	16	11·5
Motor power (kW)	33	67	135	220	350	750	1 700
Machine weight (Mg, approx.)	3	6	8	17·5	28	66	120

dictated by the use of mills and internal mixers. Examples of the continuous mixers are the Double R Mixer from Francis Shaw in the late 1940s, the Continuous Mixer from Farrel Corpn in the early 1960s, and the Transfermix from the US Rubber Co. in the late 1960s (Anon, 1962; Cargal, 1966).

The Double R Mixer is based on a three-roll mill; after mixing, the compound is forced continuously through a die to form a strip or other finished section at one end of the machine (Fig. 8.3).

The Farrel Continuous Mixer is based on a twin-screw extruder, the screws lying side by side in a single barrel. The first stage is a warming stage for strip or pellets, followed by a small set of rotors for mixing. The rotors are at the leading end of a pair of screws which force the compound through the barrel and a die to form a profiled strip (Fig. 8.4).

The Transfermix comprises a single-screw working in a barrel (Parshall and Saulino, 1967). The base of the screw helix varies in diameter, while the outer diameter is constant. The change in volume of the screw is off set by a cut-away in the barrel. The volume of the cut-away in the screw plus the volume of the cut-away in the barrel give a constant volume through which the compound must pass. Different contours and pitches of screw are available for different applications, such as mixing compound or warming compound. The objective is to provide an amount of feedback as the rubber progresses from the feed to the discharge of the machine, which results in efficient blending (Fig. 8.5 and Table 8.2).

8.2.8 CURRENT DEVELOPMENTS IN MIXING TECHNOLOGY

Internal mixers do not normally produce materials in the right physical form either for storage or for subsequent handling. In many instances, the degree of black dispersion obtained is inadequate. Thus internal mixers are always followed by additional processing equipment; either one or more mills, a tuber slitter, a roller die, or a continuous blender–mixer.

Two mills, especially if one is fitted with a stockblender, assist in improving the dispersion of carbon black. They also enable compound changes to be effected quickly and easily without contaminating one batch by another. However, the amount of manual effort required is more than with other systems, and a limitation in throughput arises at around 7000 kg/h. A roller die or tuber slitter (Section 8.3.2), on the other hand, can handle much higher

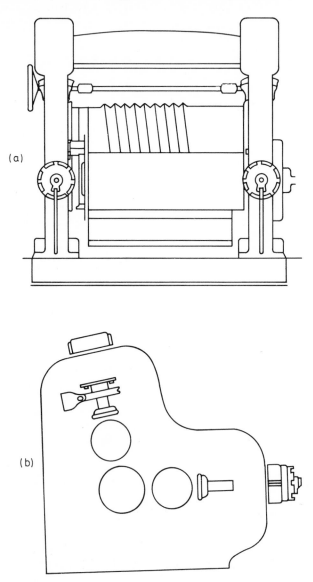

Fig. 8.3   *The Shaw Double R Mixer: (a) front elevation, and (b) side elevation. Courtesy Francis Shaw & Co. Ltd*

throughputs with much less manual effort, but both suffer from the disadvantage of not improving significantly the black dispersion in the compound.

The US Rubber Co., the originators of the Transfermix, claim that their machine gives the best of both, since it has the throughput of an extruder

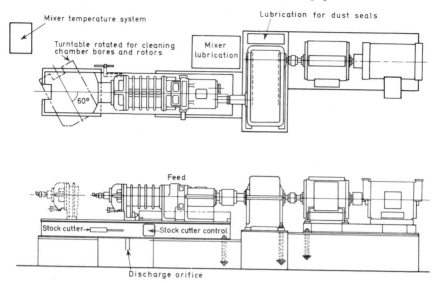

Fig. 8.4 *Bridge–Farrel Continuous Mixer*

and gives improvements in black dispersion comparable to that obtained using mills. Other factors involved are that the machines with the enclosed feed, i.e. machines other than the mills, tend to be cleaner in operation. In fact, continuous mixers of the Transfermix type may find their place as blenders, replacing mills and improving the dispersion of internal mixer output. Some of the larger companies consider that they will not purchase further mills. This seems a probable development as mills are no longer necessary for mixing operations. Cold feed extruders make them unnecessary for many production operations, and it is likely that calenders of the future will be strip fed from either cold feed extruders, a Transfermix, or some similar type of equipment.

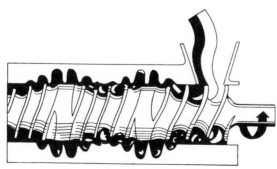

Fig. 8.5 *Diagrammatic sketch of Vickers Transfermix. Courtesy Vickers Ltd*

Mechanisation of weighing and feeding operations is proceeding slowly. The factories with high throughputs of similar compounds, such as tyre manufacturers, are in the lead in mechanisation, whereas those with highly specialised products requiring intensive labour still use mills (Nye, 1967; Jacobs, 1969; Anon, 1969-4).

8.3   Extruding

8.3.1   MACHINERY

Extruders are machines which force rubber through a nozzle or die to give a profiled strip of material. They fall into two types: those in which the pressure is produced by a ram, and those in which the pressure is produced by a screw. The latter is the type of machine most generally used in the industry and is known as an extruder, forcing machine, or tuber, whereas the ram extruder is a more specialised machine for short runs.

8.3.1.1   *Screw Extruders*

A screw extruder comprises a feed hopper, a feed screw operating within a barrel, a head, and a die or a die plate (Fig. 8.6).

The screw is rotated from an electric motor through a reduction gear, and pushes compound through the barrel into the head where it builds up a

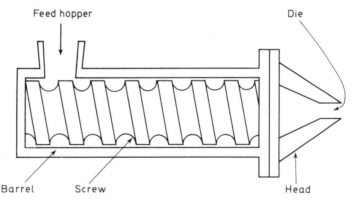

Fig. 8.6   *Section of a screw extruder*

pressure, this being relieved by allowing the compound to pass through an orifice or die to form the desired shape.

The purpose of the feed hopper is to receive the compound and pass it down into the flights of the screw. The compound may be supplied hot, in the form of an intermittent or continuous feed strip, as 'dollies', or as the discharge of an internal mixer, or it may be supplied cold in the form of a strip or pellets.

The feed hopper may be undercut to assist the feed, or may have a driven roll adjacent and parallel to the screw to give a 'roller feed'. In the case of the internal mixer discharge, a power-operated ram will be needed to push the rubber compound into the flights of the screw. The screw-type extruder is meant to be a continuous operating machine, so most systems are satisfactory with the exception of hand-feeding from dollies or strips.

The screw should preferably have a lower volume in the flights at the outgoing than at the ingoing end. This can be achieved by (a) a reduction in pitch of the screw; (b) a reduction in the depth of the base of the screw; (c) a reduction in overall diameter of the screw and barrel; or (d) an increase in the number of starts in the screw. Methods (b) and (d) are the most commonly used, the objectives being to give sufficient compression to eliminate air and also to ensure a constant pressure in the head. It is most important that an extruder screw is full at the discharge end, otherwise changes in swell of the compound will arise from surges of compound arriving in the head, resulting in dimensional changes in the extrudate which often seem to occur at the same frequency as the speed of rotation of the screw. Using hot compound, the length–diameter ratio of the screw is usually 4:1 or 5:1 whereas, with a cold feed extruder, this ratio is increased considerably, to 15:1 or 20:1 depending upon the type of compound to be extruded. The barrel of an extruder is usually of a hardened steel and is controlled so that a constant compound temperature is maintained in the extruder head. The screw rotates within the barrel and has a clearance of approximately 0·4 mm.

If the feed strip to an extruder is gradually increased, a point is reached at which the extruder chokes and will not take all the compound, which begins to back up at the feed hopper. There is an upper limit to the rate of feed for a particular extruder speed, and usually the best conditions are obtained when compound is fed at about 90% of the amount to choke the machine. The choking point increases and decreases as the speed of the screw of the extruder is increased and decreased, respectively, so each compound and screw speed demands its own rate of feed.

The purposes of the head are to equalise the pressure from the screw and barrel and to transport the compound to the die. Again, temperature control is vital. Of all parts of the extruder, the design of the head is one of the most important, and probably receives the least attention. Compound must move smoothly to the die, and ideally at equal pressures and speeds. Any points within the head where compound does not move are known as 'dead spots'. Compound vulcanises in these dead spots, then portions break away from time to time to give bits or 'nibs' of scorched compound in the extrudate.

The last stage is the die, which forms the compound into the desired shape. It should be pointed out, however, that compounds shrink along their length, so must increase in thickness and perhaps width, depending upon (a) the shape of the head and the extrudate, (b) the pressures in the head, (c) the head and compound temperatures, and (d) the compound rheological characteristics.

Extruders can be fitted with various attachments, e.g. a T-head or crosshead for mandrel covering (Section 10.3.3.1). If a wire gauze is included in the head to remove nibs or other particles of foreign matter, the machines become known as 'strainers'.

A vacuum device can also be fitted in the barrel of an extruder to remove any trace of entrapped air or other volatile matter from the compound. These vacuum extruders are useful in making articles for open steam, hot air, molten salt, and fluidised bath cures. The absence of the final traces of entrapped air and volatile matter eliminates any chance of porosity or blowing arising from this cause.

8.3.1.2 *Ram Extruders*

In the ram extruder, a quantity of warm compound is placed into the cylinder, the die head is attached to the cylinder, and the ram then pushes the compound through the die to form a profiled section.

This type of machine is intermittent in operation, and its operating costs tend to be higher than those of the screw-type extruder. The main virtue, however, is an ability to extrude compound at a lower temperature and/or speed, which can result in marked technical advantages in controlling the conditions for extruding difficult compounds. It is easier to clean and is applicable to short runs. Furthermore, it finds particular application when compounds need to be strained through gauze for high-quality products requiring completely contamination-free material. Other advantages lie in the much reduced heat build-up compared with the screw extruder, and its ability to extrude compounds with poor flow properties.

8.3.2 EXTRUSION TECHNOLOGY

The variability to be expected in the dimensions of the extrudate is at a minimum when the compound has the minimum entrapped stress. The stress can be relieved by means of a small loop immediately following the extruder die and prior to the take-away conveyor, or by means of a system of controlled shrinkage. As the depth of the loop increases, so does the variability of the section, since a large loop introduces stress. The greater the loop, the greater the stress from the compound weight that has to be supported.

Dies should always be designed to operate under conditions of minimum stress and at predetermined running speeds and temperatures, and the extrudates should always be produced under these conditions. Small adjustments in weight per unit length to accommodate minor compound variability can be made by subsequent slight stretching or compression of the section, but here again, the greater the stress in the compound after extruding, the greater the variability in the dimensions of the extrudate.

A different die will probably be needed for each compound, because of differing viscoelastic properties. The design of dies is an art, and normally requires a number of attempts to obtain the correct section with any given compound. Sections can be varied by means of different angles or 'lead-in' at the back of the die, as well as by changing the aperture of the die. Whether the die is a thick or thin one also affects the section obtained.

Whilst attempts are often made to baffle heads to give equal pressures and speeds at all points, this is rarely successful. Die design problems are more complex than this, as different swells are normally obtained at different parts of the die and vary also with the overall die aperture. It is normal practice for the die to be the hottest part of the machine, the temperature being progressively increased to this point from the feed.

Much can be done in compound design to improve the extrudability of a compound. In general terms, as discussed in Section 9.3, the more plastic the compound, and the less its elasticity, the better is the quality of finish of the extrudate. The lower the viscosity of the compound, the greater is the throughput to be expected in unit time.

Sheets of compound can be obtained from extruders as well as from calenders, one technique being to extrude a tube, cut it along its length, and open this up to form a flat sheet. Any slight eccentricity in the tube can, however, result in sheets with humps and hollows. Whilst this system has been and is still used, a type of machine known as a roller die or roller head die has been developed (B. F. Goodrich Co., 1933; Farrel Corpn, 1965; Thomson and McAlpine, 1969). In this case, the head is used to distribute the compound from the barrel to a pair of calender rolls. The roller die possesses the better features of both a calender and an extruder, namely, the high throughput and freedom from air entrapment of the extruder, coupled with the precision of gauge control and the speed of thickness changes associated with a calender. The range of thickness to be expected from such a machine is the same as that to be expected from a calender or extruder.

The two most important items in the design of a roller die are to ensure that it works at the minimum pressures, which must be kept constant as the compound leaves the head, and that the section presented to the calender is as near as possible to the ultimate shape required.

The methods of handling and vulcanising extrusions are described in Sections 8.9.1.1, 8.9.2.3, and 8.9.2.4.

8.4 Calendering

8.4.1 MACHINERY

8.4.1.1 *Types*

A calender comprises a number of rolls or bowls held in a framework. The rolls rotate to produce sheeting and, by adjusting the distance apart of the rolls, different gauges of sheeting become possible.

Whereas in industries other than the rubber industry, calenders with a multiplicity of rolls are common, the rubber industry normally uses three or four rolls, and occasionally, for rough-gauge sheeting, a two-roll calender.

Whilst two-roll calenders are normally vertical and three- and four-roll calenders used to follow the same pattern, changes are now taking place.

Modern three-roll calenders may have an offset top roll instead of a vertical configuration. The offset top roll assists the feeding of the calender from a feed strip, whereas, with older calenders, pigs or dollies were used for feeding when a vertical arrangement was preferred. Four-roll calenders can have a vertical, an inverted L, or on the most modern calenders a Z configuration of the rolls.

The vertical calender was again more easily fed manually, but a major disadvantage is that alterations to, say, the second nip down would alter the first nip also. The inverted L and the Z configuration are used principally for coating fabrics on both sides at once, with sheeting prepared in the first and third nips. Not only does the altered configuration improve the feeding of the

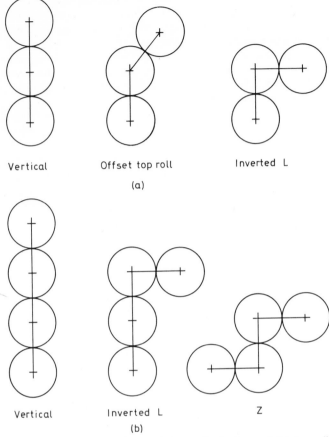

Vertical   Offset top roll   Inverted L

(a)

Vertical   Inverted L   Z

(b)

Fig. 8.7   Roll configuration of calenders: (a) three-roll calenders, and (b) four-roll calenders

calender, but it enables either the first or third nip to be altered without the second nip being affected. Sketches of these roll configurations are given in Fig. 8.7.

The loads involved in processing rubber are high, so the rolls, bearings, and supporting framework are of necessity heavy and robust in construction.

8.4.1.2   *Temperature Control*

Calender rolls are usually massive pieces of chilled cast-iron. They have sufficient thermal capacity so that, with a fixed flow of water, at some pre-set temperature, passing through an axial hole in the bowl centre, conditions can be reproduced. In any case, any temperature changes are slow, so a skilled leading hand on a calender can, from experience, maintain the temperature conditions about right by the alternate use of hot, cold, or no circulating water. Sometimes, steam is used instead of hot water.

More recent calenders use rolls with axial drillings about 50 mm under the surface, through which water at a pre-set temperature is continuously circulated. The rolls with the axial hole through the centre are known as 'cored rolls', and the latter as 'peripherally drilled rolls'. Peripherally drilled rolls without controlled-temperature water going through them are unsatisfactory in operation, since hot and cold water used alternately increase and decrease the roll temperature too rapidly. Peripherally drilled rolls are normally heated and cooled at a speed of 1°C/min. In other respects, peripherally drilled rolls give much better control than the cored rolls.

8.4.1.3 *Roll Cambering*

The calender rolls are usually cambered and are not parallel to compensate for variation in thickness across the sheet, as discussed below. Whilst this is an ideal solution for a calender which produces one gauge of sheet from one compound, a change in either of these parameters gives a departure from the desired crown (Farrel Birmingham, 1953).

Whilst calenders can be recambered in a few hours to accommodate a 'permanent' compound change or to take up wear of the rolls, if more than one compound is processed then some other device for resetting the crown will be needed. If the amount of change of crown needed is small, e.g. on a tyre-fabric processing calender, then a roll-bending device can be installed. This comprises an extra set of bearings outside the normal bearings, to which a large force can be applied hydraulically to produce roll bending. However, for compensation of more than about 0·075 mm, depending upon roll width and diameter, it is necessary to apply an axis-crossing device. This will give sheeting thicker at the edges than at the centre by as much as 0·6 mm. Thus with cross-axis and the right initial camber, great flexibility of crown control is possible, so a large range of compounds can be run (Willshaw, 1950).

8.4.2 CALENDERING TECHNOLOGY

Rubber compounds behave as viscous non-Newtonian liquids. If a uniform gauge of sheeting is to be produced, then the viscosity of the compound must be constant. If rubber is not evenly distributed across the ingoing side of a calender nip, then, in those areas where an excess of compound occurs, sheeting of a thicker gauge will be obtained at the outgoing side of the nip, i.e. the finished sheet.

In order to achieve uniform viscosity, the temperature of both the compound and the calender must be controlled, as the viscosity of rubber compounds is affected very considerably by temperature. The temperature of the compound is normally controlled by running under conditions which do not change substantially from day to day. It is not unusual, however, for the different conditions operating in summer and winter to upset the calendering operations. Fig. 8.8 shows the threading arrangement for producing sheeting between the top calender rolls. The sheeting is run round the calender and on to a liner, i.e. an interleaving material such as fabric which prevents the compound sticking to itself. If a more precise control of gauge or a reduction in the number of blisters in the sheet is needed, then a second nip is introduced. The first nip then meters compound accurately to the second nip, which produces the compound of the required gauge. This is shown in Fig. 8.9.

When an unsupported sheet is taken from a calender nip, it shrinks along its length and increases in thickness and width. This results in rubber sheets

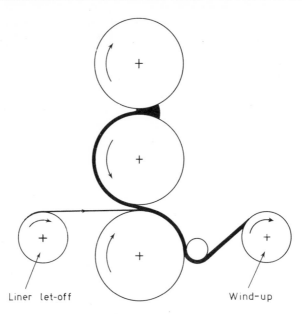

Liner let-off   Wind-up

Fig. 8.8   Threading of calender for sheeting (single nip)

having a crown, i.e. they are thicker in the centre than at the edges. Different compounds have different viscosities and elasticities even in the unvulcanised condition, so different crowns are obtained. The loads on the calender rolls amount to many megagrams, which in turn produce roll deflection resulting in some crowning. The proportion of the crown in the sheet due to roll deflection and the amount due to recovery and shrinkage depend upon the type of compound and also upon the gauge and subsequent handling. The thinner the sheet, the greater the load on the rolls. The greater the rubber content of the compound and the less the amount of milling given to it, then the greater the shrinkage to be expected, and hence the greater the crown. In order to obtain uniformity there are two methods which can be employed to overcome the problem of shrinkage. One is to chill the compound quickly, and restrain it by wrapping it tightly in a liner; the other is to allow the compound to shrink freely, or even to force shrink it, before wrapping it in a liner. Compound cooled with the strains in it will shrink lengthways when taken from the liner, especially so when heated. It depends therefore upon the ultimate use of the sheeting whether accurate gauge is obtained by cooling it and locking the strains in the material, or by allowing them to dissipate before wrapping up.

It can happen that the loads on the calender roll are approximately equal to the weight of the roll. Under these conditions, small changes in running or compound conditions can result in the upper of two rolls being a 'floating roll'. With shell bearings there must be a clearance. It is hardly likely to be less than 0·4 mm and can be 10 times as high as this in a calender requiring reconditioning. It is not possible to work to tolerances of 0·025–0·050 mm when the roll itself can be lifting under the compound pressure and falling under its own weight as the run proceeds. If a roll is floating, conditions such as temperature or running speed must be changed to maintain the roll at the top or bottom of its bearings. The other alternative is to overcome the problem mechanically by a pre-loading device which keeps the top nip roll in the up or down condition at all times.

Calenders for rubber compounds normally produce sheet of thickness in the range 0·1–1·5 mm. Both these dimensions, however, are very dependent upon the quality of sheeting required and the characteristics of the compound. Generally, the loads to produce thicknesses less than 0·1 mm become

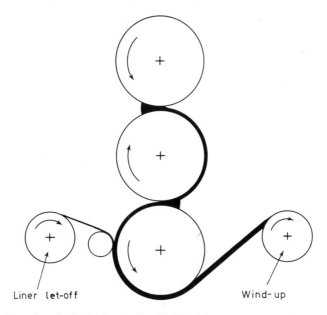

Liner let-off   Wind-up

Fig. 8.9   *Threading of calender for sheeting (double nip)*

so heavy that the normal design of rubber calender does not permit them, even with the rolls screwed tightly together before rubber is added to the nip. The upper limit arises from air entrapment in the form of 'pea blisters'. The incidence and size of blisters is reduced by reducing temperatures, but then it becomes very difficult to produce a calendered sheet without a roughness known as 'crow's feet'. Under these conditions, an alternative is to wrap the sheeting tightly in a liner in a warm state, preferably a textile liner which can 'breathe'. This causes the blisters to burst, and the holes so formed to fill in with compound over a period of time. Because, as the sheeting thickness increases, the blisters become more prevalent, thick sheeting is normally produced by 'plying up' two or more layers of compound to give material of the right overall thickness. No matter how the plying up is done, there is a potential weakness in the material and delamination sometimes occurs. Air traps between plies can also present problems. Plying up can be done as a separate operation between pairs of rolls, metal- or rubber-covered, or it can

L

be done with an extra calender roll, or with a cloth- or rubber-covered roll on the calender which applies pressure to the sheets.

8.4.3 UNSUPPORTED SHEETING

Two-, three-, or four-roll calenders are used for the production of sheeting containing no textile fabric reinforcement, different configurations of the rolls being possible.

For precision gauge control, a sheet is produced in a first nip, and this is then fed round a roll to a second nip. The second nip gives either less blistering or a thicker sheet for the same amount of blistering. The quality of sheeting is more dependent upon the quality of the feed than on any factor other than the temperature of the calender rolls. Occasionally, three nips are used for calendering, i.e. a four-roll calender is used, but one or two nips are more common. The increasing number of nips ensures an increasing quality of feed to the following nip, but it is rare to have more than two nips, the second producing the gauge, whilst the first meters a correct feed of compound to the second nip. Cross-axis or roll bending is needed on each nip if optimisation of gauge control is needed.

The compound produced may be calendered into a liner, or on to a belt (allowing some or no relaxation), or it can be taken on to cooling drums which set the compound prior to wrapping it in a liner. Most compounds are interleaved with a liner because their natural tack makes them inseparable if two wraps of compound touch each other. Sometimes they are powder dusted or are treated with a liquid antitack.

In producing a good-quality calendered sheet, the running speed is often a critical factor. Generally, the slower the running speed, the better the sheeting quality; the ultimate compromise usually becomes one of plant limitations or economics.

Whilst some compounds stick to the hotter roll and others to the cooler, there are particular compounds which will not adhere preferentially to either. Under these conditions, a change in the relative roll speeds may achieve what a change in temperature fails to achieve. For example, if there is a friction ratio of say $1.05:1$, the compound will adhere to one roll.

In addition to its use for producing unsupported sheeting, the calender is used for coating textile fabrics, and these processes are described in Sections 8.5.4 and 8.5.5.

8.5 Coating of Textile Fabrics

8.5.1 INTRODUCTION

Coating of textile fabrics is an extremely important aspect of technology. Fabric reinforcements of many kinds (Section 6.7) are used in a wide variety of products (Sections 10.1–10.4) performing a number of important functions (including structural load bearing).

The processing of fabrics has become increasingly complex with the advent of synthetic fibres, for it is necessary to treat them chemically in order to achieve a bond to the rubber and to obtain the full advantage of their reinforcement possibilities. The use of cotton, which was for many years the only suitable fabric available for rubber strengthening, presented little problem, owing to the inherent structure of the fibres providing a good mechanical key to the polymer mass.

The synthetic fibres were studied by rubber manufacturers initially as reinforcement for pneumatic tyres, and thus most of the basic research studies have been by this section of the industry. The first rayon tyre cords were produced in 1923. Du Pont started to manufacture high-tenacity rayon, Cordura, in 1933, and Courtaulds produced Tenasco in 1936. In 1947, nylon cords were first used. The next synthetic fibre to be used was polyester in 1962, and, most recently, glass fibre tyre cords have been used in a new range of tyres in N. America.

8.5.2 ADHESION OF RUBBER TO TEXTILE SUBSTRATES

Cotton presents little difficulty in securing adhesion, whether the rubber is applied from solvent dispersion or latex, or by calendering or frictioning. The degree of adhesion of rubber to the textile depends both on specific adhesion of the two materials and on the geometrical structure of the latter. The phenomenon is very complex; adhesion depends greatly on the detailed characteristics of the fibre, yarn, weave, and polymer used. Specific adhesion arises from intermolecular attraction between the polymer and fibre or fibre coating; mechanical adhesion is determined by the geometrical structure which permits interlocking through the fabric interstices of the rubber mass. The type of yarn and amount of twist affect the degree of interpenetration of the adhesive or rubber dip into the body of the yarn. Investigations by Borroff and Wake (1949, 1951) have shown that, in the absence of adhesive, the degree of adhesion is dependent on the number of fibre ends embedded in the rubber and is thus related to the surface area of the textile; the proportion of staple to continuous filament yarn in a fabric influences the degree of bond strength (Table 8.3), and correlation has been established between the adhesion and the number of fibre ends in the rubber pulled off the fabric. Because the method of application allows more even wetting and also greater interyarn penetration, the influence of fibre ends on adhesion is greater in the case of spread than of calendered fabrics.

Table 8.3 DIRECT PULL TESTS—COMPARATIVE VALUES

Rubber	Rayon			Acetate			Nylon		
	Continuous filament	20% staple	80% staple	Continuous filament	20% staple	80% staple	Continuous filament	20% staple	80% staple
Natural	88	130	165	79	117	139	75	143	240
Chloroprene	121	136	154	106	121	145	103	167	251
Nitrile	77	134	154	90	121	139	62	130	192

8.5.2.1   *Development of Adhesives*

The advent of continuous filament synthetic fibres presented problems owing to the absence of fibre ends and a lowered specific adhesion in some combinations of fibre and rubber. It was thus necessary to develop bonding agents to enable these materials to be used. The development of the bonding systems progressed as the different types of fibre became available.

One of the earliest systems for bonding rayon was based on resorcinol–formaldehyde resin and natural rubber latex, but the material suffered from a limited pot life. Butadiene–styrene and vinyl pyridine copolymer latices, in admixture with resorcinol–formaldehyde, followed. A later development was isocyanate adhesive. A modification of the latter was the formation of adducts, which were water soluble for easier application. With combinations of these systems, the range of fibres available can be pre-treated so that they will adhere to any rubber (Crocker, 1969; Takeyama and Matsui, 1969).

8.5.2.2   *Commercially Available Adhesives*

Table 8.4 lists some of the materials widely used in rubber-to-textile adhesive systems.

Table 8.4   MATERIALS FOR RUBBER-TO-TEXTILE ADHESIVES

Type	*Trade name*	*Supplier*	*Country*	*Special applications*
Synthetic latices	Gentac	General Tire & Rubber	USA	To make up RFL water-based dip for nylon, rayon, and polyesters
	Pyratex	US Rubber	USA	
	FRS 220	Firestone Tire & Rubber	USA	
Isocyanates	Desmodurs	Bayer	Germany	To make up RFL water-based dip; for nylon, rayon, and polyesters
	Hylenes	Du Pont	USA	
	Vulcabond	ICI	UK	
Modified PVC	TR 5	Canadian Industries	Canada	Used in polyester dip
Poly-epoxides	TRL-12	Du Pont	USA	Polyester dip
	Eponite 100	Shell	USA	
New materials	Pexul	ICI	UK	Polyester dip
	Pencolite Resin	Koppers	USA	To make up RFL water-based dip for nylon, rayon, and polyesters

8.5.2.3   *Recommended Treatments for Various Fibres*

Desized cotton requires only drying to achieve maximum adhesion. For rayon and nylon, water-based resorcinol–formaldehyde–latex (RFL) dips are generally used, the latex for rayon being SBR and for nylon SBR–VP (vinyl pyridine). For polyester, an isocyanate–rubber solution spread or dip followed by RFL–VP is necessary. Recently, aqueous systems based on PVC latex and a reactive polyamide, and other proprietary materials of undisclosed composition have been introduced (Chapman, 1966). Typical formulations are given in Table 8.5.

Aqueous dispersions are generally preferred to those based on solvents because of cost and the absence of fire hazard, an ever-present problem in this type of work. Solvent systems also require solvent extraction and recovery equipment (Section 8.5.3.3).

The use of the resin, apart from specific action, improves adhesion through mechanical action by spreading the stress over a greater surface area; cross-links can form between the rubber in the resin and the rubber coating, thus establishing the bond between the rubber and the fabric.

8.5.2.4 *Self-Bonding Rubber Mixes*

Considerable interest has been aroused by the development of rubbers, compounded with resorcinol, a formaldehyde donor, and silica, which will adhere to clean untreated rayon and nylon fabrics (Sections 6.6.5 and 9.5.3).

8.5.2.5 *Testing Fabric-to-Rubber Adhesion*

Conventional adhesion tests do not measure adhesion, but the resistance to stripping of the rubber–fabric interface. A direct adhesion test was devised by Borroff and Wake (1951); this was subsequently modified to the present test described by Izod and Meardon (1960).

The method consists of cementing the proofed sample of fabric between two wooden formers, which are then pulled apart at a fixed velocity. The load at break is recorded, and the elastomer area removed is measured. The adhesion is then expressed as the direct tension, in kilograms-force per square centimetre, needed to remove the elastomer. The test makes an ideal quality-control technique.

For further details of this and other tests for fabric-to-rubber adhesion, reference should be made to Section 11.3.19.

8.5.3 APPLICATION OF RUBBER FROM SOLVENT DISPERSION OR DOUGH

When it is necessary to apply a coating of rubber to a fabric which is too delicate for the calendering process, or when the compound is not suitable, the technique of spreading is used. This process consists of the application of the compound dispersed in solvent at high concentration in the form of a 'dough', as it is termed. A stationary blade (commonly called a doctor) regulates the thickness applied to the fabric as it is passed underneath the blade. The fabric is then drawn over a heated chest where the solvent is evaporated and usually recovered for re-use by adsorption on an active carbon.

Table 8.5 TYPICAL RESORCINOL–FORMALDEHYDE–LATEX ADHESIVE FORMULATIONS (PARTS BY WEIGHT). FROM BLACKLEY (1966), COURTESY MACLAREN

		RF–NR		RF–SBR		RF–SBR–VP		RF–VP		PreRF–VP		RF–CR–SBR–VP		RF–IIR	
		Dry	*Actual*	*Dry*	*Actual*	*Dry*	*Actual*	*Dry*	*Actual*	*Dry*	*Actual*	*Dry*	*Actual*	*Dry*	*Actual*
Resin solution	Resorcinol (dry)	11·3	11·3	13·0	13·0	13·0	13·0	13·0	13·0	—	—	11·0	11·0	14·7	14·7
	Formaldehyde (37%)	9·1	24·5	7·0	18·9	7·0	18·9	7·0	18·9	7·5	20·3	6·0	16·2	8·8	23·8
	Stopped resorcinol–formaldehyde condensate (75%)*									20	26·7				
	Sodium hydroxide (10%)	1·0	10	0·8	8	0·8	8·0	0·9	9·0	0·8	8·0	1·5	15	—	—
	Water	—	278	—	306·5	—	306·5	—	240	—	407·7	—	225	—	393
Latex component	Natural rubber (60% ammonia-preserved centrifuged latex)	100	167												
	29/71 styrene–butadiene copolymer (40%)†			80	200	20	50					23	62·5		
	15/15/70 styrene–vinyl pyridine–butadiene terpolymer (40%)‡			20	50	80	200	100	250	100	250	25	62·5		
	Polychloroprene (50%)§											50	100		
	Butyl rubber (55%, 1·5–2·0 mole % unsaturation)‖													100	182
	Zinc oxide (50%)											7·5	15	1·0	2·0
	Sodium hydroxide (10%)	1·0	10											To adjust final pH to 8·4	
	Water	—	157	—	381·4	—	19·3	—	44	—		—		—	
Properties	Total solids content (%)	18·6		12·3		20·0		21·0		18·3		24·5		23·5	
	Resin (p.p.h.r.)	20·4		20·0		20·0		20·0		27·5		17·0			
	Molar resorcinol–formaldehyde ratio	1:3		1:2		1:2		1:2		?		1:2		1:2·2	
	pH			8·5		8·5								8·4	
	Application	Rayon–NR		Rayon–SBR		Nylon–SBR		Nylon–SBR, nylon–NR		Nylon–SBR, nylon–NR		Rayon–CR, nylon–CR		Rayon–IIR	

* e.g. Penacolite Resin R2170.   † e.g. Gentac Latex or Polysar Latex 781.   ‡ e.g. SBR Latex 2108.   § e.g. Neoprene Latex 750.   ‖ e.g. Enjay Butyl Latex 80-21.

8.5.3.1   *Dough Preparation*

The compound, from which the dough is made, is generally formulated specially for each application and is prepared by normal mixing techniques described earlier in this chapter. After mixing and quality-approval testing, the raw stock is comminuted into small pieces or sheeted thinly from a mill and fed at once into solvent in a Z-blade mixer of the type shown in Fig. 8.10. The kneading action of the blades produces a high-viscosity dough ready for use in about 12 hours.

The solvents are selected according to the polymer type, and range from highly inflammable petroleum fractions to chlorinated hydrocarbons such

Fig. 8.10   Z-blade mixer in tilted position for emptying. Courtesy Baker Perkins Holdings Ltd

as trichlorethylene. It may be necessary in certain instances for economic or technical reasons to use a mixed solvent.

To ensure consistent behaviour and performance, tests of solids content and viscosity are carried out on the dough prior to use on the spreading machine.

8.5.3.2   *The Spreading Operation*

The main working parts of a typical spreading machine are shown diagrammatically in Fig. 8.11. A roll of dried and, if necessary, pre-treated fabric is fitted on to location A, and the 'leader' cloth is fed through the rest of the machine until finally taken up on roller J. From A, the cloth passes over a spreader bar B to ensure that all creases are removed from the fabric and to keep it under the correct lateral tension. The smooth tensioned fabric is then fed over the bearer roller C and under the doctor blade D, which is pre-set to give the correct build-up of dough on the fabric surface.

The prepared and tested dough is loaded on to the fabric lying on the bearer roll and revolves as the fabric moves, forming a rolling bank of material which is prevented from expanding laterally by means of cheeks or check

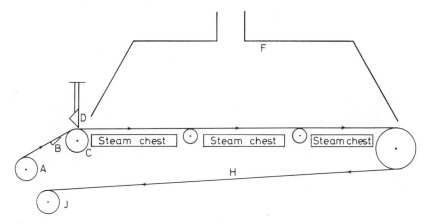

Fig. 8.11   Diagrammatic sketch of spreading machine. Courtesy ICI Dyestuffs Division

plates at the end of the spreader knife and is controlled in the direction of fabric movement by the blade itself. The angle between the blade and the fabric and the distance between them control the coat thickness and the degree of 'strike through' (degree of penetration) of the dough. The greater the angle at which the blade meets the moving fabric, the greater the degree of penetration.

The fabric then enters the steam chest area, where the solvent is driven off and removed by means of the extraction unit F; the speed of travel of the fabric is dependent on the rate of solvent removal. On emerging from the end of the steam chest, the spread fabric requires cooling before it is rolled up on roller J. This is achieved by means of the festooning device, placed at H, which may consist of a single or several rollers; if the dough is of a sticky nature, it may also be necessary to use a liner cloth or dust the surface with talc to prevent blocking together of the rubber–fabric laminate during storage.

Once a machine has been set up to run, it is necessary to carry out the spreading of the first few yards at low speed to check the coating thickness against specification, either by means of a vernier guage or electronically. Normal running speeds are of the order of 10 m/min.

8.5.3.3   *Solvent Recovery and Plant Requirements*

The solvent is recovered, in the majority of cases, by means of specially designed equipment usually employing an active carbon to adsorb the solvent vapour, which is recovered from the carbon by steam distillation.

Because of the fire hazard inherent in spreading plants, it is vital that all machinery used is fully flameproofed and that precautions are taken to ensure that no build-up of static electricity is allowed to occur. Much ingenuity has been expended to devise efficient static eliminators, some of which are extremely effective.

8.5.4 SKIM COATING OR TOPPING BY MEANS OF THE CALENDER

The operation of applying a substantial thickness of rubber to a fabric on a calender is termed 'skim coating' or 'topping'. In this method, compound is fed around a calender roll from a calender nip, and the sheeted compound is then applied to the fabric at a second nip. The rubber sheeting must be travelling at the same speed as the fabric at the point where it is laid on to the fabric; however, the sheeting can be produced from rolls which run at the same speed or with a friction ratio in the nip. It is sometimes found desirable to use a friction ratio, sometimes not, depending upon the compound. Unless the fabric used is a very open weave, it must be pre-treated with adhesive or rubber, either by frictioning (Section 8.5.5), dipping, or spreading (Section 8.5.3.2).

In tyre manufacture, or other operations demanding coating of both sides of the fabric, a four-roll calender is used. Sheeting is produced in the first and third nips, and the sheets are brought together and laid on both sides of the fabric at the same time in the second nip.

The second and third rolls, which apply the coats to the fabric, must be even speed, whilst the first and fourth can be even or odd speed. Various configurations of rolls are used, from the vertical four of many years ago to the Z configuration on modern calenders. The configuration must take into account the feeding of the compound to the rolls, and the interactions of such items as roll bending or cross-axis on the adjoining nips. It is an unsatisfactory configuration if adjustment of cross-axis on one nip affects a second nip.

8.5.5 FRICTIONING

If there is an appreciable difference in the speed of the fabric and the rubber sheet at the point of contact in the nip of the calender, rubber is forced into the fabric weave. If the compound is tacky so that a proportion sticks to the bowl and is returned to the first nip, the process is known as 'frictioning', and, if all the compound drops from the roll, it is known as 'layering' or 'friction coating'.

For the frictioning operation, a three-roll calender is almost always used, the top and bottom rolls going slowly and the middle roll fast; the friction ratio is between 1·5 and 2. The temperatures of the rolls are adjusted on a trial and error basis to give the required viscosity–adhesive characteristics of the compound, but there is an upper limit of temperature which must not be exceeded as this results in the scorching of the compound.

Frictioning fabric on both sides on a four-roll calender cannot be carried out because of the problem of obtaining a speed difference between fabric and compound. Under these conditions, two three-roll calenders in tandem have to be used. This arrangement can be used for coating as well as frictioning, or for coating one side and frictioning the other in a single pass. It is more usual, however, to make two separate passes through a single calender, as there is not usually the demand for frictioned fabric in sufficient quantity to warrant setting up a twin calender unit.

8.6 Moulding

8.6.1 INTRODUCTION

Moulding is the operation of shaping and vulcanising the plastic rubber compound, by means of heat and pressure, in a mould of appropriate form. Fundamentally, all processes of moulding are similar, the ways of introducing the material into the mould distinguishing one technique from another. The basic processes are compression, transfer, and injection, and these are illustrated diagrammatically in Fig. 8.12.

For the satisfactory large-scale production of components, it is necessary to use carefully designed and well-constructed steel moulds suitably hardened

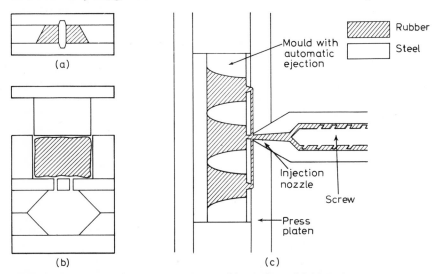

Fig. 8.12   Moulding methods: (a) compression, (b) transfer, and (c) injection

and finished, depending on the surface quality required in the product; small runs can be achieved using moulds of cheaper materials such as aluminium alloys. With compression moulding, to ensure dimensional consistency it is necessary to allow the excess material to move away from the edge of the cavity so that the 'lands' can contact with minimum thickness of rubber (flash) between them. Spew grooves and channels are provided of sufficient dimensions to accommodate this excess, and also to allow the escape of air from the mould cavity. In some cases, where the shape is complex, it may be necessary to provide extra venting to allow air to escape from a 'blind' area, where it is likely to be trapped.

Injection and transfer moulds do not require any provision for excess material flowing out of the mould, simply an escape route for the volume of

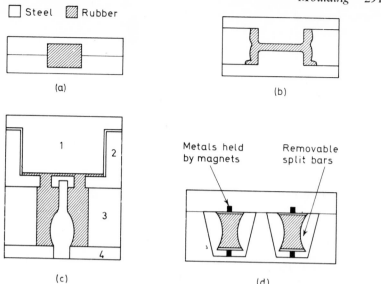

Fig. 8.13   Types of mould design: (a) simple two-piece compression, (b) three-piece compression, (c) four-piece transfer, and (d) split-bar multi-cavity compression

air in the cavities, which the parting faces provide. The mould is closed under pressure and held so while the rubber is forced into the cavities. These two latter methods give parts which carry little or no flash adhering to them at the parting lines of the mould (Fig. 8.13).

8.6.2   BLANK PREPARATION

The type of blank preparation is dependent on the moulding technique employed. Compression moulding requires a blank of accurate weight and shaped so that it will fill the cavity when the mould closes. It can vary from a simple geometrical shape cut from sheeted or extruded rubber to one built up of several pieces.

Transfer moulding requires the compound to be either in pad form or as pellets in sufficient quantity to fill the mould cavities plus an allowance for the transfer pot. Injection moulding machines will accept feed either as strip or pellets.

8.6.3   BLANK HEATING

In order that moulding cycles can be kept to a minimum, it is desirable that the compound be presented to the mould at a temperature as near as possible to that of the mould, consistent with the avoidance of prevulcanisation or scorch.

Several methods of pre-heating moulding blanks are available, ranging from the simple but crude water bath or oven, to the far more controllable microwave or radiofrequency heater.

8.6.3.1 *Water Baths and Ovens*

Dwell times to achieve uniform pre-heat temperature are long and, therefore, with thick blanks, which are most in need of pre-heating, overheating of the outside is very likely to occur before the centre is heated.

8.6.3.2 *High-Frequency and Microwave Heating*

On the other hand, high-frequency and microwave techniques, by exciting the molecules throughout the mass, rapidly achieve uniform heating of the whole blank. The intensity of the force field used governs the degree of pre-heat obtained. The earlier high-frequency method suffered from the defect that inevitable variations in the uniformity of the black dispersion in the rubber led to hot spots. The newer microwave techniques are not so sensitive to this variability and are most successful (Section 10.2.1.3).

8.6.4 COMPRESSION MOULDING

Compression moulding, the oldest and still most universally used technique, is in many respects and for many products the cheapest process because of its suitability for short runs and because of the low mould costs.

The press used for compression moulding is substantially constructed and has two or more platens which are heated either electrically or, more commonly, by saturated steam under pressure. The platens are brought together by pressure applied hydraulically, either by water or oil, to give a loading of from 75 kgf/cm^2 to 150 kgf/cm^2 of projected mould cavity area. Mould designs are generally based upon the availability of such pressures, which are necessary to achieve closure with acceptably thin flash.

Many types of component are produced by this method with little difficulty, but, unless very extreme measures of mould design are employed, some components, of an intricate nature, defy moulding by compression techniques. The reasons are usually the difficulty or cost of preparing a suitable blank shape, the impossibility of flowing the required amount of rubber into the cavity, or the inability to retain loose metal inserts for bonding.

8.6.4.1 *Compression Clamp Moulding*

In some sections of the industry, it is still common practice to use the hydraulic press only to close the mould, which is fitted with clamps while it is under pressure and is then removed from the press; vulcanisation is carried out in an autoclave. This technique is often used for bulky components requiring long low-temperature cures.

8.6.5 TRANSFER MOULDING

Compression moulding is a relatively slow process and, therefore, attention has been directed to means of increasing cure rates. Limited success can be achieved by compounding techniques and/or increase in curing temperatures, but difficulties arise from prevulcanisation or scorch and from the low thermal conductivity of rubbers creating uneven cure, the outside being degraded while the centre remains incompletely cured.

Two methods of introducing a heated mix to the cavity are common. The first of these, the use of outside pre-heaters, has been dealt with previously (Section 8.6.3).

8.6.5.1 *Within-Mould Transfer*

The other method is to use a transfer mould where the blank is pre-heated in the one part of the mould as it passes into the cavity [Fig. 8.12(b)]. This method is frequently adopted for rubber-to-metal bonded, large, and complex components.

The mould consists of four or five parts. The mould cavity is made in a conventional manner, but, mounted above and connected to it by a series of holes, is another cavity. The prepared rubber is placed in this upper chamber, and the integral ram is used to transfer the rubber into the mould cavities

Table 8.6 ADVANTAGES AND DISADVANTAGES OF TRANSFER MOULDING

Advantages	*Disadvantages*
1. Single large blank, giving simplified preparation 2. Shorter cure times 3. Better rubber-to-metal bonds 4. Lower level of rejects made 5. Easy location of metals	1. High mould costs 2. Extra weight in mould is cumbersome 3. Longer changeover time between cures 4. Usually fewer cavities than compression moulding

under pressure. The energy required to force the rubber into the cavities increases somewhat the temperature of the rubber mass as it flows through the connecting holes. As a result, curing times are significantly reduced; other advantages and disadvantages are stated in Table 8.6.

8.6.5.2 *Transfer Press Moulding*

In commercial use is a wide variety of machines which transfer fixed volumes of compound to clamped moulds, by the operation of a ram in a cylinder fed with pre-heated slugs. These machines are capable of high-speed semi-automatic cycling and are a great advance over the use of loose transfer moulds in conventional presses. The components which are produced are to a large degree flash-free, it only being necessary, in the majority of cases, to remove the injection feed and runner system.

8.6.5.3   *Flashless Moulding*

Several patented techniques are available which give a completely flash-free component requiring no further handling except for a statistical quality inspection (Daubenberger, 1961; Jurgeleit, 1962, 1964; Gilette, 1967).

The systems used are very costly with respect to moulds and presses, but once in operation these machines are capable of large volume output and the blank is a simple small extruded pad or disc.

8.6.6   INJECTION MOULDING

The main feature of injection moulding, in which it differs from compression moulding, is in the presentation of rubber to the mould at or near moulding temperature. A large part of the cure time necessary to allow the whole of the rubber mass in the cavity to reach the vulcanising temperature by conduction and convection, as in the case of compression moulding, is eliminated (Wheelans, 1966, 1968, 1969; Geschwind, 1969; Treida, 1969).

The types of injection machinery available differ basically in the method of heating and preplasticisation of the compounded rubber. The main types of machine are ram (Section 8.6.5.2) or reciprocating screw.

The screw machine generates its own preplasticising heat by work on the rubber, which is fed into it in strip form, between the screw and the barrel; this heat is governed by a temperature-controlled fluid medium (water, glycol, or oil) which circulates around the barrel (and, in larger machines, also through the centre of the screw). Progression of the rubber up the screw both heats and preplasticises it. The rubber, collected in front of the screw, pushes the screw back until a trip switch is operated, and the preplasticisation phase is completed. Final heating of the stock takes place during passage through a small orifice (die) into the runner system of the mould.

Machines are available which combine both basic concepts. Coupled to any type of injection moulding machine can be single or multi-stage press units to give intermittent or continuous moulding (Anon, 1968-3). Compounding of the rubbers for injection moulding requires special attention (Penn, 1969-2).

8.6.7   MOULDING SHRINKAGE

Shrinkage is the term used to describe the difference in the dimensions of the mould and the article produced from it, when both are measured at ambient temperature. Shrinkage factors must be determined for individual compounds if strict dimensional accuracy is required, and can be obtained by curing a standard test specimen at production temperatures. In general, linear shrink-age figures fall within the range 1·5–3·0%, depending on polymer type and filler loading. Shrinkage takes place because the coefficient of thermal ex-pansion (volume) of steel is 0.3×10^{-4} per °C, while that of rubbers is from 4×10^{-4} per °C to 7×10^{-4} per °C (Juve and Beatty, 1954, 1955); in addition, anisotropy may be present in the material and will often influence the dimensions of the moulding.

8.6.8 MOULD LUBRICATION

It is usually necessary to provide lubrication at the surface of the mould to facilitate easy stripping of the component and to allow unrestricted flow of the rubber stock. The substances used are termed 'mould release agents' or lubricants, and comprise surface-active materials such as detergents, soaps, wetting agents, silicone emulsions, aqueous dispersions of talc, mica, and fatty acids, applied by spray or brush. Alternative 'dry' lubricant types are based on PTFE or polyethylene, usually carried in solution and applied from an aerosol.

The amount of lubricant required is dependent on several factors: mould complexity, type of lubricant, and polymer type or grade.

The main disadvantage of the use of release agents is the build-up of a solid deposit on the mould surface, which results from the constituents of the mould release, stock residues, and decomposition products. This deposit must be removed frequently otherwise pitting of the surface will result.

Release agent application should always be minimal, otherwise a number of moulding faults, due to poor knitting together of the stock, will result. In addition, the presence of silicones and similar materials can be extremely detrimental to rubber-to-metal bonds.

8.6.9 MOULD CLEANING

As mentioned above, frequent mould cleaning is essential in long-run production to ensure good part appearance. The cleaning frequency is dependent on the application of the component being manufactured.

The methods used are manual scouring with steel wool and abrasives, wire brushing, grit blasting, vapour blasting, and electrolytic oxidation in salt baths. The first two methods are suitable for simple moulds, while the remainder are applicable to any size or configuration of mould. The types of material used for grit blasting are ground vegetable matter, such as nut shells, of 10–30 mesh; the particles after use are cleaned and recycled. Vapour blasting consists of high-pressure application of a detergent solution at high velocity to the mould surface. The abrasive and detergent action effectively removes the deposit.

Salt baths are used in large-scale plant and are extremely efficient but are hazardous in use, requiring the usual safety precautions associated with caustic chemicals (Bitting, 1969).

8.7 Bonding of Rubber to Metals

8.7.1 INTRODUCTION

For many engineering applications of rubber, it is necessary for them to be intimately and strongly attached to supporting metal structures. Historically, the successful uniting of rubbers to metals is of recent origin. The early attempts to unite rubber and metals were by mechanical means, but inevitably there proved to be severe limitations on the use of products of this type. As early as 1869 an ebonite interlayer was being used commercially for soft rubbers, and this method continues to be used for some applications, such as solid tyres for castors, and rollers for the printing industry. The ebonite by its rigidity adheres to the metal surface, which is usually roughened, and to the rubber during vulcanisation by chemical crosslinks. The union fails at about 100°C as the ebonite layer softens.

In 1862, Sanderson submitted a British patent application for the use of electrodeposited brass as an intermediary for rubber-to-iron and rubber-to-steel adhesion. Thus began the era of what has come to be known as the bonding of rubber to metal. However, it was not until the 1920s that the process became successfully commercialised.

8.7.2 COMMERCIALLY AVAILABLE ADHESIVES

In more recent years, there has been an almost complete change to adhesives, of which there are today a great many. The early cements or adhesives were based on cyclised rubber, rubber hydrochloride, and chlorinated rubber, in blends with resins and rubber. The discovery of the value of isocyanates marked a major step forward in the immediate post-war period, but their moisture sensitivity and tendency to be wiped off the metal during moulding have proved somewhat of a disadvantage.

Table 8.7 RUBBER-TO-METAL BONDING AGENTS

Country	Producer or supplier	Trade name	
W. Germany	Bayer	Desmodur R Pergut	Isocyanate
USA UK	Dayton Chemical Gills & Hollings	Thixon	Range available for press, hot air, and open steam cure
USA UK	Hughson Chemical Durham Raw Materials	Chemlok	Types available for most metals and polymers
UK	ICI	Vulcabond TX	Isocyanate
W. Germany	Kautschuk Gesellschaft	Megum	Range available for most metals and polymers
USA UK	Marbon Anchor Chemical	Ty-Ply	Range available for most metals and polymers
USA	Vanderbilt	Braze	Suitable for bonding NR, IIR, SBR, and CR to metal

The majority of the commercially available adhesives in use today are complex mixtures of undisclosed composition and are specific in many instances for particular rubbers and substrates. A list of some of the suppliers, their main types and grades, and particular applications appear in Table 8.7. The bonds obtained with these materials show a high degree of reliability and reproducibility.

8.7.3 METAL PREPARATION

Whatever bonding system or adhesive is adopted, it is necessary to ensure that the metals to be bonded are free from all oil, grease, scale, and contaminants.

The most common degreasing technique is the solvent vapour bath, and both chemical methods and mechanical grit blasting are employed to prepare the surface. In the former, systems based on either acids or alkalis are used, and it is common practice to follow with a phosphate conversion and passivation. Special circumstances may demand other techniques, such as rumbling to clean, and zinc or cadmium plating to protect from corrosion, prior to the application of the adhesive, which must always be made as soon as possible after the metal has been cleaned.

The aim is to coat the surfaces to be bonded with an even thin consistent film of the bonding agent, and brushing, spraying, and dipping are all suitable methods to which some degree of automation can be introduced (Section 10.5).

8.7.4 SELF-BONDING COMPOUNDS

A considerable amount of work has been carried out to obtain a direct bonding system avoiding the treatment of the metal apart from cleaning.

Developments have been along two main lines. The addition of cobalt salts to the rubber mix is covered by several patents (US Rubber Co, 1957; Renault, 1958; Pirelli, 1966; Glagoler, Il'm, and Kosheler, 1967). The other development is the incorporation of resorcinol and a formaldehyde donor in the compound to form a resin *in situ*; the addition of fine particle silica enhances the degree of adhesion considerably. This latter technique has also been applied to textile-to-rubber adhesion (Sections 6.6.5 and 9.5.3). Mention should be made also of the silicone rubber compounds which are offered as being self-bonding to metals (Section 4.13.5).

The use of these compounds has limitations and presents problems with conventional moulds because the bond is indiscriminate and the rubber adheres to the mould unless liberal application of a release agent is made.

8.7.5 MOULDING AND FINISHING

Once the bonding agent has been applied to the metal and dried, it is important that steps are taken to keep it free from contamination until moulded. Furthermore, when the metal is being transferred to the mould, handling must be minimal and operatives must wear gloves. Since no adhesion can occur unless there is intimate contact of the unvulcanised mix with the adhesive-coated metal and there is mutual wetting, the compound must be correctly formulated both to have the necessary delayed-action vulcanising system and to be of suitable rheological properties (Sections 9.5.1. and 10.5.5). The fundamentals of adhesion have been discussed by deBruyne (1953), Alstadt (1955), and Iyengar (1963, 1969).

With some combinations of rubber and metal, care must be exercised in removing the component from the mould as it is frequently the case that full bond strength is not developed in the hot moulding, and, in fact, may not reach its maximum for some hours or days.

8.7.6   CONTROL TESTING FOR ADHESION

To ensure a satisfactory service life for the component, it is necessary to carry out some form of test of the bond strength. It has been the practice to stress the rubber in tension to a percentage of its anticipated breaking strength or to a strain in excess of its expected service condition by a factor of 2–4. Although it is possible that this excessive testing might be the ultimate cause of failure later in service, it has been the accepted method of quality testing for many years (Section 11.3.20).

Recent developments have been concerned with the use of ultrasonics in the testing of the bond. The ultrasonic machine gives a picture of the attachment of the rubber to the metal by examining the continuity of the interface. Any break alters the ultrasonic reflection and causes an alteration of the trace on the viewing screen. This test is carried out while the rubber–metal interface is very lightly stressed. Evidence has been collected which shows this to be an acceptable control technique for components of the type discussed in this section.

8.8   Hand Building and Forming

Reference to Chapter 10, in particular Sections 10.1 and 10.4, will show that, in spite of the development of automatic processes, hand assembly steps are involved in the production of many components, prior to their moulding and vulcanisation. There are, in addition, a wide variety of products, which are hand built and formed because they are either very difficult or very costly to produce by any other means.

The rubber mix used for these products is prepared in a conventional manner and supplied to the operator as calendered sheet, with reinforcing textile, if required, spread fabric, or extrusion. Compounds must be selected to give smooth-surfaced materials with the requisite tack for adhesion to themselves and other substrates. The basic techniques applied are best discussed by considering some typical products made by hand building and forming.

8.8.1   TANK AND PIPE LINING

The lining of vessels, tanks, and pipes of all sizes including large industrial reactors, of many gallons capacity, is a hand operation. The tank or pipe to be lined has to be thoroughly cleaned to remove grease and scale, either by pickling or grit blasting. After degreasing, the surface of the metal is coated with a bonding agent and usually a tie coat of material is applied to ensure that an adequate bond is achieved between the final layer of rubber and metal.

Sheets of rubber are next applied to the prepared metal surface, the rubber surface being lightly freshened with solvent to ensure that it has the tack for adhering to the tie coat. The rubber is hand rolled on to the metal to expel trapped air, and any joints are made firm by the same means.

When lining pipes it is necessary to use a 'dolly' of the correct size, which is pulled through the pipe to press the rubber firmly into contact with the pipe. These dollies are usually made of wood or of metal.

Vulcanisation of this type of lined product is carried out in an autoclave if of a suitable size. It is usual to use a water cure (Section 8.9.1.4) to ensure an adequate pressure being maintained on the rubber–metal interface. In cases where components are large, such as reactor vessels, it is necessary to use the vessel as its own steam or hot-water pressure retainer.

The lined parts are tested, after cure, for pinholes, etc., by an electrical discharge.

8.8.2 ROLLER COVERING

Many rubber-covered rollers of the larger sizes, used on a variety of machines, are built by hand. Plies of calendered sheet are rolled on to the prepared metal core, great care being taken to ensure that no air is included during the rolling process. Cloth wrapping is usually applied before vulcanising in a steam pan. The surface will generally require to be ground to the correct size and surface finish.

8.8.3 HOSES

Many types of hose are produced by semi-automatic processes (Section 10.3). Other types, and those special purpose ones required in relatively small quantities, are made by hand building on mandrels.

Unreinforced hoses, used in low-pressure or zero-pressure applications, are built from calendered sheet or extrusion directly on to the shaped mandrel. If building from calendered sheet, it is necessary to roll the plies together and finally wrap the hose with a wet cotton fabric to ensure adequate pressure to consolidate the mass. If extruded stock is being used, the necessary length of uncured extrusion of suitable bore and wall thickness is cut off and blown on to the mandrel, a cushion of air facilitating its progress.

The fabric reinforcement for hoses for high-pressure use may be square woven, flat, or tubular knitted, and is generally pre-treated with rubber to ensure adequate adhesion to the rubber components. Square woven fabrics are frictioned on a calender, while stockinette can be either spread with a dough or alternatively hand painted with a rubber solution subsequent to positioning on the hose.

A knitted reinforcement in the form of a continuous cylindrical sleeve, if applied to a lining on the mandrel, can be covered with an outer layer of rubber by means of a cross-head extruder. Owing to its flexibility, a knitted construction hose of this type allows the manufacture of curved and other shaped hoses.

Where high burst pressures are required, it is necessary to use a single heavy ply of square woven fabric or multiple plies of a lightweight fabric. Some degree of flexibility is achieved by using the material on the bias, but this construction is limited to less complicated shapes because of removal of the mandrel.

Knitted tubular stockinette can be applied, as an external means of re-inforcement, for limited applications. The technique adopted is to apply the coated stockinette on to the exterior of an uncured hose, using a straight mandrel, and subsequently to transfer the built-up hose to its final shaped mandrel.

Wire spirals are sometimes fitted inside finished structurally unreinforced hoses to enable them to be used for light vacuum work. Hoses are cured in an autoclave, and, after removal from the mandrels, are cut to the specified dimensions.

8.8.4 INFLATABLE PRODUCTS

Two main types of inflatable products are manufactured by hand forming techniques: (a) bladders and air cushions from calendered sheet; and (b) a larger group, constructed from processed and coated fabrics—dinghies, air beds, dracones, storage tanks, etc. The methods of construction for both classes of products are similar in that die or machine cut shapes are assembled into the final shape either on supporting formers or entirely by operative skill.

All joints of such products have to be prepared with solvent and adhesive, and firmly hand rolled to ensure complete joint adhesion.

When constructing inflatables from coated fabrics, it is normally necessary to tape all the seams to ensure that there is no possibility of leakage by wicking along the fibres of the fabric.

8.9 Vulcanisation by Methods Other Than Moulding

It is convenient to group these vulcanisation methods into those in which one batch at a time of material or components is cured and those involving continuous vulcanisation of lengths of material.

8.9.1 BATCH CURING METHODS

8.9.1.1 *Autoclave or Steam Pan*

The autoclave is used for the vulcanisation of extrusions, sheeting, and components which are of an unsuitable size for convenient mould curing. Examples of these components are rollers and hand-formed goods. The temperature and pressure required for cure are achieved by the use of steam.

An autoclave is a cylindrical pressure vessel, normally used in the horizontal position but, for certain products, manufacturers prefer to use it vertically (often sunk below floor level). Autoclaves can be of two main types:

1. A *jacketed* autoclave consists of two large pressure vessels, one inside the other, constructed so that it is possible to fill the inner vessel with an inert atmosphere, such as nitrogen (to prevent polymer degradation through oxidation), and the outer vessel with high-pressure steam to act as the heating medium.
2. An *unjacketed* autoclave consists of a single cylindrical pressure vessel, the pressure medium being the high-temperature steam.

One of the advantages of an autoclave is that a large number of components can be vulcanised at one time, provided that the curatives used in different materials do not interact. It is usual to support the extrusions and small components by embedding them in talc on trays. Products formed on mandrels and sheeting, for example, are cloth wrapped to prevent distortion. In some instances, clamped-up mandrels are used.

For bulky components, it is necessary to step the cure to raise the component to the required vulcanisation temperature slowly, ensuring adequate heat transfer.

8.9.1.2 *Gas Curing*

Gas curing is an extension of the technique used with the conventional autoclave. A jacketed pan is used, the jacket being filled with high-pressure steam, while the interior of the pan is filled with high-pressure gas, usually nitrogen, although air or coal gas can be used.

The use of nitrogen prevents oxidation of the rubber surface during vulcanisation and is less of an explosion risk during blow-down of the pan. Often such pans have a circulatory fan inside them to eliminate hot spots and to ensure even temperatures throughout the cure cycle.

8.9.1.3 *Oven Curing*

The majority of rubber compounds, if heated at atmospheric pressure, sponge as a result of dissolved air and moisture, or degrade owing to oxidation. Oven curing is not, therefore, generally applicable as a curing procedure. It is, however, possible in the case of certain heat-resistant polymers, e.g. silicone and fluorocarbon; indeed, it is desirable to supplement the initial press cure with an oven cure to obtain the required physical properties and to eliminate active residues.

Ovens for this purpose must be vented to atmosphere and adequately ventilated to ensure the complete removal of evolved gases.

8.9.1.4 *Water Curing*

Water curing techniques usually employ an autoclave as a pressure vessel, heating being either from a steam jacket or from steam injection. This method of curing finds application in the vulcanisation of very large hand-built rubber-lined machinery, such as pumps, pipes, and chemical tanks. The water supplies the pressure means for ensuring adequate adhesion of the rubber to metal, in addition to acting as the heat transfer medium.

8.9.1.5   *Lead Curing*

The formation of a lead sheath around such products as long-length hose has been a long-established method of vulcanisation (Section 10.3.3.1).

8.9.1.6   *Cold Curing*

Thin rubber products can be vulcanised by immersion in a carbon bisulphide solution of sulphur chloride or by exposure to its vapour (Sections 1.2.2 and 5.2).

8.9.1.7   *The Peachey Process*

Peachey and Skipsey (1921) patented a process of cold curing thin rubber products, such as proofings and rainwear, by exposing them first to sulphur dioxide gas and then to hydrogen sulphide. This process had some success because no heat or steam was necessary and bright-coloured products could be vulcanised in this way without loss of brilliance.

8.9.2   CONTINUOUS VULCANISING METHODS   (Iddon, 1965; Gregory, 1966)

8.9.2.1   *High-Pressure Steam (Cables)*

A continuous vulcanising technique is possible when covering cable with rubber: the wire core provides a means of drawing the rubbered cable from the extruder through a long steam-filled tube (pressures of 15 kgf/cm^2 and higher are used), where vulcanisation takes place. This tube, approximately 60 m long, is sealed to the extruder at one end; the other end is sealed to retain the steam pressure, the cable passing through a series of mechanical seals or a pressurised water seal, to atmosphere.

8.9.2.2   *Hot Air Tunnel*

It is possible to use a single gas-heated tunnel for the vulcanisation of cellular extrusions and carpet underlay, which do not have to meet tight dimensional tolerances.

The method is very cheap to run, particularly if several extrusions are traversing together, but the cost of installation of this type of tunnel is very high. Hot air tunnels, usually electrically heated, are used for the vulcanisation of silicone rubber products, in which case the exit temperature is 250°–300°C.

8.9.2.3   Molten Salt Bath

In principle, this method, which is applicable to solid and cellular extrusions, employs, as heating medium, an eutectic mixture of inorganic salts in the molten state, at a temperature of from 150°C to 250°C.

Owing to the high density and non-turbulence of the salt mixture, it is necessary for the extrusion to be held under the surface by means of a conveyor. Furthermore, the density of the salts obviates the necessity for the use of the dessicant necessary with the fluidised bed (Section 8.9.2.4). Any salt adhering to the extrusion is removed by means of a water wash tank.

8.9.2.4   Fluidised Bed

The principle upon which the fluidised bed (Geldart, 1969) operates is the flotation of particles of an inert material, such as sand or very small glass balls (ballotini), in an air stream, giving fluidity to the system.

Heat can be supplied initially by means of steam, which also performs the flotation. Once the operating temperature has been obtained, air is substituted for steam and the heating is continued electrically. The two-phase system behaves as a simple fluid and has heat transfer characteristics similar to those of a liquid (Fig. 8.14).

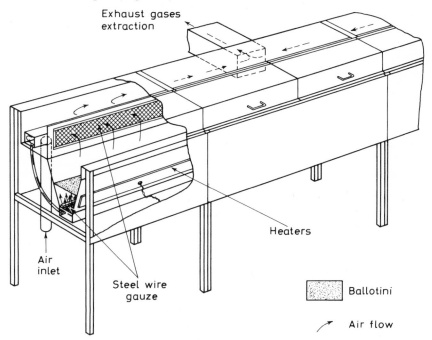

Fig. 8.14   Diagrammatic sketch of fluidised bed for continuous vulcanisation. Courtesy RAPRA

The product to be vulcanised floats near the surface of the bed, and turbulence ensures that heat transfer to the rubber is sufficient for an adequate cure to take place, in a short time.

It is necessary to use a dessicant (Section 9.3.5) in the rubber compound to ensure minimum porosity, which would otherwise occur owing to the presence of water in the polymer and fillers and to lack of pressure during the vulcanisation.

The fluidised bed is used for the vulcanisation of continuous extrusions of many types, and is cheap and efficient for long runs.

8.9.2.5   *Continuous Drum Cure*

Continuous drum cure is used for the vulcanisation of sheet material, flooring, and belting (Section 10.2.1.3). Two machines in common use are the Berstoff and the Rotocure.

The sheeted material is fed between a large heated curing drum and a tension belt which holds the sheet around the circumference of the drum.

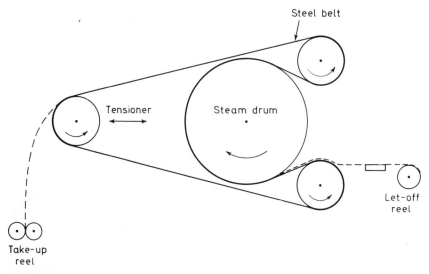

Fig. 8.15   *Drum machine for continuous vulcanisation*

On emergence, it is fully cured. The speed of rotation of the curing drum is variable and is adjusted to achieve the full vulcanisation of the material in the time of its passage through the machine (Fig. 8.15).

8.10   Finishing of Rubber Components

Once a vulcanised product of any process is obtained, a finishing operation is usually necessary to make the part acceptable for consumer use.

It is not possible to discuss the finishing of each individual product, and, therefore, the subject will be discussed under the headings of the techniques employed; namely, flash removal, punching, grinding, painting, and surface treatments.

8.10.1 FLASH AND SPEW REMOVAL

8.10.1.1 *Hand Trimming*

It is still necessary to hand trim many components owing to their awkward or complex configuration or their small quantity production. The usual tools are scissors, both straight and curved, and knives of various types suited to the product being trimmed.

8.10.1.2 *Machine Trimming (Roller Trim)*

Various machines are available which trim components, such as discs, lids, or mats, by the action of two rollers or wheels running under power, the rubber being sheared between the two wheels and effectively removed.

8.10.1.3 *Buffing*

This process uses a buffing wheel, which may be hard felt, flannel, cotton cloth, or wire brush, according to the type of work. The textile wheels are used to remove flash and also to polish hard rubbers, e.g. ebonites. The wire buff is used to prepare areas of rubber which have to be bonded to other components, to remove rubber from the metal parts of bonded components, and in some instances to render them suitable for application of paint to prevent metal corrosion.

8.10.1.4 *Low-Temperature Tumbling*

Low-temperature tumbling is an efficient and widely used means of flash removal in large-scale production. There are three basic types of machine available: the dry ice tumbler, the liquid nitrogen tumbler, and the 'wheelabrator'.

The dry ice (carbon dioxide) and the liquid nitrogen tumblers are similar in function but differ in the refrigerant used. The action consists of freezing the flash, by means of the refrigerant, until it becomes brittle; steel shot, pebbles, or other media are tumbled in a revolving drum with the frozen parts and break off the flash. The wheelabrator is a modern development in which the flash is removed by shot blast action on frozen parts. Separation of good components from the shot and detritus is usually carried out on vibrating sieves, which may be incorporated in the deflashing machine (Anon, 1968-4, 1969-3).

8.10.2 PUNCHING

Independent of quantity, there are some products which require hand punching to clear the flash from an orifice inaccessible to a machine tool. It is now common to find punching being done semi-automatically in reciprocating presses, which only require the operator to locate the sheet of components into the appropriate jig. Punch and die sets are used with this type of machine, which is particularly suitable for thin diaphragms, which require extreme accuracy of punching.

Considerable use is still made of clicker and fly presses with simple cutters, although the latter are now disappearing as more automated equipment is used. The main advantages of the machine punches are the greater throughput possible, and the ability to use multi-gang punches.

8.10.3 GRINDING

For some products it is necessary to obtain a superior finish to that achieved by extrusion, moulding, or wrap curing. The surface may be marred by the presence of flow marks, cloth impression, sprue lines, or injection pips, for example. It is then necessary to produce the parts oversize and grind to the correct dimensions. Such products are typewriter platens, printing rollers, and tubing from which, for example, parted off washers are to be made.

It is necessary to use a lathe grinder, often of the centreless type, in order that the product can be machined successfully. Rubber deforms under the action of the grinding wheel, and careful selection of grinding wheels and speeds is necessary for fine finish work.

8.10.4 SHOT BLASTING

For some applications, it is desirable that sheets of rubber have a clean rough surface. This can be achieved by shot blasting the surface of the rubber with fine aluminium oxide particles of controlled size.

8.10.5 PAINTING AND LACQUERING

It is sometimes necessary to treat the exposed unprotected metal surfaces of rubber-to-metal bonded components. There are many types of protective finishes in use for this purpose, subject to consumers' requirements.

Whereas it is not now general practice to apply paints or lacquers to rubber products, there are exceptions, notably such products as footwear and toys, which depend for their market appeal on a high glossy finish. The lacquer, based on rubber, shellac, chlorosulphonated polyethylene, or polyurethane, is applied either prior to the final cure stage or as a final operation.

8.10.6 CHEMICAL SURFACE TREATMENT

The majority of rubber compounds show a high coefficient of friction, and it is sometimes necessary to lower this for certain applications. This is achieved by the halogenation of the surface of the vulcanisate and usually follows upon the final inspection of the part. Both chlorination and bromination are in general use, and the process is carried out by immersing the moulded article in a solution of chlorine or bromine in water for a few minutes. The parts are subsequently rinsed thoroughly to remove all traces of the halogen.

Other methods of achieving a low coefficient of friction are the use of a thin coating of polyurethane or polytetrafluorethylene, applied from solvent or aqueous dispersion or as a film. Reference may also be made to the Lubrin process developed by Quantum Inc. (Doede, 1961).

NINE

Principles of Compounding

BY J. E. JACQUES

9.1 Introduction

When designing a mixing formulation for a specific end use, a compounder must take account not only of those vulcanisate properties essential to satisfy service requirements but also costs of the raw materials involved and the production processes by which these will be transformed into the final product.

The principal task of compounding is therefore concerned with securing an acceptable balance between the demands arising from these three considerations: in brief, final vulcanisate properties, price, and processability. For any particular product, these requirements are not necessarily fixed. Over a period of time the vulcanisate properties may change owing to altered service requirements. Raw material prices fluctuate and, ironically enough, natural rubber itself is one of the most variable in this respect as reference to any copy of the *Rubber Statistical Bulletin* will show. Processing requirements, too, may not remain static for any length of time. Mixing, process operations, and vulcanisation cycles tend to shorten as production efficiency increases, and at the same time process temperatures and vulcanisation temperatures tend to rise.

It will be appreciated from these remarks that a compounder finds himself walking frequently on shifting sands in attempting to achieve his task. The time element in formulation development work can sometimes be considerable and preclude, for example, rapid reaction to change in material cost. Certain features of factory processability and service performance cannot be assessed quantitatively by laboratory scale tests alone. Extensive factory trials and sometimes service testing may have to be resorted to before a formulation can be finally proven, and these may be both costly and time consuming.

In Chapters 4 and 6, the sources, manufacture, and principal properties and uses of the various ingredients of a mix are discussed. In Chapter 5 the chemistry of vulcanisation and the vulcanising recipes applicable to the various polymers are described in some detail. In this chapter an outline is given of the principles adopted in arriving at a mix formulation to produce particular patterns of properties in both the unvulcanised material and the final vulcanisate.

In consulting the considerable literature dealing with formulations, the newcomer to rubber compounding may often find to his surprise that nominally identical formulations are quoted but appear to possess appreciably different property patterns, the differences being beyond those attributable to interlaboratory testing or to test method variations. It must be admitted at once that this surprise is not always restricted to those unfamiliar with the technology.

The reasons for this, however, are not difficult to understand. A typical formulation may consist of perhaps a dozen individual constituent materials which are brought together by a physical mixing process that will never achieve perfection. Although a written formulation may quote 'zinc oxide, 5·0 parts', there are many commercial grades of zinc oxide, each having its own particle size distribution and characteristic levels of impurities. Similarly, nominally identical polymers produced on different plants or by different manufacturers will have their own minor variations on the basic characteristics. These considerations apply to a great many compounding ingredients, few of which are restricted to a single source, and, when such differences are summed up for perhaps most or all of the constituents of a formulation, some variations in properties must be anticipated. If to these variations are added the effects produced by differences in the mixing process (e.g. mix cycle times, shear rates, order of additions) and in the total heat history before the final vulcanisate is obtained, then appreciable property shifts are not difficult to understand.

For this reason, any published or quoted formulation should always be crosschecked using the grades of polymers and compounding ingredients which will be employed at the place of manufacture, and the properties determined on materials prepared in this way should only be accepted when they can be reproduced allowing for any reasonable experimental error.

The choice of raw materials is steadily increasing, new polymers and compounding ingredients jostling for a place among those already established and requiring critical evaluation before their worth can be assessed. This review has been drawn up around established materials available at the present time. The picture is, therefore, only transitory, and in the future, with a wider choice, both the range of application and the property–cost ratios will shift and a number of the recommendations may no longer apply.

In particular, it should be noted that only the most widely used rubbers are discussed, but, in most cases, the general principles will apply. The very specialised rubbers, such as acrylic rubbers, silicones, and fluorocarbon, have been covered in Chapter 4, and points to be considered in compounding them can be found there.

9.2 The Ingredients and Formulation of a Mix

Today, a technical vulcanisate is made up from the following constituents:

1. Base polymer or blend of polymers.
2. Crosslinking agents.
3. Accelerators of the crosslinking reaction.

4. Accelerator modifiers (activators and retarders); zinc oxide and stearic acid form a common activator system for sulphur-vulcanised systems.
5. Antidegradants (antioxidants, antiozonants, inhibitors of metal catalysed oxidation, protective waxes).
6. Reinforcing fillers (black, mineral, and organic).
7. Processing aids (chemical peptisers for polymers, softeners, plasticisers, dispersing aids, tackifiers, factice, and lubricants).
8. Diluents (inert mineral fillers, organic materials, and extending oils).
9. Colouring materials (organic and inorganic).
10. Specific additives, e.g. blowing agents, fungicides, fibrous materials.

In addition to the above, reclaimed rubbers or vulcanised rubber crumb may be included and, according to the manner of their use, function under groups 1, 7, or 8.

9.2.1   RELATIVE RATING OF TYPICAL POLYMER VULCANISATES

A summary of comparative properties of a range of representative polymer vulcanisates has been drawn up in general terms to show the pattern of properties that may be expected from the types of polymer listed. Table 9.1 deals with general purpose and non-oil-resistant polymers, and Table 9.2 with oil-resistant and the other special purpose polymers.

It should be recognised that appreciable shifts in properties are sometimes made possible by special compounding variations—for instance, the heat resistance of NR vulcanisates may be improved considerably by variation of the vulcanising recipe (Sections 5.2 and 5.6)—but such exceptional compounding is ignored.

Most of the terms used in the tables are self-explanatory. The ASTM D1418 nomenclature for rubbers based on the chemical composition of the polymer chain has been referred to. ASTM D2000, entitled *Classification System for Elastomeric Materials for Automotive Applications*, is a classification system drawn up specifically with the automobile industry in mind. It is based on the premise that the properties of rubber products can be arranged into material designations based on types relating to heat-ageing resistance and classes based on resistance to oil swelling. A two-letter (type and class) designation is followed by three figures, the first indicating hardness, and the second and third tensile strength. Special suffix letters and numbers are used to refer to other characteristics for which a particular mix has been formulated. The type and class requirements are indicated by the group of letters shown in Table 9.3.

In this ASTM D2000 system, recently reviewed by Stoeck (1969), a normal NR vulcanisate not particularly compounded for heat resistance would be designated AA. A special purpose fluorinated polymer intended for extreme heat and oil resistance, on the other hand, would be rated HK. Other typical polymers occupy intermediate positions, as indicated in the tables. For further details reference should be made to the most recent edition of ASTM standards.

Solubility parameter data have been included because of its importance in compounding. It gives an indication of the likely compatibility of a

Table 9.1 GENERAL PURPOSE AND NON-OIL-RESISTANT RUBBERS

	Natural	Polyisoprene	Styrene–butadiene	Polybutadiene	Isobutylene–isoprene	Ethylene–propylene terpolymer
ASTM D1418 class	NR	IR	SBR	BR	IIR	EPDM
ASTM D2000 rating	AA	AA	AA/BA	AA/BA	AA/BA	CA
Polymer specific gravity	0·93	0·93	0·94	0·93	0·92	0·86
Relative volume cost* of polymer only	1·50	1·58 (high-cis)	1·00	1·08	1·45	2·02
Glass transition temperature (°C)	−75	−70	−60	−85	−79 (2% isoprene)	−58
Solubility parameter (cal$^{1/2}$/cm$^{3/2}$)	8·25	8·25	8·29	8·20	7·60	8·0
Normal compounded hardness range (IRHD)	25–95	25–95	40–95	45–95	40–80	40–95
Resilience at 20°C	4†	4	3	4	1	3
Resilience at 100°C	4	4	3	4	3	3
Tensile strength (gumstock)	3	3	1	1	2	1
Tensile strength (reinforced)	4	4	4	3	3	3
Dielectric properties	3	3	2/3	3	3	3
Bonding to substrates	3	3	3	3	2	2
Resistance to gas permeation	2	2	2	2	4	2
Resistance to abrasion	3/4	3/4	3/4	3/4	2/3	2/3
Resistance to tearing	3	2	2	2	2	2
Resistance to cutting and cut growth	3	2/3	2	2	2	2
Resistance to flexing and fatigue	3	3	3	3	3	3
Resistance to heat	2/3	2/3	2/3	2/3	3	3
Resistance to flame	2	2	2	2	2	2
Resistance to fluids:						
Aqueous	3	3	2	3	3	3
Acids	2	2	2/3	2/3	3/4	3
Hydrocarbons:						
Aliphatic	1	1	1	1	1	1
Aromatic	1	1	1	1	1	1
Animal and vegetable oils	1	1	1	1	3	3
Oxygenated organics	2/3	2/3	2/3	2/3	3	3
Resistance to light (staining and discolouration)	2/3	2/3	2/3	2/3	2/3	2/3
Resistance to ozone	1	1	1	1	3	4
Resistance to atmospheric exposure	2/3	2/3	2/3	2/3	3	4

* Volume costs of raw polymers relative to SBR 1500 at 15½ p/kg (SBR 1500 taken as unity).
† Vulcanisate properties are rated: 1, poor; 2, moderate; 3, good; 4, outstanding.

Table 9.2 OIL-RESISTANT AND SPECIAL PURPOSE RUBBERS

	Chloroprene	Butadiene-acrylonitrile	Chlorosulphonated polyethylene	Silicone	Fluorocarbon
ASTM D1418 class	CR	NBR	CSM	Si‡	FPM
ASTM D2000 rating	BC/BE	BF/BG/CH	CE	FC/FE/FK/GE	HK
Polymer specific gravity	1·23	1·00	1·18 (35% chlorine)	0·98	1·85
Relative volume cost* of polymer only	3·22	3·14	4·83	29·6	128·0
Glass transition temperature (°C)	−50	−22 (40% acrylonitrile)	−28	−120§	−22
Solubility parameter (cal$^{1/2}$/cm$^{3/2}$)	9·26	9·92 (40% acrylonitrile)	9·1	7·30	8·60
Normal compounded hardness range (IRHD)	30–95	30–95	40–90	30–90	50–80
Resilience at 20°C	3†	1	2	3	1
Resilience at 100°C	3	2	3	2	2
Tensile strength (gumstock)	3	1	2	1	3
Tensile strength (reinforced)	3	3	2/3	1/2	3
Dielectric properties	1	1	3	3/4	3
Bonding to substrates	3/4	3/4	2/3	2	2
Resistance to gas permeation	3	3/4	3	2	3
Resistance to abrasion	3/4	2/3	3	1/2	3
Resistance to tearing	3	2/3	2	1/2	3
Resistance to cutting and cut growth	2/3	2/3	2/3	1/2	2/3
Resistance to flexing and fatigue	2/3	2/3	2/3	2/3	2/3
Resistance to heat	2/3	2/3	2/3	4	4
Resistance to flame	3/4	2	2/3	2	3/4
Resistance to fluids					
Aqueous	2	3	3	3	2/3
Acids	2/3	2/3	3/4	1	3/4
Hydrocarbons:					
Aliphatic	3	3	2	2	3
Aromatic	1	2	1	2	3
Animal and vegetable oils	3	4	2	3	2
Oxygenated organics	1	1	2	3	2
Resistance to light (staining and discolouration)	2	2	3/4	3/4	3/4
Resistance to ozone	3/4	1	3/4	3/4	3/4
Resistance to atmospheric exposure	3/4	2/3	3/4	3/4	3/4

* Volume costs of raw polymer relative to SBR 1500 at 15½ p/kg (SBR 1500 taken as unity).

† Vulcanisate properties are rated: 1, poor; 2, moderate; 3, good; 4, outstanding.

‡ Methyl polysiloxane; other classes are FSi, lluoropolysiloxane; PSi, phenyl methyl polysiloxane; PVSi, phenyl methyl vinyl polysiloxane; VSi, methyl vinyl pol·siloxane.

Table 9.3 THE ASTM D 2000 CLASSIFICATION SYSTEM: TYPE AND CLASS REQUIREMENTS

Type requirements		Class requirements	
After 70 h at stated temperature, the tensile strength change shall be < ±30%, the elongation at break change < 50%, hardness change < ±15 IRHD		70 h at temperature as for type test (max. 150°C) in ASTM reference oil no. 3*	
Reference letter	Test temperature (°C)	Reference letter	Maximum volume swell (%)
A	70	A	None
B	100	B	140
C	125	C	120
D	150	D	100
E	175	E	80
F	200	F	60
G	225	G	40
H	250	H	30
J	275	J	20
		K	10

* ASTM D471.

polymer with other polymers or compounding ingredients or with fluids encountered in service (Section 9.2.3).

Relative polymer volume costs have been quoted based on those published in mid-1969. Since the specific gravities of polymers vary so widely, it is important to remember that any comparison must be made on a volume basis, and this applies also to the fully compounded mixes.

Vulcanisate properties have been rated on a 1–4 scale corresponding to the following property levels: 1, poor; 2, moderate; 3, good; 4, outstanding. It must be emphasised that these general levels are capable of some modification by special compounding.

9.2.2 CROSSLINKING SYSTEMS

For the general purpose polymers (NR, IR, and SBR) and for IIR, BR, NBR, and EPDM, sulphur is the common and most generally used crosslinking agent. Its action is controlled by organic accelerator systems, of which there is a wide choice, and an activating system commonly supplied by zinc oxide in the presence of stearic acid.

The normal sulphur vulcanisation system is capable of many variants which will govern the chemical nature of the sulphur crosslink, i.e. whether it is essentially a mono-, di-, or poly-sulphidic type. The nature of the sulphur crosslink can have a considerable influence on the heat resistance of the vulcanisate (Section 5.6.1).

For CR, crosslinking is induced using combinations of zinc oxide with light calcined magnesia, with or without an additional organic accelerator according to the type of CR. Sulphur is used only occasionally for special

M

effects (Section 5.4). CSM can be crosslinked by means of a number of systems, utilising either a metal oxide with organic accelerator or an epoxy resin combined with organic accelerators and retarders (Section 5.4.1.2). FPM can be crosslinked by means of metallic oxide–amine systems (Section 5.4.2.3).

Silicones and fluorocarbons and certain other saturated polymers are effectively crosslinked using organic peroxides, and for silicone rubbers a wide choice of peroxides is available (Section 5.4.3.2).

9.2.3   SELECTION OF INGREDIENTS

For a particular application, the basic polymer, or when desirable and possible the polymer blend, has first to be selected according to the broad pattern of final properties required. This must then be combined with an appropriate crosslinking system, and, whether black or non-black, the reinforcement, if any, has next to be decided, sufficient processing aids being included to ensure satisfactory mixing and processing. It should be remembered that reinforcing fillers of finer particle size are, as a rule, progressively more difficult to incorporate during mixing, yield stiffer stocks, and are also generally more expensive; excessive reinforcement should, therefore, be avoided when not critical. Any permissible diluents or special additives for specific effects are then added to the formulation list.

Protective antidegradants are selected according to the severity and type of service exposure and to their relative effectiveness in the particular base polymer used.

The stages in which the formulation will be mixed in production must also be borne in mind, and the appropriate ingredients grouped accordingly. For instance, protective agents are commonly added early—crosslinking agents last.

9.2.3.1   *Compatibility—The Solubility Parameter Concept*

The range of possible ingredients is now so large that problems associated with compatibility of polymers with each other or with other compounding ingredients, such as plasticisers, extenders, process aids, or solvents, may often be encountered. In resolving such problems, the concept of the solubility parameter is a useful one (Section 9.2.1, Tables 9.1 and 9.2) (Beerbower, Pattison, and Saffin, 1964; Corish, 1967). The solubility parameter may be regarded as a measure of intermolecular attraction and is numerically equal to the square root of the cohesive energy density. Polymers and compounding ingredients having closely similar solubility parameter values are likely to be compatible, and this is often of considerable value when dealing with unfamiliar materials.

The concept must be applied with care; no indication of the relative rates of solution will be given since this will be influenced by molecular size, viscosity, and temperature. Where two polymers are concerned, vulcanisation rates may be quite different so covulcanised blends are impracticable. Methods of determination of solubility parameters have been reviewed

Table 9.4 TYPICAL VALUES OF SOLUBILITY PARAMETERS

	Solubility parameter $(\text{cal}^{1/2}/\text{cm}^{3/2})$		Solubility parameter $(\text{cal}^{1/2}/\text{cm}^{3/2})$
Process aids		*Solvents*	
Petroleum jelly	7·5	Propane	6·0
Paraffin wax	7·7	N-Hexane	7·3
Low molecular weight		N-Octane	7·6
polyethylene	7·8	Methyl isobutyl ketone	7·8
Stearic acid	8·5	Cyclohexane	8·2
Polyethylene glycol		Xylene	8·8
($M = 400$)	9·0	Toluene	8·9
Polyethylene glycol		Benzene	9·2
($M = 1500$)	8·8	Methyl ethyl ketone	9·2
Diethylene glycol	10·4	Acetone	10·0
Plasticisers (ester type)		Ethylene dichloride	9·8
Dioctyl sebacate	8·7		
Dibutyl sebacate	8·9		
Dioctyl phthalate	8·9		
Dibutyl phthalate	9·4		
Tritolyl phosphate	9·8		
Trixylenyl phosphate	9·9		

Petroleum oils	*Viscosity gravity constant*	*Aniline point* (°C)	*Solubility parameter* $(\text{cal}^{1/2}/\text{cm}^{3/2})$
Paraffinic	0·78	115	7·2
Naphthenic	0·89	71	8·2
Aromatic	0·96	21	9·2

(Burrell, 1955), and a calculative method based on summation of molar attraction constants where the molecular formula is known (Small, 1953) has also been devised. The method however has certain limitations. Typical values for a range of plasticisers, process aids, and solvents are given in Table 9.4.

9.2.4 DESIGN OF COMPOUNDING EXPERIMENTS AND EVALUATION OF DATA

In compounding studies involving more than one variable, e.g. studies of the effect of a range of concentrations of more than one compounding ingredient, it is important that the maximum amount of information is wrung from the evaluation of a given number of mixing formulations. A graphical method has been described (Bertsch, 1961) involving the construction of contour graphs for each property of interest, using as coordinates the concentrations of the compounding ingredients concerned. By plotting the contour curves for each property on a transparent overlay, several properties may be examined simultaneously with respect to variations in the concentration of any two compounding ingredients, e.g. carbon black and process oil or primary and secondary accelerators. By suitable modification, up to four variables may be examined in this way. The utility of this method is increased by careful choice of ingredient concentrations to be evaluated, disposed around the most likely areas of interest.

A more sophisticated approach to this type of problem involves the setting up of a statistically designed experiment, followed by the use of a digital computer to analyse the experimental results. Examples of compounding studies demonstrating typical procedures have been described by Weissert and Cundiff (1963), Buckler and Kristensen (1967), and Hartmann and Beaumont (1968).

The object is to set up a mathematical model or response equation, as it is usually known, which relates a particular formulation property to all the variable factors (in this instance the compounding ingredient concentrations) under examination. The form of the general response equation used is an empirical polynomial one, usually quadratic.

The range of concentrations of each compounding ingredient to be varied is first decided, and a statistically designed pattern of formulations involving selected ingredient concentrations within these ranges is chosen. These formulations are mixed, and the appropriate properties of both unvulcanised and vulcanised stock are determined. The data are evaluated, using a regression analysis computer programme, to determine coefficients of the polynomial response equation relating a particular property and the concentrations of all the variable ingredients of the mix giving that property.

The coefficients having been established, the data can now be used, if necessary in conjunction with an optimisation programme, either to predict the property pattern of a given formulation or to give a formulation possessing a stated pattern of properties. One of these properties could, for instance, be required at an optimum level, subject to certain restrictions on permitted levels of others. Since the response equation used is an empirical one, predictions outside the range of concentrations of compounding ingredients used in the study should not be attempted.

As a practical example of the technique, it is instructive to consider the evaluation described by Buckler and Christensen (1967). This involves the examination of an experimental polybutadiene polymer in blends with natural rubber, four variables being investigated: Mooney viscosity of the polybutadiene, percentage by weight of polybutadiene in the blend with natural rubber, and levels of carbon black and oil respectively in the formulation. The mathematical model assumed was the generalised quadratic in four variables having the form:

$$Y = B_0 + B_1 X_1 + B_2 X_2 + B_3 X_3 + B_4 X_4 + B_{11} X_1{}^2 + B_{22} X_2{}^2 + B_{33} X_3{}^2 \\ + B_{44} X_4{}^2 + B_{12} X_1 X_2 + B_{13} X_1 X_3 + B_{14} X_1 X_4 + B_{23} X_2 X_3 \\ + B_{24} X_2 X_4 + B_{34} X_3 X_4$$

where Y equals any value of a property or response related to the factors being studied; letters B represent numerical coefficients in the quadratic response equation; and letters X represent any value of the factors (here compounding ingredient concentrations or Mooney value) under examination.

Each variable was employed at three levels, as shown in the experimental design, which gives 81 possible combinations of the levels. Since the above equation contains 15 coefficients, it is only necessary to mix and evaluate a slightly greater number of combinations. A one-third order fractional factorial design was therefore employed, requiring 27 mixings. The levels of ingredients were chosen so that each level of each factor was used only once in combination, as illustrated in Table 9.5.

The mixed stocks were evaluated for a range of properties of both unvulcanised and vulcanised materials. For each property 27 substitutions

Table 9.5 EXPERIMENTAL DESIGN. FROM BUCKLER AND KRISTENSEN (1967), COURTESY RUBBER & TECHNICAL PRESS LTD

Experimental polymer in blend (wt %)		P_1			P_2			P_3		
Mooney viscosity of experimental polymer		M_1	M_2	M_3	M_1	M_2	M_3	M_1	M_2	M_3
Oil concentration (p.p.h.r.)	Black concentration (p.p.h.r.)									
O_1	C_1	X	—	—	—	—	X	—	X	—
	C_2	—	X	—	X	—	—	—	—	X
	C_3	—	—	X	—	X	—	X	—	—
O_2	C_1	—	—	X	—	X	—	X	—	—
	C_2	X	—	—	—	—	X	—	X	—
	C_3	—	X	—	X	—	—	—	—	X
O_3	C_1	—	X	—	X	—	—	—	—	X
	C_2	—	—	X	—	X	—	X	—	—
	C_3	X	—	—	—	—	X	—	X	—

X = formulation actually mixed and evaluated.

were made in the above mathematical model, each one containing a particular property determination and the corresponding concentration or value of the four variables employed in that particular mix.

These equations were next subjected to the computer programme to give the range of numerical coefficients required. Knowledge of these then permits calculation of formulations having certain specified property levels, or alternatively property levels which may be expected from a given formulation. In addition, investigations into possible relationships between properties and analysis of property trends with respect to compounding ingredient concentration changes or other factors systematically varied in the study (e.g. Mooney viscosity changes of the polymer as examined here) become possible. For more detailed discussion, the original paper should be consulted.

The compounder will need to establish suitable records of the formulation of mixes and their properties. With a suitable system, well arranged and planned, such as a punched card system (Garvey and Rostler, 1968), quick retrieval of a mix to suit or nearly suit a particular requirement or specification is easy. Considerable time in the development of new compounds can be saved in this way.

9.3 Compounding to Meet Processing Requirements

Before compounding procedures for the various specific vulcanisate properties are described, some general remarks will be made about the requirements for mixing and processing characteristics. This is necessary as a compromise has often to be made to permit a formulation, which has been developed for particular vulcanisate properties, to be mixed and processed in a given manner.

9.3.1   VISCOSITY CONTROL

For a raw polymer type, choice of correct viscosity level is important to ensure acceptable mixing and processing characteristics. Viscosity levels for polymers are commonly expressed in terms of Mooney viscometer readings taken at 100°C (Section 11.2.1.1). This instrument, however, operates at very low shear rates, and, with presentday trends towards faster mixing and processing operations, viscosity data obtained with much higher shear rate instruments may prove ultimately to be more realistic (Krol, Grijn, and Henten, 1964; Collins and Oetzel, 1969, 1970). In the meantime, Mooney data remain the generally accepted criteria, since manufacturers rate their polymers this way. Most types of synthetic polymers are available at various viscosity levels, a nominal Mooney viscosity [ML $(1+4)$ 100°C] of 50 being a common one. For easy mixing and processing, e.g. for formulations with high filler loading or for sponge production where particularly uniform blowing is required, a viscosity of the order of 30 is necessary; at the other end of the scale, a viscosity level of 100 or so should be selected when loading with oil diluents, for economical compounding, or when processing demands a certain level of strength ('green strength') in the unvulcanised material.

Natural rubber produced by conventional means has a typical Mooney viscosity of the order of 90, which is too high for normal direct mixing except by special compounding where oil extension is used. Its viscosity is, therefore, usually reduced to a suitable level by mechanical working or mastication before mixing. For more efficient operation, small additions of chemical peptisers are made (Section 6.3.1). In the case of standard Malaysian rubbers (SMR), viscosity-stabilised grades (known as CV types) with a Mooney viscosity of 55–65 are available and may be suitable for some mixings directly without further mastication (Section 4.1.5.1).

A number of low molecular weight synthetic polymers can be obtained which are in viscous liquid form. These can be used as part replacement of solid polymer, functioning as general process aids and softeners. After vulcanisation they become part of the crosslinked structure and are, therefore, not extractable by liquids which would extract other softeners or processing aids. Specially depolymerised NR is also available in fluid form and behaves similarly.

9.3.2   CONTROL OF 'NERVE'

Some rubbers exhibit a degree of elastic recovery on deformation, even before crosslinking. This 'nerve', as it is usually called, if excessive, must be reduced during mixing processes to a controlled level, so that processing will be reproducible and shaped stocks will exhibit dimensional stability.

Depending on the polymer type, elastic recovery can be reduced by a suitable degree of mastication, by increasing filler loadings, by choice of filler type (e.g. use of high structure furnace blacks in place of normal structure carbon blacks, Section 6.1.6), or by use of additions of materials such as mineral rubber or factice (Sections 6.3.4 and 6.3.8). Another method is to add a proportion of specially produced polymers which have been partially cross-linked during manufacture. These are particularly advantageous in improving shaping operations and ensuring dimensional stability of un-vulcanised stocks. Polymers, modified in this way, are produced from NR, SBR, IIR, and NBR : examples are SP (superior processing) grades of NR, and types 1009 and 1018 of SBR (Sections 4.1.4.6 and 4.2.1.6).

9.3.3 ADHESION TO MILL ROLLS, ETC.

During mill mixing, mill sheeting, or calender sheeting operations, sticking to roll surfaces can sometimes be a problem, and additives to aid release are necessary. Certain soft tacky stocks or ones containing intermediate loadings of mineral fillers are most likely to present this problem. Fatty acids or their derivatives, microcrystalline waxes, or low molecular weight polyethylenes are useful additives to control this, but must be used with particular care in NBR, where, because of limited compatibility, other problems due to bloom or lack of stock coalescence may arise.

9.3.4 CALENDERING AND EXTRUDING

In calender frictioning operations, the stocks used must be soft so that they penetrate the interstices of the fabric easily and tacky so that ready adhesion to the fabric is ensured. Inevitably, such stocks can give rise to handling problems and call for compromise compounding. NR and CR compounded with resins such as coumarone–indene or petroleum types produce good frictioning stocks (Section 6.3.9); refined reclaimed rubbers are used frequently in blends for lower-cost stocks (Section 6.4).

To obtain good extrusion characteristics, compounding requirements are similar to those described previously for reduced elastic recovery. These minimise undue die swell on shaping and contribute to shape retention until this is made permanent by crosslinking. Internal lubricants assist in reducing die drag, promoting a smooth surface. Waxes, fatty acids, or fatty acid derivatives are also useful additives. For extrusion work generally, it is just as important for easy processing to avoid compounding to too low a stock viscosity as to one that is too high. Too low a viscosity can lead to loss of extrusion efficiency and poor shape retention.

9.3.5 CONTINUOUS VULCANISATION

Continuous vulcanisation of extrudates by passage through liquid salt baths or fluidised beds requires special compounding techniques because of the high vulcanisation temperatures and normal atmospheric pressures involved. Volatile softeners or plasticisers and moisture-containing ingredients must be avoided, any moisture traces being eliminated by use of scavengers, such as a quicklime dispersion in mineral oil. Rapid rates of crosslinking are desirable since these ensure a build-up of strength throughout the mass of the stock, causing entrapped air to diffuse away rather than to form expanding pockets which make the material tear internally (Hughes *et al.*, 1962). Compounding for high-temperature injection moulding operations requires a similar approach with respect to elimination of volatile or unstable compounding ingredients, although moisture scavengers are not normally required.

9.3.6   TACK

Forming operations which involve building layers of stock together require the mix to have tack, the property of auto-hesion or adhering rapidly and strongly to itself when surfaces are brought into contact. This surface condition may be required to last for considerable periods of time, e.g. in tyre retreading stocks or adhesive tapes. Excessive filler loadings are best avoided since they may have a drying effect on the stock surface. NR and CR can be compounded readily to give good tack levels. Other general purpose polymers require suitable resins, e.g. coumarone–indene, rosin derivatives, petroleum resins, or phenolic types to give acceptable tack. EPDM is at present difficult to tackify. Waxy compounding ingredients which may form a bloom should be kept at a minimum, and, where necessary, freshening of the surface with a solvent wipe is beneficial. Excessive use of solvent is, however, undesirable since, if traces remain trapped, vapour blows or porosity may result during vulcanisation (Mueller *et al.*, 1969).

9.3.7   AVOIDANCE OF PREVULCANISATION OR 'SCORCH'

Vulcanisation systems are normally designed to be as fast as practicable, subject to certain limitations. These are the cost of the vulcanisation system, and the heat unavoidably generated during mixing and processing which causes prevulcanisation or scorch problems.

Delayed-action vulcanisation systems must be used to minimise risk of premature vulcanisation, and, for sulphur-cured NR or SBR, sulphenamide accelerators are widely used, with an additional chemical retarder such as nitrosodiphenylamine in critical applications. For CR, increased levels of magnesia, additions of sodium acetate, or small amounts of mercaptobenzthiazyl disulphide are effective.

9.4   Compounding for Vulcanisate Properties

In this section, the means by which individual vulcanisate properties can be varied will be considered one by one.

9.4.1   HARDNESS AND MODULUS

Hardness is probably the property first specified and, of all vulcanisate characteristics, the one most frequently determined. It gives a measure of the modulus at low strains. In the following section, hardness values are expressed in international rubber hardness degrees (IRHD), measured according to the ISO method R48 (Section 11.3.5).

Hardness and modulus of a vulcanisate are normally increased by the use of particulate fillers, the greatest effect being obtained by the reinforcing types or by the use of crosslinked resin systems. These latter may be phenolic types capable of crosslinking (Sections 6.2.4.6 and 6.3.9.4), or monomers capable of being grafted *in situ*, e.g. acrylic monomers used with peroxide crosslinking systems. Relatively unreactive resins, such as high-styrene resins (Sections 6.2.4.5 and 6.3.9.3) are used with NR and SBR for certain hard vulcanisates. PVC may be incorporated into NBR (Section 4.10.5).

Carbon blacks are the most common reinforcing agents used, and the relative influence of typical types in four common polymers is illustrated in Table 9.6. This indicates approximately the parts of the various types of

Table 9.6 EFFECT OF PARTICLE SIZE ON HARDNESS OF VULCANISATES

Carbon black type		*Average particle size* (nm)	*Parts of filler per point increase in hardness*			
ASTM	*Code*		*CR*	*IIR*	*NR*	*SBR*
N-220	ISAF	28	1·3	1·5	1·7	2·0
N-330	HAF	32	1·5	1·7	1·9	2·3
N-660	GPF	70	2·0	2·2	2·5	3·0
N-990	MT	300	3·3	3·7	4·2	5·0
Non-black						
Hydrated silica		20	1·5	1·6	2·0	1·6
Calcium silicate		25	2·2	1·9	3·0	3·3
Hard clay		2×10^3	4·5	5·9	5·0	4·9
Soft clay		10×10^3	5·0	9·1	7·7	5·6
Whiting		12×10^3	5·0	9·1	6·4	8·4

carbon black required to increase hardness by 1 IRHD. It will be noted that the blacks of finer particle size exert relatively more effect and that the quantity required varies from one polymer to another, being least for CR and most for SBR of the four polymers illustrated. The values stated should only be applied over the central part of the hardness range, say between 45 IRHD and 80 IRHD. At a given particle size of black, increase in structure level produces a small increase in hardness.

With non-black fillers, the pattern is less well defined, the varying chemical natures of the materials concerned, in addition to their varying particle shape and surface state, evidently influencing the relative hardening effect. The broad picture of finer particle size resulting in greater hardening effect remains, however, as will be seen from Table 9.6.

For a given non-black filler type, decreasing particle size gives a progressively increased hardness. Data for a range of whitings of known particle sizes have been given by Gibson and Pratt (1969).

In addition to the low strain modulus, of which hardness is a measure, the compounder is interested in modulus at higher strains, which is expressed not as true modulus but as the stress required to produce a stated extension, e.g. 100%, 300%, or 500% (Section 11.3.2). Compounding which produces increases in hardness also produces increases in the higher strain moduli values. The effects of particle size variation and structure level in black are shown for NR and SBR in Tables 9.7 and 9.8. In the case of CR, the effects

Table 9.7   EFFECT OF BLACK PARTICLE SIZE ON PHYSICAL PROPERTIES

Carbon black type	*Code ASTM Particle size* (nm)	HAF N-330 32	FEF N-550 47	GPF N-660 70
*NR**				
Tensile strength (kgf/cm^2)		265	250	230
Stress at 300% strain (kgf/cm^2)		150	140	115
Elongation at break (%)		475	500	530
SBR†				
Tensile strength (kgf/cm^2)		233	209	197
Stress at 300% strain (kgf/cm^2)		140	125	91
Elongation at break (%)		500	525	560

* NR formulation:		† SBR formulation:	
NR-RSS	100	SBR 1500	100
Zinc oxide	4	Zinc oxide	4
Stearic acid	3	Stearic acid	2
Carbon black (as shown)	50	Carbon black (as shown)	50
Process oil	5	Process oil	10
Antioxidant	1	PBN	1
CBS	0·5	CBS	1
Sulphur	2·5	Sulphur	2
Vulcanisation, 30 min at 148°C.		Vulcanisation, 30 min at 148°C.	

of progressive loading with FEF (N-550) black and fine particle silica (20 nm) are given in Table 9.9.

Increases in extension modulus in a given polymer–filler system are also sometimes possible by using heat treating cycles during mixing, with or without the use of a chemical promoter, These treatments modify the polymer–filler interaction. The effect is illustrated in Table 9.10 by three examples, in which vulcanisate properties are compared in stocks prepared in a normal mixing technique with those obtained when the polymer–filler system is first subjected to a heat treatment cycle in an internal mixer, as indicated, before conversion to the final mixing. In the preparation of SBR–silica masterbatch by heat treatment, 1·5 parts of a trimethyl dihydroquinoline antioxidant was

Table 9.8   EFFECT OF BLACK STRUCTURE (NR FORMULATION AS FOR TABLE 9.7)

	Carbon black type		
	HAF-LS (N-326)	*HAF (N-330)*	*HAF-HS (N-347)*
Structure level (DBP absorption) (cm^3/100g)	70	105	125
Tensile strength (kgf/cm^2)	284	260	255
Stress at 300% strain (kgf/cm^2)	115	154	171
Elongation at break (%)	575	480	450

added to protect the polymer from undue degradation. From the results quoted, it will be seen that, in addition to a reduction in mix viscosity, the vulcanisate properties are modified appreciably, extension modulus and resilience being increased, hardness and compression set reduced, and the rubbery nature of the vulcanisate is thereby enhanced. Because of the magnitude of the property changes, heat treatment cycles must be carefully controlled to achieve reproducible results on a production scale (Gessler *et al.*, 1953; Ecker, 1959; Reissinger, 1960; Wolf, 1960).

Examples of phenolic resin and high-styrene resin additions and the use of *in situ* resin formation with acrylate monomer to increase hardness and

Table 9.9   EFFECT OF FILLER LOADING FOR CR MIX

N-550 black	Filler loading		
	30 p.p.h.r.	*50 p.p.h.r.*	*70 p.p.h.r.*
Tensile strength (kgf/cm^2)	205	235	225
Stress at 300% strain (kgf/cm^2)	50	105	134
Elongation at break (%)	290	160	150
Reinforcing silica			
Tensile strength (kgf/cm^2)	232	220	215
Stress at 300% strain (kgf/cm^2)	48	77	112
Elongation at break (%)	810	746	450

Formulation:
CR (non-sulphur-modified)	100
Filler	as shown
Light calcined magnesia	4·0
Stearic acid	0·5
Process oil	2·0
Zinc oxide	5·0
Ethylene thiourea	0·5
Antioxidant (PBN in black mix, octylated diphenylamine in silica mix)	2·0

Vulcanisation, 30 min at 148°C.

modulus appear in Tables 9.11–9.13. It is interesting to note that all these three systems produce similar increases in hardness at equivalent resin addition. High-styrene resin (or PVC addition to NBR), because of its relatively unreactive nature, has a more adverse effect on resilience and set than phenolic resin crosslinked with HMT, as shown in Table 9.14. The formulation is based on 100 parts SBR 1500 and 50 parts of HAF (N-330) carbon black.

Changes in accelerator systems produce relatively small differences in hardness and modulus. Sulphenamides tend to produce slightly harder vulcanisates than thiazoles, and boosted accelerator systems small hardness increases over those produced by straight systems.

Softeners and plasticisers reduce hardness and, at the levels normally used to secure acceptable processing, require additional filler or reinforcing agent to counteract their effect. In this respect, where their cost can be accepted, acrylic monomers or phenolic resins can act as effective aids during mixing and processing without recourse to additional softeners. Mineral process oils, which are widely used in general purpose polymers as low-cost softeners, produce a hardness drop of approximately 1 IRHD for each two parts added.

Table 9.10 INFLUENCE OF HEAT TREATMENT ON POLYMER–FILLER MASTERBATCHES

	IIR (1·6% unsaturation)		IIR (1·6% unsaturation)		SBR 1502	
Polymer						
Reinforcing filler (50 p.p.h.r.)	Carbon black (EPC, S-300)		Carbon black (HAF, N-330)		Synthetic silica (20 nm)	
Mixing method	Normal	Masterbatch treated (10 min at 165°C)	Normal	Masterbatch treated (10 min at 160°C)	Normal	Masterbatch treated † (10 min at 170°C)
Promoter	—	—	*p*-Quinonedioxime*		—	—
Mix viscosity [ML(1+4)120°C]	75	60	80	67	97	69
Vulcanisation	45 min at 148°C	45 min at 148°C	45 min at 148°C	45 min at 148°C	20 min at 148°C	20 min at 148°C
Physical properties						
Tensile strength (kgf/cm²)	182	185	156	186	202	207
Stress at 300% strain (kgf/cm²)	56	91	79	167	41	98
Elongation at break (%)	680	560	510	340	690	470
Hardness (IRHD)	54	50	64	57	80	67
Resilience at 50°C (%)	49·9	55·6	50·8	54·0	55·0	61·1
Compression set (25%, 24 h at 70°C)	29·8	24·1	33·6	28·8	17·7	10·8

* The masterbatch was made in concentrated form from 100 parts polymer, 80 parts carbon black and 1 part *p*-quinonedioxime using the method described by Edwards and Storey (1957); it was then diluted back to 50 parts black equivalent with fresh polymer.
† The subsequent processing of the heat-treated SBR–silica masterbatch is enhanced by special metallic soaps used at levels of 2–4 p.p.h.r.

Table 9.11 ADDITION OF PHENOLIC RESIN TO SBR

Resin addition (p.p.h.r.)	0	5·0	10·0	20·0
HMT addition (p.p.h.r.)	0	0·5	1·0	2·0
Physical properties				
Tensile strength (kgf/cm^2)	200	210	175	150
Stress at 300% strain (kgf/cm^2)	90	100	90	90
Elongation at break (%)	500	525	525	500
Hardness (IRHD)	56	60	64	72

Formulation:

SBR 1500	100	CBS	1
N-330 black	30	PBN	1
Process oil	5	Sulphur	2
Zinc oxide	4	Phenolic resin	as shown
Stearic acid	2	HMT	as shown

Vulcanisation, 30 min at 148°C.

Table 9.12 ADDITION OF ACRYLIC MONOMER TO PEROXIDE-CROSSLINKED NBR

Physical properties	Monomer addition			
	0	5·0 p.p.h.r.	10·0 p.p.h.r.	20·0 p.p.h.r.
Tensile strength (kgf/cm^2)	190	195	195	200
Stress at 100% strain (kgf/cm^2)	80	105	140	165
Elongation at break (%)	250	200	150	120
Hardness (IRHD)	70	75	80	86

Formulation:

NBR polymer (36% acrylonitrile)	100	Stearic acid	1
N-774 black	50	Monomer	as shown
Zinc oxide	5	Dicumyl peroxide	1·75

Vulcanisation, 30 min at 148°C.

Table 9.13 ADDITION OF HIGH-STYRENE RESIN TO SBR

Physical properties	Resin addition			
	0	5·0 p.p.h.r.	10·0 p.p.h.r.	20·0 p.p.h.r.
Tensile strength (kgf/cm^2)	140	144	149	135
Stress at 300% strain (kgf/cm^2)	33	42	50	54
Elongation at break (%)	750	660	600	550
Hardness (IRHD)	60	63	67	74

Formulation:

SBR 1502	100	MBTS	1
Zinc oxide	3	DPG	0·5
Stearic acid	1	Sulphur	2
Aluminium silicate	30	Resin (85% styrene)	as shown

Vulcanisation, 20 min at 148°C.

Table 9.14   ADDITION OF RESIN TO SBR–CARBON BLACK MIX

	Phenolic			High-styrene	
Resin addition (p.p.h.r.)	0	10	20	10	20
HMT addition (p.p.h.r.)	0	1	2	0	0
Physical properties					
Hardness (IRHD)	67	80	90	78	88
Resilience at 50°C (%)	55	49	44	46	41
Compression set (25%, 24h at 70°C)	25·5	30·5	36·2	33·7	42·0

Formulation:
SBR 1500	100	Antioxidant	1
N-330 black	50	Process oil	5
Zinc oxide	4	CBS	1·25
Stearic acid	2	Sulphur	2
Resin	as shown		

Vulcanisation, 30 min at 148°C.

Relatively less softening is produced by factice (vulcanised vegetable oil), which not only acts as an efficient process aid but is also capable of enhancing the crosslinking system in some instances (Flint, 1962).

9.4.2   ELASTICITY

As discussed in Chapter 3, vulcanised rubbers show viscoelasticity, and the departures from perfect elasticity are evaluated by measurement of resilience (Section 11.3.12), creep and stress relaxation (Section 11.3.10), and set (Section 11.3.11). Compounding which contributes to a more tightly knit crosslinking system occupying the maximum possible volume proportion of the vulcanisate will enhance the elastic properties, as displayed by resilience. If this crosslinking system is stable, with regard to any degradative condition which may be encountered in service, then the associated properties of set, creep, and stress relaxation will also be minimised. Appropriate antioxidant protection of the polymer will give further improvement.

As discussed in Chapter 5, polymers such as NR, IR, CR, BR, and certain types of EPDM may all be crosslinked to give high resilience levels, under ideal conditions in excess of 90% with some polymer systems compounded in gumstock form (Table 9.15).

Loading with particulate fillers progressively reduces resilience, particularly with the more reinforcing types of carbon black and silica, the effect being more marked at a given loading level as the ultimate particle size is reduced (Table 9.16). At the same time creep, set, and stress relaxation characteristics are worsened.

Softeners and plasticisers at the levels normally used have relatively small effects but, at higher concentrations, generally produce reductions in resilience, high-viscosity petroleum softeners of increased aromaticity being more detrimental than lower-viscosity paraffinic types (Table 9.17). Some ester plasticisers in NBR rubbers can increase resilience. In NBR, SBR, and IIR, increases in the proportion of butadiene or isoprene in the copolymer give increased resilience (Table 9.18).

Table 9.15 EFFECT OF LEVEL OF VULCANISING INGREDIENTS

	Sulphur			CBS		
Accelerator level (p.p.h.r.)	0·4	0·6	0·8	0·5	0·5	0·5
Sulphur level (p.p.h.r.)	2·5	2·5	2·5	1·0	2·0	3·0
Vulcanisation	25 min at 148°C			35 min at 148°C		
Physical properties						
Tensile strength (kgf/cm^2)	240	260	250	230	255	260
Stress at 300% strain (kgf/cm^2)	125	140	150	90	130	150
Elongation at break (%)	530	500	450	550	510	450
Hardness (IRHD)	60	63	64	57	63	64
Resilience at 50°C (%)	68	70	73	64	69	72

Formulation:
NR-RSS   100   Process oil   5
N-330 black   50   CBS   as shown
Zinc oxide   4   Sulphur   as shown
Stearic acid   2

Table 9.16 COMPARISON OF INCREASINGLY REINFORCING CARBON BLACK

Carbon black type	Code ASTM Particle size (nm)	HAF N-330 32	ISAF N-220 28	SAF N-110 22
Vulcanisation, 20 min at 148°C	Tensile strength (kgf/cm^2)	253	258	268
	Stress at 300% strain (kgf/cm^2)	220	205	200
	Elongation at break (%)	360	400	420
	Hardness (IRHD)	72	75	77
	Resilience at 50°C (%)	63·2	58·9	54·8
Vulcanisation, 35 min at 148°C	Compression set (25%, 24 h at 70°C)	15·8	17·2	17·8
	Goodrich flexometer (10·5 kgf/cm², 6 mm stroke, 30 min run at 24 Hz):			
	Dynamic compression drift (%)	6·2	7·6	12·6
	Permanent set (%)	4·4	7·0	18·2
	Temperature rise (°C)	51	58	65

Formulation:
NR-RSS   100   Process oil   5
Carbon black   60   Antioxidant   2
Zinc oxide   4   CBS   1·1
Stearic acid   2   Sulphur   1·5

Table 9.17 EFFECT OF SOFTENER TYPE—PETROLEUM OILS IN SBR

Oil type	Aniline point (°C)	Resilience (%)	
		At 20°C	At 50°C
Highly aromatic	15	34·2	44·6
Aromatic	30	36·2	46·1
Naphthenic	80	41·1	49·2
Paraffinic	100	41·8	50·8

Formulation:

SBR 1500	100	Resin tackifier	2
N-220 Black	70	Antioxidants	2·5
Extending oil	42·5	CBS	1·25
Zinc oxide	5	Sulphur	2
Stearic acid	2		

Vulcanisation, 45 min at 148°C.

Table 9.18 EFFECT OF COPOLYMER CONSTITUTION OF NBR

Physical properties	Acrylonitrile in polymer			
	28%	33%	40%	50%
Resilience at 50°C (%)	66	63	56	39
Compression set (25%, 24 h at 70°C) (%)	13·3	16·3	17·8	28·0
Tension set (75% elongation at break, 10 min stretch, 10 min recovery) (%)	8	9	10	28

Formulation:

Polymer as shown	100	Brown factice	2·5
N-774 black	50	MBTS	1·5
Stearic acid	1	DPG	0·25
DBP	7·5	Sulphur	2
Zinc oxide	5		

Vulcanisation, 45 min at 148°C.

9.4.3   STRENGTH

At relatively low hardness levels, say below 50 IRHD, highest tensile strength levels are most easily obtained using high gum strength polymers which crystallise on stretching, such as NR, IR, CR, or isoprene–acrylonitrile co-polymer. Other polymers, such as SBR, NBR, and BR of low gum strength, require fine particle size reinforcing fillers to develop maximum strength, and this reinforcement is accompanied, as indicated in Section 9.4.1, by increased extension modulus and hardness.

In both groups of rubbers the highest strengths are given by using the fine particle size carbon blacks or reinforcing silicas, and to obtain highest levels it is essential that these are well dispersed. Optimum levels vary according to the polymer–filler system but usually fall in the range 30–60 p.p.h.r. (Table 9.19). Certain resins are also capable of producing increased strength levels in some conditions. Thus high-styrene resins can produce strength increases in SBR, and crosslinkable phenolic resins can also produce strength increases, particularly with NBR (Table 9.20).

Table 9.19   COMPARISON OF CARBON BLACK TENSILE REINFORCEMENT IN RUBBERS OF HIGH AND LOW GUM STRENGTH

Gum strength		High			Low		
Carbon black type	Code	GPF	FEF	HAF	GPF	FEF	HAF
	ASTM	N-660	N-550	N-330	N-660	N-550	N-330
Physical properties							
Tensile strength (kgf/cm^2)		225	245	265	190	215	235
Stress at 300% strain (kgf/cm^2)		115	143	150	85	120	145
Elongation at break (%)		540	510	480	600	530	490

Formulation:

	High	*Low*		*High*	*Low*
NR-RSS	100	—	Process oil	5	5
SBR 1500	—	100	PBN	1·5	1·5
Carbon black as shown	50	50	CBS	0·5	1·0
Zinc oxide	4	4	Sulphur	2·5	2·0
Stearic acid	2	2			

Vulcanisation, 30 min at 148°C.

Table 9.20   ADDITION OF PHENOLIC RESIN TO NBR

Resin addition (p.p.h.r.)	0	10	20	40
HMT addition (p.p.h.r.)	0	1	2	4
Physical properties				
Tensile strength (kgf/cm^2)	51	85	110	145
Elongation at break (%)	400	340	295	260
Hardness (IRHD)	44	60	68	81

Formulation:

NBR polymer	100	HMT	as shown
Zinc oxide	5	MBTS	1·5
Stearic acid	1	Sulphur	1·75
Resin	as shown		

Vulcanisation, 45 min at 148°C.

The rapid fall in the tensile strength values of many high-strength vulcanisates at elevated temperatures has to be borne in mind when compounding for high-temperature service (Fig. 4.29).

9.4.4   RESISTANCE TO ABRASION AND TEAR

9.4.4.1   *Abrasion*

The resistance of vulcanised rubbers to abrasion is a difficult measurement to carry out since ratings depend on the precise conditions of wear, and a number of features are not understood. As a result, compounding aimed at improving abrasion resistance sometimes produces confusing results.

In general, as a rubber is compounded with progressively smaller particle size carbon blacks, say FEF (N-550), HAF (N-330), or ISAF (N-220), the resistance to abrasion increases, optimum resistance occurring in the region of 50 p.p.h.r. With non-black fillers a similar pattern is seen, maximum resistance being given by reinforcing silica of ultimate particle size around 20 nm.

For vulcanisates based on NR, SBR, and CR, abrasion resistance can generally be improved appreciably by replacing a minor proportion of polymer by polybutadiene. The high abrasion resistance of this polymer is best utilised in this way since, for a number of reasons, it is not normally used as the sole polymer in technical vulcanisates. This practice cannot normally be applied to NBR or IIR because of lack of compatibility and inability to covulcanise respectively.

Table 9.21   RELATIVE ABRASION RESISTANCE OF POLYMER–CARBON BLACK COMBINATIONS: AKRON LABORATORY MACHINE, RATINGS RELATIVE TO HAF (N–330) IN EACH POLYMER

Polymer	Carbon black type		
	FEF(N-550)	*HAF(N-330)*	*ISAF(N-220)*
NR-RSS	85	100	112
SBR 1500	67	100	125
CR	75	100	115
IIR (2% unsaturation)	80	100	130

Table 9.22   INFLUENCE OF SEVERITY OF WEAR CONDITIONS ON RELATIVE ABRASION RESISTANCE OF NR–BR BLEND COMPARED WITH 100% NR. COURTESY NRPRA

Test condition	Slip angle (degrees)	Rating*
Wet	4	78
Dry	1	96
Dry	1·5	118
Dry	2	132
Dry	4	168

* Rating = $\dfrac{\text{wear resistance of 50–50 NR–BR}}{\text{wear resistance of 100\% NR}}$

In Table 9.21 the three carbon blacks already referred to are compared for relative abrasion resistance in four different polymers. These are representative laboratory ratings obtained under the severe conditions operating in the Akron abrader using a slip angle of 15° (Section 11.3.7). In each mix the black loading was 50 p.p.h.r. in the presence of 5 p.p.h.r. of a process aid appropriate to the polymer. In each polymer the N-330 black rating is taken as 100. Comparisons between polymers should not be made from these data because of the conditions of test. Under average service conditions in pneumatic tyres, NR or SBR 1500 reinforced with N-220 black gives abrasion resistance ratings of the order of 10% above those for N-330.

The effect of substituting solution polybutadiene for part of the NR in an HAF (N-330) reinforced tread stock can be judged from the data in Table 9.22 taken from NRPRA Tech. Info. Sheet No. 72 (1964). This also illustrates the influence of test severity on the abrasion rating obtained. Results were obtained from tyres on trailer wheels operating at the stated slip angles.

Under mild abrasive conditions NR is superior, whereas under the most severe conditions the BR blend is markedly more resistant. A good level of antioxidant protection, particularly of certain p-phenylenediamines, has a favourable influence on abrasion resistance. Data relating to NR have been given for a number of antioxidants and antiozonants (NRPRA Tech. Info. Sheet No. 81, 1965).

9.4.4.2 *Tear*

As with abrasion resistance, the determination of tear resistance presents a number of problems which render compounding variations intended to improve this property sometimes difficult to interpret. The nicked crescent test-piece is widely used (Section 11.3.6), and the data presented here are based on this test. It suffers from the disadvantage that some of the energy applied to the test-piece is dissipated in stretching the arms of the sample rather than tearing it. Methods have been proposed to avoid this difficulty (Greensmith and Thomas, 1955; Veith 1965-1).

Unlike abrasion resistance, reinforcement with carbon blacks of decreasing particle size does not always produce a noticeable effect on tear resistance. With NR and IIR improvement is evident, but with other polymers, such as CR or SBR 1500, the effects are much less marked (Table 9.23). In low-hardness stocks the high gum strength polymers, such as NR or CR, give

Table 9.23 TEAR RESISTANCE IMPARTED BY REINFORCING CARBON BLACKS IN OESBR 1712*

Loading per 137·5 parts SBR 1712	Tear resistance (kgf/cm)		
	FEF(N-550) black	HAF(N-330) black	ISAF(N-220) black
55	61	64	66
70	66	71	68
85	71	70	68

* Within these loadings and black types there is relatively little effect.

Table 9.24   EFFECT ON TEAR RESISTANCE OF 40 VOL. % LOADINGS OF NON-BLACK FILLERS IN CR

Filler	Loading (p.p.h.r.)	Tear resistance (kgf/cm)
Whiting	90	26
Soft clay	85	40
Talc	90	55
Reinforcing silica	65	70

higher tear levels than low gum strength types, such as SBR or NBR. In CR small particle size non-black reinforcing agents such as synthetic aluminium silicate or reinforcing silica give high tear strength figures, silica being capable of producing levels superior to that given by reinforcing carbon blacks.

Coarse fillers of mineral origin, particularly those with relatively large particle fractions, generally detract from tear resistance, but fine clays give some improvement (Table 9.24).

Certain resinous processing aids, e.g. coumarone–indene resins, petroleum resins, and phenolic types, are capable of improving tear resistance values when added to a formulation in small amounts of the order of 5%. This is probably due to their ability to enable the polymer to wet out the surface of the reinforcing filler more effectively, thereby improving the polymer–filler interaction.

Tear resistance falls appreciably with increasing temperature (Table 9.25).

Table 9.25   EFFECT ON TEAR RESISTANCE OF RUBBER TYPE AND TEMPERATURE: REINFORCEMENT WITH 50 P.P.H.R. HAF (N–330) BLACK

Rubber type	Tear resistance (kgf/cm)	
	At 20°C	At 100°C
NR	125	75
NBR (40% acrylonitrile)	70	45
SBR 1500	70	46
IIR (1·5% unsaturation)	85	63
BR (92% cis)	45	25

9.4.5   RESISTANCE TO CYCLIC STRESSING, FLEX CRACKING, CUT GROWTH, AND FATIGUE

Where rubber products are subjected to cyclic stresses in service through either repeated or intermittent flexing or compression, the initiation and development of cracks is a frequent cause of failure. Furthermore, at high rates of flexing, e.g. in rapidly rotating tyres or drive belts, heat build-up can be an important factor and the compounding involved must allow for this.

Apart from the choice of base polymer, compounding variations can have quite complex effects on relative crack development. The nature and stability of the crosslinking system, the choice of protective agents, and the degree of dispersion of curatives and reinforcing and other fillers, leading to localised strains, can all influence performance on flexing. If compounding reduces resilience, additional heat will be generated on flexing, and, if severe, an increased rate of cracking will result.

The Goodrich flexometer data quoted in Table 9.16 illustrate the influence of reduced resilience on heat build-up during rapid cyclic flexing, the reduced resilience in this example having been produced by increasingly reinforcing carbon blacks (Section 11.3.9).

The effect of polymer choice can be illustrated by the data reported by Beatty (1964) (Table 9.26). Results were obtained on moulded test-pieces stressed at 12 kHz. They show that polymer ranking depends on the strain

Table 9.26 INFLUENCE OF RUBBER TYPE ON FATIGUE RESISTANCE. FROM BEATTY (1964), COURTESY AMERICAN CHEMICAL SOCIETY

Rubber type	Cycles to failure × 10⁶		Time for × 4 crack growth (h)
	Strain from −75% to 0%	Strain from +50% to +125%	
OESBR	30·0	5·8	38
OESBR–BR blend	13·0	11·3	41
EPDM (DCPD)	30·0	23·5	13
SBR	22·0	0·2	—
NR	4·0	30·0	—

cycle used and that good resistance to crack initiation is not always accompanied by good crack growth resistance (see, for example, the data for EPDM).

For exacting service, the precise operating conditions must, therefore, be considered before choosing the base polymer. For NR the importance of the grade chosen for severe service has been emphasised by Heal (1963).

Where a risk of physical cutting is involved, then the use of high gum strength polymers such as NR or CR is desirable, reinforced to give a balance between low modulus, high extensibility, and optimum tear resistance. For this purpose, carbon blacks such as HAF-LS (N-326), channel blacks (S-330), reinforcing silicas, or silica–carbon black blends are used.

When the anticipated flexing is of constant amplitude, extension modulus should be kept as low as possible to minimise stresses. The data in Table 9.27 illustrate this for an SBR 1500 formulation reinforced with HAF (N-330) black. Auer, Doak, and Schaffner (1958) have reported more extensive data.

The influence of sulphur crosslink variants may be anticipated from the fact that, for many years, the fatigue resistance obtained using the so-called 'sulphurless' systems, e.g. using TMTD alone, has been regarded as inferior to that of a conventional system. This is presumably because of the relative lack of mobility of elastomer chains joined by monosulphide links as opposed to polysulphidic ones. Results for NR show that, although this is true if the vulcanisate is cycled through zero strain, efficient vulcanisation systems can be superior under non-relaxed cycling, and also under elevated temperature conditions, because of the greater thermal stability of monosulphidic links (NRPRA Tech. Bull., 1968).

The degree of cure is also important. If vulcanisation cycles are prolonged,

Table 9.27 INFLUENCE OF EXTENSION MODULUS ON CUT GROWTH RESISTANCE: DATA FOR SBR 1500 BASED FORMULATION LOADED WITH HAF (N–330) BLACK

Stress at 300% strain (kgf/cm²)	De Mattia cut growth (kc)	
	Growth from 2 mm to 4 mm	Growth from 4 mm to 8 mm
136	6·3	13·2
147	2·9	7·9
199	1·0	2·4

thermally susceptible polymers such as NR or IR may be affected by degradation, whilst SBR can suffer from high modulus effects due to continued crosslinking with the consequences on cut growth resistance indicated above.

Antioxidants vary in their influence on fatigue (Section 6.5.4). At elevated temperatures, crack growth rate increases comparatively rapidly for SBR and CR but less so for NR (Beatty, 1964).

9.4.6 RESISTANCE TO DEGRADATIVE AGENCIES

9.4.6.1 *Heat*

The relative ratings of typical polymers with respect to heat resistance have already been given in Tables 9.1 and 9.2. Attention to crosslinking systems and choice of other compounding ingredients can increase appreciably the maximum service temperature of a polymer.

Reference has already been made to the enhancement of heat resistance of NR by using a thermally stable monosulphidic crosslinking system as opposed to the normal polysulphidic type. In the past such a crosslinking system was achieved with TMTD alone, but this method is limited because of prevulcanisation risk and objectionable bloom in the final vulcanisate. The new systems avoid these disadvantages and, when combined with a good heat-resistant antioxidant system, are capable of increasing the typical service temperature of NR from around 70°C to 100°C, and even above this for intermittent exposure. An example is given in Table 9.28. A number of other systems have also been described (Section 5.6.1) (NRPRA Tech. Info. Sheet Nos 78, 79, 86, and 94).

When sulphur crosslinking is not employed, as with CR, other procedures are used. Non-sulphur-modified CR shows greater heat resistance than the sulphur-modified types and is, therefore, preferred. It is used with a crosslinking system based on increased levels of zinc oxide and ETU in the presence of a selected retarder to minimise prevulcanisation problems. At the

Table 9.28 VULCANISING–PROTECTIVE AGENT SYSTEM FOR HEAT-RESISTANT NR MIX

Ingredient	p.p.h.r.	
Polymerised trimethyl dihydroquinoline	2·0	Antioxidants
2-Mercaptobenzimidazole	2·0	
TMTD	0·66	Efficient crosslink system
2-Morpholinothiobenzthiazole	1·40	
Sulphur	0·35	

same time, a high level of an octylated diphenylamine antioxidant is added to improve the polymer stability. A typical system incorporating these principles is given in Table 9.29. Further data on choice of antioxidants for other synthetic rubbers will be found in a recent paper (Hill, 1969).

Softeners, processing aids, and plasticisers should be chosen so that they have long-term thermal stability at the anticipated service temperature and are of low volatility. High flash point petroleum oils, or monomeric or polymeric ester plasticisers are used frequently, choice being governed by cost and compatibility with the polymer concerned. Somewhat unexpectedly, rapeseed oil has been found to be particularly effective in CR.

Comparative data on the resistance of vulcanisates at temperatures above 100°C have been published by Thomas and Sinnott (1969).

Table 9.29 FORMULATION FOR HEAT-RESISTANT MIX (NON-SULPHUR-MODIFIED CR)

Ingredient	p.p.h.r.	
Octylated diphenylamine	4·0	Antioxidant
Dodecyl mercaptan	1·0	Retarder
Light calcined magnesia	4·0	Vulcanising system
Zinc oxide	10·0	
Ethylene thiourea	1·0	

9.4.6.2 *Flame*

By using halogen-containing polymers such as CR, it is possible to produce vulcanisates which, although not non-inflammable, are self-extinguishing when an applied flame is removed (Thompson, Hagman, and Mueller, 1958). In such formulations tricresyl phosphate and liquid chlorinated paraffin wax should be used as processing aids in place of oils, waxes, or factice, and the carbon black level should be kept at a minimum, the black being used only as a pigment (Section 6.6.2). Reinforcing mineral fillers are used, along with antimony trioxide and the intumescent zinc borate, which assists by forming a hard surface crust on exposure to flame. Recently, certain chloroalkyl phosphates have become available which may prove to be of value.

The flame resistance of NBR can be improved by blending with PVC and using tricresyl phosphate plasticisation.

9.4.6.3 *Liquids*

The action of liquids on vulcanisates may be regarded as the net effect arising from physical swelling of the polymer, the degradation of the polymer and fillers, and the leaching of, or attack on, plasticising agents. Tables 9.1 and 9.2 give a summary of the general action of liquids.

For optimum resistance to a particular liquid which will have primarily a swelling action on the polymer phase, the polymer should be chosen so that its solubility parameter is as far removed from that of the contact liquid as the required processing and other vulcanisate properties will allow (Beerbower, Pattison, and Saffin, 1964). Chemical attack on the polymer is most likely to arise from oxidising substances, and in this event polymers having a relatively inert backbone structure should be chosen, e.g. IIR, CSM, or EPM.

In addition to the intrinsic swelling and chemical resistance of the polymer, its volume content in the vulcanisate should be kept as low as practicable since this will minimise its contribution to any volume change. Fillers should also be chosen to be as inert as possible under the service conditions, e.g. precipitated barium sulphate (blanc fixe) is useful under acid immersion conditions. Plasticiser stability is also important, and ester plasticisers should be avoided where hydrolysis could occur, e.g. under hot alkaline conditions. Extractable plasticisers or additives are undesirable where shrinkage effects cannot be permitted, for instance in certain seals, unless swelling of the polymer by the liquid in which it operates compensates for this. Polymeric plasticisers, low molecular weight polymers which will crosslink on vulcanisation, and oil-resistant factices all have applications in such cases. A high state of crosslinking in the final vulcanisate will also assist in reducing ultimate swelling.

Where resistance to water or aqueous solutions is required, the presence of water-soluble salts in the vulcanisate is detrimental as they facilitate the ingress of water. Thus in SBR manufactured by salt–acid coagulation methods, water swelling is greater than that given by specially finished types (e.g. 1503, 1509, and 1708) and solution-polymerised types, which are for the same reason preferred for electrical applications. CR which is crosslinked using magnesia–zinc oxide systems produces traces of soluble chlorides; so, for maximum water resistance, lead oxide cures are preferred since lead chloride has only limited solubility. Magnesia-crosslinked CSM reacts similarly. Mineral fillers containing traces of soluble salts can also produce these effects. NR and the water-resistant grades of SBR have useful levels of resistance to a wide range of aqueous solutions and to chemicals which do not have any marked oxidising action.

IIR and EPM are resistant to animal and vegetable fats, oxidising chemicals, and some oxygenated organics, e.g. ketones, but not to aliphatic or aromatic hydrocarbons. CR is also resistant to animal and vegetable fats and waxes, as well as aliphatic hydrocarbons (Table 4.17). The general oil and solvent resistance of NBR is of a high order (Fig. 4.26). CSM has a high level of chemical resistance, particularly towards oxidising solutions. When very high temperatures are involved, then the best general chemical and solvent resistance is given by the fluorinated polymers (Table 4.26).

9.4.6.4 *Light*

When considering the influence of light, it should be remembered that the ultra-violet region of the spectrum has a much higher potential energy than the visible or infra-red regions, the level being of an order capable of breaking down organic molecules (Gysling and Heller, 1961). In exposed vulcanisates, this manifests itself as discolouration, surface embrittlement, or break-up in which oxygen also plays a part.

To protect rubbers, particularly from intense sunlight, which is rich in ultra-violet radiations, compounding with materials which will either reflect it or absorb and transform it into innocuous wavelengths is necessary. This involves the use of suitable pigments, stabilisers, or possibly ultra-violet absorbing materials. Carbon black is fortunately an outstanding absorber of ultra-violet light and is effective at quite low loadings of a few parts per hundred parts of rubber. Smaller particle size blacks are particularly effective. Among non-black fillers, barium sulphate and calcium carbonate have good ultra-violet reflectance characteristics, whereas zinc oxide and titanium dioxide are relatively poor. The rutile form of titanium dioxide shows greater absorption than the anatase form. Among coloured pigments, phthalocyanine blues and greens, quinacridone reds, iron oxide reds, yellows, and browns, and chromium oxide green have good ultra-violet absorption characteristics. Specific absorbents of ultra-violet light include a wide range of benzophenone derivatives.

For non-black vulcanisates destined for long-term light exposure, CSM and NBR–PVC blends are frequently used as base polymers.

Processing aids (such as low aniline point aromatic process oils, aromatic resins, and extenders) which can degrade under exposure conditions should be avoided in compounding non-black formulations. Staining antioxidants are not acceptable, and the use of non-staining phenolic types is essential (Table 6.11). Some accelerator residues, e.g. those from certain sulphenamides and from guanidines, can also produce discolouration and may not be acceptable under critical conditions.

Black formulations which will be used in contact with light-coloured paints or lacquers or other coloured surfaces should be compounded similarly if migrating or contact stains are unacceptable.

9.4.6.5 *Ozone*

When stretched, vulcanisates are exposed to ozone, even at very low concentrations, and, particularly under dynamic conditions, cracking develops at right angles to the direction of strain, the rate of cracking varying markedly from one polymer to another and being influenced by the compounding employed.

With susceptible formulations, additions of materials, such as petroleum wax blends, of limited compatibility result in the formation of surface blooms which give some physical protection because the flexible protective barrier prevents ozone attack at the polymer surface provided it is not disrupted. Chemical protective agents, notably the substituted *p*-phenylenediamines, give additional protection without suffering from this defect (Table 6.10).

The ozone resistance of non-protected polymer vulcanisates increases through the series: SBR, NR, CR, IIR (increases as the degree of unsaturation decreases), CSM, EPDM. In addition, NBR, which is susceptible to ozone attack, can be improved considerably by blending with PVC, suitably stabilised. At NBR–PVC ratios of the order of 60:40, ozone resistance is of the same order as that given by IIR but resilience levels are low at such ratios because of the proportion of non-crosslinked PVC. The ozone resistance of any of the above polymers will be improved by compounding with small additions of selected wax blends or *p*-phenylenediamine antiozonants or both, common levels being in the range 1–5 p.p.h.r. Ozone resistance of EPDM is so high that additional protection is rarely necessary. Levels used with any particular polymer must be chosen with care, because of limited compatibility, to avoid excessive blooms. Heavy wax blooms which crack on flexing are undesirable.

9.4.6.6   *Atmospheric Exposure*

When vulcanisates are exposed to outdoor weathering, the material is not only subject to the combined effects of light and ozone but also to fluctuating temperatures, rain, perhaps hail and snow, and the eroding action of any airborne dusts. Conditions will be aggravated if the service is a dynamic one. Rain has a leaching action on the degraded surface, and, when mineral fillers are present, these gradually become exposed and may be rubbed off—the phenomenon known as 'chalking'. Compounding which minimises the effects of ultra-violet radiation will reduce this. Geographical location has a marked effect on rates of weathering. Ultra-violet effects will be most noticeable in regions of considerable sunshine, whilst ozone effects will vary since atmospheric ozone concentrations at the earth's surface increase from equatorial regions towards the poles, and with increasing altitudes. An account of ozone distribution in the atmosphere has been given by Jones (1968). Exposure to direct sunlight can give rise to considerable heat build-up at the surface of

Table 9.30 SURFACE TEMPERATURES OF EXPOSED COLOURED VULCANISATES: SAMPLES WERE FACING DUE SOUTH AND WERE EXPOSED TO HAZY SUNSHINE IN MID-AUGUST IN THE UK

Test no.	Surface temperature (°C)		
	Black	Yellow	Medium blue
1	51	43	45
2	58	$47\frac{1}{2}$	47
3	61	49	$52\frac{1}{2}$
4	62	49	56
5	61	50	55

vulcanisates. The build-up is greatest for black rubbers and least for white, with colours giving intermediate values. In extreme conditions, surface temperatures of black vulcanisates can exceed 100°C, and conditions can become particularly severe inside parked motor vehicles. In such circumstances, protective wax blooms may melt and redissolve in the polymer. Examples of surface temperatures, measured with a low heat capacity probe and reached on vulcanisates exposed in the UK during mid-August with ambient temperatures in the range 19°–22°C are given in the Table 9.30. It will be seen that black gave the highest temperature, yellow the lowest, with blue intermediate.

For any severe outdoor weathering application the most satisfactory results will be obtained by combining the most resistant polymer, protective agent system, and filler system appropriate to the conditions.

9.4.7 LOW-TEMPERATURE RESISTANCE

If high gum strength polymers which crystallise easily on stretching, such as NR or CR, are exposed to moderately low temperatures for considerable periods of time, a time-dependent reversible stiffening phenomenon known as crystallisation occurs. For NR this is most rapid at about −26°C, and at −12°C for CR. It is a reversible effect, and the relative rates of crystallisation vary from one CR type to another. A special crystallisation-resistant grade of NR has also been developed. Crystallisation may be minimised by using very tight states of crosslinking and, where possible, by compounding with increased amounts of filler and petroleum plasticisers, including resinous types. In CR, ester plasticisers should be avoided since they tend to increase crystallisation rates—the reasons for this are not fully understood. Exposure of all rubbers to low temperatures also produces a reversible stiffening, under static conditions, at a temperature dependent on the glass transition temperature T_g of the polymer (Tables 9.1 and 9.2). Under dynamic conditions, cracking may occur if the stiffening is severe.

For optimum results with respect to low-temperature flexibility and with regard for other essential properties, formulations should be chosen to combine the polymer having the lowest T_g and, if possible, being free from crystallisation tendencies and having a high degree of crosslinking, with compatible plasticisers, also having the lowest possible freezing point (e.g. adipates, sebacates, paraffinic process oils, or butyl oleate, as appropriate).

Silicone rubbers have extremely good resistance to low temperatures (Section 4.13.4), but, where their cost or mechanical property level cannot be entertained, blends of NR with low-*cis* variants of BR give good results. The low-temperature dependence of NBR on the acrylonitrile content has been shown in Fig. 4.27. Compounding with ester plasticisers of the sebacate or adipate type is frequently necessary.

9.4.8 ELECTRICAL PROPERTIES

Electrical properties of vulcanisates depend not only on the choice of polymer but also on the level and nature of filler and the type of plasticiser. Compounds of NR, those based on specially finished types of SBR with low water extracts (e.g. 1016, 1503, and 1708), and IIR are all capable of giving products characterised by high resistivity. CR, CSM, and NBR–PVC blends are suitable as insulants in low-voltage work but are more applicable as outer protective coatings over insulating layers where weathering resistance or solvent-resistant characteristics are necessary (Table 9.31).

Carbon blacks detract from electrical resistivity, their precise effect depending on ultimate particle size, degree of structure, and level of dispersion in the

Table 9.31 EFFECT OF RUBBER TYPE ON ELECTRICAL RESISTIVITY

Formulations*	Rubber type			
	IIR (2% unsaturation)	NR-RSS	SBR 1503	CR (non-sulphur-modified)
Rubber as shown	100	100	100	100
Zinc oxide	5	5	5	5
Stearic acid	1	0·5	1	0·5
Activated calcium carbonate	50	50	50	50
China clay	50	50	50	50
Phenolic antioxidant	1	1	1	1
Paraffin wax	2	2	2	2
Process oil	5	5	5	5
Light calcined magnesia	—	—	—	4
ETU	—	—	—	0·5
MBT	1	—	—	—
MBTS	—	1	1	—
TMTD	1	1	1	—
Sulphur	2	1	0·75	—
Electrical resistivity (Ω cm)	25×10^{15}	9×10^{15}	$1·5 \times 10^{13}$	1×10^{12}

* Vulcanisation, 30 min at 148°C.

vulcanisate. The lowest loadings of the least reinforcing types will detract least from insulating characteristics. Of the reinforcing blacks, channel types give highest and high structure furnace give lowest resistance, particularly if compounded into normal grades of SBR or NBR (Table 9.32). The super conductive grades of furnace blacks and acetylene blacks are special purpose materials for those applications requiring antistatic or conducting properties in the final product, e.g. for the dissipation of static electricity.

Mineral fillers, such as clays, whitings, calcined clays, and coated silicas, especially those with low water-soluble contents, give appreciably higher electrical resistance and dielectric characteristics than carbon black. Some synthetic aluminium silicates are detrimental and can give non-black anti-static products.

When considering choice of plasticisers or processing aids, petroleum oil softeners will be found to give relatively high volume resistivity of the order of 10^{13} Ω cm, paraffinic oils giving higher resistance than naphthenic or aromatic types. Ester plasticisers vary considerably in their effects, dioctyl phthalate and dioctyl sebacate having resistivities of 10^{11} Ω cm, whereas those of phosphate plasticisers can be as low as 10^{8} Ω cm. Special plasticisers are available for antistatic and conductive rubbers. Possible migration effects

Table 9.32 EFFECT OF CARBON BLACK TYPE ON ELECTRICAL RESISTIVITY

Carbon black type	Volume resistivity (Ω cm)	
	SBR mix*	NR mix†
ISAF (N-220)	$2{\cdot}5 \times 10^3$	$1{\cdot}25 \times 10^3$
HAF (N-330)	$4{\cdot}0 \times 10^6$	$4{\cdot}5 \ \times 10^4$
EPC (S-330)	$4{\cdot}5 \times 10^6$	$3{\cdot}5 \ \times 10^7$

```
* SBR formulation:                      † NR formulation:
  SBR 1500              100                NR                   100
  Carbon black as shown  50                Carbon black as shown 50
  Zinc oxide              4                Zinc oxide             4
  Stearic acid            2                Stearic acid           2
  Process oil            10                Process oil            5
  CBS                     1                PBN                    1
  Sulphur                 2                CBS          0·5 (0·75 for EPC black)
  Vulcanisation, 30 min at 148°C.         Sulphur               2·5
                                          Vulcanisation, 30 min at 148°C.
```

of plasticisers into insulation stocks should be borne in mind when compounding outer protective stocks. Compounding for minimum water absorption will assist in maintaining good electrical characteristics (Section 9.4.6.3), and the use of copper-inhibiting antioxidants, e.g. di-β-naphthyl-p-phenylenediamine, in insulation stocks, particularly those based on NR, is desirable. For all insulation applications, the best possible dispersion of ingredients is essential. When considering insulating compositions, the dielectric strength (the voltage difference per unit thickness at which electrical breakdown occurs) is an important characteristic. Mechanical flaws, filler agglomerates, or other inhomogeneities can have a marked effect on such measurements, and hence the need for a high degree of ingredient dispersion. Typical comparative values for a number of rubberlike polymers have been given by Mcpherson (1963) (Table 9.33). The author points out that the range of strengths quoted for various polymer formulations illustrates the importance of choice of compounding ingredients, polymer purity, and manufacturing method, these sometimes overshadowing the influence of the polymer itself. The original paper should be consulted for further details since it contains a comprehensive review of the electrical properties of rubberlike polymers.

In general terms, volume resistivity values above 10^8 Ω cm are regarded as insulating, between 10^4 Ω cm and 10^8 Ω cm as antistatic, and below 10^4 Ω cm as conducting.

Table 9.33 TYPICAL VALUES OF DIELECTRIC STRENGTH OF RUBBERS. FROM MCPHERSON (1963), COURTESY AMERICAN CHEMICAL SOCIETY

Material	Apparent dielectric strength (MV/m)
IIR	14–30
NR	24–31
SBR	10–19
CR	4·5–30
NBR	9–11
SI	4–26

9.4.9   RESISTANCE TO GAS PERMEATION

Permeation of a gas through a vulcanised polymer is influenced by the nature of both the polymer and the gas, and is governed by the diffusivity and the solubility levels of the gas in the polymer concerned.

9.4.9.1   *Solubility of Gases*

For a particular polymer, gas solubility increases as the molecular weight of the gas increases and is also greater for gases of increased polarity. Solubility is influenced by temperature in accordance with the Clausius–Clapeyron equation. In different polymers, the solubility of a given group of gases follows a similar pattern but tends to decrease as the solubility parameter of the polymer increases.

9.4.9.2   *Diffusivity of Gases*

The influence of polymer type on diffusivity depends on the presence or absence of polar groups, such as acrylonitrile, and also whether methyl groups are present. Both polar and methyl groupings decrease diffusivity appreciably, and this accounts in part for the low permeability levels both of high-nitrile NBR and of butyl rubbers.

Diffusivity in a given polymer is influenced by the molecular volume and shape of the gas concerned, the state of crosslinking within the normally used range having a relatively smaller influence.

9.4.9.3   *Permeability*

In addition to diffusivity, solubility, and temperature considerations, permeation rate varies with the pressure differential initiating the permeation and with the thickness of the polymer barrier involved. By reducing the relative volume content of polymer in the vulcanisate, increase in filler content reduces permeation; fillers having lamellar particles, such as talc, mica, or graphite, produce a proportionally greater reduction in permeability than an equivalent volume of more regular fillers. Other mechanical properties, however, are adversely affected by such fillers, tear resistance being particularly influenced. Wherever thin vulcanisate layers are used to act as permeation barriers, it is also important to ensure that all particulate fillers are well dispersed and free from agglomerates and gritty impurities, otherwise mechanical imperfections in the sheeting can seriously impair gas retention properties. Permeability of various polymers to a wide range of gases is given in Table 2.1.

9.4.10   CONTACT WITH FOODSTUFFS, DRUGS, ETC.

Where rubber products may come into contact with foodstuffs, drugs, or drinks, certain restrictions on choice of polymers or compounding ingredients arise. Comprehensive lists of materials which are acceptable in the USA and W. Germany have been published (Anon, 1968-7) together with details of permitted concentrations and test methods where appropriate. These should be consulted when formulations for this type of application are being considered. A useful review of toxicity problems in the rubber and plastics industry has been published by Estevez (1969).

9.5 Compounding for Bonding to Non-Rubber Substrates

Although the bonding of compounded rubbers to various substrates is not fully understood, it is usually accepted that both chemical and physical agencies are involved. Furthermore, it is found in practice that enhancement of the chemical activity or improvement in the degree of physical contact of the surfaces concerned will give improved adhesion levels. At the same time, any contamination of either prepared surface is detrimental.

In rubber technology the two most common bonding requirements encountered involve bonding vulcanisates to metals (most frequently steel) or to fabrics. The bond is normally developed at the same time as the vulcanisation reaction is carried out.

9.5.1 BONDING TO METALS

For successful rubber-to-metal bonding, it is found that highly polar polymers, such as CR or NBR, may be bonded more readily than NR or SBR of lower polarity. These in turn bond more readily than IIR. In the past, bonding to metals was performed either via an ebonite tie layer or by brass plating the metal where this was practicable. The brass-plating technique requires stringent control to obtain consistent results. Nowadays, these methods have in general given way to the use of chemically reactive cements, of which a wide range is available commercially, according to the type of polymer base and metal to be bonded (Sections 8.7.2 and 10.5.3.2).

In compounding the rubber stock to be applied to a cement film, carbon black loading is normally preferred for high bond strengths, and the use of ingredients capable of blooming to the surface of the unvulcanised stock should be avoided. If this is not possible, freshly prepared material must be used. Delayed-action acceleration systems are preferred to ensure that optimum contact between stock and cement surface has been achieved before the onset of crosslinking. A number of aspects of rubber-to-metal bonding have been discussed by Cox (1969-1, 1969-2, 1969-3).

9.5.2 BONDING TO TEXTILE MATERIALS

When bonding compounded rubbers to fabrics, it is found that for many applications mechanical interaction between the surface of cotton fibres and the compounded rubber is sufficient to ensure acceptable adhesion. This does not, in general, apply to synthetic fibres, which require treatment with a chemically active coating to secure maximum adhesion (Section 8.5.2).

When chlorine-containing polymers are being used to coat fabrics, particularly when elevated service temperatures or outdoor exposure of thinly coated fabrics are involved, compounding with acid acceptors and antioxidants must be adequate to minimise formation of traces of hydrochloric acid on ageing. Failure to do this may result in serious loss of fabric strength during service, particularly when cotton, rayon, or nylon is used.

9.5.3   BONDING SYSTEMS INCORPORATED DIRECTLY INTO THE RUBBER STOCK

A recent development is the compounding of a reactive resin system directly into the rubber stock, which then bonds to untreated fabric. A formaldehyde donor is included in the formulation, and reinforcing silicas are found to enhance the adhesion and also the retention of adhesion on ageing (Creasey and Wagner, 1968; Klötzer, 1969). The best results are achieved with rayon and nylon fabrics. Table 9.34 gives the adhesion levels obtained by direct vulcanisation to crosswoven nylon for 25 min at 145°C. The formulation is based on

Table 9.34   EFFECT OF SILICA LEVEL ON SELF-BONDING RUBBERS. FROM KLÖTZER (1969), COURTESY RUBBER & TECHNICAL PRESS LTD

Carbon black (p.p.h.r.)	Reinforcing silica (p.p.h.r.)	Adhesion (kgf/3 cm)	
		Unaged	Aged for 3 days at 100°C
45	0	6	16
40	2	10	30
35	7	25	53
30	12	33	59
25	17	33	50
20	22	27	40

NR and contains carbon black and reinforcing silica levels as shown in the table, with 6 p.p.h.r. of a 50–50 blend of resorcinol and reinforcing silica and 1·5 p.p.h.r. of HMT (as a formaldehyde donor). In this formulation, optimum adhesion is obtained at about 15 p.p.h.r. of silica. The technique can also be applied to other common polymers and blends, and is reported to improve adhesion to metals.

TEN

Manufacturing Techniques

10.1   Pneumatic Tyres

BY G. F. MORTON AND G. B. QUINTON

10.1.1   INTRODUCTION

During the last decade, changing attitudes in the motor industry together with the rapid developments in the field of synthetic textiles and elastomers have had a considerable impact on tyre manufacturing plant and operating techniques.

Recent legislation in Europe and N. America has placed emphasis on safety, and this has brought about the development of more complex tread patterns to assist water displacement and thereby to minimise the tendency to aqua-plane. The 'wet grip' requirement has also led to the introduction of special rubber compounds. These features have tended to complicate factory processing of the tread elements, including the steps of vulcanisation and extraction of the finished tyre from the mould.

Improvements to road networks and, in particular, the construction of motorways have led to higher sustained operating speeds. Concurrently, suspension systems and chassis designs have altered significantly, resulting in more sensitive structures. The combined effect of these developments has tended to highlight the ill effects of vibrations and resonances excited by forces generated by lack of uniformity within the tyre structure. As a direct result, vehicle manufacturers insist upon rigid specifications for tyre uniformity particularly in the passenger car field and, to a growing extent, for the truck ranges. This requirement has become more critical with the adoption of the radial ply tyre and the increasing use of high-modulus materials in the carcass. Variables that require to be controlled include radial and lateral force variations, static unbalance, and radial and lateral run-out. It should be appreciated that the force or unbalance actually measured on the tyre is a vectorial summation of an infinite number of small forces or moments about the central axis. A complete reappraisal of previously accepted manufacturing practices and operating tolerances has been made, leading to the raising of quality acceptance standards for raw materials, in-process components, tyre manufacture, and finished tyre grading.

N

Tyre manufacture has also been complicated, particularly in the passenger car range, by a trend towards the adoption of squat section tyres. Whereas a few years ago the section height–width ratio was approximately 100%, in present-day tyres it is nearer 80%, and the trend is still downwards. From the car manufacturer's point of view this results in vehicles with a lower centre of gravity and consequently improved control characteristics. Often the opportunity is taken to increase wheel diameters to permit the fitting of more efficient brakes while still maintaining the low vehicle line. Such changes tax the designer's art in producing mould shapes and manufacturing processes to ensure that the inflated raw tyre shape, immediately prior to mould closure in the vulcanising press, is reasonably in keeping with the tyre mould profile. Failure to achieve this condition can create undesirable rubber flow, which affects dimensional accuracy in the vulcanised product and also promotes moulding defects.

In Europe over the past 10 years, there has been a progressive swing from diagonal ply or cross ply tyre constructions to radial ply in the passenger car and truck ranges. This is likely to continue and may extend to earth-mover and other specialised tyre types. This revolutionary concept demanded a complete change in manufacturing methods and factory control practices. Not only did it involve the introduction of entirely new tyre building plant and far more sophisticated tyre moulding techniques, but also created the need for a much tighter overall quality control system. By virtue of its basic construction, this type of tyre is far more sensitive to in-built variables than the conventional diagonal ply tyre. The implications of the flexible sidewall structure and the relatively rigid tread zone will be outlined later. The alternative belted bias construction, which is the N. American counter to the radial tyre, has the virtue that the carcass can be assembled using the standard type of tyre building and vulcanising plant. Economic considerations undoubtedly promoted this development, which only goes part way towards the advantage in tread life which the radial holds over the conventional cross ply tyre. Furthermore, the improved vehicle handling imparted by the radial tyre is not realised with belted bias constructions.

10.1.2 MATERIALS

Reference should be made to the changes that have taken place over the years in the materials used in the tyre carcass. For a long period, high-extensible cotton cord was the universally used textile. Steam stretching techniques were introduced to achieve lower extensibilities, improved strength, and a lower cord denier, thereby yielding cord of higher tenacity. This development was quickly outmoded by the general adoption of rayon, with nylon coming into use for specialised applications. Both these latter fibres required the use of adhesives to achieve adequate bonding to rubber compounds, as discussed in Section 8.5. Recently, polyester, glass fibre, and steel have been added to the carcass materials available to the designer; in the future, carbon fibre may find applications in this field.

The physical properties of materials used in tyre structures influence, to a marked degree, the initial rubbering techniques, handling and storage of the

Table 10.1 FEATURES OF GENERAL PURPOSE ELASTOMERS FOR USE IN TYRES. COURTESY DR G. J. VAN DER BIE

Property*	NR	IR	SBR	BR	IIR	EPM/EPDM
Wear resistance	2	2	3	4	2	3
Road holding	2	2	3	2	4	2
Low heat build-up	3	4	1	1†	1	1
Tear resistance	4	3	1	1	2	1
Gas and vapour imperme- ability	1	1	1	1	4	1
Ageing resistance	1	1	2	2	3	4
Ozone resistance	1	1	1	1	3	4

* Ratings: 1, poor; 2, good; 3, very good; 4, excellent.
† See Section 4.4.3.

processed sheet, and tyre building and vulcanising procedures. Features which are of importance include moisture regain, extensibility, thermal stability, flexibility, and acceptability of adhesives. These and other properties of textile materials in relation to the requirements of the tyre designer are discussed in Section 6.7.

The scorch rate, plasticity, modulus, shrinkage factor, ease of processing, tack, and shelf life of rubber compounds likewise affect manufacturing procedures. Equally, such properties as heat build-up, abrasion resistance, flex cracking and cut resistance, and gas permeability determine the selection of compounds for the various components of the tyre. The raw polymers used are NR, SBR, IIR, BR, IR, and, most recent, EPDM. The properties of these materials are discussed in Chapter 4 and the general principles of their compounding in Chapter 9. The differences in their properties in relation to their use in tyres is shown in Table 10.1.

10.1.3 TYRE DESIGN

The tyre designer is faced with an impossible task in trying to satisfy all the needs of the vehicle manufacturer and the user, and is, therefore, forced to seek a compromise with emphasis on the important factors of safety and tread life (Clifton, 1969).

In determining the tyre size and type, the vehicle and tyre designers give prime consideration to:

1. Vehicle weight distribution, which will determine the load-carrying capacity of the tyre and the operating inflation pressure.
2. Axle height and clearance for the chassis, suspension, and braking system; these decide the overall diameter, section width, and bead diameter of the tyre.
3. The vehicle suspension system, which will influence the basic tyre construction including the selection of either radial or diagonal ply construction.
4. The speed capability and operating conditions, both of which will have to be considered in relation to construction, compounds, and tread pattern design.

Having established the basic tyre dimensions, which must conform to agreed industry standards, the tyre designer is in a position to decide on the tyre mould dimensions, tread pattern details, rubber mixes, textile reinforcement, and the form of the structure. From the vehicle weight and speed requirements, mathematical calculation of the casing and bead wire strengths can be made; the materials can then be selected and their make-up settled.

When formulating a tyre design it is essential, from a manufacturing standpoint, to pay due regard to the cost and availability of materials, factory plant, and production methods, so that a viable product is produced at an economic price and meeting the needs of the consumer.

10.1.3.1 *Tyre Sizing*

The generally accepted system for indicating tyre dimensions is to quote the approximate cross-sectional width of the tyre, followed by the nominal diameter of the bead seat of the wheel to which the tyre must be fitted. These dimensions are either expressed in inches or millimetres; dual imperial and metric unit markings are sometimes employed.

The load-carrying capacity of a tyre is dependent on its internal volume and its inflation pressure. It therefore follows that, for a fixed section height–width ratio, the load-carrying capability will increase with increasing wheel diameters. This arrangement has applied in the past for standard passenger car and truck tyre ranges, but recently, in the car group, there has been a trend towards designing tyres of common load-carrying capacity for all wheel diameters. This means that the cross-sectional area reduces as the nominal bead diameter increases, i.e. the internal volume remains constant regardless of the nominal bead diameter. In this latter circumstance, the convention is to identify a tyre by a letter of the alphabet indicating the load-carrying capacity, followed by the percentage height–width ratio, and finally the wheel diameter (Table 10.2).

Although the basic principles of tyre manufacture are similar for all tyre ranges, it will be appreciated that the method and plant used to produce a

Table 10.2 EXAMPLES OF TYRE SIZING

5·20-13	A 5·20 inch section width, and nominal 13 inch wheel diameter
145-14	A 145 mm section width, and nominal 14 inch wheel diameter
H.70-15	H indicates the load-carrying capability, 70 the section height–width ratio, and 15 the nominal wheel diameter in inches

simple motorcycle tyre made up of approximately 10 components will differ radically from that necessary to handle an extra large earthmover tyre embodying around 175 individual units. A typical tyre of this type will contain as many as 40 layers of textile in the carcass plies and a composite tread weighing 600 kg giving an overall tyre weight of 1·25 Mg. Although semi-automatic building machinery is employed, much of the assembly is a skilled manual operation requiring more than 24 hours to complete, and followed by a vulcanising time of 8 hours. Other specialised ranges include tyres for aircraft and racing cars.

It is only possible in this chapter to cover one area in detail, and passenger car tyres have been selected for this purpose. Reference will be made where appropriate to special manufacturing procedures involved in other groups of tyres.

10.1.4   TYRE CONSTRUCTIONS

Figs 10.1, 10.2, and 10.3 illustrate the three basic well-established tyre constructions.

10.1.4.1   *Standard Diagonal Ply Tyre Construction*

It will be seen (Fig. 10.1) that the internal structure comprises layers of cords, normally two or.four in number which run diagonally from bead to bead, with adjacent layers assembled at opposite bias. (The operation of coating

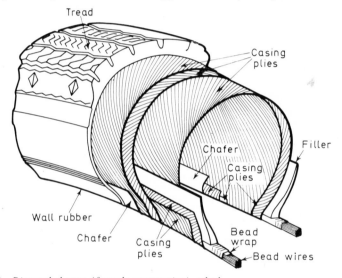

Fig. 10.1   Diagonal ply tyre (four-ply construction), tubed

with rubber compound has been outlined in Section 8.5.) Each layer of rubbered cord is known as a ply. It will be noted that the outer edges of the plies are interlocked around the steel wire bead coils in order that reorientation of the casing angles, during the vulcanisation process, will take place in a controlled manner. It is imperative that individual cords are evenly tensioned in the finished product, in order to contain tyre growth, obtain the optimum structural performance, and achieve an acceptable level of uniformity.

10.1.4.2   *Belted Bias Tyre Construction*

Again the casing plies are cut at an angle approximating to that used for standard diagonal ply tyres and are assembled together in a similar manner to produce a balanced and uniformly tensioned structure (Fig. 10.2).

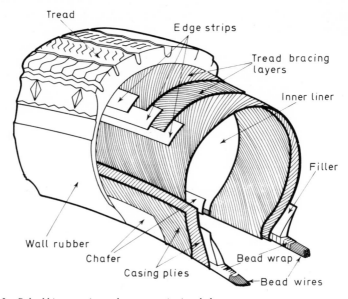

Fig. 10.2   *Belted bias tyre (two-ply construction), tubeless*

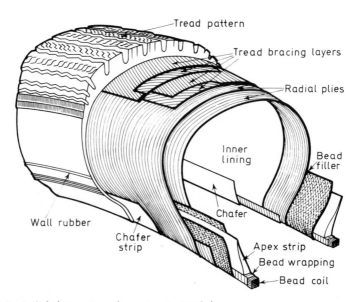

Fig. 10.3   *Radial ply tyre (two-ply construction), tubeless*

The fundamental difference in the construction is that the plies are sur-
mounted by two or more layers of rubbered cord material, cut at a lower
angle than the carcass plies. These are in the form of circumferential strips
extending across the full width of the crown area of the tyre, and fitted
with an opposed bias relationship to the casing plies, and to each other. These
form a restraining belt which raises the modulus of the tread area, thereby
controlling the inflated tread profile and also reducing tread pattern
movement.

In N. America, where this construction is rapidly replacing the diagonal
ply tyre, glass fibre is widely used for belt components.

10.1.4.3   *Radial Ply Tyre Construction*

As the terminology implies, the essential difference between the radial ply
tyre and the two alternative constructions just described lies in the disposition
of the casing ply cords (Fig. 10.3). In the radial ply tyre the cords lie at approximately 90° to the circumference. This results in an extremely flexible sidewall
which acts independently of the tread bracing belt, thus further reducing the
distortion which takes place in the belt and the tread block as the tyre
passes through the road contact zone. Advantages include increased tread
life, improved comfort at high speed, and improved security on cornering.

The belt construction, in general use, comprises four low-angle plies of
textile cord; some manufacturers have adopted steel or glass fibre cords for
the belt. The choice depends upon performance requirements, practical
manufacturing facilities, and economic considerations.

Table 10.3 gives casing ply and belt angles typical of those used in the
three constructions. To a large extent it is the reorientation of angles and the

Table 10.3   TYRE CARCASS ANGLES (INCLUDED ANGLE FROM THE CIRCUMFERENTIAL CENTRELINE
OF THE TYRE)

Tyre condition	Cross ply		Belted bias		Radial	
	Casing	*Belt*	*Casing*	*Belt*	*Casing*	*Belt*
Raw	54°–63°	—	58°–61°	52°	80°–90°	16°–21°
Vulcanised	29°–36°	—	32°	26°	75°–90°	12°–18°

required differential between the casing ply and belt angles which dictate
tyre building procedure.

10.1.5   TYRE COMPONENTS

Before dealing with the detailed operation of raw tyre assembly and ancillary
operations, prior to tyre vulcanisation, it is necessary to outline the basic
components forming a tyre structure and to indicate their function and
method of preparation. Fig. 10.4 details these for a radial ply tyre.

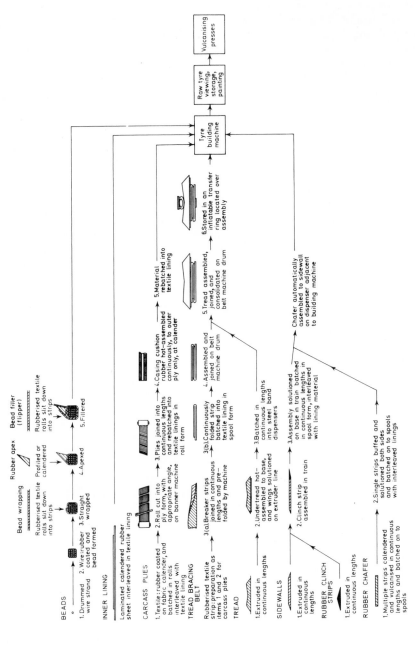

Fig. 10.4 Component flow diagram for radial ply tyres (single-stage process). Courtesy The Dunlop Co. Ltd

10.1.5.1   *Beads*

The bead coils are a combination of multi-strand copper-coated high-tensile steel wires. They have the function of providing rigid, practically inextensible units, which retain the inflated tyre on the rim under all conditions of loading.

The appropriate number of wires, formed into a flat layer and uniformly separated, are coated with rubber compound by means of a T-head extruder. The layer of wires is coiled to form a ring, and the free wire ends are taped or stapled. The wire treatment and use of fast-curing compound ensure good bonding and a regular bead shape in the finished product. For some purposes, the bead coil is covered with a light cross-woven rubberised textile to contain the wires and preclude any possibility of looseness in service.

10.1.5.2   *Bead Apex*

The bead apex is a fibrous or rubber compound strip located on the top face of the bead; its primary function is to pack out the area of the structure immediately above the bead coil and so provide a steady gradation of thickness between the latter and the sidewall zone, the thickness of which is only that of the casing plies and sidewall rubber.

This component has been eliminated from many of the conventional passenger car tyres with small bead coils, there being sufficient compound and compound flow during moulding to fill the void adequately. It has to be retained, however, in the radial ply tyres, in which high sidewall deflections occur, and in large tyres of multi-bead construction. The stress-carrying zone in the low wall region has to be spread to avoid rim chafing or structural break-up.

The component, which is generally triangular in shape with a finely tapered upper edge, is formed either on a profile calender or, preferably, by extruding from a multi-head die. Application and consolidation of the base of the apex to the outer circumference of the bead is a semi-automatic machine operation. The bead is gripped and revolved while the apex is fed forward manually to be located centrally and consolidated by means of angled pressure rollers.

10.1.5.3   *Bead Fillers (Flippers)*

In those instances where further reinforcement of the tyre bead area is required, the already wrapped and apexed bead may be enclosed by a strip of rubberised textile or, in some applications, by steel cords. Emphasis is on the avoidance of localised circumferential stress lines which could promote looseness or cord break-up under constant flex conditions. High-grade flexible heat-resistant compounds are essential for this region of the tyre.

10.1.5.4   *Carcass Plies*

It is the carcass plies that give the tyre its strength. These consist of cords of cotton, rayon, nylon, or polyester, woven as the warp of a fabric with only very light yarns, widely spaced, as the weft. These weft strands serve to maintain the uniformity of cord spacing during handling but play no part in the performance of the product. The fabric is treated with adhesive, rubberised to a thickness of approximately 1·0 mm, and interleaved with a low moisture regain textile lining. Steel is produced in weftless form from a creel feeding directly through a rubber calender. The large rolls of rubbered textile approximately 1·5 m wide and 300 m long are cut into strips, termed plies, on a horizontal Banner machine. This ply cutting plant has facilities for mechanically unwinding large rolls of fabric and simultaneously rewinding the interleaving lining. The fabric sheet is fed forward, through a festoon unit to allow for continuous operation, and then guided along a horizontal multi-belt conveyor to a cross-beam rotary cutting knife. This knife is complete with its own drive motor, and the entire unit traverses across the support beam when cutting the material to pre-determined width and bias angle; the latter is variable between 45° and 90°, and the operation is controlled by photoelectric cells.

Cut plies are placed manually on adjacent batching tables, where they are joined, end to end, into a continuous length and batched into roll form, interleaved with a textile lining to prevent self-adhesion. Suitable devices are provided at all stages to prevent distortion of the material.

10.1.5.5   *Tread Bracing Components (Breakers or Belts) for Radial or Belted Bias Tyres*

Tread bracing components raise the modulus of the tread area, thereby maintaining the inflated tyre tread profile and reducing tread pattern movement as it contacts the road.

The method of strip cutting the rubbered textile is identical with that used for cutting plies, but, in this case, the travelling beam on the bias cutting machine is adjustable for angles of 15°–25°.

The method of converting the single-layer low-angle cut strips into the form of the final belt varies. The most widely used construction, for radial ply tyres, employs four layers, made from folded strips, adjacent layers being of opposite bias angle. The method of assembly is shown in Fig. 10.4. In practice the two strips are slightly offset to achieve a graduated step-down in thickness at the belt edges, thereby reducing stress concentration and minimising the development of looseness in service.

10.1.5.6   *Insulating Components (Undertread, Breaker Cushion Insulations)*

These insulating components are calendered strips of rubber compound usually of 1 mm gauge or less. They are located in positions within the structure where additional insulation is required between components to prevent chafing. The compounds are similar to those used for coating the carcass plies.

10.1.5.7　*Tread*

The tread is the wearing surface of the tyre. It is applied in the raw state as an extruded slab of rubber compound. In cross-section it is substantially rectangular across the centre portion, tapering down to very fine edges. The thickness must be calculated to accommodate the pattern fragmentation in the tyre mould and to allow an adequate residual thickness beneath the pattern grooves. The tread width is such that the tapered edges extend to a position slightly above the maximum flex zone in the upper sidewall region.

The extruding operation is continuous, and the extrudate is either batched as a continuous length into a band dispenser, for subsequent cutting to length at the tyre building machine, or pre-cut into individual lengths and stored on flat metal trays in multi-leaf stillages. The former method is preferred; length variation of pre-cut treads often occurs as in-built stresses in the compound are released.

10.1.5.8　*Sidewall*

The sidewall is an extruded rubber compound layer which serves to protect the carcass structure from weathering and chafing damage. Together with the tread, which it overlaps in the buttress region, it forms the outermost layer of the tyre.

As with the tread, the extruding operation is continuous, and sidewalls are normally batched into spools interleaved with a textile or polyethylene lining material.

For conventional diagonal ply and belted bias tyres, built on relatively flat drums, frequently a common formulation permits extrusion of tread and sidewalls as one piece. Two separate compounds are used either to achieve performance or for reasons of economy. Modern dual extruders, feeding through a common Y-box head, produce a combined tread–sidewall unit; by joining the two stocks under high pressure and at high temperature, interface failures are eliminated.

10.1.5.9　*Chafer*

The chafer is a narrow circumferential strip of material which encloses the completed bead area. Its upper edge is located slightly above the rim flange height and extends downwards and around the bead base. This arrangement provides some protection from rim chafing and, in the case of tubeless tyres, serves to prevent air leakage either into the tyre or through the tyre in the bead area. To meet these conditions, the material used is either a rubber-coated wick-proof cross-woven textile cut at 45° bias or a strip of calendered compound. In the latter case, the strip, of approximately 1 mm gauge, is generally fully cured, buffed, and solutioned, prior to assembly into the raw tyre. In this way, stock flow during vulcanisation is avoided and the retention of an adequate rubber covering over the casing ply edges is assured.

Processing of cured rubber strips is a continuous operation of multi-strip calendering and vulcanising by drum cure (Section 8.9.2.5). After surface buffing and solutioning, individually on an ancillary unit, the strips are spooled in continuous lengths suitably interleaved with polyethylene.

10.1.5.10   *Inner Lining*

For tubed tyres, the inner lining is a thin layer of compound usually calendered direct on to the underside of the first ply after the latter has been joined and batched in continuous lengths. The component serves to insulate fully the tyre cords from the inner tube and thereby prevent tube chafing damage. It also protects the cords from possible degradation due to atmospheric moisture absorption.

In tubeless tyres, the inner lining is the air-retaining member and is usually calendered as a two-layer laminate having stepped edges. The overall gauge may be as high as 2·5 mm, and the width must ensure that the edges are overlapped by the inner edges of the chafer. This provides a low permeability layer from bead to bead.

10.1.5.11   *Clinch Strip*

The clinch strip may be considered as an extension to the lower edge of the sidewall, introduced as a further anti-abrasion measure on radial ply tyres, which are more subject to rim chafing than the cross ply tyre. It is a narrow extruded strip, about 25 mm wide and tapered at both edges, which is normally hot-assembled to the sidewalls by operating extruders in train.

It should be appreciated that the above elemental breakdown only deals with a comparatively simple passenger car tyre construction. Truck tyres and allied ranges are far more complex, involving many additional components, inserts, and compounds. Also, it will be obvious that preparation and handling techniques have to be adjusted to deal with the sheer dimensional size and weight of components used in these products. Steel ply radial giant, earthmover, and aircraft tyres are particular examples.

10.1.6   TYRE BUILDING

10.1.6.1   *The Building Drum*

Throughout the conventional diagonal ply and belted bias tyre ranges, including passenger car, truck, and earthmover tyres, the tyre building drum in general use is a plain segmental open-ended metal drum which is mounted on a driven shaft. The internal mechanism is designed so that the segments can be collapsed radially to permit the removal of the completed tyre carcass. The ends of the drum are flanged to suit the bead configuration of the tyre to be built. In all cases, the drum overall diameter exceeds the nominal tyre bead diameter. This difference, known as the drum crown height, varies from 25 mm for single bead tyres to 300 mm for large truck and earthmover tyres. Furthermore, in the latter cases, the flange profile has an appropriate undercut to achieve a balanced ply-to-ply tensioning in the finished product (Figs 10.5 and 10.6). The raw tyre can be completed on the building drum

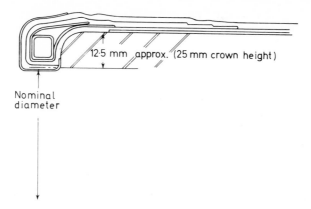

Fig. 10.5   Low crown height building drum. Courtesy The Dunlop Co. Ltd

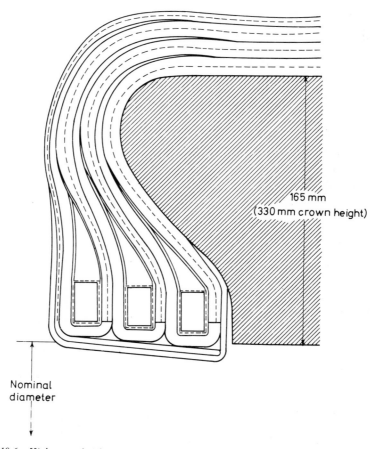

Fig. 10.6   High crown height building drum. Courtesy The Dunlop Co. Ltd

and is transformed from the cylindrical to a toroidal shape during the mould closing operation, the circumferential stretch being of the order of 60%.

The fact that the tread bracing components or 'belt' of radial ply tyres form a relatively inextensible unit and must be fitted at a diameter within 5% of its final diameter in the moulded tyre, means that for this type the partly built casing has to be inflated into a toroidal shape. Radial ply tyres necessitated, therefore, new complex and costly machinery incorporating inflatable textile-reinforced diaphragms, overlying a skeletal metal drum shell, to shape the carcass plies and other foundation components up to the diameter for belt fitting. It will be understood that, to achieve this, the shoulders of the drum move inwardly in a controlled manner so that the correct tension is maintained in the carcass plies.

Because it can still be accommodated on standard tyre building machinery, the belted bias tyre has come into being.

10.1.6.2   *Diagonal Ply Tyres*

Modern building plant, for passenger car tyres, is highly mechanised and constructed for maximum accuracy and productivity. For diagonal ply tyres, including the belted bias version, the standard form of building machine is designed specifically for operation with low-crown building drums; the uncured raw tyre is completed in a one-stage operation.

The machine (Fig. 10.7) consists basically of an accurately machined and balanced collapsible segmental metal drum mounted at the end of a driven shaft. One bead carrier ring is located on either side, and concentric with the drum. These rings can be traversed inwardly to provide an interference contact with the drum shoulders and thereby transfer and consolidate the tyre beads against the partly built casing structure. Built within the carrier ring frame, spring steel fingers, which form a circle, turn the ply material down the shoulder of the drum immediately before the beads make contact. The outer carrier ring is retractable to permit the removal of the uncured tyre.

Attached to the building machine base-plate and to the rear of the drum are two pneumatically operated component consolidating assemblies, each comprising two pairs of shaped disc rollers. The rollers of one assembly are located on either side of the drum and pivot around the tyre bead area for the purpose of turning and interlocking the various components around the bead. The pair of rollers forming the second assembly traverse laterally, outwards from the centre line of the drum to consolidate the tread–sidewall elements to the carcass.

Mounted at the rear of the machine and on the centre line of the drum is a multi-station component servicer [Fig. 10.7(b)]. This permits the dispensing of continuous length plies, via photocell-controlled festoon loops, guides, and feed trays, to the face of the building drum.

A gravity roller track is incorporated above the ply dispensers and furnished with guides on to which individual pre-cut tread lengths are placed, manually, immediately prior to running on to the carcass. Facilities are also provided for feeding forward and guiding the two spooled chafer components. The combined building machine and servicer functions are controlled by means of a series of foot-operated pedals and push buttons.

Each building drum is designed for one tyre bead diameter, e.g. 10 inch (254 mm), 13 inch (330 mm), or 16 inch (406 mm), but is adjustable laterally and one drum will usually accommodate several tyre sections, e.g. 5·20 inch, 5·60 inch, 5·90 inch, or 6·00 inch.

The following is a description of the building operation for a simple two-ply construction. With the building drum collapsed, the two beads are located on their respective bead carrier rings, the drum is expanded, and

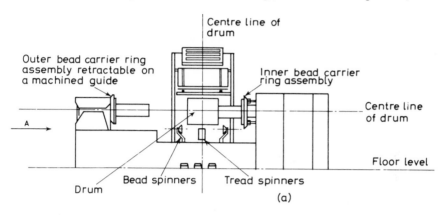

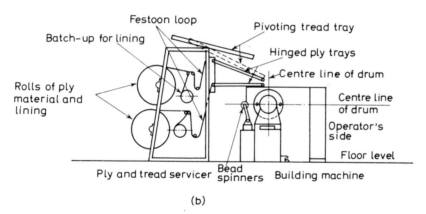

Fig. 10.7 *Typical cross ply tyre building machine: (a) view from operator's side, and (b) end view in direction 'A'. Courtesy The Dunlop Co. Ltd*

rubber-based adhesive is applied to the drum shoulders. The leading edge of the first ply, with its attached inner lining, is applied to the drum face, which is rotated slowly once. With allowance for a small overlap, the ply is cut by running a hot knife between two adjacent cords, and the join is consolidated manually. The ply widths are such that they extend beyond the edges of the building drum.

The bead carrier ring assemblies are automatically traversed inwardly, turning down the ply edges round the drum shoulders and consolidating the beads against the ply in one continuous operation. The ply edges are auto-matically turned up around the beads by the bead consolidation rollers. The second ply is fitted in a similar manner, and then the chafer strips, overhanging the ply edges, are run on and joined with a small overlap.

Both chafers and ply edges are turned down simultaneously around and under the bead bases by the rollers which, in this case, follow the conforma-tion of the beads in the reverse direction. This interlocks the beads within the tyre structure.

A tread length is now run on by drum rotation, the bevelled ends being matched and hand consolidated to ensure overall symmetry. Final consolida-tion is performed mechanically by traversing the tread rollers from the centre to the beads. Automatic collapse of the drum allows removal of the raw tyre, which is conveyed to the storage area.

10.1.6.3   *Belted Bias Tyres*

The procedure outlined for the building of a conventional diagonal ply tyre is applicable to a belted bias construction but with the addition of the breaker belt layers. These strips are assembled centrally to the top ply prior to fitting the tread unit.

10.1.6.4   *Radial Ply Tyres*

No standardised type of building machine has yet emerged for radial ply tyres, but two general methods of tyre building are currently employed.

First Method. Plies, beads, and chafers are assembled in the manner already described on the same machine as used for diagonal ply and belted bias tyres. The partly built carcass is then transferred to another machine, which com-prises a driven shaft carrying a shaping drum. The first operation on this

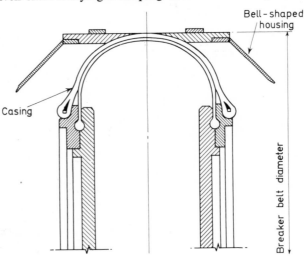

Fig. 10.8   Shaping of tyre casing: the casing is shaped and the bells are positioned for belt assembly. Courtesy The Dunlop Co. Ltd

machine is to inflate the flexible drum sleeve and shape the casing [Figs 10.8 and 10.9(a)]. During the operation the drum flanges and bell-shaped housings traverse inwardly in a controlled sequence partly to enclose and also to contain the tyre casing.

A pre-folded breaker length is applied manually to the exposed top carcass ply, the recessed inner edges of the bell surfaces serving as guides. A marginal overlap joint is made, and a second breaker length is applied with the fold on the side opposite to that of the first length. The assembled belt is then consolidated by roller pressure. A pre-cut tread length is run on by drum rotation [Fig. 10.9(b)], joined, and hand consolidated.

Fig. 10.9   Tyre shaping machine: (a) the casing is shaped and the bell-shaped housings are shown in retracted position; (b) tread is fitted to the completed belt. Courtesy Pirelli and Akron Standards

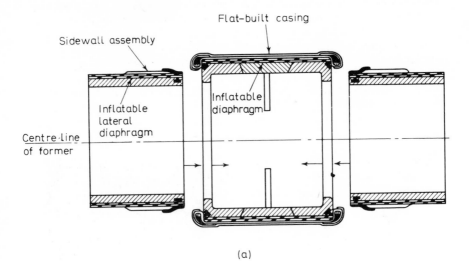

Sidewall assembly

Flat-built casing

Inflatable
lateral
diaphragm

Inflatable
diaphragm

Centre-line
of former

(a)

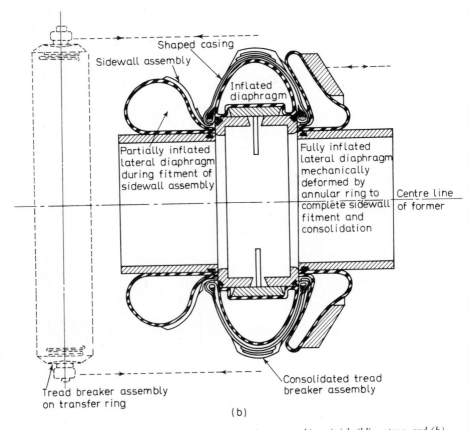

Shaped casing

Sidewall assembly

Inflated
diaphragm

Partially inflated
lateral diaphragm
during fitment of
sidewall assembly

Fully inflated
lateral diaphragm
mechanically
deformed by
annular ring to
complete sidewall
fitment and
consolidation

Centre line
of former

Tread breaker assembly
on transfer ring

Consolidated tread
breaker assembly

(b)

Fig. 10.10 Automatic radial ply tyre building and shaping machine: (a) building stage, and (b) shaping stage. Courtesy The Dunlop Co. Ltd

At this stage the shaping bells are withdrawn to allow pivoting disc rollers to traverse and consolidate the tread. Sidewall rubbers, extending from the tyre beads to overlap the centre tread tapered edges, are fitted, cut to length, joined, and consolidated, first manually and then by rollers. In some variants of the machine, inflatable rolling bags automatically apply sidewalls to the otherwise completed tyre. Vacuum is applied to collapse the flexible drum centre and permit removal of the finished tyre.

In the above method, the initial partly built casing is in a very unstable condition and precautions have to be taken to avoid contamination and distortion.

Second Method. One of the most modern building machines carries out the building and shaping operations automatically. All components are prepared on ancillary plant designed to place material accurately in rechargeable cartridges so that no operator intervention to correct for misalignment is necessary.

The central inflatable drum portion is used in its collapsed state for the flat building operation [Fig. 10.10(a)] and serves to shape the casing when inflated. Lateral diaphragms, on either side, turn up the ply material edges around the bead and also apply the preassembled sidewalls and chafers to the casing. Bell-shaped housings control the shape of the inflated casing and incorporate bead setting rings, ply turn-down fingers, and in-built diaphragms for consolidation of the outer ply edges.

The breakers and tread are assembled on a separate machine (Fig. 10.11) and are transferred to the building operation supported internally by a

Fig. 10.11   Belt making machine: A, first layer of pre-folded belt about to be joined; B, tread feed tray. Courtesy The Dunlop Co. Ltd

narrow split metal ring and externally by an inflatable transfer ring [Fig. 10.10(b)]. The operator's function is confined to fitting beads, joining components, fitting the transfer ring, initiating the automatic sequences, and removing the finished tyre. Inevitably a high degree of skill is required for setting up and maintaining this equipment.

10.1.7　PREPARATION OF RAW TYRES FOR VULCANISATION

Tyres from the building machine lines are generally transferred by line conveyors, either to a holding stock area or direct to the vulcanising department.

Intermediate operations include viewing and the application of a lubricant to the inside surface of the casing. The latter is necessary to facilitate the introduction of the BagoMatic press diaphragm or the Autoform bladder (Section 10.1.8.4) at the outset of the vulcanising process.

In some circumstances, an external coating of a thin rubber solution extended with a heavy loading of black and with talc is also employed. This prevents adherence of the tread and sidewall to the hot mould face and provides a coarse medium through which air, trapped between the casing and the mould, can wick to channels and vents incorporated in the mould profile.

A variety of single or dual purpose semi-automatic machines are used for these operations. With the tyre rotating, the paints are applied by means of strategically positioned spray guns.

10.1.8　MOULDING AND VULCANISATION

10.1.8.1　*Tyre Moulds*

For most applications, tyre moulds are basically of two-piece construction, in steel or aluminium, with the split being circumferential and either central or slightly off centre, dependent upon tread pattern complications. The use of male and female registers and dowels ensures both circumferential and radial location during mould closure. Each mould half is recessed just above the bead diameter to accommodate rings, termed 'slip rings', which incorporate the tyre bead profile.

The tread pattern in each mould half may be either cut from the solid, produced in the form of a ring which is then bolted to the mould shell, or built up from a series of aluminium die cast segments. The latter method is advantageous for complex pattern designs, incorporating sipes, tunnels, buffer bars, and variable-pitch arrangements, because segments from a few master dies, in the requisite number of pitches, can be utilised for a range of tyre diameters and sections by making small machining adjustments. Larger truck tyres, and tractor and earthmover tread patterns are either sand cast (simple bar designs) or produced as tread rings.

For the vulcanisation process, heat is applied externally to the tyre within the mould either (a) by locating the mould in a suitable form of pressure vessel with direct circulation of steam around it, or (b) by mounting the mould between two steam-heated chests or platens.

10.1.8.2　*Curing Bags, Bladders, and Diaphragms*

To avoid porosity, to consolidate the components, and to force the uncured tyre into contact with the mould profile, internal pressure must be applied. Advantage is taken of this requirement to introduce a secondary source of heating. The media are normally contained within a pre-cured form of tube, termed a curing bag, made from a resin-cured butyl rubber; this component is introduced into the uncured tyre as an ancillary operation prior to loading the combined unit into the mould. With modern vulcanising plant, the curing bag has been largely replaced by a curing diaphragm or bladder which remains permanently *in situ* in the vulcanising unit (Section 10.1.8.4). For internal inflation, either high-pressure steam, within the range from 12·6 kgf/cm^2 (193°C) to 19·7 kgf/cm^2 (213°C), or circulating hot water is used; use of the latter enables pressure and temperature to be varied independently. High internal pressures up to 28 kgf/cm^2 are generally employed for steel casing tyres, and for aircraft, earthmover, and large truck tyres, having thick cross-sections.

Thus heat is applied externally to the tyre mould to cure the outer layers of the tyre (tread, sidewalls, etc.), whilst the secondary source of internal heating cures the inner layers of the structure. The internal heating is less direct than the external because the former is transmitted by conduction through the thickness of the containing curing bag or diaphragm.

10.1.8.3 *Cure Temperatures and Times*

The thickness of the tyre to be vulcanised determines the highest temperature that can be used.

At one end of the range, cycle tyres, having a thickness of 5 mm, are cured for 4 min at 170°C mould temperature, whereas very large earthmover tyres, at the opposite extreme, with a thickness of, maybe, 150 mm, may require 8 hours with a mould temperature of 120°C.

The cure requirements are derived from laboratory tests on slab mouldings, but it is not always practicable to set the ideal cure in tyre manufacture. From a technical aspect a long cure at a low temperature is preferable, but for productivity reasons maximum temperatures must be employed to reduce curing cycles to a minimum. Where bonding of metal to rubber is involved, cure setting requires to be precise for consistent high fatigue performance and bond strength.

10.1.8.4 *Vulcanisation Plant*

Autoclaves and Pans. Originally the general purpose curing plant, for conventional diagonal ply tyres, was in the form of a deep circular pan or autoclave, having a hydraulically operated centre ram and capable of containing a vertical stack of up to 20 individual moulds, dependent upon their cross-sectional depth.

The curing bag, painted with lubricant to facilitate centralisation and prevent adhesion, is inserted in the raw tyre by machine, and the tyre is transformed to an approximation of the toroidal shape of the vulcanised product. Bagged-up tyres are placed manually in the lower mould half, the bag valve is located, and the top mould half is positioned; the assembly is gravity fed along the track, via the mould closing press to the autoclave, into which it is hoist loaded. The steam inlet supply is connected to each curing bag. After the lid of the fully loaded autoclave has been locked, hydraulic pressure is applied to the central ram, closing the moulds tightly against the lid. External steam services are admitted through the side of the autoclave, with the steam circulating around the moulds during the curing sequence. Some cooling of moulds, prior to unloading, is effected by a short cold-water flood introduced after exhausting the steam. After disconnection of inflation fittings, moulds are transferred, by hoist, to the roller track and opened; the vulcanised tyre with bag *in situ* is ejected. Condensate is syphoned from the bag, which is extracted from the tyre and put back into circulation.

Autoclave moulding is a flexible but inefficient system in curing time, because moulds require a lengthy warming up period at the commencement of each cure, and is now considered obsolescent. For very large tyres, the principle is retained and a relatively shallow 'pan' or 'kettle' accommodating two or three moulds is used, the hydraulic ram often being replaced by an inflatable rubber diaphragm located beneath the base plate on which the moulds rest. To obtain the higher internal pressures, hot water rather than steam is used, and cold water is circulated prior to pressure release at the end of the cure to provide some degree of cooling.

Presses. Various forms of press, generally accommodating a pair of moulds, have replaced the autoclave or kettle. The mould halves are bolted to the lid and base of the press and remain *in situ*. At the end of the curing cycle, the press opens automatically.

For supplying external heat, in the case of small tyres, e.g. cycle tyres, a steam chest may be an integral part of the tyre mould. For passenger car tyres, use is made of steam chests or 'cavities' which remain permanently in the press and into which the mould halves are fixed in intimate contact. In other circumstances, mould halves may be fixed directly to flat upper and lower platens. For press operation with relatively deep section truck tyre moulds, steam circulates directly around the moulds using the sealed press dome as a pressure vessel. As with autoclave curing, these early press processes used inflatable solid-base curing bags previously inserted into the uncured tyre.

The BagoMatic Press. BagoMatic and Autoform are two designs of press in which separately fitted curing bags have been eliminated. They are capable of accepting both diagonal ply and radial ply tyres; the uncured shape of the former being that of an open-ended barrel and the latter toroidal. In both instances the internal heating medium is contained with a butyl rubber diaphragm of some 7 mm thickness which remains permanently in the press.

In the BagoMatic press, the diaphragm (otherwise termed bladder) is of open-ended barrel shape having bead formations at either extremity for attachment to ring assemblies carried at the head of a central sleeved piston rod, projecting from the press base. The arrangement is such as to permit relative movement so that the bladder can be extended, by raising the piston, to facilitate tyre loading and can be lowered to permit entry of the bladder into the tyre during the shaping and mould closing operations (Fig. 10.12). At the end of the cure, the internal pressure is released and vacuum is applied to the bladder. The tyre is released automatically, first from the top mould half and then from the lower half by movement of the sleeve and clip rings.

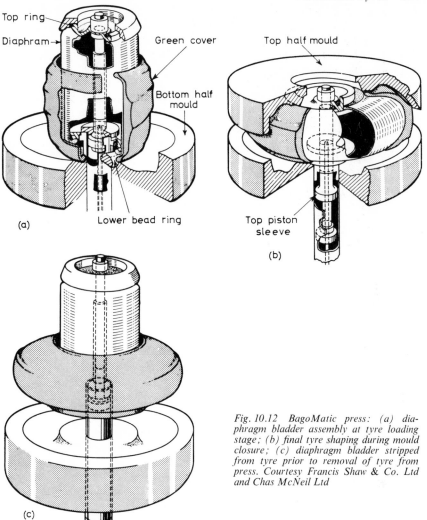

Top ring
Diaphram
Green cover
Top half mould
Bottom half mould
Lower bead ring
(a)
Top piston sleeve
(b)
(c)

Fig. 10.12 BagoMatic press: (a) diaphragm bladder assembly at tyre loading stage; (b) final tyre shaping during mould closure; (c) diaphragm bladder stripped from tyre prior to removal of tyre from press. Courtesy Francis Shaw & Co. Ltd and Chas McNeil Ltd

Finally, the bladder is stripped from the tyre, which is supported on arms and ejected from the press rearwards (Fig. 10.13).

Autoform Press. With this design of press, the bladder is again barrel-shaped but the upper end is closed. The single bead is clamped between the lower clip ring, which in this case is fixed in the mould, and the lip of a central submerged cylinder or 'well'. At the time of tyre loading, the bladder is contained within the well. As with BagoMatic moulding, uncured tyres are automatically picked up from stands located in front of the press and are positioned centrally over the lower mould half with the lower bead contacting the bottom mould half clip ring (Fig. 10.14). With the tyre held in this manner, the bladder is forced from the well into the casing by the introduction of low-pressure steam.

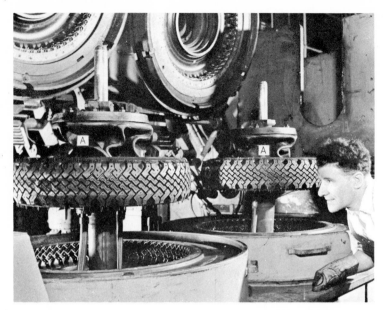

Fig. 10.13   *BagoMatic press: A, bladders stripping from tyres as centre piston rod rises after tyre extraction from moulds. Courtesy The Dunlop Co. Ltd*

Fig. 10.14   *Autoform press: A, radial ply tyres held in position over bottom mould halves during insertion of bladders from the submerged press wells; B, further tyres located on stands in readiness for automatic pick-up by loading chucks; C, loading chucks. Courtesy David Bridge & Co. Ltd and National Rubber Machinery Co.*

The loading chucks move out, and several detailed final shaping operations are carried out before press closure. On completion of curing, and after exhausting the bladder, an air-actuated piston rod, located in the press lid, extends downwards to strip the bladder from the tyre and transfer it back into the well. Chuck sector plates, mounted on an ejector mechanism, are expanded to register under the top tyre bead and strip the tyre from the bottom mould half. As press opening continues, the top half mould moves upwards away from the clip ring, and the tyre strips from the top mould half to remain suspended by the upper bead between the clip ring and the expanded sector plates. Finally, the latter are withdrawn and the tyre is forced off the clip ring seat to drop on to a track for take-away. External heating is similar to that for BagoMatic curing. The Autoform principle has merit in that it eliminates hydraulic and vacuum services. Furthermore, replacement of bladders is a quick operation. With the BagoMatic principle, the bladder assembly, with associated clamping rings, has to be removed from the press, dismantled, and reassembled as a separate operation.

10.1.8.5 *Instrumentation and Control Systems*

The total cure time, and the sequence of service requirements during this time, are controlled by a cycle timer, having a series of manually adjustable cams. According to their setting, these cams produce a pneumatic servomechanism to open or close the appropriate diaphragm-actuated valve, and so permit or terminate the flow of the inflation media.

Press platen temperatures are individually controlled and recorded for each press by means of a complex instrument using solid state electrical circuits.

Where high-pressure circulating hot and cold water is used as the curing medium, instrumentation requires to be more elaborate, e.g. flow rates need to be known, also differential temperatures between internal inlet and outlet.

10.1.8.6 *After-Treatments*

When employing textile casing materials which are thermally unstable, e.g. nylon or polyester, it is frequently necessary to prevent heat retraction and tyre distortion by cooling the tyre to below the critical temperature while still inflated. Two alternative methods are as follows:

1. At the conclusion of the curing cycle, the internal steam or hot water is replaced by cold water under pressure. This method is somewhat inefficient and may increase cure cycles appreciably.
2. Immediately after extraction from the mould, the tyre is mounted on a rim and allowed to cool under an air inflation pressure of approximately $2 \ \mathrm{kgf/cm^2}$. Fully automatic plant for this purpose is located immediately behind modern presses. Some units have dual mountings which rotate in order that each pair of tyres from a twin press can be held for the duration of two curing time cycles.

10.1.9 PRACTICAL MANUFACTURING PROBLEMS

A few of the key problems, which are common to the industry, are worthy of note. These arise mainly from the intrinsic properties of textiles and rubber stocks and from the semi-plastic condition of materials in the unvulcanised state.

Although the rubber coating applied to textiles provides protection from moisture absorption, it is important to limit exposure time to a practical minimum. A high moisture regain in rayon, for example, can result in separations within the casing structure when internal pressure is released at the conclusion of the vulcanisation process. For this reason, and also to prevent self-adhesion, 'in process' stocks must be interleaved at all stages in dry low-moisture absorbent textile linings.

The instability of rubbered materials and compounds creates handling problems. For example, relatively minor distortion to cut ply materials will alter the bias angle and regularity of cord spacing. This can affect ultimate casing tension and uniformity of the tyre. In critical areas, like radial ply tyres, all plant is designed with festoon appliances and air flotation systems to minimise this danger.

Care is necessary to ensure that compounds are free from incipient scorching and that the desired surface condition is retained at the time of tyre building. Clearly, over-tacky or over-dry materials present handling difficulties and due attention has to be paid to maturing times and shelf life. It is imperative that interface bonding, during assembly, is good, and cleanliness is a key requirement at all stages of manufacture. The development of surface blooming, with some compounds, can be counteracted by light application of a solvent-soaked textile pad during the tyre casing assembly operations.

One of the anathemas of tyre manufacture is included air. Even after all precautions have been taken in the fitting and progressive consolidation of individual components, it is inevitable that minute channels of air will persist at the stepped edges within the structure. During the power stitching operations, this air may collect in the form of small blisters which require to be released by manual pricking.

Regularity of the outside contour of the raw tyre is important in the avoidance of vulcanised tyre surface defects. Considerable movement or flow of stock can take place in the initial stages of vulcanisation owing to its thermoplastic properties and to the influence of the high internal pressure applied to the casing. Any surface irregularities induce flow which can result in moulding defects in the form of splits or blisters. The need for fine component edges and graduated stepping-down of components, particularly in the bead zone, has been emphasised previously.

10.1.10 FACTORY CONTROLS

Apart from the normal laboratory checking of incoming raw materials, it is essential to apply quality controls, on a statistical basis, throughout the manufacturing process and to the finished product.

The mixed compounds are subject to specific gravity checking and to curemeter testing for scorch, plasticity, and modulus. Water hardness, viscosity, alkalinity, and total solids of impregnants, used for textile treatment, must be controlled, and a routine check of static rubber-to-cord adhesion is desirable with certain critical materials before release of processed rolls for use in manufacture. Rubbered ply and belt materials must be examined for extensibility, regularity of cord spacing, and, in some instances, percentage moisture regain. The continuous auditing of all key operations in component preparation and tyre building is essential, including the dimensional accuracy of components and machine settings. Finally, tyre uniformity and grading machines are used to check static or dynamic balance, radial and lateral run-out, etc, of the finished product.*

10.2 Belting

BY W. C. NELLER

The main type of belting manufactured is for the conveyance of materials. Belting is also produced for power transmission purposes in the form of flat transmission belting and V-belting.

10.2.1 CONVEYOR BELTING

Conveyor belts are constructed of a reinforcement, either textile or steel cord, which is protected from mechanical damage, arising from impact or the ingress of moisture, by a rubber or plastics cover. The rubber conveyor belt is usually manufactured by laminating a number of plies of a rubberised textile reinforcement, and consolidating and vulcanising them in a special press or rotary curing machine (Kyle, 1968).

The emphasis, in recent years, has been to design conveyor belting for a given application or purpose. As examples of different constructions, mention may be made of heat-resistant belting for carrying hot sinter and coke, oil-resistant types, and others specially designed to be non-toxic for food carrying. Belts are also made with a moulded rough or chevron surface for incline conveying. The nationalised industries, such as the National Steel Corporation, the Central Electricity Generating Board, and the National Coal Board, are the biggest users. Following the Cresswell Colliery disaster in 1950, the National Coal Board in conjunction with the conveyor manufacturers developed fire-resistant antistatic belting for underground use in the mines. Approximately $1\frac{1}{2}$ Mm/year of this type of belting is currently being produced.

10.2.1.1 *Material Requirements*

Textile Reinforcement. The use of the well-known British Standard cotton ducks of 81 g/m^2, 93 g/m^2, 122 g/m^2, and 175 g/m^2 is decreasing in favour of synthetic fibres. The latter have better physical properties for conveyor belting and a significantly lower cost per unit strength than cotton.

* *Rubber Chemistry and Technology*, **41**, part 4, contains several papers on many aspects of tyre design, performance, and properties.

The main synthetic fibres used are continuous filament fibres of rayon, of polyester, and of nylon (Table 10.4 gives their physical properties in comparison with cotton). Rayon and polyester are both high-modulus materials,

Table 10.4   PHYSICAL PROPERTIES OF TEXTILE FIBRES

Physical property	Polyester	Nylon	Rayon	Cotton
Tenacity (g/den.)	8·0	8·5	5·0	1·5–2·0
Elongation at break (%)	8·5	13·5	11·0	3–10
Initial modulus (g/den.)	110–130	40–50	90	12–70
Moisture regain (%)	0·4	4·2	13·0	8
Specific gravity of fibre	1.38	1·14	1·5	1·54

being particularly suitable for use in the warp or longitudinal direction for long-haul high-tensile belting which requires to undergo the minimum amount of stretch in service. Rayon is lower in strength than polyester and has the great disadvantage of losing strength when wet.

Nylon is the most widely used synthetic fibre in belting today. It has the highest tenacity or strength of the three fibres, a comparatively low modulus, and a high elongation at break compared with rayon or polyester. The principal application is in low to medium strength belting, where its high strength and extensibility give very good energy absorption characteristics and, therefore, excellent resistance to impact damage. However, when used as a warp member for high-tensile long-haul conveyor belt installations, problems arise owing to its extensibility.

The physical properties of the synthetic fibres can be modified during manufacture by heat stretching and heat relaxing processes. The former process increases the modulus and gives lower stretch characteristics for use as a warp member, the yarn losing some tenacity. Heat relaxing the yarn gives a lower thermal free shrinkage, which is particularly advantageous for a weft member, to give maximum stability to the fabrics during processing.

Continuous filament fibres such as nylon and polyester are more difficult than staple cotton to bond to rubber and have a far lower weight for equivalent strength; this factor can lead to load support problems of the belt in service. In the low to medium strength type of belting, a proportion of cotton or staple fibre rayon may be woven with continuous filament yarn to aid adhesion of the laminated belting by normal frictioning methods (Section 8.5.5) and to provide the bulk required. However, with the development of improved bonding techniques for synthetic fibres, a significant growth has taken place in the use of 100% synthetic fibre, e.g. nylon, in belting. For the lower strength products, the bulk is maintained by introducing additional rubber between the plies.

High-tensile very long-haul belts are manufactured from fabrics with a nylon weft and a polyester warp, as mentioned above; alternatively, a steel cord reinforcement, which gives the required very small stretch in service, is employed.

Fabric Designs. Simple interlacings, such as plain or oxford weaves, are normally used for fabrics for plied belting. Special weaves are used to enhance such physical properties as fastener holding, stretch characteristics, edge wear, and tear resistance. Solid woven or compound weaves are also used, whereby two- or three-layer fabrics are combined together during weaving to produce a monoply construction.

Rubber Formulations for Carcass and Cover. Choice of rubber mixes will depend on the type of belt being manufactured, i.e. general purpose, heat resistant, oil resistant, or food quality.

The carcass compound for the general purpose belt is usually a high-quality natural rubber with low-modulus fillers and a high-scorch-time vulcanising system.

There are two main types of cover compound for the general purpose belt: for abrasion purposes, BS 490 Grade N; and Grade M for maximum resistance to cutting and gouging since it is a more resilient stock usually based on natural rubber. The compounds generally have flat cure characteristics and must be compatible with the carcass stocks. Typical formulations are given in Table 10.5.

Compounds based on SBR, IIR, and EPDM are used for heat-resistant belts, and on CR, NBR, or PVC–NBR for oil resistance.

For underground mining, conveyor belting is mainly based on PVC to provide flame resistance, which will prevent the danger of spread of fire, and

Table 10.5 TYPICAL FORMULATIONS OF CARCASS AND COVER COMPOUNDS

Ingredient	Carcass compound	Cover compound	
		Grade N	Grade M
Natural rubber (1st grade)	100	—	80
SBR 1500	—	100	20
Black	—	HAF, 50	ISAF, 45
Whiting	20	—	—
Zinc oxide	5	5	5
Stearic acid	1	1·5	1·5
Process oil	3	3·5	5
Pine tar	3	3·5	—
Coumarone resin	—	5	—
Accelerator	{ MBT, 0·4 MBTS, 0·4	HBS, 0·75	HBS, 0·50
Sulphur	3	2	3
Retarder	0·5	—	—
Antioxidant	1·5	1·5	1·5
Antiozonant	—	1·5	1·5

has antistatic properties to minimise the risk of the explosion of mine gases due to static discharge.

10.2.1.2 *Manufacture*

Drying. Drying of the fabrics is essential to avoid blowing of the laminate occurring during the vulcanising operation. The fabric is dried by passage over a multiple steam-heated drum drier or hot plate at a speed of 15 m/min at a surface temperature of 115°C. A moisture level of 1·0% for cotton-containing fabrics is required. It is desirable to carry out the frictioning operation as soon as possible after drying; alternatively, temporary storage boxes must be available at a temperature of 70°C.

Frictioning and Topping. To ensure good frictioning, a hot fabric is essential. Frictioning on each side is carried out on a three- or four-bowl rubber calender. The lighter-weight fabrics are friction coated, and the heavier fabrics are also topped or skim coated to give additional rubber between plies and between the outer plies and covers.

To obtain good adhesion, synthetic fibre fabrics are either pre-spread with isocyanate-based solutions or RFL-dipped prior to calendering (Section 8.5.2). A topped skim coating of rubber only is then required; with a four-bowl calender, this can be carried out on both sides in one operation. The rubberised fabrics are stored in cotton liners to prevent sticking.

Recently, techniques have been developed whereby no pre-treatment of the fabric is required, the bonding being achieved by special bonding agents added to the friction or topping compounds (Section 9.5.3).

Belt Building. Usually full-width fabrics to the optimum width of the calender are used for the calendering operation. Fabrics are then cut accurately to the width required on a cam-cutting machine which has multiple circular cutting knives. The actual cutting widths will depend on the construction required; typical constructions used are shown in Fig. 10.15.

The cut plies are generally plied together on long building tables equipped with two-bowl consolidating rolls. Semi-automatic tables, letting off a

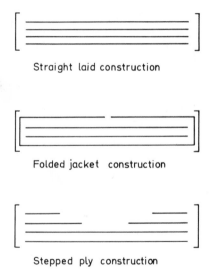

Straight laid construction

Folded jacket construction

Stepped ply construction

Fig. 10.15   Conveyor belt constructions

multiple number of plies in one operation with individual liner rewind for each roll and tension control, are also used.

Transverse and longitudinal joints are allowable within certain prescribed limitations, the joints in adjacent plies being staggered to avoid any weakness occurring in the belt.

The separate face and back rubber sheeting calendered to the required thickness are applied to the carcass on the building table to provide the rubber cover. The completely built belt is then consolidated and passed through pricking rollers to remove any trapped air. Alternatively, the covers may be calendered directly on to the belt carcass using a three- or four-roll calender. The finished raw belt is stored in a cotton or polyethylene lining, or chalked to prevent sticking prior to the vulcanising operation.

10.2.1.3   *Vulcanisation*

Vulcanisation of conveyor belting is carried out in either a press or a continuous drum curing machine (Berstoff or Rotocure) (Section 8.9.2.5).

Pre-heating by Microwave Techniques. The latest development is the introduction of pre-heating, by means of microwaves, into production press and continuous vulcanising operations, with up to 50% saving in vulcanising time.

Because materials, such as rubber, are difficult to heat uniformly without degrading their structure, the many attempts in the past to reduce vulcanising cycles by pre-heating by conduction, convection, or radiation methods have met with little success. The development of microwave heating has opened up new possibilities. Rubber has high dielectric characteristics and thus can absorb energy of very high frequency, generating heat uniformly within the material structure. Microwave heating is basically similar to dielectric heating but, with the frequency increased from 100 MHz to 2000 MHz, the disadvantages and difficulties of the latter are overcome (Section 8.6.3.2).

The microwave heating system consists essentially of a power supply to raise the mains voltage to approximately 7 kV, which is then fed to a magnetron oscillator. The magnetron oscillator contains within its vacuum envelope a tuned circuit and delivers the energy via an aerial and waveguide to the applicator.

One form of applicator is a metal chamber (Dunlop, 1967) so designed that, for the frequency generated, the chamber becomes a resonant cavity. The laminate is placed inside the cavity: no direct contact with metal is required as in dielectric heating, and the material can be heated irrespective of the product shape. Even distribution of energy is obtained through cavity design and the provision of a specially designed rotary deflector system mounted at each entry point.

The illustration (Fig. 10.16) shows the pre-heater in series with a 10 m press, a system of approximately 20 multiple $1\frac{1}{2}$ kW power units and magnetrons being used to generate the microwave energy. This makes it possible to distribute energy evenly over the wide face of the belt and so eliminate non-uniformities.

The pre-heater is worked in conjunction with the press by automatic electronic controls; pre-heat times, power input, and press conditions are carefully selected. The reduction in overall cure time (Brooke and Robinson,

1964) depends on the particular type of belt but reductions of *up to 50%* are achievable.

In addition to reduced cure time the reduction in strain during vulcanising and the more consistent vulcanising conditions lead to a better-quality product.

Fig. 10.16   General view of belt-making shop with microwave unit in foreground. Courtesy The Dunlop Co. Ltd

Press Operations. Various types of large flat multi-ram presses are used, of frame or column construction, both single and double daylight.

The raw belt is unreeled from a braked 'let-off' station to the press, and a section is vulcanised. The presses have cool areas at the ends to prevent over-cure between successive sections of the belt. Each press is equipped with stretching gear, usually consisting of flat hydraulically operated clamps, the belts being stretched a given amount prior to closing the press. This stretching process is essential to prevent excessive lengthening of the belt occurring in service. A moulding frame is made from flat 50–75 mm wide metal irons placed along each side of the belt. Lateral pressure to form the belt edges is usually applied by hydraulically operated cams which move the metal irons in a fixed amount after the press has been closed on low pressure. The thickness of the metal irons is selected to give 10–12% compression on the raw belt thickness. A specific pressure of the order of 21 kgf/cm^2 is used.

The length of cure depends upon the thickness of the belt, and an average cure time would be 17 min at 145°C. On completion of the cure of a section, the next length is indexed into the press, the small semi-cured end section being brought to the exit end of the press and its cure completed on the next press operation.

Because synthetic fibres, such as nylon, are not affected by the ingress of moisture or by microbiological attack and also have good edge wear, belting is now being manufactured from these materials to the full width of the press without a moulded rubber edge. The slab of belting produced is slit to the required width after vulcanising, two or three belts being obtained from one making and one pressing operation.

Continuous Vulcanisation of Belting. The belt is passed between a rotating steel drum and an endless high-tensile-steel band, pressure being applied by tensioning the latter hydraulically; heat is applied to both sides of the product, from the internally heated drum and also through the steel band by contact with steam-heated shoes (Section 8.9.2.5). Typical dimensions of one such machine (the Shaw–Boston 80 inch Rotocure) are as follows: width of steel band, 2 m; width of main drum, 2·3 m; approximate maximum product width, 1·9 m; drum diameter, 1·5 m; maximum pressure, 4·8 kgf/cm^2. The maximum product thickness is 32 mm, but the maximum product width varies with the application. The curing speeds are variable between 65 m/h and 4 m/h, giving cure times of from 3·5 min to 53 min, which adequately covers the range normally required for rubber belting. The belt being fed through the vulcaniser is subjected to an initial tension accurately set and maintained, and also to a predetermined stretch.

10.2.2 PASSENGER CONVEYOR BELTING

Belts for passenger carrying have been used for some time. However, the problems associated with handling large numbers of people at airports, railway stations, and coach terminals have led to considerable developments recently.

The most widely used type of belt today is designed to run on short (50 mm) edge-support rollers so that the area on which the passengers stand is completely free from hard and uncomfortable supports and a smooth ride is obtained. To achieve this result, the belt is designed to have high lateral rigidity supporting the load 'bridge wise', coupled with good longitudinal flexibility enabling it readily to negotiate small terminal pulleys.

Essentially, the belt consists of a rubber–textile reinforcement sandwiched between two outer layers of transverse steel cords (Robinson, 1960) which are at opposite sides of the sandwich, the whole being enclosed in a rubber casing. The textile reinforcement provides longitudinal strength to take the drive tension, and the combined action of the upper and lower steel cords creates lateral rigidity across the belt to limit deflection. Such a belt 1·1 m wide, supported on 50 mm edge rollers and loaded at the centre with 68 kgf, will deflect no more than 9·5 mm at the centre.

Travelling speeds are dictated by the ease and safety with which passengers can enter and leave the belt. It has been found in practice that a maximum speed of 36 m/min suits the variety of passengers travelling.

O

The surface of the belt is ribbed along its length, the ribs approximately 6 mm wide and 6 mm apart meshing in and out of a comb-like platform at the entry and exit points. The ribs are moulded to very close tolerances in order to ensure that they move smoothly through the teeth of the combs without wear or damage. The ribbed surface sliding into a comb removes the danger of clothing being trapped between the moving belt and the stationary platform at the exit point. If any obstruction is encountered, the comb plate is designed to trip out the conveyor drive. Moving handrails are an integral part of the passenger conveyor installation.

Passenger conveyor systems of this type are already in use at many airports, railway stations, docks, shopping precincts, and sport stadia.

10.2.3 TRANSMISSION BELTING

10.2.3.1 *Flat Transmission Belting*

Flat transmission belting is still mainly manufactured from plain weave $1 \cdot 05$ kg/m^2 and $1 \cdot 15$ kg/m^2 cotton fabrics, known as hard driven types, giving the required high modulus characteristics combined with good edge wear and fastener holding efficiency. Belts are also manufactured from polyester- or rayon-containing fabrics.

The fabrics are friction and skim coated with natural rubber compounds, or, in case of belts operating in contact with excessive oil, with chloroprene or nitrile compounds, formulated to have low heat build-up, high strength, and good adhesion to fabric. The finished belts can range from 3 to 10 plies, and in manufacture widths of calendered fabric are plied together to form a slab. The slab is then vulcanised in a press or drum curing machine, a high percentage stretch being applied during vulcanising to ensure that the minimum amount of stretch occurs in service.

The slabs are usually from 200 m to 400 m long by 2 m wide and are then slit on a multi-slitting machine to the widths and lengths required for stocking. This type of flat transmission belting, known as 'cut edge', is the main type manufactured today. Folded-edge type belts which are made in the narrow finished widths, with the outer plies folded round a centre core to form a jacket, show better edge wear but their production is slow.

Transmission belting is also manufactured from single ply or compound weave fabrics, usually made from specially latex-impregnated and coated yarns.

10.2.3.2 *V-Belts*

Endless V-belts are manufactured in various types including fan belts for automobile use, fractional horse power drives for domestic and light industrial applications, and larger industrial belts.

The standard industrial belts are manufactured in five section sizes, usually from rayon cord material. More recently, premium industrial belts, known as 3 V, 5 V, and 8 V belts, or wedge-type belts, have been introduced. In these, the ratio of depth to the top width is greater than in the standard sections, which results in a more compact drive. These constructions are made possible by the use of high-tenacity materials such as polyester cord.

V-belts are generally constructed of a top filler layer of rubber, a cord section, situated at the neutral axis, and a base or cushion rubber, the whole assembly being surrounded by a fabric jacket. The shorter belts are manufactured on rotatable collapsible drum formers. The required number of layers of rubber are applied to the drum, and the cord is wound on at touch pitch. The individual belts are then parted off by means of circular knives and transferred to a skiving machine, which roughly forms the V shape; finally, according to the size of the belt, one or two jackets are applied by machine. The belts are subsequently vulcanised in multi-ring moulds in open steam.

The long-length belts are made similarly, except that they are built on twin-drum building machines using weftless cord fabric instead of an individually wound cord. The belts are vulcanised endlessly by moulding in a press under controlled stretch conditions. After manufacture, the belts are checked for length and made up into matching sets.

A link form of V-belting is also manufactured, with the advantage that any endless length required can readily be made up. The links are stamped out of slabs of transmission-type material and joined together with metal components.

10.2.3.3 *Positive Drive Belts*

Positive drive belts, for precision drive and timing equipment applications, are moulded with a toothed inner surface which meshes with mating axial grooves in the pulley to give positive non-slip engagement. The belts are required to have very low stretch characteristics. Steel or glass fibre cord is, therefore, used and is continuously helically wound at a set pitch on to a backing of suitable compound which encases the load-carrying members. The teeth are moulded integrally with the belt, and a tough nylon fabric is also moulded on to the wearing surface to protect the teeth. The nature of the backing compound and teeth can be varied according to the application, to be resistant to oil, heat, or fire.

10.3 Hose

BY C. W. EVANS

10.3.1 INTRODUCTION

Within the rubber industry it is generally accepted that, as distinct from a plain tube, which is solid-walled throughout and contains no reinforcement, a rubber hose consists of the lining or tube, the reinforcement or carcass, and the cover for outer protection.

The lining, which is the innermost section, has to be specially compounded to meet the conditions and resist the action, either chemically or physically, of the material passing through. This may be in any form (solid, liquid, or gaseous) and either acid, alkaline, or neutral. A solid may be extremely abrasive and flow at a fairly high velocity, as in the case of sand or shot blasting operations. Furthermore, certain liquids have also active solvent properties, and operating temperatures may vary from as low as $-65°C$ to $+200°C$ and higher, added to which there may be a requirement for some flame or fire resistance.

Very fortunately, not all of these properties are required simultaneously; nevertheless, on occasion, the ingenuity of the compounders and hose designers is stretched to say the least!

10.3.2 MATERIALS

Nearly all of the extremely wide range of available rubbers are used by the hose industry, either singly or in various blends.

All the following fibrous materials have applications as the reinforcement or carcass in current and future hose manufacture, and are selected according to properties for each service requirement (Evans, 1969-1):

Asbestos fibre	Linen	Polyester
Carbon fibre	Rayon	Polypropylene
Cotton	Nylon	Steel wire
Flax		Glass fibre

The reinforcement is applied to the lining in yarn form by knitting, braiding, spiralling, or circular loom weaving, or as cut woven fabric wrapped straight or on the bias. If more than one ply of textile is applied, it is usual to include a layer of insulating rubber between them to prevent chafing in service.

Generally speaking, the operating pressure of the hose is the deciding factor in the selection of material. For a low-pressure application, such as a garden hose, steel wire would not be used, but rather a weaker material such as cotton or flax. Nylon has little use as a reinforcing material, mainly because of its low modulus and high extensibility, but it is these very properties that make it eminently suitable for the expanding hose portion of the power steering system of an automobile or truck.

The cover, generally speaking, is the outer protective layer of the hose, and has to be compounded to resist environmental conditions such as weathering (including ozone in certain circumstances), temperature extremes, oil and chemical contamination, as well as cutting and abrasion when the hose is dragged around.

It is essential for optimum service performance that the lining, reinforcement, and cover of the hose are fully consolidated and bonded together.

10.3.3 MANUFACTURE

The method and machine used in the production of a particular hose, as a general rule, depend upon the type of reinforcement necessary for the service conditions to be met and upon such practical considerations as the bore size and length required. The manufacture of hose will be described under five headings which form a convenient breakdown by type, use, process, and production method (Bradshaw and Welsby, 1966; Anon, 1968-5; Evans, 1968-1, 1968-2, 1969-2, 1969-3; Harkleroad, 1969).

10.3.3.1 *Moulded Hose*

In the manufacture of moulded hose, as the name implies, the hose is formed during the vulcanisation stage by being moulded against a lead sheath.

The lining (tube) is normally extruded in lengths of up to 500 m, depending on bore size. The extrusion may be carried out on conventional equipment, either cold or hot fed or a combination of the two, and in either strip or pellet form. To meet the subsequent processing requirements, the compound has to be fairly hard and firm in the green uncured state.

After cooling and maturing, the linings are reinforced by either braiding or spiralling the textile around them. As a rule, in the UK the braid is mainly of cotton for the general purpose hose (air, water, garden) and in the USA it is of rayon, whereas alternate left- and right-hand laid textile reinforcement (mainly of rayon) is used throughout Europe.

A braider is a machine which applies the textile yarn to the hose lining in a manner very similar to a maypole dance. In fact they are often called maypole machines, and in this instance are usually installed in a vertical position.

The outer cover coat of rubber is next applied to the reinforced hose by means of a cross-head extruder. After covering, the hose is passed through either a lead press or a lead extruder (Section 8.9.1.5).

After lead covering and winding on to drums, the lining is filled with water (in some instances, air is used also), pressure is applied, and the ends of the hose and lead are clamped. The drum and contents are placed in large pans for vulcanising. The water inside the hose in fact expands and becomes super-heated, and the hose is pressed against the lead, which acts as a mould (hence the name of the process). If the inside of the lead is fluted, then such a finish is imparted to the hose. If smooth dies are used to form the lead, then, of course, a smooth finish is produced.

After vulcanisation and cooling, the clamps are cut from the hose, and then the lead is removed by slitting along its length in a stripping machine. The cured hose is coiled up, tested, and inspected, and the scrap lead is returned to the melting pot for re-use.

Generally speaking, moulded hose is used for comparatively low-pressure applications and is manufactured in lengths of up to 500 m, depending on size, of which the range is inclusive to 38 mm bore.

10.3.3.2 *Hydraulic Hose*

Another interesting process is that used for the manufacture of hydraulic hoses. Basically, hydraulic hoses, depending on pressure ratings, are reinforced with either textile (usually rayon) or steel wire, applied by either braiding or spiralling techniques; in this process, the braiders and spiralling machines are usually mounted horizontally.

The process consists basically of extruding the lining tube (which must be

hard and firm uncured, and must also have good compression set characteristics when cured), blowing it on to a steel mandrel with compressed air, and
then braiding and/or spiralling with reinforcement. Covering, usually with
CR, and cloth wrapping is then carried out. After vulcanisation, the finished
hose is stripped from the mandrel.

In this method, the production length of the hose is limited to 40 m by the
steel mandrel length. The introduction of the flexible mandrel process has
made lengths up to, and even longer than, 300 m possible, depending on bore
size. The limiting factor in this case becomes the handling weights and the
wire capacity of the braiding machine carriers and bobbins.

Vulcanisation under lead is often used instead of cloth wrapping, particularly in the flexible-mandrel long-length process. In this process, the steel
mandrels are replaced by long lengths of flexible cores of suitable polymeric
material, on which the lining is applied. Braiding is carried out conventionally,
and then the cover is extruded into position. After the tubing has stood for
some time for maturing purposes, the lead is applied through either a lead
ram press or a lead extruder, the whole is vulcanised, and then the lead and
mandrel are removed from the hose. In modern hose plants, the curing may
be carried out in steam pans, continuous cateneries, fluidised beds, or salt
baths, or by high-frequency methods.

Flexible hydraulic hoses are used in almost every industry, the main ones
being aircraft, automobiles, earthmoving, materials handling, and mining.

In the majority of hydraulic hoses, the linings consist of either NBR- or
CR-based compounds, which are suitable for most of the hydraulic fluids in
common everyday use. Because of the extremely low temperatures encountered during N. European winters, hoses made for outside use in these
latitudes are specially compounded for temperatures below the normal
specification limits of $-35°C$ to $-40°C$. Special liquids, such as the phosphate ester type, require other materials, e.g. IIR or EPM, to be used
(Daugherty and Kehn, 1968).

Over the years, service requirements and conditions have increased in
severity, particularly for burst pressures and for impulse performance. Considerable research and development in the field of metallurgy has been
directed to discovering the best type of steel alloy and wire finish to be used,
the correct tensile strength of wire, the pressure of application during braiding,
and the angle of braids and/or spirals, to ensure that all layers of wire are
working together in harness and unison.

As a general guide, the universally recognised operating factor for burst
pressure to working pressure is 3:1 for static conditions and 4:1 for pulsating
conditions.

Considerable progress in design has been made over the past year or two,
and wire-reinforced braided hoses capable of well over 500 000 impulse
cycles are now commonplace, whereas such performance was only possible
in the past with hose of all-spiral-wire construction. Under certain conditions, with attention to the operating and safety factors, cycles of well over
1 000 000 can be achieved with braided constructions (Fig. 10.17) (Evans and
Melsom, 1968).

It can be seen from Fig. 10.17 that the impulse test has been carried out
with a square-wave format, and the plotted points were obtained by carrying
out a large number of impulse tests on the same piece of hose at various

pressures on each bore size in the range. A useful and practical diagram has thus been obtained, from which the hose capabilities can be seen at a glance. This provides a good guide to circuit designers when considering what size of hose to use to meet known impulse conditions and at the same time giving thought to volume delivery requirements.

The impulse characteristics of a completed hose are dependent on many factors, including the properties of the steel wire used in its construction. As

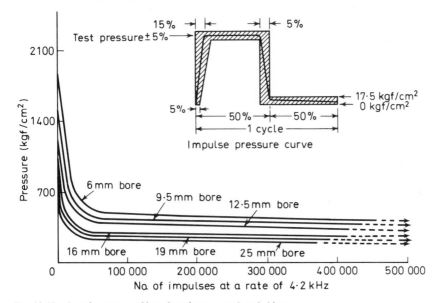

Fig. 10.17   *Impulse testing of long-length two-wire braided hose*

a generalisation, the lower the tensile strength of the steel, the more flexible is the hose, whereas the higher the tensile strength, particularly at the upper end of the range, the less flexible and more brittle it is. Much work has been carried out to obtain the point at which all the desirable properties are at a maximum and the undesirables at their lowest influence. The thickness of the wire is also important, and the wire finish may consist of galvanising, dry or wet drawn copper, bronze, or brass. The last is the one most widely used, and the composition must be correct to obtain the maximum bonding with the various adhesive systems.

Finally, an extremely compact hose is needed for maximum impulse performance with the minimum amount of movement under pressure surges. This is achieved by the correct design of the braiding carriers, by ensuring even and consistent tensions during the braiding operations, and by attention to the consolidation of the completed hose.

10.3.3.3   *Machine-Made Wrapped Hose*

As the name implies, this type of hose is made on machines, consisting of three rollers whose centres are on the corners of an equilateral triangle, the base being formed by two of these rollers (Fig. 10.18).

The process consists of extruding the lining tube in the straight position, and, after cooling, blowing it on to a circular-section mandrel treated to prevent sticking. The mandrel, the outside diameter of which is the nominal bore size of the hoses, may be made either of steel for the smaller-diameter

Fig. 10.18   Three-roll wrapping machine showing rollers (centre roller in raised position). Courtesy The Dunlop Co. Ltd

hoses or of light alloy for the larger sizes to facilitate ease of handling and movement; its length is that of the available three-roll making machine, which can be as much as 40 m. The mandrel, together with its blown-on lining tube, sits nicely on and between the two bottom rollers, the third top roller applying pressure as the fabric reinforcement is applied lengthways in strip form into the revolving rollers. The woven fabric (or duck as it is sometimes called), previously frictioned on a calender or spread coated on a spreading machine, is supplied cut to the required width, at a bias angle of 45° to give the hose flexibility.

The cover rubber is now applied lengthways and is pulled into position and consolidated by the rollers, after which a wrapping cloth, usually of nylon, is applied by spiralling it through the rollers. After vulcanisation, the hose is stripped of the nylon cloth wrapping (which is used for future cures) and removed from the mandrel.

A new concept is the development of a machine which laps the lining, reinforcement, and cover around the mandrel; the textile fabric is usually straight cut and may be applied at any angle as required. There is, therefore, greater flexibility in angle control than with the normal 45° bias cut and jointed fabrics. Furthermore, weftless cords can be laid in position by this machine, and, if so desired, armouring wires can also be incorporated in one operation and on the run during manufacture. The hose is completed by applying a wrapping cloth in line with the other taping heads, and thus all the hose manufacturing stages are completed, ready for cure, in simultaneous operations.

Hoses made by either of these methods are used, for example, for water, compressed air, and steam delivery, for welding and shot blasting equipment, and for conveying beer, wine, and food.

10.3.3.4 *Hand-Made Hose*

Many special large-size hoses required in the oil industry are made by hand on large lathes or making tables which rotate as necessary, each layer of lining, carcass, and cover being hand applied and rollered into position. Steel-wire helices are incorporated by spiralling them into position when it is required to prevent collapse of the hoses under suction conditions.

Oil suction and discharge (OS & D) hoses are currently produced with bore sizes up to a nominal 600 mm (24 inch), and development work is active on even larger sizes to keep pace with the requirements of the oil companies.

Furthermore, with the advent of the super oil tankers, and the need for loading and discharging on the open sea to avoid the cost of very expensive port and docking facilities, buoy mooring systems have been introduced. In addition to conventional OS & D hoses used in the system, there is a need for some of the hoses (those between the buoy and the ship) to be capable of flotation. This is achieved by one of two methods. The first is to place flotation beads around the hose. While this is very effective, it has the disadvantages that assembly is difficult and time consuming and the outfit is prone to damage. The other method is to build into the hose an integral flotation system by the use of expanded rubber, with closed-cell structure, introduced into the carcass of the hose.

10.3.3.5 *Circular Woven Hose*

The carcass of this types of hose consists of long lengths of fabric woven on circular looms. The warp threads run in the longitudinal or lengthways direction, and the weft threads are woven into the warps to produce a seamless circular weave.

One type of kerbside petrol-pump hose is produced by weaving around an oil-resistant lining tube, with wire helices included in the weft. The largest volume user of this form of reinforcement is, however, fire hose, and, briefly, such hoses are made by weaving the jacket (which also serves as the cover), as a separate operation, on vertical looms.

The lining tube is extruded or hand built from calendered material, and either in an uncured or in a semi-cured state (backed with uncured compound) inserted into the woven jacket. Consolidation and vulcanisation of the hose are completed by admitting live steam to each length, the ends of which have been suitably closed. The fabric is usually treated to prevent mildew, etc.

10.3.3.6 *Other Processes*

The five processes just described cover the vast majority of manufacturing methods, but, of course, there are others which are only in very small volume production by comparison. These include open steam and water curing methods, where the consolidation pressures necessary during vulcanisation are applied externally to the hoses in autoclaves by either steam or super-heated water.

Another interesting hose is that containing convolutions and which has applications as flexible connections in low-pressure systems, such as are encountered in the cooling systems of internal combustion engines. This type of hose is usually made in discreet short lengths by moulding an extruded tube (with or without fabric interply) between shaped metal formers. Provision is sometimes made for the application of a low negative pressure during vulcanisation to assist the forming operation.

10.3.4 HOSE DESIGN

Apart from the general characteristics of the rubber mixes used in the linings and covers of the various hoses, the most important design feature of a hose is probably its carcass construction. Reference has been made to the various textiles which are used and to some of the safety factors. The angle of application of textile yarn cord or fabric to the hose lining (the braid angle) and the ultimate bursting pressure achieved are of paramount importance in hose design. With regard to the former, generally speaking, the designer aims at obtaining the so-called neutral angle, at which there is no change in length or diameter of the hose, under internal pressure, assuming no elongation of the reinforcement.

The effect of any departure from the neutral angle can be predicted. When the braid angle is greater than neutral, the hose will increase in length and its diameter will decrease. When the braid angle is less than neutral, the hose will shorten in length and its diameter will increase.

10.3.4.1 *Braid Angle*

From Fig. 10.19, the braid angle θ is given by

$$\tan \theta = \frac{\pi D}{L} \tag{10.1}$$

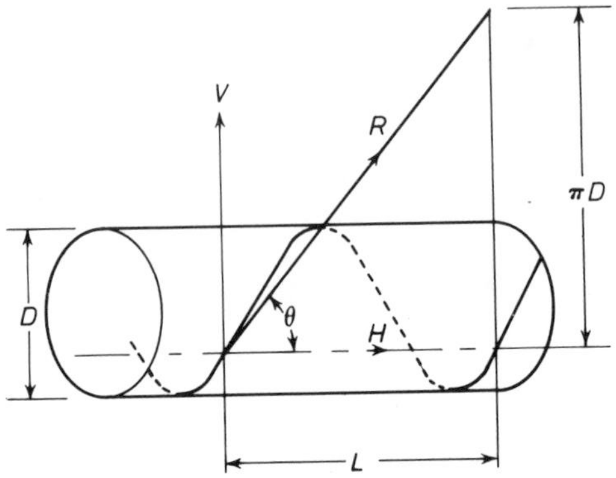

Fig. 10.19   *Diagram for hose design calculations*

where D is the mean diameter of reinforcement and L is the pitch or lead of the spiral.

10.3.4.2   Derivation of Neutral Angle

When the resultant of the hoop force V and the end force H is at an angle equal to the braid angle, the neutral angle is achieved and is derived as follows:

$$H = \frac{\pi P D^2}{4} \tag{10.2}$$

and

$$V = \frac{PDL}{2}$$

where P is the internal pressure. Substituting for L from equation 10.1,

$$V = \frac{\pi P D^2}{2 \tan \theta} \tag{10.3}$$

Therefore,

$$\tan \theta = \frac{V}{H} = \frac{\pi P D^2 / 2 \tan \theta}{\pi P D^2 / 4} = \frac{2}{\tan \theta}$$

Thus

$$\tan^2 \theta = 2 \tag{10.4}$$

or

$$\theta = 54° \, 44'$$

10.3.4.3   Burst Formula for Braided or Spiralled Hose (Hoop Force Burst)

For a hose with a total of N ends of yarn or cord in both directions and of an individual breaking strength of R, the burst pressure is

$$P_b = \frac{2NR \sin \theta}{DL} \tag{10.5}$$

10.4   Rubber Footwear

BY D. G. JONES

10.4.1   INTRODUCTION

The many differing types of footwear which include rubber compounds in their make-up can be classified as heavy industrial boots, light protective wellingtons and fashion boots, sports shoes and casuals with canvas uppers, slippers, and rubber-soled leather footwear.

Rubber has in the past been associated with utilitarian type footwear, but latterly it has become an established commodity in fashion footwear. Traditionally rubber footwear was manufactured by hand assembling pre-shaped components parts on a last or tree and vulcanising in an autoclave. This method is still used, but, during the post-war period, there have been great advances in machinery, equipment, and automated processes, thereby reducing the high labour content of the hand-assembled product. The rubber footwear industry has also taken advantage of new developments in polymer science and technology to formulate compounds which meet the new process requirements and give products with practically any desired feature.

10.4.2   MATERIALS

The rubber footwear industry has to meet a wide diversity of quality, depending on the end use of the product, the method of manufacture, and the price, with the result that compounds vary considerably and there is no typical footwear formulation.

Compounds for hand-assembled products are normally based on natural rubber because good building tack is essential and excessive shrinkage can cause distortion of shaped component parts. Adhesives for hand-built products are also based on natural rubber. The grades of natural rubber selected depend on the end use. For good-quality thin upper compounds, where flexing is a factor, a rubber with minimum dirt content is necessary, but lower grade rubber can be used for soles and heels. In moulded products synthetic rubbers are used extensively, such as non-staining low-Mooney SBR, oil-extended SBR, IR, and BR. Where heat- and oil-resistance properties are required, compounds are based on nitrile rubbers, blends of nitrile rubbers and PVC, or polychloroprene, the last of these being used mainly in hand-assembled products because of its superior tack. Whole tyre reclaim is used extensively in black solings and in some moulded black upper compounds. Finely ground vulcanised crumb can be used for cheapness and to help reduce porosity.

Carbon black is the reinforcing ingredient for black industrial boots, but, to meet the demand for non-black compounds, reinforcing siliceous fillers are used. A high-quality non-black soling will have fine particle size silica as a reinforcing agent, and the medium grade solings either aluminium or calcium silicate. Good-quality clays or activated calcium carbonate will give reasonable reinforcement with natural rubber, but to obtain equal reinforcement with SBR and particularly with oil-extended SBR, the addition of some siliceous fillers is required. The main diluents are whiting and cheap clays, and for whiteness titanium dioxide.

The accelerator system is normally a thiazole with a guanidine, thiuram, or dithiocarbamate as a secondary, depending on the rate of cure required. For white and light-coloured products an antioxidant must be selected which is non-staining. (Section 6.5.4).

Four examples of compound formulations are given in Table 10.6.

Cotton is still the most commonly used fabric in the rubber footwear industry. Blends of cotton and man-made fibres are sometimes used where additional resistance to abrasion is required. The leg and foot lining of an industrial boot is normally a square-woven fabric of 0·30–0·40 kg/m^2 in weight, rubberised by frictioning and topping, and with sufficient elongation to enable it to be lasted on to the shape of the boot tree. On moulded boots, knitted fabrics (either plain knit or ribbed) are also used, giving the boot leg a more flexible feel.

The upper material of canvas-topped footwear is prepared from a square-woven face fabric combined to a twill backing fabric by means of a thick

Table 10.6 FOOTWEAR MIXES

Ingredient	Hand-built boot upper	Good-quality direct-moulded sole (leather upper)	Industrial boot soling (compression moulded)	Medium-grade sole and heel unit (injection moulded)
Smoked sheets	80	27·5	45	30
SBR	—	25·5	55	30
IR	20	—	—	30
High-styrene resin MB	—	12	—	40
BR	—	35	—	—
Whole tyre reclaim	—	—	40	—
Zinc oxide	8	3	3	3·5
GPF black	20	—	—	—
HAF black	—	—	30	15
Precipitated silica	—	55	—	—
Aluminium silicate	—	—	20	50
Calcium carbonate activated	40	—	—	—
Ground whiting	30	—	—	—
Coated clay	—	—	20	—
Stearic acid	1	1	1·5	1
Coumarone resin	0·5	5	6	5
Process oil	—	30	3	10
Paraffin wax	0·5	—	—	—
Antioxidant	0·5	1	1	1
MBT	1	1·5	—	—
MBTS	—	—	1	1·5
TMTD	0·15	—	—	—
HMT	0·3	—	—	—
DOTG	—	1	1·5	1
Sulphur	2·5	2·5	2·5	2
30 mesh crumb	—	—	20	10
Colour	—	As required	—	—
Polyethyleneglycol	—	6·5	—	—
Properties				
Mooney viscosity [ML(1 + 4) 121°C]	52	—	50	45
Mooney viscosity [ML(1 + 4) 100°C]	—	55	—	—
Mooney Scorch* (121°C)	7 min	—	3 min	6 min
Mooney Scorch* (100°C)	—	10 min	—	—
Tensile strength (kgf/cm^2)	175	130	130	95

* 10 points rise.

rubber solution or compounded latex. For top-quality sports shoes, the combined weight of the fabrics is approximately 0·50–0·60 kg/m^2. For slippers, which are very much a fashion article, a variety of fabrics are used such as cotton, wool, nylon, suede, and lurex.

10.4.3   MANUFACTURE

10.4.3.1   *Hand-Assembled and Hot Air Vulcanised Products*

Although this method is labour intensive, it still has its advantages in that it allows for greater flexibility in changing product design without the need for acquiring new and sometimes expensive equipment. Some all-rubber industrial and wellington boots, and canvas-topped sports shoes and plim-solls are still made by this method.

A pair of all-rubber industrial boots may consist of 40 rubber or rubberised fabric component parts, each prepared and cut to the required shape prior to assembly. These parts consist of the outer rubber leg, outer rubber vamp, fabric leg lining, fabric vamp lining, anklet, insole, fillings, toe and heel reinforcements, back strips and strengthening pieces, foxing strips, top binding, sole, and heel (Fig. 10.20).

The calendered sheet for the production of outsoles is produced on a four-roll calender where the fourth roll is engraved with the necessary sole

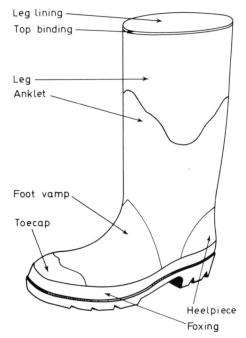

Leg lining
Top binding
Leg
Anklet
Foot vamp
Toecap
Heelpiece
Foxing

Fig. 10.20   Rubber boot components

pattern and profiled to give the variations in thickness. Any required sole design and thickness is obtained by changing the fourth roll.

The heel of an industrial or wellington boot which is hand assembled is usually pre-moulded to the required pattern, size, and shape. The insole, heelpiece, and fillings are made from unvulcanised rubberised fabric waste which has been ground on a tight mill and sheeted to the required thickness or topped on to fabric.

The components, freshened by a solvent wipe or thin rubber solution, are assembled, in a specified sequence, tightly on the metal last and then thoroughly rolled to ensure good bonding.

Another method of manufacture is to omit the outer rubber leg and vamp and to apply a coating of compounded latex by dipping and coagulating to provide the outer surface. After thoroughly washing and then drying off the moisture, the foxing strip, outsole, heelpiece, and top binding are attached. The latex forms a seamless, impermeable layer of rubber around the carcass.

For canvas-topped footwear, the fabric components are die cut and machined together to form the upper, on to which bindings have been sewn and eyelets or gussets and heelpieces fitted. This is lasted on to the last, and the necessary rubber components attached.

Vulcanisation is carried out in a large cylindrical autoclave heated internally by banks of steam coils fitted in the base and around the sides. The boots are loaded on to carriers and wheeled into the autoclave. In some factories, the carrier peg on which each boot is fixed is attached to a vacuum pump so that, as the boot is vulcanised, the component parts are drawn tightly around the last, thus ensuring good adhesion. Immediately the autoclave doors are closed, air, at a pressure of around 3 kgf/cm^2, is admitted to offset any blistering or blowing of the outer rubber parts. The temperature gradually rises to 135°C, where it is then maintained for the remainder of the vulcanising period, which is normally 60–65 min in total.

Because of the different thicknesses of rubber components and the hot-air method of vulcanisation, the compounds need to have a plateau-type curing system to avoid degradation. Accelerator systems based on thiazoles with TMTD as a booster have proved effective. A good heat-resisting antioxidant is usually added.

10.4.3.2   *Compression Moulded Industrial Rubber Boots*

A complete boot is moulded in one operation using vulcanising presses consisting basically of two lasts, two side moulds, suitably profiled to give the required pattern and thicknesses, and an engraved sole mould. Five components only are required by the moulder, a pre-formed fabric lining shaped to the last, a sole blank, a rubber leg, a rubber vamp, and a filler. A one-piece rubber leg and vamp is sometimes used. The lining fabric is first coated with rubber compound, and then pre-cured to avoid any strike-through of the outer rubber leg during moulding under high pressure. Correctly shaped legs, vamps, and insole socks, die cut from the pre-cured roll, are machined together to form the boot lining. This lining must have sufficient stretch to enable the moulder to fit it on to the last, and yet, when on, it must be a reasonably tight fit to avoid pinching during the moulding operation. Knitted fabric linings may also be used, suitably coated with rubber compound. The rubber leg and vamp compounds are calendered on a profiled roll and die cut to shape. The filler compound can be calendered or sheeted off a mill and die cut to shape. Correct distribution of the rubber in the mould is essential to give good mould definition. An economical method of preparing the sole blank is by the use of the Barwell Precision Preformer, which can produce a pre-shaped sole and heel unit of accurate weight and dimensions.

The heel portion of the sole, being the thickest in the boot, governs the vulcanisation time. Desma-Werke of Bremen have incorporated in their Desma 600C press (Fig. 10.21) a method whereby the sole and heel can be given double the vulcanising time set for the completed boot, reducing cure times by nearly half. While one boot is being vulcanised, a sole and heel blank is being moulded and semi-vulcanised in a second sole mould for the next boot. Overcuring the sole must be avoided, otherwise the bond of sole to

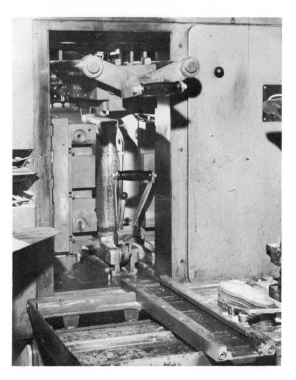

Fig. 10.21   Desma 600C press for compression moulded boot

upper could be impaired. Two lasts, normally collapsible or split to ease the fitting of the lining, are provided so that, whilst one last is in the press vulcanising a boot, the other is having the lining and rubber parts assembled on to it. The press is fitted with automatic temperature controls and timing devices.

To obtain efficient economical operation, compounds have to be rapid curing yet with delayed action to give good mould flow and perfect definition of pattern. With sole plate and mould temperatures of from 150°C to 155°C, a complete boot can be obtained every 4 min, the sole and heel having had two 4 min cures. Compounds with Mooney viscosity and scorch values at 120°C of 45–50 and 3–4 min (for a 10 points rise) respectively, are satisfactory for such conditions.

The vulcanised leg compound must be very flexible to avoid any discomfort in wear, have good flex cracking resistance, good atmospheric ageing resistance, and reasonable tear strength. The vulcanised sole and heel com-

pound must have adequate abrasion resistance and good cut growth re-
sistence. Moreover, the specific gravity of the compound should be no more
than 1·2–1·25 to avoid excessive weight in the boot.

10.4.3.3   *Direct Moulded Process for Shoe Bottoming*

As the name implies, direct moulded footwear is produced by vulcanising
rubber compounds direct to lasted uppers. Initial developments in this field
date from the early 1900s, but it is in the post-war era that the direct moulded
process has come to the fore. More shoes are now made this way than by any
other method (Fig. 10.22).

Moulding presses can be classified into two types: for cellular rubber
soled footwear, internal pressure machines which rely on pressure created

Fig. 10.22   Direct moulding of canvas shoe on Nova press

by the blowing agent in the compound; and external pressure machines,
either pneumatically or hydraulically operated, for moulding solid compounds.

Cellular Rubber Soled Footwear. By far the largest volume of cellular
rubber footwear produced is the sponge-soled slipper.

There are several different types of internal pressure machines but each consists essentially of (Fig. 10.23):

1. A base plate, having the contour of the sole and heel.
2. A ring which sits on the base, split longitudinally, hinged at one end and forming the sole cavity.
3. A last on which the upper is fitted and which acts as the top plate of the mould.

The sole plate and last are fitted with electrical elements, the sole plate being heated to 150°–160°C, the last to 125°–130°C.

The slipper upper, already prepared and shaped, is fitted on to the inverted metal last and held tightly to it by string or a sock lining which has been

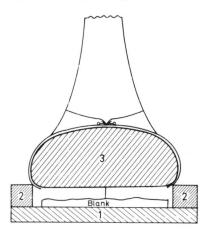

Fig. 10.23   Schematic diagram of internal pressure moulding machine: 1, base plate; 2, ring; 3, upper fitted on last

sewn to the base of the upper. A semi-vulcanised non-cellular outsole cover is placed centrally in the mould cavity, followed by a shaped blank of the un-vulcanised sponge compound and a filling, usually a felt. The last is then lowered until it is clamped tightly over the mould cavity. The total volume of material placed in the mould cavity is usually 60–70% of the volume of the cavity. The gas generated by the blowing agent expands the sponge compound, creating internal pressure sufficient to fill the cavity and form satisfactory bonds between the upper, cellular sole, and outsole cover. Vulcanisation time is from 6 min to 8 min.

Both preparation and presentation of the sponge compound require exacting control. Inorganic blowing agents are used, with a small addition of an organic to give a finer overall cell structure. The polymers are given extended mastication with the addition of chemical plasticisers. The compound must be highly plasticised to allow for ease of expansion by the blowing agent, and vulcanisation must be delayed sufficiently to allow this expansion to take place. After mixing, several remills may be necessary to attain the required plasticity, and the finished batch is then calendered to the required thickness, well dusted to give ease of handling, and die cut to the required shapes. The outsole cover compound is also calendered and pre-moulded into sheets 1·0–1·5 mm thick by approximately one metre square from which shapes are cut. This cover compound must have good ageing properties as the cut soles are to be given further vulcanisation in the slipper press.

Solid Rubber Soled Footwear. The attachment of solid rubber soling to uppers of either canvas or leather by the direct moulded process has been a revolution in footwear manufacture. Essentially, it is a mass-production process and most efficient when allowed to operate as such. The operation is carried out by the shoe manufacturer in individual presses with electrically heated moulds.

External pressure machines operating at low pressures, between 10 kgf/cm^2 and 20 kgf/cm^2, on the sole area are used for canvas-topped footwear, leather bootees, and casuals. For walking shoes, work boots, and heavy footwear, high-pressure machines operating at 55–85 kgf/cm^2 on the sole area are used. The pressure is normally applied by the sole plate acting as a ram.

Machines consist essentially of (Fig. 10.24):

1. A base plate, contoured to the shape of the sole and heel and carrying the required sole pattern.
2. Movable side plates which open and close automatically.
3. The last, which, when fitted with the prepared upper, fits exactly on to the cavity formed by the base and side moulds when closed.

The sole plate, side moulds, and lasts are fitted with electrical elements, and the machines include automatic temperature, pressure, and time controls.

As in the case of the moulding of cellular soled footwear, the operation consists of assembling the upper on to the last and loading the mould cavity

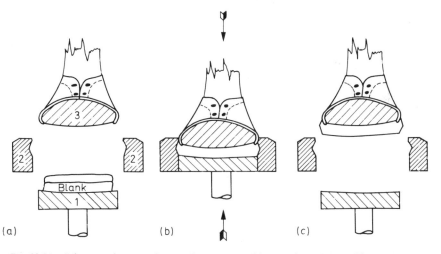

Fig. 10.24   Schematic diagram of external pressure moulding machine; (a) mould open, upper on last; (b) mould closed, under curing pressure; (c) mould open, cure completed

with the unvulcanised rubber sole and heel blanks. The side moulds are first closed tightly around the upper, and the sole plate is then forced up, acting as a ram, until it reaches a set position to give the required finished gauge of outsole. The time control is set to the required cure time, and, when this is reached, the press automatically opens. The footwear is stripped off the last and allowed to cool, and the spew is trimmed off.

With fabric uppers, the pressure applied is sufficient to force the rubber compound into the interstices of the fabric to give a satisfactory bond of rubber to fabric. In the case of leather uppers, the bonding area is roughened to allow better penetration of adhesive into the leather fibres. Two coats of a fast-curing natural rubber cement, sometimes activated with an isocyanate, are applied to the roughened leather of the upper. A wooden or fibre-board heel filler is also included with the rubber blank. The lasted upper must fit correctly within the bite line of the side moulds so that a good edge seal is formed without cutting into the upper. This bite line follows the insole edge. A good bond of sole to upper is essential, and, on the SATRA toe testing equipment, adhesion tests on finished shoes usually register figures of 32 kgf and over.

Accurate control is essential of the preparation and volume of the rubber blanks which are in cut shapes, strip, or granules, depending on the type of finished product and the sole pattern. The uncured rubber compounds must have fast-curing delayed-action characteristics and a shelf life of at least 3–4 weeks is essential. To utilise the equipment efficiently, the cycle times must be as short as possible. Where leather uppers are used, the lasts are normally cool, thus heat transfer into the rubber is from the sole and side moulds only and temperatures at the leather–rubber interface will not exceed 90°C. Temperatures of the sole and side moulds are 155°–165°C. With canvas uppers, the lasts are heated to 130°C so that shorter curing cycles are possible. Typical cure times for heavy boots are 8–10 min, for lightweight men's and children's shoes 5–6 min, and for canvas plimsolls 3–4 min. To reduce cure times, pre-heating of the unvulcanised sole and heel blank in an electrically heated oven at around 95°C is commonly practised.

10.4.3.4   *Injection Moulded Rubber Sole and Heel Units*

Injection moulded rubber sole and heel units are produced on multi-purpose moulding machines (Hall, 1967). These machines are basically a rubber injector which consists of a reciprocating screw working within a pre-plasticising cylinder, the screw and cylinder being fitted with independent temperature regulators. The moulding stations vary depending upon the machine design, but four-, six-, and ten-station machines are commonly used, the mould stations being carried on a rotary table. At the beginning of each cycle, the screw rotates and simultaneously moves to the back end of the cylinder, and the uncured compound in strip or pellet form is fed into the injection unit (it is generally found that pellets are more satisfactory for harder stocks). The rotary action of the screw plasticises the compound, and the movement backwards of the screw leaves hot plasticised material in front of the screw. The entire injection unit then advances to form a tight seal between the nozzle and the mould sprue. The screw, acting as a ram, moves forward without rotating, forcing the compound through the nozzle into the closed mould. Pressure is maintained for several seconds after the mould is filled, to allow the compound to set up. The injection unit then retracts, the rotary table moves the next station in front of the unit, and the cycle is repeated. At the same time, the station prior to the injection point opens, allowing the moulded units to be removed from the mould cavities.

The number of mould cavities per station depends upon their volume and the maximum shot size of the machine. In addition, the total mould surface area must be such that the upward thrust is not in excess of the closing pressure of the presses themselves, which is in the region of 300–500 Mg, depending upon the machine.

Moulds of varying capacity may be used in any one run, since the volume of compound injected into each mould is determined by the back pressure developed in the mould. Effective flash-free mouldings are not produced from this type of machine, but, by designing the moulds for tear trimming, minimum spew waste can be achieved.

Each station carries individual temperature controls, so that cure temperatures for mouldings of different sole and heel thicknesses can be achieved. Typical output from a six-station injection moulding machine would be:

Number of stations	6
Pairs per station	2
Mould temperature	177°C
Cycle time	35 seconds
Cure time	$2\frac{3}{4}$ min
Output per hour	206 pairs

Whilst the process for bottoming footwear can be adapted as above, the manufacturer of full-length rubber boots presents difficulties in both engineering and rubber technology. Boots of this type are, however, being injection moulded, using PVC, which lends itself readily to this method and has proved to be a very satisfactory material for lightweight wellington boots.

Thermoplastic rubbers (Section 4.7) can be satisfactorily injection moulded and are now used for some casual footwear.

10.4.3.5 *Resin Rubber and Microcellular Soling*

Rubber soling can be supplied to footwear manufacturers either in the form of soles cut from pre-moulded sheets or as pre-moulded sole and heel units. The advent of high-styrene resin (Section 4.2.1.5) rubber soling in the late 1940s was the most significant development in this field. It is light in weight, attractive in appearance, combines high hardness with good flexibility, and is 4–5 times as durable as leather. Because of these properties, it has practically replaced leather as a soling medium in fashion footwear. Microcellular soling containing minute discreet air cells and having a specific gravity as low as 0·3 is also used extensively. It is made by the addition of an organic nitrogen-generating blowing agent to a soling compound, usually based on NR–SBR blends, with the addition of high-styrene resin to obtain a satisfactory hardness and of siliceous-type fillers for good resistance to abrasion.

The manufacture of microcellular soling requires very careful control in mixing and moulding. Expansion occurs only after release from the press, and, therefore, the mould is loaded with a blank of volume slightly in excess of that of the mould. As soon as the press is opened, the sheet of vulcanised rubber immediately starts to expand and must not be restrained in any way, otherwise permanent distortion of the sheet occurs. The moulds are usually bevelled along the edges, and the press is made to open as rapidly as possible. The degree of vulcanisation affects the finished density, overcure resulting in near-solid sheets. Accurate control of vulcanising time and temperature is therefore essential. After initial expansion has occurred, some degree of shrinkage will continue indefinitely. Stabilisation is, however, achieved by heating the sheet in an autoclave at 120°–130°C for approximately 30 min. Sheets can be moulded up to 20 mm thick and subsequently slit to any required thickness. Sole shapes are die cut from the finalised sheet.

10.4.4 ADHESIVE ATTACHMENT

The present-day method of attaching rubber soles and heel units is by the use of adhesives, based on CR or NBR. The adhesive is applied to both the surface of the upper and the sole, the sole then accurately placed in position, and pressure applied. Where it is necessary to apply the adhesive some time in advance to meet production requirements, the adhesive-coated sole surface is reactivated by heat from infra-red lamps immediately before contact is made with the uppers.

10.4.5 SPECIAL PURPOSE FOOTWEAR

10.4.5.1 *Safety Footwear*

The need for added protection of the foot in the vulnerable areas has led to the manufacture of safety footwear, incorporating a steel toecap in the toe area and a spring steel midsole between the base of the foot and the outsole. The demand for this type of footwear in industry has increased considerably during the last few years, and approximately $2\frac{1}{2}$ million pairs are sold annually. The inserts are fitted during the assembling of the product, which can be hand built, compression moulded, or direct moulded.

The steel toecap is shaped to fit over the toe area of the last and is fitted after the boot lining has been lasted. It is essential that the toecap is a tight fit so that no movement occurs during manufacture. The toecap must meet the impact test requirements as laid down in BS 953. The steel midsole needs to cover as much of the sole area as possible, and BS 1870 specifies that the margin between the protective midsole and the edge of the insole must not exceed 6·35 mm. To ensure that no rusting occurs during service, the steel parts are given a chemical anti-rust treatment and enveloped between two thin layers of rubber compound. Depending on the service requirements of the product, safety footwear is manufactured with steel toecaps only, protective midsoles only, or incorporating both toecaps and midsoles.

10.4.5.2 *Conductive and Antistatic Footwear*

Where it is necessary to eliminate the build-up of static charges in the body to avoid sparking, as in factories handling explosives and in hospital theatres,

conductive or antistatic footwear is used. Conductive footwear should only
be used where there is no danger from electrical equipment. Where potential
electrical defects can occur, then it is essential to have a lower limit on
resistivity to give adequate protection against dangerous electrical shock and
this type of footwear is referred to as 'antistatic'.

In the manufacture of conductive and antistatic footwear, the upper parts
are of standard materials but the base of the footwear is made of conductive
materials. The insole, outsole, and any fillings required are prepared from
compounds possessing the required resistivity (Section 9.4.8).

For conductive footwear, BS 3825 specifies an upper resistance limit of
1.5×10^4 Ω, and, for antistatic footwear, BS 2506 specifies a lower limit of
7.5×10^4 Ω and an upper limit of 10^7 Ω.

10.5   Rubber-to-Metal Bonded Components

BY J. F. POWELL AND G. HOPKINS

10.5.1   INTRODUCTION

Bonding of rubber to metal is usually carried out during vulcanisation of the
rubber. Although direct bonding of rubber to metal is possible, this entails
the use of special compounds and/or metals, and the resulting bonds are not
always as satisfactory as those obtained by more conventional means
(Sections 6.6.5, 8.7.4, and 9.5.3). In normal practice, a bonding medium is
applied to the metal surface after all surface contamination has been
removed.

10.5.2   CLEANING THE METAL OR OTHER SUBSTRATE

The soils to be removed consist of oils and greases used in fabrication and
protection of the metal, corrosion products, oxides from heat treatment, and
possibly burnt-on carbon. Oil and grease may be removed with an organic
solvent, preferably neutral-stabilised trichlorethylene used in a vapour de-
greaser. Hot aqueous alkali mixtures containing wetting agents are sometimes used as an alternative to solvent-type degreasing agents.

Mechanical methods are frequently employed to remove the remaining
contaminants, and, of these, blasting with metallic or non-metallic abrasive
is the most popular. On ferrous metals, crushed iron grit is widely used, whilst
on aluminium, blasting with aluminium oxide grit is generally preferred. It is
important to avoid the use of dissimilar metals in order to eliminate the
possibility of galvanic action resulting, for example, from the use of iron
grit on aluminium. A final degrease in trichlorethylene vapour is a useful
safeguard in removing any abrasive debris from the metal surface.

Chemical cleaning methods are sometimes more convenient to use than
mechanical methods, particularly with automatic handling of the metal
components. Removal of metal oxides is achieved by means of dilute acids
or other chemical means appropriate to the metal being treated. Several
proprietary treatments are available in which removal of oil and oxide films

is combined with the deposition of a thin adherent film of iron or zinc phosphate. Thick powdery phosphate films must be avoided, and it is also important to include a final chromic passivating rinse. Ultrasonics can be used to improve the cleaning efficiency at any stage—in particular, the removal of inert soils such as carbon. (Fiske, 1961; Peterson, 1965; Smith, 1967; Cox, 1969-1.)

Many rigid plastics, some capable of replacing metals in certain applications, can be bonded to rubber using techniques similar to those for metals. Cleaning presents less of a problem, and in many cases (e.g. nylon), degreasing in trichlorethylene is the only cleaning required. However, acetal resin (Delrin) does require a special etching treatment (Bruner and Baranano, 1961).

10.5.3   APPLICATION OF BONDING MEDIUM TO SUBSTRATE

The substrate having been cleaned free of contaminants, the next stage is the application of the bonding medium. Nowadays, it is the usual practice among rubber-bonding firms to employ proprietary bonding agents, which are formulated specifically for this purpose.

10.5.3.1   *Brass-Plating*

Until the 1950s, the common method of bonding involved the use of a thin layer of electrodeposited brass. Although the brass-plating method is rarely employed nowadays for general bonded work, it is still used for specialised products, such as the wire reinforcement in tyres. In the brass-plating process, the metal part, which will have been previously cleaned by chemical methods, is made the cathode in a plating bath consisting mainly of a solution of copper and zinc cyanides in sodium cyanide. The plating conditions are adjusted to give a brass deposit of the correct composition for bonding, usually in the region of 68–75% copper. Rubber compounds for brass bonding require to be specially formulated as the presence of sulphur is essential to bond formation. The mechanism of bond formation has been discussed by Buchan (1959) and Stuart (1962).

The ability of rubber to bond to brass is taken advantage of in tyre-valve manufacture. Brass-plating is, of course, unnecessary, but great importance is attached to the composition of the brass used in fabrication of the valve stem. Etching treatments based on ammonium persulphate are used to bring the brass surface into optimum condition for bonding. In order to obtain greater reliability in bonding, the tendency nowadays is to use a proprietary cover cement over the brass, and it is therefore doubtful whether this can be classed as a direct bond between brass and rubber.

10.5.3.2   *Bonding Agents*

Proprietary bonding agents are used almost exclusively in modern manufacturing processes for bonded products. Advances in adhesive technology in the 1950s have enabled a range of reliable materials to be marketed, which perform at least as well as, and in many cases better than, the brass-plating technique. These materials are employed as one-coat or two-coat systems. Various grades (see Section 8.7.2) are marketed, depending on the polymer to be bonded and whether the material is intended as a primer or as a cover or secondary coating. The primer coating is formulated to have good adhesion to metal and consists of polar materials, such as chlorinated rubber, dissolved in a suitable solvent. The cover coat may consist of a mixture of several polymers, usually of intermediate polarity and rubberlike rather than resinous in nature, with the addition of materials to form chemical bridges between the adhesive polymer and the rubber (Cox, 1969-3). The formulation of these materials is said to involve considerable 'know-how', and their complexity may be judged from a study of the patent literature (Lord Manufacturing Co., 1958, 1959; Borg Warner, 1961; DeCrease and Schafer, 1966).

Application of bonding agents to the metal is achieved by any method appropriate to solutions and dispersions of organic polymers and pigments. Brushing is the simplest method of application and might well be used in a factory where bonded products do not form a large proportion of the production. Automated or semi-automated lines would employ spraying, dipping, or roller-coating techniques. Automatic spraying is commonly used for long runs of the same item. The method chosen depends largely upon the shape of the job, the area to be covered, throughput, etc., bearing in mind that adhesive wasted or applied to non-bonded areas represents a considerable financial loss. Each coat must, of course, be dry before the next coat is applied, and, in the case of spraying, it is often the practice to pre-heat the metals so that drying takes place in seconds rather than minutes, thereby economising on the size of plant required.

The dried adhesive coating is normally non-tacky, and coated metals may be stored in tote-tins for several weeks if needs be, provided the metals are protected from dust and other contamination.

Special tacky cover coats are employed where building tack is required, as in tank lining.

10.5.4   BONDING OF VULCANISED RUBBER TO METAL

Bonding of vulcanised rubber to metal is practised far less frequently than the technique of bonding during vulcanisation. Where the size of the metal part would be inconveniently large for the mould, it would be easier to adhere the moulded and vulcanised piece of rubber in a separate operation. Epoxy-type adhesives can be used for this purpose, although special treatment of the rubber surface, e.g. by cyclising with sulphuric acid, is necessary in order to promote adhesion of the rubber to the epoxy resin. Mention should also be made of the cyanoacrylate type of adhesive, in which polymerisation of the adhesive takes place when the two surfaces are brought together under pressure, but high material cost precludes its general use (Eastman 910 adhesive, marketed by Eastman Kodak).

10.5.5   COMPOUNDING OF RUBBER FOR BONDING

Reference has already been made to the need for special compounding of rubber for bonding to brass plating (Sections 9.5.1 and 10.5.3.1). Sulphur is the key ingredient in the formation of the bond, which is considered to take place through reaction between the copper in the brass and the sulphur in the rubber. A sulphur level of around 3% on the rubber is used commercially. The accelerator may also have an influence, and those types that inhibit the copper–sulphur reaction are to be avoided. It has been shown (Blow and Hopkins, 1945) that thiurams, in particular, have a deleterious effect on bond formation with brass, owing presumably to their ability to react with copper (to give a dithiocarbamate), thereby rendering it unavailable for the copper–sulphur reaction.

Preferred accelerators are the thiazoles and their derivatives; N-cyclohexyl-2-benzthiazylsulphenamide (CBS) is a popular accelerator for bonded work. Antioxidant selection is not normally critical, but mercaptobenzimidazole (MBI) has been known to inhibit brass bonding. Stearic acid and processing aids should be kept at reasonable levels, and care taken to avoid incompatibility. The effect of carbon black and other ingredients on bond strength has been discussed by Hicks and Lyon (1969). Synthetic rubbers, as well as natural rubber, can be bonded to brass provided these general rules on compounding are followed, although bond strengths to synthetic rubbers tend to be lower. Rubbers not normally cured with sulphur (e.g. polychloroprene) can be bonded to brass by inclusion of sulphur, but the effect on properties, such as heat resistance, needs to be considered.

One of the reasons for the decline in popularity of the brass-bonding process was the restriction placed on the rubber compounder. It was virtually impossible to bond heat-resisting compounds produced, for example, by curing with a thiuram in place of elemental sulphur. The advent of the modern adhesive has removed many of these restrictions, although certain difficulties may still be encountered.

Dealing with the polymer itself, there is little or no restriction on its choice, which obviously will be dictated largely by the end use of the product. Practically all of the commercially available rubbers can be bonded to metal by the correct choice of bonding agent (Section 8.7.2). On the more theoretical side, the polarity of the rubber has been considered as a factor in its bonding capability (Alstadt, 1955; DeCrease, 1960). The more polar rubbers, such as NBR and CR, can be bonded to metal more easily, that is to say, a simple one-coat bonding agent, such as chlorinated rubber or a phenol–formaldehyde resin, being adequate in many cases. Rubbers of low polarity, such as NR and SBR, often require a two-coat bonding system, although certain one-coat systems are effective, probably because they rely on cross-linking rather than a polarity effect.

The vulcanising system can be chosen to suit the properties required rather than the need for bonding. Nevertheless, thiuram disulphide/sulphurless and sulphur-donor-type curing systems have been known to give rise to difficulties with certain bonding systems, particularly if a low curing temperature is employed (Cox, 1969-2).

Antiozonants must be chosen with care as quite a number of these, even when used at normal levels, have a deleterious effect on bonding. The more effective antiozonants appear to have the greatest effect on bonding. This problem can be overcome by choosing a bonding agent with maximum resistance to antiozonants, and by careful selection of the antiozonant system (Cox, 1969-2).

The use of waxes in combination with reasonable levels of chemical anti-ozonants can result in satisfactory bonding properties. Moderate loadings of fillers, particularly carbon black, result in an increase in bond strength. Where low modulus is required, it is better to use a moderate loading of carbon black and add oil to reduce the modulus to the required value. Excessive amounts of processing aids or plasticisers, particularly if the limit of compatibility is exceeded, can be harmful to bonding, naphthenic oils being preferred to the aromatic type.

10.5.6 MOULDING

The moulding of rubber-to-metal bonded parts introduces complications over and above those experienced in the manufacture of all-rubber items. Three points that are worthy of attention are: the necessity to accommodate the metals in the mould, the need to ensure clean surfaces of adhesive and rubber, and the dangers of indiscriminate usage of mould lubricant.

So far as metal location is concerned, it will be obvious that, whereas rubber will accommodate quite a degree of undercutting, this is not the case with metal. Thus, the mould has to be designed so that the metal can be released easily: this may mean a multi-part mould. Alternatively, the manufacturer will discuss the design with the customer to ensure that metals of an acceptable shape are used. It will, of course, be realised that any post-cure metal assembly such as welding must be undertaken with great care, as neither the rubber nor the rubber–metal interface must be allowed to approach degradation temperature.

It must also be realised that comparatively tight tolerance on metal dimensions is called for. When located in the mould, the metals may be regarded as forming part of the cavity in which the rubber is shaped. Thus, leakage between metal and mould must be reduced to a minimum to avoid not only wastage and extra trimming costs, but also the possibility of the rubber flow wiping heat-softened adhesive from the edge of the metal surface, leading to poor bonding. In this same connection, mould design must be such that spew relief does not coincide with a bonding edge, thereby reducing pressure in this area to an unacceptably low level. The mould designer must also bear in mind the fact that parts are normally demoulded hot, at which time the rubber-to-metal bond strength could well be low. Undue strain on the interface must not be imposed when the part is removed from the mould.

The second point mentioned above is the need to present a clean rubber surface to the adhesive, plus the need to preserve the adhesive in a clean condition, which usually involves supplying the treated metals to the press line in covered containers or wrapped in polyethylene or similar sheeting. Brass-plated metals must similarly be kept clean, and should be used promptly before atmospheric tarnishing can set in, or they should receive a protective coating of, for example, the compound to be bonded applied as a solution. The moulder must be made very conscious of the need to avoid handling the treated surface: as he will normally be wearing heavy and probably dirty protective gloves for his work, he must be encouraged to remove these, possibly replacing them by clean cotton gloves, when he is locating the metals. Alternatively, it may be preferable, particularly where multi-cavity moulds involving the location of many metal pieces are concerned, to use a service man for metal location. Pre-location of metals in a jig for subsequent transfer to the mould will sometimes meet with success, but, owing to the need for precise and close-fitting metal location, this has limited application. An advantage of jig loading is that the possibility of pre-cure of the bonding agent on the first metals to be located is eliminated: neither the bonding agent nor the rubber must be allowed to start crosslinking prior to their coming together (Cox, 1969-3).

So far as cleanliness of the rubber is concerned, compression moulding will often be acceptable provided that more than usual care is taken in keeping the blanks clean: it may be necessary to brush or swab them. Transfer moulding may be preferred in that a clean surface of rubber is automatically presented to the metal. By the same token, injection moulding has advantages, but these must be set alongside the disadvantage of slowing down mould servicing because of the need to handle metals. Vertical injection rather than horizontal may be favoured, to overcome difficulties of retaining metals in their location: in this connection, a two-station vertical injection machine that enables one mould to be serviced whilst the other is filling and curing is worthy of note (Anon, 1968-1).

So far as mould lubrication is concerned, it is evident that metal and rubber surfaces that are to be brought together must not be contaminated with lubricant. On the other hand, owing to possible demoulding difficulties imposed by restrictions of mould design mentioned above, lubrication of the mould itself may be very desirable. This normally means that the moulder must be properly trained: the lubricant must be applied by brush rather than spray, and due precautions must be taken if any adjacent presses are using sprayed lubricant. The possibility of using a baked-on release agent, e.g. PTFE, can be considered. This must be set alongside the realisation that such non-stick surfaces have generally been extremely susceptible to mechanical damage, which can easily occur when metals have to be assembled in the mould. However, recent improvements in these treatments, resulting in scratch-resistant surfaces, suggest that a wider use of them might be envisaged in the future.

So far, the conventional manufacturing process of introducing metals and then rubber into a suitable mould cavity has been considered. Related concepts, such as the use of multi-cavity moulds fed from a common transfer pot, and the use of specially designed stripping tools to assist in removal of the semi-rigid component from the mould, will be understood as logical and reasonable extensions of the basic concept. However, the possibility of using the metals themselves as a mould, or a substantial part of the mould, should not be overlooked. For example, the inner and outer metals of a typical bush may be assembled in a suitable jig and filled with uncured rubber from an extruder nozzle (Woodcock and Goodman, 1945). This procedure presents a ready-made mould, apart from top and bottom plates, which may be designed to work between the platens of a conventional press and to apply the necessary pressure at the bush ends during curing.

The difference in coefficients of thermal expansion of metals and rubber compounds plays a dual factor where rubber–metal composites are concerned. The shrinkage of the rubber to something less than cavity size is well-known for all-rubber articles and is readily allowed for (Section 8.6.7). In addition, however, since the metal of the composite will shrink less than the rubber on cooling from moulding temperature, the bond is thereby placed under some strain. This feature has to be catered for in design, and, provided an adequate bond strength is achieved, some additional compensation may sometimes be obtained by pre-compression of the rubber—as, for example, by expanding the inner metal of a bush—before the part is put into service. This counteracting of shrinkage by pre-compression is in the nature of a bonus effect, since the main virtue of pre-compression is that in service the rubber will not pass through zero strain and will thereby (particularly in the case of natural rubber) gain a many-fold increase in fatigue life (Lake and Lindley, 1964).

10.5.7   TESTING OF FINISHED PRODUCTS

In testing bonded products, it is important not to approach so near to the ultimate bond strength (which may range from 50 kgf/cm^2 to 90 kgf/cm^2 in tension) as to run the risk of causing internal flaws that may lead to service failure. A typical test schedule for a high-quality product would involve 100% testing to 70% of the minimum specified value for bond strength, and this in turn would be of the order of 50% of the expected value. It is important to realise that very few bonded products are made to such flowline techniques that sample testing will give a true picture of bond strengths throughout a batch of parts: a consideration of the possible causes of bond weaknesses in a product where the adhesive is brushed on and the metal is handled to locate it in the mould will show how little a sample can be representative of a batch. Methods of non-destructive testing involving lower stresses combined with some type of probe (e.g. acoustics) to seek separation between rubber and metal have been proposed (Matusik, 1968; Peer and Fountain, 1969).

10.6   Cellular Rubber

BY J. G. WEBSTER

10.6.1   INTRODUCTION AND TERMINOLOGY

Cellular rubber products are probably more varied than any other rubber product. In addition to the normal technological differences between one rubber formulation and another, the degree of expansion causes significant changes in the properties and these are further varied by the cellular structure, which may be open or closed. To help appreciate the differences between one cellular material and another, reference may be made to the glossary of terms in BS 3558.

Cellular rubber is defined as a mass of cells in which the matrix is rubber. The three main classes of cellular rubber are foam, sponge, and expanded rubber. Foam rubber is particularly defined as a product made from liquid starting materials; the best example is latex foam, in which the cellular structure is intercommunicating.

On the other hand, both sponge and expanded rubber are made directly from solid starting materials; the essential difference is that the production method for sponge gives rise to an open or intercommunicating structure, whereas the term 'expanded rubber' is applied to those materials which have a substantially closed cellular structure. It is unfortunately common in the literature for the terms 'foam' and 'sponge' to be used to describe closed cell products; this confusion can only be resolved by studying the processing details or the properties of the end product.

10.6.2   EXPANSION TECHNOLOGY

To appreciate the refinements in technological formulations and the control of processing conditions in all types of cellular rubber production, it is imperative to have some understanding of the expansion process itself.

The expansion technology of cellular rubber is conveniently divided into two sections, covering foams on the one hand and sponge and expanded rubbers on the other. The link between the two is the gas phase, which may be air, carbon dioxide, nitrogen, or ammonia, each of which has special functions in individual circumstances. Furthermore, they are not always interchangeable. It is essential, for example, in closed cell products, that the entrapped gas does not diffuse rapidly from the cells and allow collapse of the structure. The most suitable gas, next to air, is nitrogen. The gas phase may be introduced by beating in air, by dissolving nitrogen in the rubber under high pressure, or by the generation of various gases from the decomposition of inorganic or organic chemicals dissolved or dispersed in the mix. The timing of the expansion step is critical, and careful control has to be exercised over the rheological properties of the mix if high-quality products are to be achieved at an economic level of production.

10.6.2.1   *Blowing Agents*

The earliest chemicals employed as blowing agents were inorganic, such as sodium bicarbonate, which gave off carbon dioxide gas and was used to produce sponge rubber. Nowadays, particularly for expanded rubber production, organic substances are used, primarily because they are more readily dispersed in rubber and because the major gaseous decomposition product is nitrogen, which has a lower diffusion rate than carbon dioxide. The essential features of blowing agents are the yield of gas per unit cost, the decomposition temperature, the composition of the gasses given off, and their effectiveness in producing uniformly fine cellular structures (Reed, 1960; Lasman, 1965). The principal organic blowing agents commercially available in the UK and W. Europe are given in Table 6.12.

AZDN has the disadvantage that one of its decomposition products is toxic. Both DNPT and BSH are more commonly used in expanded rubber production, but DNPT gives off an unpleasant fishy odour which is masked to some extent by the incorporation of urea; BSH is almost twice as expensive (volume for volume of gas) as DNPT. OB and AC both have higher decomposition temperatures than BSH and are used in applications where premature decomposition of the blowing agent is a problem.

The use of pure nitrogen in a pressure solution process is described later, and, although this process requires high capital expenditure, it is still economically attractive because of the high cost of blowing agents. A number of gas-liberating chemicals, including sodium carbonate, petrol, and metal hydrides and peroxides, have been tried in latex foam, but all commercial processes rely on mechanical air-foaming techniques.

10.6.2.2   Foam Rubber

The production of a foamed latex depends upon the inclusion, in a latex, of gas bubbles which are initially discrete and non-communicating to enable an increase in volume to take place but which, in the end product, are open and interconnecting and give the resulting foam its unique properties.

As foaming takes place, competing conditions exist between the formation of further bubbles, the breakdown of existing bubbles, and the densification of the rubber phase by flocculation. When conditions are optimised, it is necessary to gel the latex and, at the same time, control the breakdown of the air phase such that an interconnecting cellular system is produced. Finally, the gelled interconnecting foam is vulcanised.

The expansion technology of the competitive polyurethane foam closely resembles that of latex foam, but with the important differences that a highly regulated set of chemical reactions is used to produce the gelled polymer and that one of these reactions may be used to generate carbon dioxide as the pneumatogen. The high rate of diffusion of this gas through rubber makes it necessary that cell rupture is guaranteed; this is sometimes achieved by physically crushing the cell walls.

10.6.2.3   Sponge and Expanded Rubbers

Literature references to the production of sponge and expanded rubbers indicate a requirement for a low-viscosity formulation. Fig. 10.25 shows the qualitative relationship between the degree of expansion of a particular mix

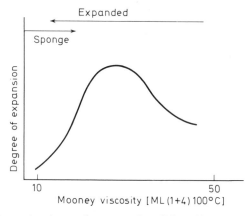

Fig. 10.25   Viscosity against degree of expansion for cellular rubber

and its rheological properties as measured by Mooney viscosity (Section 11.2.1.1). For the sake of simplicity, it has been assumed that the expanding gas is completely available for expansion at the appropriate level of viscosity. In practice, it is generated progressively over a period of time and the viscosity increases as the rubber is crosslinked by the vulcanisation reactions.

At low levels of viscosity, the low strength of the rubber matrix allows the gas bubbles to coalesce and escape from the matrix, leaving a coarse open structure with a high density or low degree of expansion. If a more viscous compound is used, the matrix, first of all, withstands the expansion forces of the gases enough to increase the degree of expansion, but eventually the cells break down. At a higher level of viscosity, the forces within the rubber are large enough to prevent sponge formation and a fine closed cell product results. In compounds with a still higher viscosity, the gas will be restrained and contained within the matrix at pressures higher than atmospheric pressure, leading to a reduction in the degree of expansion.

The non-quantitative character of Fig. 10.25 is stressed, and it should be used only as a starting point in determining the specific parameters necessary to control a particular mix. Mooney viscosity has been used here as an illustration; other measurements such as Wallace or Williams plasticity may also be used (Section 11.2.1.2); the important consideration is that the parameters must be determined and rigidly adhered to. A further point about Fig. 10.25 is that it is representative of a formulation with one level of blowing agent, and the change in the degree of expansion is due entirely to viscosity change. If a second level of blowing agent were used, the basic character of the curve would be retained but its quantitative aspects would differ.

The choice of blowing agent is important. Reference has already been made to the use of bicarbonates for sponge rubber production, but nitrogen blowing agents may also be used. All closed cell products are expanded using nitrogen gas. In both sponge and expanded rubber production, the gas must be available at the viscosity level best suited to the product being expanded. A careful balance must be struck between the decomposition temperature of the blowing agent and the temperature relationships of the vulcanising system adopted.

To summarise, it can be stated that the optimum rheological condition for a mix depends upon the polymer, its viscosity, the amount of softener used, the level of fillers, the curing system, and the blowing agent, not omitting to take account of the interaction of the latter with the curing system.

10.6.2.4 *Density Variation*

Cellular products are usually heterogeneous to a much higher degree than other materials, and it is not unusual to find variations in the degree of expansion throughout a moulded product leading to a $\pm 20\%$ spread in its properties. When a mass of cells are produced, the shaping of the cellular matrix within the mould is aided by the internally generated expansion forces; the gas pressure in the cells causes the matrix to move progressively to the surfaces of the mould, where gas diffusion and partial or complete collapse of the structure occur. This situation allows more matrix to move outwards from its centre to the surfaces until equilibrium is attained. The effect is more limited with sponge because of its open structure, but, nevertheless, the qualitative density variation represented in Fig. 10.26 can be detected to a greater or lesser extent in all cellular products. A high-density

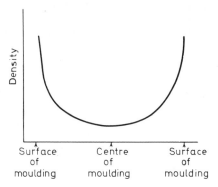

Fig. 10.26   *Density against cross-section for cellular rubber*

structure with elliptical cells parallel to the moulded surface is produced at the surface, and a lower density with spherical cells exists in the centre. Complex mouldings obviously have very complicated density contours.

The density differential effect is not so important in whole mouldings because it is the integrated property which is under consideration. It assumes greater importance in slab stock which is sliced into thin sheet for later fabrication work; dimensional changes take place as the stresses due to the differential are relieved. These changes are not always instantaneous and are often aggravated over a period of time by a further dimensional change due to gas diffusion. The total change varies considerably with the product but can be up to 10%.

10.6.3   MANUFACTURE

10.6.3.1   *Sponge Rubber*

Sponge rubber production follows normal mixing practices except that the highly plasticised formulations create sticking problems. Mixing in internal mixers is preferred to open mill mixing, and peptisation is commonly adopted to reduce cycles to an economic level. This procedure, however, creates some problems because of the difficulty of obtaining precise control over the level of viscosity required. Synthetic rubbers, particularly SBR, are widely used because they are supplied at low viscosity levels and are more consistent than natural rubber. To assist in achieving the low viscosity, it is common practice to incorporate high proportions of oil (50–100 p.p.h.r.) of suitable compatibility with the rubber (Sprague, 1968). The information on the formulation of compounds issued by the suppliers of rubbers and blowing agents will serve as a good starting point for the development of suitable production formulae.

Sponge products include carpet underlay, industrial sheet, mouldings, soleing, and automobile sealing strips. The mixed batches are processed on

P

extruders or calenders to give the required section or sheet stock from which the blank is prepared. In applications requiring a tough outer skin, such as door seals or slipper soleing, a mix without blowing agents in calendered sheet form is applied to the sponge profile. This process has limitations because of the differential expansion behaviour of the two materials during the moulding process, but this is overcome, for example, in better quality soleing, by first placing the solid compound in the mould followed by the expansible blank.

The curing and blowing of the sponge is carried out either freely or in a mould which is only partially filled with the moulding compound. In the free blow method, the section is hot air or steam cured and the resultant product usually shows considerable dimensional variation. In the moulding operation, a trial and error procedure is adopted to determine the mould volume loading necessary to produce a product with the desired properties. Sponge compounds are not usually highly blown, and a volume loading of 40–60% is common. Adequate venting of the mould and liberal use of talc or other dusting agent enable air to escape freely from the surface of the matrix and so avoid the formation of surface blisters.

10.6.3.2   *Expanded Rubber*

Closed cell products are manufactured as sheets, profiles, and mouldings, and in many ways follow the sponge manufacturing methods except that the viscosity control values are higher; the fluid bed processes are used to manufacture profiles and moulding machines to produce footwear (Section 10.4.3.5). A fuller description is given below of two methods of manufacturing industrial sheets by the nitrogen gassing and chemical blowing techniques.

Nitrogen Gassing. Normal processing methods are used to produce a calendered or extruded slab of rubber. Expansion of the slab is not usually isotropic, and careful control is exercised over the dimensions, residual strain in the slab, and its viscosity, to ensure that the dimensions of the finished expanded sheet are those required. The value of a typical linear expansion ratio for an industrial sheet of expanded rubber suitable for slicing and punching into gaskets is 2·0, giving a cubical (volume) expansion ratio of 8·0. In other words, a natural rubber formulation with a solid density of 1200 kg/m^3 is expanded eightfold to a density of 150 kg/m^3.

The calendered or extruded slab is placed in a high-pressure vessel, and nitrogen gas at a pressure between 140 kgf/cm^2 and 700 kgf/cm^2 is introduced. The vessel is heated for up to 24 hours at temperatures between 100°C and 130°C to effect solution of the gas in the mix. After cooling, the gas pressure is released, causing the slab to expand partially. The expansion is completed by further heating of the slab in hot air or on a hot plate. The initial expansion of the slab takes place under relatively low temperature conditions and is an important difference from the chemical expansion procedure where expansion takes place hot. The fully expanded sheet is then cured either in hot air or in a frame mould to obtain more accuracy in the final sheet dimensions.

Gases other than nitrogen have been used in this process, but the products suffer from considerable shrinkage. Nitrogen is preferred because it is an environmental gas and there will be the minimum interchange between the enclosed gas and the external atmosphere.

Use of Chemical Blowing Agents. The processing also follows the normal methods, except that a blowing agent is added to the formulation. It is important that mix ingredients are adequately dispersed to prevent the production of coarse cellular structures. For sheet production, the extruded or calendered slab is partially vulcanised in a pre-form mould; during the heating cycle, the blowing agent decomposes to give off nitrogen and other gases, which remain dissolved in the rubber because of the high specific pressures used in the pre-forming operation. The dimensions of the pre-form mould are as important as the extruded dimensions in the nitrogen gassing process: when the press is opened and the sheets 'explode' hot from the mould, the expanded dimensions relate to the original pre-form mould cavity dimensions. In some pre-form mouldings, specific pressures as high as 350 kgf/cm^2 are used to prevent premature opening of the press. In formulating the compound, a balance has to be achieved between the rate of evolution of gases and the partial vulcanisation of the rubber which takes place in the pre-form operation. The expanded sheet is finally cured in a frame mould where the specific closing pressure is as low as 0·35 kgf/cm^2.

Reference has been made earlier to the dimensional stability of expanded rubber. The size of a fully expanded sheet is typically 1 m × 1 m × 30 mm, and, when this sheet is subdivided into thinner sheets by a splitting operation, the split sheets can shrink by as much as 10%. One of the methods used to minimise this problem is to heat mature the vulcanised sheet in an air oven at elevated temperature. As a major part of the maturing treatment requires gas diffusion to take place, the cycle is of necessity a long one and can be up to 24 hours. In some cases, an alternative lengthy maturing cycle of 4–8 weeks at room temperature is used.

10.6.3.3 *Foam Rubbers*

Detailed discussion of latex technology and polyurethane product manufacture is outside the scope of this book. Reference is made in Chapter 2 to these two subjects (Sections 2.4.3.4 and 2.5.3).

10.6.4 TESTING

Reference should be made to Section 11.3.21.

10.7 Sports Goods

BY R. C. HAINES

10.7.1 INTRODUCTION

Many items of sports equipment incorporate rubber in some form; the grips on golf clubs and cricket bats, the bladders in footballs, and the pimpled surfaces of table tennis bats are examples. The largest quantities of rubber are, however, used in the manufacture of balls for the games of tennis and golf.

10.7.2   TENNIS BALLS

A tennis ball consists essentially of a hollow rubber core covered with a cloth usually composed of wool and nylon. The International Lawn Tennis Federation requires that the following specification is met at a temperature of 20°C and a relative humidity of 60%:

1. Diameter ('go-no-go' gauges), 2·575–2·700 in (65·4–68·6 mm).
2. Weight, $2–2\frac{1}{16}$ oz (56·70–58·47 g).
3. Rebound from 100 in (2·54 m) on to concrete, 53–58 in (1·35–1·47 m).
4. (a) Deformation under 18 lbf (8·2 kgf) load, 0·230–0·290 in (5·85–7·35 mm).
   (b) Deformation under 18 lbf (8·2 kgf) load on recovery after ball has been compressed through 1 in (25·4 mm), 0·355–0·425 in (9–10·8 mm).

The test in 4(a) measures the 'compression' or 'hardness' property of the ball, and that in 4(b) measures hysteresis after the ball has been compressed through 1 in (25·4 mm). The tests are carried out on a special 'Stevens' machine.

Two distinctly different types of tennis ball are now made. The type that is most common has a core pressurised with air or gas to about 0·7 kgf/cm² above atmospheric pressure. The second type has a non-pressurised or 'pressureless' core, having the advantage that its physical properties are retained over indefinite periods of storage. Permeation of the pressurising gas causes the pressurised type to become softer and less resilient in storage over periods of a few months unless it is kept in pressurised containers.

The two types are made by similar methods, but different rubber compounds are used for the cores and the non-pressurised ball has a thicker-walled core. Compound formulations are given in Table 10.7; that for the non-pressurised is given by Dunker (1952). The high-styrene resin (Sections 4.2.1.5 and 6.3.9.3) imparts the required stiffness to the core compound to achieve the necessary rebound and compression characteristics.

The compounds are normally formed into blanks of controlled size by an

Table 10.7   TENNIS BALL MIXES

Pressurised core		*Non-pressurised core*	
Natural rubber	100	Natural rubber	100
GPF black	30	High-styrene resin	30
Clay	32	Kaolin	20
Zinc oxide	9	Stearic acid	2
Sulphur	3·5	Sulphur	2·5
DPG	2	Accelerator	1
HBS	1		
Cure	$2\frac{1}{2}$ min at 150°C	Cure	4 min at 150°C

extrusion and pelletising process, and hemispherical shells are compression moulded and vulcanised at a temperature of 150°C. The edges of each shell are abraded and coated with a vulcanising rubber solution. Two shells are then joined together to form each core and are subjected to a further press moulding operation to vulcanise the joining solution. One of two alternative methods is used for pressurisation of the pressurised type of core:

1. Pellets of sodium nitrite and ammonium chloride are inserted inside the core before it is closed. During the subsequent press moulding operation, the chemicals are activated by heat to form nitrogen, which pressurises the core.
2. The shells are mounted separately in the platens of a clamshell press, and compressed air is introduced to each core by the following sequence of operations: the press closes until the pairs of shells are almost touching, when a rubber seal around the periphery of the platen area is actuated to isolate the platen area from the outside atmosphere; compressed air is then introduced into this area, and the two halves of the mould are brought together so that the compressed air becomes sealed inside each core. Cooling is necessary at the end of each cycle, as in the case of all inflated products.

These two methods of pressurisation offer different advantages. The 'chemical' method gives higher uniformity in practice, but, on occasion, the byproducts of the chemical reaction crystallise and produce 'rattlers'. The compressed air method is somewhat cheaper, but cure times tend to be longer owing to the more complicated press arrangement and consequent greater mass of metal involved in the heating–cooling cycle. Typical cure times and temperatures are:

Chemical pressurisation	4 min at 150°C
Compressed air	16 min at 150°C

After this stage, the further processing is identical for both pressurised and non-pressurised balls. The moulded core is first subjected to surface abrasion in preparation for adhesion of the cloth, and solution is then applied by a barrelling operation.

Tennis ball cloth is made from weft yarns of blended wool and nylon woven into a cotton warp in a manner such that the weft appears predominantly on one side of the fabric. This side of the fabric is then given a teasling or 'raising' treatment so that a fluffy texture is produced on what is going to be the wearing surface of the cloth. The cloth is then subjected to a felting operation in which the raised surface is consolidated by the natural felting properties of the wool to give the cloth its final texture which is necessary for satisfactory flight and wearing properties. For adhesion to the core, the reverse side of the cloth is spread with a vulcanising solution and dumb-bell shaped blanks are then stamped out. These are arranged into packs and their cut edges coated with a vulcanising solution by a dipping process. It is this solution which subsequently forms the seam on the surface of the ball.

Two dumb-bell shaped pieces are applied to each ball, the first by a hand-operated jig and the second purely by hand (Fig. 10.27). This operation must be carried out with extreme precision if rejects are to be avoided in the subsequent press moulding operation, which is carried out at 130°C for 12 min to vulcanise the cover to the core and form the seam. The moulded ball is steamed to raise the nap and to remove the moulding line. It is then tested for compression before applying the transfer marking and boxing.

It is interesting to note that the operation of making a tennis ball involves three separate moulding operations: (a) the hemispherical shells, (b) the core,

Fig. 10.27   Covering a tennis ball: the spatula is used for smoothing down the joint between the cover segments. Courtesy The Dunlop Co. Ltd

and (c) the covered ball. Moulding conditions and compound formulations take recognition of this fact.

10.7.3   GOLF BALLS

Golf balls were originally made by stuffing feathers into spherical leather containers; they were first made from solid gutta percha in 1845. The present-day core-wound ball dates from the beginning of the century, from which time it has remained basically unchanged. During the last few years, solid balls have been reintroduced using modern materials, and these are now made alongside conventional wound balls. The latter are, however, still the only type acceptable to the majority of golfers.

Golf balls are made to meet specifications laid down by two controlling bodies, which are the Royal & Ancient Golf Club of St Andrews (R & A) and the United States Golf Association (USGA). Both bodies specify that the ball must weigh not more than 45·9 g (1·62 oz) but the specification for diameter differs as follows:

R & A   Not less than 41·15 mm (1·62 in)
USGA   Not less than 42·67 mm (1·68 in)

The USGA additionally specifies that a golf ball must not leave a projection apparatus of their own design at a speed greater than 77·8 m/s (255 ft/s), and this sets an upper limit on resilience for the 42·67 mm ball.

Compared with a tennis ball, the specification for a golf ball is relatively simple and arises from the fact that, other things being equal, the smaller and heavier a ball the further it will go. Manufacturers (in particular, those making 41·15 mm balls) are therefore encouraged to make balls as near as possible to maximum weight and minimum size allowable so as to achieve maximum flight length.

Certain properties cannot be adequately covered by specifications. Such properties are feel and 'click' (which are subjective assessments of the club–ball contact), flight pattern (which depends upon the rotational inertia and aerodynamic qualities of the ball), and durability. Such qualities come into the category of 'know-how' perfected by individual manufacturers.

10.7.3.1 *Conventional Wound Golf Balls*

A 'wound' golf ball consists of four components: (i) the centre, (ii) the windings of rubber thread, (iii) the cover, and (iv) the paint and markings. Each component has an important contribution to make to the properties of the finished ball and these will be considered in turn (Fig. 10.28).

Centre. The centre must be the nucleus for the winding operation. It must be large enough for this purpose, but not so large that the resilience of the core is adversely affected. It must contribute a significant amount of mass to

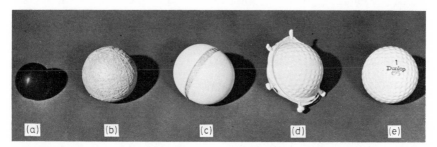

Fig. 10.28   The components of a golf ball: (a) the centre, (b) the wound core, (c) the wound core with cover shells in position prior to moulding, (d) the moulded ball and (e) the finished ball. Courtesy The Dunlop Co. Ltd

the ball because the overall ball weight and rotational inertia are largely dictated by those of the centre. A further most important property is that it should absorb the minimum amount of energy at club–ball impact. For this reason, centres of the best-quality balls consist of a liquid or suspension enclosed in a deformable container. Centres of this type are also found to give an acceptable 'feel' to the ball owing to the fact that the centre does not restrict free distortion of the ball on impact.

Initially, liquid centres were made from a suspension of barytes and bentonite in water and glycerine contained in a bag made from latex and tied at its neck. Modern methods use a similar suspension or, alternatively, oil contained in a moulded rubber jacket. In one method of centre manufacture, the suspension or 'paste' is dispensed into a series of hemispherical cavities in a rubber mould; two similar moulds are located together to form spheres of paste, which are then frozen in a refrigerator so that they can be demoulded and handled individually. These spheres are then encapsulated in rubber by the following process. A sheet of unvulcanised rubber is placed over cup-shaped cavities fixed in the bottom platen of a press and pushed down slightly by hand into each cavity; a frozen pellet is placed in each depression. A further sheet of unvulcanised rubber is laid over the top of the assembled pellets, and the press is closed so that mating cup-shaped cavities in the top platen join with those at the bottom. The rubber is stretched and joined around each pellet, the excess being cut off by the sharp cavity edges. The rubber is then vulcanised to produce a small rubber-covered liquid ball, which, before winding can commence, must again be frozen. This is done by a barrelling operation in the presence of dry ice so that a hard true sphere is formed.

Centres for second-quality balls are often made from solid rubber compounded to provide the necessary mass. The high loading of filler required affects the ball resilience. Some manufacturers have overcome this problem by using a lower-gravity compound with a metal or glass bead to make up the necessary mass fitted in the geometrical centre of two separately moulded half-centres.

Windings. Rubber thread used in the windings of a golf ball must produce the highest possible resilience in the ball core. The thread normally based on natural rubber or a blend of natural rubber with *cis*-polyisoprene is made by dry rubber or latex techniques on the cut thread principle.

The percentage of *cis*-polyisoprene is limited by the requirement of good chafe resistance necessary for the winding operation. A typical 'dry rubber' thread formulation based on natural rubber is given in Table 10.8.

The thread is cut to dimensions of 0.5 mm $\times 1.5$ mm for modern direct winding, in which one continuous thread is wound on to the centre. Older

Table 10.8   GOLF BALL THREAD FORMULATION

Natural rubber	100
Sulphur	6
Aldehyde ammonia	0·5
Cure	150 min at 130°C

methods employed the use of thread of several different gauges, the thickest being applied nearest the centre. Such methods were particularly applicable to centres of the latex bag type.

A modern core-winding machine (Hurst and Jenkins, 1954) consists essentially of two horizontal rollers on which the core sits and which both rotate and oscillate axially in anti-phase. The core is held in position on these rollers by two further small rollers attached to a carriage which moves up and down so that contact with the core is periodically released. The function of this mechanism is to rotate the core in a completely random manner so that the thread is randomly wound on; the action of the machine ensures that the thread is always wound on to the smallest diameter of the core and produces a spherical product.

The thread is tensioned during winding either by friction pads applied directly to the thread or, preferably, by running the thread around a braked roller. High compression is imparted by high tension associated with thread elongation approaching 1000% and this leads to high ball resilience. The winding operation is automatically stopped as soon as the required core size has been reached.

Cover. The cover of a golf ball provides protection for the core and carries the surface pattern necessary to produce aerodynamic lift. Covers have been traditionally made from natural gutta percha and balata, until recently the only materials having suitable properties—in particular, mouldability at 80°–100°C, which is necessary to avoid degradation of the rubber core by heat. Synthetic gutta percha (*trans*-polyisoprene) is now also used for golf ball covers.

In conventional procedure, hemispherical shells are compression moulded by putting heated blanks into a cold mould. The shells are then mounted on a core, and the assembly is subjected to a compression moulding process at 90°C, in which the shells are moulded to the core and to each other and the surface pattern is moulded in. The moulds are cooled by refrigerated water before extraction takes place. The spew lines are ground by rotating frozen balls against a revolving emery disc.

Vulcanisation of the cover is then carried out by exposing the moulded balls to carbon bisulphide vapour for several hours at a temperature of 32°C. The mechanism of the process is that of the Geer process, in which the carbon bisulphide reacts with a secondary amine and zinc oxide to form zinc diethyl dithiocarbamate. Vulcanisation is completed by putting the balls in ovens for 48 hours at a temperature of 32°C.

With other plastics materials, special moulding techniques must be used to avoid the problem of thread degradation. One such process is the direct injection of the cover on to the core; one using a polyester polyurethane is referred to by Spalding (1963).

Painting and Marking. The balls are carefully graded for weight and compression, before being given a paint pre-treatment. White polyurethane paint is then applied by rotating the balls, held by metal prongs, in the path of a spray gun which oscillates through the arc of a circle and covers the whole ball surface (Fig. 10.29). The name and numerals are stamped on, and the ball is finished with a clear polyurethane coat, wrapped, and packaged.

10.7.3.2   *Solid Golf Balls*

Many attempts have been made in the past to produce acceptable golf balls by methods other than the complicated process of core winding. These have generally failed due to inadequate flight length and poor 'click'. A patent filed by James Bartsch (1963) described a method of making homogeneous solid balls without the shortcomings of the previous attempts. The problem is to devise a compound which combines high resilience with extreme hardness

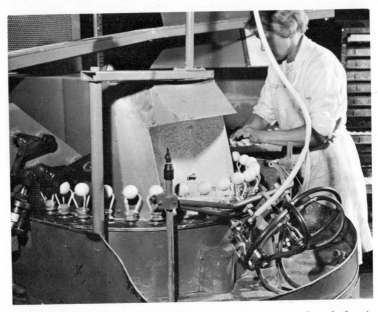

Fig. 10.29   *Painting golf balls: the balls are located on jigs which rotate in the path of a paint spray shown in the bottom right-hand corner of the photograph; the gun reciprocates in a circular arc so that paint is evenly applied to the ball surface. Courtesy The Dunlop Co. Ltd*

and toughness. Bartsch discovered such a material in a combination of certain monomers with an elastomer and a polymerisation initiater of the free radical type. The monomer is selected so that it is capable both of crosslinking the elastomer and of itself undergoing further polymerisation. A series of suitable monomers is given in the patent referred to (Bartsch, 1963). Silica is used to adjust the gravity and to reinforce the compound. Typical formulations for a 42·67 mm ball are given in Table 10.9.

Although the performance of these balls is extremely good by the standards

Table 10.9   TYPICAL MIXES FOR SOLID GOLF BALLS

cis-Polybutadiene	100		*cis*-Polybutadiene	100
Butylene glycol dimethacrylate	56·2		Divinyl benzene	62·5
Silica	37·5		Silica	62·5
Dicumyl peroxide	3·13		Dicumyl peroxide	3·13
Cure	10 min at 150°C		Cure	10 min at 150°C

of earlier solid balls, their flight distance is about 10% inferior to that of top-quality wound balls. However, being much more damage resistant, they appeal to the relatively inexpert golfer, and are now made in substantial quantities, particularly in the USA.

It is interesting to note that, because the solid ball is homogeneous in construction, its rotational inertia differs from that of a conventional core wound ball, which has much of its weight concentrated at its centre. For this reason, the solid ball accepts less spin from a lofted club and so has a flatter trajectory than a conventional ball.

10.8 Cables

BY D. POLLARD

10.8.1 INTRODUCTION

The modern elastomer-based electric cable consists generally of a metallic conductor covered with a crosslinked elastomeric compound formulated primarily with regard to its electrical properties, and a sheath or jacket formulated primarily with regard to its mechanical properties; in both formulations, the primary considerations are influenced by other environmental conditions.

In the period when natural rubber was the commonest alternative cable insulation to paper, its use developed in those areas where the smaller conductor sizes were appropriate—as in lighting and general low-power final distribution. These are areas where its greater manufacturing convenience in covering small wires and the greater flexibility of the resultant cable shows to the maximum advantage.

As the use of electrical power developed, the basic rubber cable construction extended into larger cables, particularly into areas where their flexibility was a salient advantage.

The availability of polychloroprene next brought improvements in cable sheaths since it provided outstanding technical advantage over natural rubber compounds under three main headings, namely flame retardance, weathering, and oil resistance. Consequently, it was adopted universally as the sheath material for mining cables, quarry cables, and ships cables, and to a major extent in, for instance, flexible cables in workshops and in farm wiring. Its inferior electrical properties were a deterrent to its use as an insulant, although it did see some use in this role in low-voltage cables, notably in aircraft (Pren cables, BS 2E 21), where size, weight, and flame retardance are primary considerations. Chlorosulphonated polyethylene (CSM) is now replacing CR, particularly over EPM and IIR insulations.

In applications where flame retardance and electrical characteristics are both primary factors and are equally important, as in ships cables and railway signalling cables, a layer of natural rubber insulation compound is used immediately on the conductor, followed by a layer of polychloroprene compound bonded and covulcanised with it. This type of insulation system is now paralleled in the modern EPM–CSM and IIR–CSM combinations.

10.8.1.1 *General Purpose Cables*

The general purpose category is defined as that where the environment, or end use, is typical of everyday modern life and such requirements as there are, can be met by most, if not all, elastomeric materials. In this group of applications, economics largely determine the choice of material.

In this field, natural rubber holds sway; only the cleanest grades are satisfactory for insulation, although less stringency is necessary for sheaths. The growth of bulk production methods, e.g. continuous vulcanisation, has put emphasis on uniformity of quality and properties, and reliable grading

Fig. 10.30   Representative range of cables. Courtesy British Insulated Callender's Cables Ltd

schemes, such as the standard Malaysian rubbers, have been of considerable assistance. As yet, no polymer has been developed which behaves in all its aspects exactly as does the natural product; hence, strong economic pressure is the only incentive to use the general purpose synthetic rubbers. The extent of use of these synthetics is therefore indicative of the vagaries of natural rubber prices. In the UK, SBR and IR, generally blended as a minority component, rarely as the base polymer, are most commonly used in this area. In the USA and to some extent in European countries, the general purpose synthetics are often used alone.

The cable specifications in the UK have tended to be particular about the polymer, but successive revisions have given some freedom. The tendency to write specifications in terms of the performance required of the cable is increasing.

General purpose formulation economics are subject to the same broad considerations as is the selection of polymer. Not only the prime cost of the individual compound must be taken into account, but also the processing and handling economics. Because material costs exceed labour costs by several times, compound formulation economics, along conventional lines, subject only to limitations from gross effects on electrical and mechanical properties, receive primary attention. Minimum requirements for these formulations are indicated in BS 6899. It should be appreciated that, in cables, the cost per unit volume is the criterion because definition of the product is dimensional.

Illustrative of the general purpose cables are building wires and cables (BS 6007) and flexible cords for portable appliances (BS 6195). The wiring of domestic, commercial, and general industrial buildings is also an outstanding example of the dominance of economics in the selection of materials. It is in this area that the use of the thermoplastic PVC compounds has largely replaced that of elastomers.

10.8.1.2 *Special Purpose Cables*

The principal applications and specifications and the salient design requirements for special purpose cables are given in Table 10.10, together with the elastomer type used.

Heat resistance is defined as resistance to degradation and/or permanent deformation as the result of a few temperature excursions to $1\frac{1}{2}$–2 times the maximum continuous operating temperature and of duration not exceeding a few hours. Ageing is related to the life of the cable coverings at the maximum continuous operating temperature.

Under the heading of electrical requirements are included dielectric strength, volume resistivity, permittivity, dielectric loss, and corona resistance.

Mechanical requirements cover tensile and compression stresses, abrasion, cut, tear, shearing, and flexing.

Oil resistance also includes resistance to solvents.

Fire resistance is meant to be the ability of an insulation system to operate for a limited time during, or after, a fire involving the cable, as distinct from flame retardance, which is the capability of a covering to prevent the propagation of flame.

In the special purpose category, electrical, environmental, and mechanical performance requirements can limit the choice of polymer to a small number, leaving economics as the ultimate arbiter, or can narrow the choice to one polymer with very limited scope for compound variation, irrespective of price.

10.8.2 COMPOUND FORMULATION AND EVALUATION

Reference has already been made to some of the factors determining the selection of polymer and formulation for general and special purpose cables. Typical insulation and sheathing formulations for general purpose cables are given by Evans (1953) and Tew (1956). The more recent work by Skinner and Watson (1967) has particular relevance to NR cable compounds. In this section the main factors having peculiar interpretations related to cables, their influence on formulations, and their evaluation in cables, are discussed.

10.8.2.1 *Ageing and Heat Resistance*

The maximum continuous operating temperature of cable is the sum of the maximum continuous ambient temperature and the temperature rise in the

Table 10.10 SPECIAL PURPOSE CABLES

Cable application	Specification	Insulation polymer	Sheath polymer	Main design features					Additional features
				Heat resistance	Ageing	Electrical	Flame retardance	Mechanical	
Ship wiring	BS 6883	EPM, IIR	CSM, CR, NBR–PVC } HOFR (BS 2899)	X	X		X		
Mining	NCB 188, NCB 504, NCB 505,	EPM–CSM, IIR–CSM, CSM	CR				X	X	
Quarrying	BS 6116	NR, IIR, EPM	CR, CSM, NBR–PVC } HOFR (BS 2899)					X	Weather resistance
Railway signalling	BR 972	NR–CR	CR				X		Weather and moisture resistance
Loco and rolling-stock wiring	BRB/LAMA 7, BRB/LAMA 9, BRB/LAMA 11	EPM, IIR	CSM		X		X		Oil resistance
Railway track feed	BRB 704-REN4	EPM, IIR	CSM		X			X	Weather resistance

Table 10.10 *continued*

Application	Specification								Properties
Naval ship wiring	MOD(N) DGS211, MOD(N) DGS212	SI, EPM, CSM, IIR	CSM	X	X	X	X		Fire resistance
Lift wiring	BS 6977	NR	—					X	
Welding electrode leads	BS 638	NR	NR, CR					X	
		IIR, EPM	CSM CR NBR–PVC } HOFR (BS 2899)	X					
			CSM					X	
Medium voltage (15 kV) power distribution	American IPCEA S-19-81	EPM, IIR	CSM, CR	X	X	X			Weather and moisture resistance
X-ray tube supply		NR–CR, EPM	CSM			X			
Ignition (car, internal combustion engine, air-craft)	BS 6862, MOT EL1864, MOT EL1895, MOT EL2107	NB–CR, EPM	CSM	X		X			Oil resistance
Coil leads	BS 6195		EPM–CSM, IIR–CSM, CSM, SI	X					Oil resistance

conductor due to the maximum current rating of the cable. For general purpose cables, the maximum is approximately 60°C, and for the special purpose cables 85°–90°C. NR, its alternatives, and CR are used for the former; EPM, IIR, CR, and NBR–PVC for the latter. A much smaller temperature category of the special purpose cables with operating temperatures in the range 100°–150°C is satisfied by silicone rubber compounds. These temperature categories are generally understood to be associated with a life of at least 10 years, and usually 20–30 years.

The correlation between the accelerated ageing and natural ageing of NR has been obtained by experience. From this knowledge and the application of the Arrhenius reaction equation to data obtained from extended accelerated ageing programmes (Cloutier, 1963), it has been found possible to establish a logical basis for the minimum short-term accelerated ageing requirements of the new elastomers. Typical ageing curves for cable insulation and sheathing compounds are shown in Fig. 10.31. Idealised Arrhenius plots are also

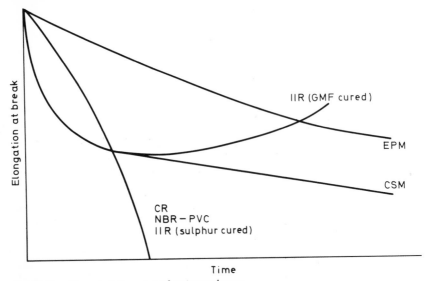

Fig. 10.31   Characteristic ageing of various polymers

given in Fig. 10.32 for IIR and EPM insulation compounds in current use. Mechanical criteria for the end of life are used, usually in terms of elongation at break, since electrical failure of a cable results from the mechanical consequence of ageing degradation. The different shapes of the ageing curves (Fig. 10.31) render the old two-point test inadequate, and a three-point test to define the maximum slope once the rate of ageing has stabilised has been adopted (BS 2899, Part 4; BS 6899, Sections 4 and 5).

Because of the different mechanisms involved in the ageing behaviour of the different polymers and compounding systems, and because of their temperature sensitivity, the linear Arrhenius relationship (reciprocal of temperature proportional to the logarithm of time) holds reasonably well only over the lower part of the temperature range, representing continuous operating conditions. Hence, the results of short-term tests used to specify heat ageing behaviour at normal operating temperatures are inadequate; where heat resistance is a design factor, tests to simulate service conditions, consisting primarily of conventional ageing procedures carried out at temperatures considerably above the maximum continuous operating temperature, have

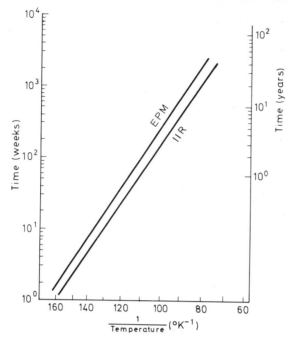

Fig. 10.32   Arrhenius plot of life against temperature for IIR and EPM insulation

been introduced. In ageing tests generally, cable samples are usually included so that studies can be made of the catalytic interference of the conductor metal, shown to occur with some polymers, and the interaction between polymer compounding systems used in insulation and sheath. These tests are supplemented by, for example, current overload tests, the so-called 'threshold of damage' tests for power distribution cables described by Carlson (1961), the varnish stoving test for coil leads (BS 6195; CSA C22.2, No. 116), or the heat endurance test for navy cables [MOD(N) Specification SES21, Part 1].

10.8.2.2   Electrical Properties

The electrical properties of polymers and their compounds and the factors affecting them have been reviewed by McPherson (1963). Most of the information given is relevant to compound characteristics and design for cables, and no attempt is therefore made here to deal with primary dielectric characteristics.

In low-voltage applications (up to 1 kV), electrical parameters are not very critical and are usually limited to volume resistivity and dielectric strength. Adequate levels, when measured dry, are usually readily obtained with the hydrocarbon polymers (Section 9.4.8). Minimum levels for the various categories of insulation and sheathing compounds are given in cable specifications (BS 6899; American IPCEA S-19-81; Underwriters' Laboratories No. 44). With this elastomeric type of cable, mechanical and manufacturing considerations dictate the thickness of the coverings and this is usually more than adequate electrically.

Only barely adequate electrical properties are achievable with CR compounds, which are rarely used alone as insulants. With CSM, a balance can be struck with other design requirements so that 'single shot' coverings are feasible.

Electrical requirements, particularly in general purpose and low-voltage special applications, are often specified not as an indication of the performance necessary in service, but as a criterion of compound formulation or of manufacturing processes. For example, volume resistivity or insulation resistance can be used to dictate polymer–filler ratios in hydrocarbon polymer compounds.

Although the mechanisms involved in the ageing of elastomers result in molecular changes which can be traced by the changes in electrical properties (Jandial and Spade, 1967), the changes in physical properties are much more significant.

Higher voltage cables are much more demanding with regard to compound design. Cables for power distribution have been manufactured for voltages of the order of 30–35 kV a.c., and x-ray cables operate up to 150 kV d.c. The normal criterion of performance of an elastomeric high-voltage cable insulant is the apparent dielectric strength against time characteristic, measured on cable samples. A typical example is given in Fig. 10.33.

It will be noted that the curve tends eventually to zero slope. It is the electrical stress represented by this part of the curve, corresponding to a

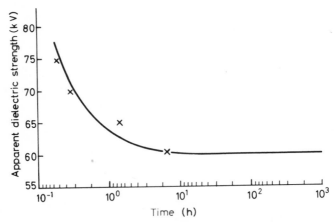

Fig. 10.33   *Typical relationship between apparent dielectric strength and time, for 50 kV x-ray cable*

time $\geqslant 1000$ hours, which is used for cable design purposes. The early part of the curve, in which dielectric strength decreases rapidly with time, is indicative of the effect of degradation of the polymer due to corona discharge and thermal effects. Corona discharge is the result of the ionisation of a a gaseous atmosphere and the acceleration of the ions in the electric field at the surface of the insulant, or in voids associated with filler agglomerates, or introduced accidentally within its body during manufacture. If the gas contains oxygen, one of the products of ionisation is ozone, which chemically attacks some polymers in addition to the erosion resulting from bombardment by other charged particles.

Within the insulant subjected to a high-voltage electric field, heat is generated owing to energy loss in the dielectric. This energy loss (loss index) is a function of volume resistivity, dielectric constant, and dissipation factor. Which of these parameters is dominant depends on whether the electrical field is unidirectional (d.c.) or alternating (a.c.) and on the frequency of the alternating field.

The degradation effects due to corona discharge and to thermal effects are highly temperature sensitive and interact; also, the intensity of the electric field at any particular point in the dielectric is affected by the dispersion of compound constituents and manufacturing flaws. Consequently, a considerable scatter is obtained in the determination of individual points and the shape of the early part of the curve is subject to considerable variation from compound to compound.

In practical compounds, at the higher electrical stresses the thermal effect is cumulatively unstable and, consequently, predominates. As the stress is reduced, the corona discharge effect becomes dominant. The influence of this decreases with the energy in the discharge, which is proportional to voltage, until a stress level is reached which is insufficient to produce ionisation. For compounds for the insulation of high-voltage cables it is hence necessary to formulate, mix, and process for minimum loss index, maximum dispersion, and elimination of voids.

Natural rubber has been used successfully over a number of years as a high-voltage cable insulant and tends to give somewhat better practical short-term voltage breakdown values than the synthetic elastomers; this is probably because of the greater experience and better understanding of the compounding chemistry for electrical applications. However, NR is rapidly degraded by corona discharge and needs protection on both inner and outer surfaces by a thin layer of a compound of a more resistant elastomer, such as CR. Such protective layers, based on the halogen-containing polymers, with high dielectric constants, modify favourably the electric stress distribution in the dielectric, particularly near to the conductor where the stress is highest and where discharge attack is most damaging.

IIR and EPM, with their low degree of molecular unsaturation, can be compounded to yield almost the same levels of practical dielectric strength and have the added advantage of considerably enhanced corona resistance and heat resistance. EPM has better corona resistance than IIR, as indicated by the results obtained on two typical high-voltage insulation compounds given in Table 10.11. Schwartz (1954, 1955) gives the basic principles of compounding butyl rubber for these purposes, and Marchesini and Gazzana (1962) have published similar information for EPM.

None of the hydrocarbon rubbers has yet been shown to be completely resistant to corona discharge, and hence cables are designed to be normally free from discharge at the working voltage. Corona resistance of a compound is therefore valuable as insurance against the accidental occurrence of discharge in service, or as the result of a minor manufacturing defect. Corona resistance is measured in terms of time to electrical failure by submitting a sample of cured compound, in either sheet or cable form, to an electrical stress such that an ionised air gap is included at the surface of the sample, between the electrodes. The compound is also usually submitted to mechanical stress during the test. Typical of such tests are those in specifications

Table 10.11   LONG-TIME BREAKDOWN TESTS CAUSED BY CORONA: U-BEND TEST TO AMERICAN SPECIFICATION IPCEA S-19-81

Compound	Temperature (°C)	Breakdown stress (kV/cm)	
		10 hours	*100 hours*
EPM	20	>450	>450
	90	360	275
IIR	20	125	90
	90	110	80

BS 6862 for motor car ignition cables, IPCEA S-19-81 U-bend test for power cables, and VDE 0472 ozone test for power cables.

For the sake of flexibility, the conductors of cables often consist of a number of small wires, a formation which leads to local increases in electrical stress and the formation of air gaps at the inner surface of the insulant; corona discharge is thereby encouraged. To overcome this, a layer of conducting elastomer (Section 9.4.8) is commonly included immediately over the conductor and bonded to the insulating compound.

The mechanical properties required, and hence the formulation of elastomeric compounds for cable sheathing, are similar to those of tyre compounds. The polymers are, for the general purpose cables where economics are important, natural rubber and, for the more special applications, the halogen-containing materials CR, CSM, and NBR–PVC. Many of these are inherently poor electrically and, being reinforced with carbon black, cannot be expected to have good electrical properties; nor are they necessary. However, if the volume resistivity, for instance, is too low and the cable sheath contacts a terminal, current leakage can take place through and along the sheath. To avoid the risk of shock, a reasonable minimum value of volume resistivity is 10^8 Ω cm. In most cases, the upper limit of volume resistivity is unimportant. An exception is the sheath on an x-ray cable, which may be used in an operating theatre; the value in this case should be low enough to discourage the accumulation of static charges.

10.8.2.3   *Moisture Resistance*

Perhaps of more concern than the initial electrical properties of compounds is the stability of electrical parameters which are most affected by the absorption and permeation of water. Access of an elastomeric insulant to free water leads fairly quickly to some deterioration in electrical properties and eventually to its electrical failure. The concept of 'life' with electrical failure as the criterion is thus suggested.

The mechanism of the absorption of water by compounded rubber depends on the structure of the base polymer, the distribution of fillers and other components, the presence of water-soluble or hydrophilic constituents, and whether water is present in the liquid or vapour phase (Briggs, Edwards, and Storey, 1963). The distribution, rather than the amount, of water in a compound is important for electrical properties, and in the presence of moisture small quantities of some substances have a quite disproportionate effect.

Elastomeric compounds containing very little ionic material can absorb 2–3 % of water without serious effect on electrical properties; in this case, an increase in dielectric constant can be detected without a corresponding deterioration in dissipation factor, volume resistivity, or dielectric strength. Disturbing substances, such as crosslinking agents and antioxidants, can be introduced knowingly in designing the compound. Impurities, such as the residues of catalyst and initiators and regulators of polymerisation in synthetic polymers, impurities in fillers, and proteins and other water-soluble materials in natural rubber, also occur unintentionally. For this reason, the more highly purified grades of rubbers are used for insulation compounds; the fillers are purified, or are treated to neutralise any polar impurities adsorbed on their surfaces or to promote polymer–filler bonding, e.g. coated calcium carbonates (Section 6.2.2) and the use of silane coupling agents (Schwaber and Rodriguez, 1967). The less 'clean' grades are sometimes compounded with substances which combine with the polar residues from the

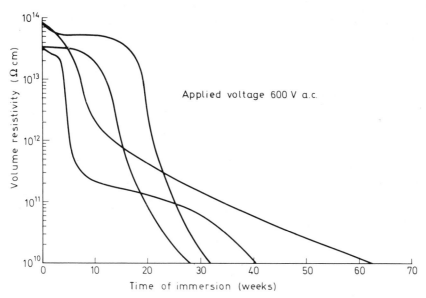

Fig. 10.34 *Water stability test: insulation resistance against time of immersion for various IIR insulation compounds*

polymerisation reaction to form non-polar substances, e.g. the use of metallic soaps with the corresponding oxide in EPM formulations (Marchesini and Gazzana, 1962).

During curing in steam, water in the vapour phase is forced into the compound and is released in a time dependent on the vapour pressure gradient, the temperature, and the diffusion coefficient. The effect of the retained water, which may be in the liquid or vapour phase, is to cause a deterioration in electrical properties relative to those obtained with the dry cured compound by a factor of two or three. The volume resistivity of a steam cured sample can be several powers of 10 different from a press cured equivalent. In the case of insulated wire cured in high-pressure steam or in a continuous vulcanisation process, this effect can persist for several weeks after processing, during which a steady improvement in electrical properties occurs.

Polar sites in the molecular structure of polymers created by oxidation contribute to the osmotic and diffusion effects. In the halogenated polymers, the products of degradation are particularly actively polar and hydrophilic. It has not been found practical, therefore, to apply physical laws even to a limited extent, as in the case of heat ageing.

With natural rubber compounds as a base, compound evaluation is by comparison of standard samples of insulated wire immersed in water, deterioration being traced by measuring insulation resistance with time. A typical plot is given in Fig. 10.34. To accelerate degradation, the test temperature is of the order of the maximum service temperature and an electrical stress of about twice the working value is applied across the dielectric. Significant changes can be detected in about six weeks (Extended Insulation Resistance Test in UL Specification No. 44). No satisfactory short-term test is available, and full evaluation of a compound's performance may take several years. Resistivity has been found to be a more sensitive criterion of performance than any other electrical parameter.

10.8.3 MANUFACTURE

10.8.3.1 *Compound Production*

The variety of polymers and corresponding compound constituents used in cable manufacture leads to some complication in maintaining segregation of non-compatible materials. The automation of the cable factory mill room and mechanical compound-handling arrangements are also influenced by this large complex range of materials (Nye, 1967). Mixing is commonly by internal mixer, with mill mixing the exception. Because of the importance, electrically, of good dispersion, particular attention is paid to the particle size of fillers in respect of both level and range. Dispersions, in oil, of the components of curing systems (e.g. red lead, GMF in butyl compounds) are generally used, when appropriate, for the same reason. In addition, insulation compounds are passed through a straining machine to minimise electrical failure due to powder agglomerates or contamination.

10.8.3.2 *Cable Covering*

Application of rubber cable coverings is carried out with plant designed for one of two basic methods, namely batch curing (discontinuous) or continuous curing. The older one involves the application of rubber as an operation separate from its cure. In the continuous process, the curing operation is integral with the application of the covering and thus labour cost is reduced.

Application of Insulation and Sheath. The three methods of fabricating the rubber coverings of cables are longitudinal, lapping, and extrusion. All three are used for insulating wire; only the latter two are appropriate to sheathing.

The two older methods—longitudinal and lapping—of rubber-covering wire have now been almost completely superseded by extrusion techniques. Both involve the application of calendered tapes of uncured compound; longitudinal covering consists of passing the wire, sandwiched between two rubber tapes, between grooved rollers so that a seam is formed on each side. A machine has three or four sets of rolls in tandem. The requisite thickness of covering is built up by applying the appropriate number of tapes of suitable thickness. Up to 36 wires are covered simultaneously, the covered wires being separated between each station by parting blades. Lapping consists of spirally wrapping tapes on to the wire with an overlap. Again, the appropriate thickness is obtained by using several tapes from heads in tandem.

The two methods require compounds capable of being handled in roll form with the conflicting properties of good building tack, good green strength, and low stretch. Both processes are very susceptible to ambient temperature variations and contamination (Evans, 1953; Tew, 1956).

Extrusion of insulation and sheath is carried out using conventional machines fitted with cross-heads. The conductor to be covered, or the core or cabled cores to be sheathed, as the case may be, pass through a core tube supported coaxially within the head and located concentrically with the die by a tapered extension to the core tube—the 'core point'. Cold feeding of compound, either in chip or strip form, to machines with length–diameter ratio of the scroll from 12:1 to 15:1 has been widely adopted.

The use of dual extruders (see Fig. 10.35) permitting the simultaneous application of the two components of composite insulation, e.g. EPM–CSM,

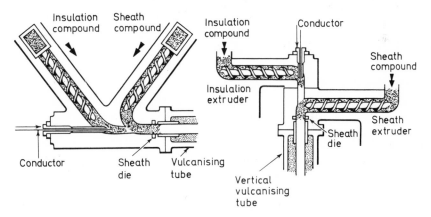

Fig. 10.35 Principle of synchronised dual extrusion machines

NR–CR, is now common. Insulation and sheath, or stress control layer and insulation, are dealt with similarly. With appropriate combinations of compounds, the ensuing curing promotes bonding of the two layers. (It should be noted that, with such an arrangement, balancing of the cure characteristics and rheological properties of the two compounds is necessary.)

Line equipment consists of a capstan, or capstans, either of the drum type or fitted with caterpillar tracks (caterpuller), together with pay-off and take-up stands. In modern extrusion lines, the two capstans, one behind the extruder, one in front, are interlinked—one with the other and with the extruder drive—and co-ordinated so that the line can be run at low speed for setting up and then run up to full production speeds with the minimum of extruder readjustment. Tension control is effected by adjusting the speed differential between the two capstan units.

10.8.3.3   *Curing Techniques*

The choice of curing method, either discontinuous or continuous, is determined mainly by economics, dependent on the length and size of cable to be produced. In some cases, technical requirements dictate the method to be used, e.g. the use of continuous vulcanisation for the insulation of high-voltage cables.

Batch Curing. In the batch, or discontinuous, method, the insulation or sheath applied by one of the processes outlined above is cured in an autoclave in steam at pressures corresponding to temperatures of 130°–160°C; curing cycles vary between 15 min and 90 min overall. Where a rubber sheath is applied over thermoplastic-insulated cores, curing temperature is below the softening point of the insulant.

The cable or core, in its uncured state, is coiled in trays or reeled on drums, precautions being taken to avoid adhesion and deformation by dusting with talc, or other anti-tack agent. Cooling before take-up and some form of restraint during curing are also necessary. The smaller cables, particularly those covered by extrusion, are sometimes coiled spirally in a large-diameter shallow tray and packed in talc.

To provide support during handling and restraint during cure, spiral over-lapped cotton tape is applied, in the case of the lapping or extrusion methods, as part of the rubber-covering operation; or separately, in longitudinal covering. The steam during the cure causes the tapes to absorb moisture, to shrink, and so to consolidate the covering. This is reinforced by the expansion forces within the rubber in the later stages of the curing cycle. Other low-stretch tapes, such as woven glass fibre and polyethylene terephthalate fibre (Terylene), or the latter in film form (Melinex), have been used under particular circumstances where cost is not the main concern. Choice of tape strength is dependent on the volume of rubber per unit length. In the case of core, the cotton tape is often 'proofed' with a coating of coloured rubber and left in place after cure is complete to serve as circuit identification. For sheath cures, however, the tape is dusted with dry talc and removed as a separate operation.

Where maximum consolidation is required, or to improve the transfer of heat from the curing medium, a metal sheath is applied either over a tape, as above, or direct on to the rubber. This technique has been used in the USA for core insulation by enclosing it in a longitudinal tinned-copper tape with a folded seam at the overlap. More commonly, it takes the form of a lead tube extruded on to the cable by either a ram or screw type press (Sections 8.9.1.5 and 10.3.3.1). The lead is stripped off subsequent to cure and re-used. Lead is used directly on extruded coverings, particularly for sheaths, where a smooth surface finish is required (e.g. cables to pass through pressuretight glands). Close control of the uniformity of the extrudate is required, otherwise air pockets are trapped under the lead, resulting in 'pock marking'. Some limitation is also imposed on compound formulation since coloured sheaths, especially those which contain sulphur, are stained by interaction with the lead.

Cables which are taped or lead sheathed for cure are wound on to large-diameter steel drums, having dimensions designed to present maximum area (minimum number of layers) to the steam. These drums are placed in the autoclave with their major axis vertical (for small core or cable) or horizontal. Differential expansion between conductor and rubber insulation produces a disadvantageous effect, which becomes serious on core with thick walls of rubber, and which results in de-centralisation of the conductor towards the centre of the curing cylinder. With large insulated conductors cured on horizontal cylinders, there is also a tendency for eccentricity of the covering to occur owing to gravity. For this reason, the largest curing cylinders are sometimes arranged to be rotated during the cure cycle. This also tends to offset any non-uniformity in the temperature distribution.

Continuous Curing. Of the fluid heat transfer media which have been suggested, the simplest is steam because it is an extension of an older method and pressure is an essential feature. This system is referred to as CV—continuous vulcanisation (Tunnicliffe, 1953). The earlier plants were arranged horizontally (HCV), but later, vertical (VCV) and catenary or 'vector' (CCV) arrangements to cater for the larger cables were introduced. A typical modern layout of each type is shown in Fig. 10.36.

Essentially, a CV plant consists of a high-pressure steam tube attached to the die face of the extruder head. At the remote end of this tube, the steam pressure is balanced by water pressure in a cooling tube, maintained against a controlled leak at the cable exit gland. The length of the steam section is of the order of 50–80 m, that of the water section 4–10 m. The ratio of steam length to water length is proportional to the size of cable and the thickness of the covering. Curing temperatures vary from 170°C to 200°C. Single gland tubes are now used universally.

The HCV plant has the obvious disadvantage that, for reasonable tube diameters and with practical tension, the cable must touch the bottom of the steam tube at some point. This is unimportant if the point of touch-down is such that the cure is well advanced. These conditions can be achieved with the smaller cables and cores, and no significant damage occurs with extrusions up to a diameter of about 15 mm. For cables larger than this, the CCV and VCV arrangements are more appropriate. In the catenary form, the path of the steam tube conforms to the natural catenary of a cable suspended from two points and under a given tension. With a tension appropriate to its weight per unit length, the cable can be made to conform to the longitudinal axis of the tube. CCV gives higher throughput speeds for a given cable size

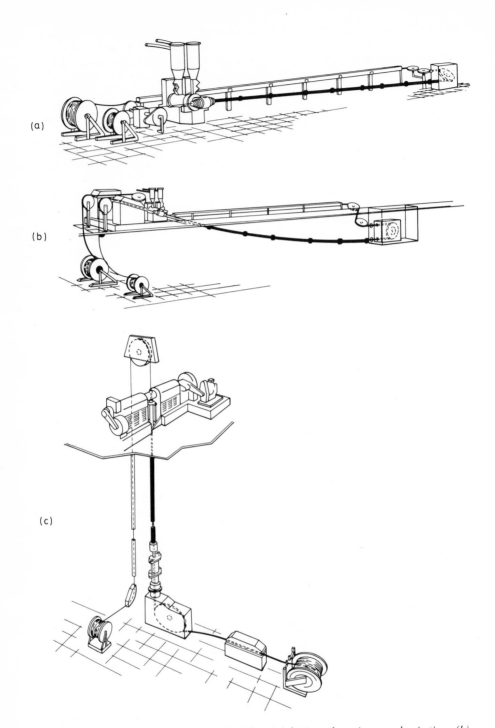

Fig. 10.36 Continuous vulcanisation of cables: (a) horizontal continuous vulcanisation; (b) catenary continuous vulcanisation; (c) vertical continuous vulcanisation, showing VCV tower

than VCV, since the height and consequently the cost of the latter's supporting tower limits the length of tube. Tension control of the catenary is, however, more critical than with VCV. Consequently, the largest cables are made
by the vertical plant.

The considerable length of cable required to fill a CV line and the consequent high scrap at each start-up, together with the high capital cost of the
installation, make this method more appropriate to long production runs.
Batch curing techniques have hence tended to be perpetuated for short-length
manufacture and jobbing work.

The above circumstances relative to CV encourage the highest practical
line speeds, demanding the formulation of compounds with the shortest
practical curing times (Mason, 1953). Table 10.12 gives examples of practical
conditions.

Obviously, to achieve optimum cure times on a given CV line, the maximum practical steam pressures are used, subject to the thermal characteristics

Table 10.12   PRACTICAL CONDITIONS FOR THE CONTINUOUS VULCANISATION OF CABLES

Covering	Diameter under covering (mm)	Covering thickness (mm)	Line speed (m/min)	Steam pressure (kgf/cm^2)	Steam length (m)	Cure time (s)
General purpose NR insulation	2·6	0·8	200	14·5	80	24
High-voltage EPM insulation (11 kV)	5·2	67·0	9	13·0	22·5	150
NR sheath	9·5	1·3	65	14·5	55	50
CSM sheath	9·5	1·3	30	14·5	55	110

of the compound and the geometry of the covering, which determine rate of
flow of heat and, consequently, the radial cure gradient. Reversion must not
occur on the outer surface of the covering before an acceptable state of cure
is achieved at the inside.

Adequate cooling is also an important factor influencing optimum line
speed since, otherwise, the high internal vapour pressure, resulting from CV
conditions, results in 'blowing' of the covering.

Since equilibrium thermal conditions are not generally achieved during
CV cures, the theoretical estimation of curing conditions is complex. From
initial work by Heap and Norman (1966), electrical analogue techniques have
been developed (Atkin and Nye, 1969) which can be computerised (Le Nir
and Dodwell, 1968) to make it rapid and practical. It will be evident that
compounds for CV must have a 'flat' cure characteristic at the curing temperature, and a compromise between minimum cure time and scorch time
must be achieved.

Both horizontal and vertical fluidised beds, integrally coupled with extrusion, are attractive for short-length jobbing cable production (Section 8.9.2.4).
Liquid salt baths also have considerable promise (Section 8.9.2.3). The short
individual lengths require the use of shorter curing units than CV (Hughes
et al., 1962). Since these media are operated at, or near, atmospheric pressure,
porosity of the extrudate from the occlusion of air and moisture in compounds
has to be combatted by the use of vented extruders and anhydrous lime

(Caloxol) in the compound, respectively (Hughes *et al.*, 1962; Shaw and Nelms, 1967).

10.8.4   SPECIALISED ELASTOMERIC CABLE COMPONENTS

Apart from the main elastomeric cable components of insulation and sheath, a number of other components are used which call for different compound formulations.

Cured conducting compounds of NR, IIR, and EPM, formulated to have resistivities of $<10^3 \, \Omega$ cm, are used for electrical stress control in high-voltage cables. NR and CR compounds with resistivities of $<10 \, \Omega$ cm have been proposed (Frost and Liverpool Electric Cable Co., 1939) for protective circuits in flexible mining cables and are now coming into general use, as defined in NCB specifications.

Radiofrequency interference from petrol engines is suppressed by the use of CR compounds, with resistivities of the order of $1 \, \Omega$ cm, for the conductor instead of a metallic wire in cables for ignition circuits. Formulations of these conducting compounds follow the lines indicated by Norman (1957).

Paints based on CR and CSM are used to impart flame retardance and weather resistance to cable coverings, e.g. textile braids. They are also used for core identification, particularly CSM which gives better colours, and to add ozone resistance and oil resistance to insulants based on hydrocarbon elastomers.

10.9   Mechanical and Miscellaneous Products

BY C. M. BLOW

10.9.1   INTRODUCTION

The products, the manufacture of which has been described in the previous eight sections of this chapter, consume probably 95% of the natural and synthetic rubbers produced. In this section, attention is turned to the large number and variety of other products consuming relatively little rubber because of their small size, their small quantity usage, or their low content of new polymer. Many of them are commercially very remunerative to the manufacturer; others, because of their technical importance for the engineering applications in, for example, the aircraft, electrical, and electronic industries, receive a great deal of effort in government and co-operative research laboratories, as well as in individual manufacturer's laboratories.

Because of their variety, it is difficult to group or classify them; furthermore, it is not the intention to attempt to discuss or to list every product. Many are produced by one or other of the methods described in general terms in Chapter 8, no particular technique or technology being involved. It is proposed to consider the more important and interesting items under the headings of the five processes, spreading, calendering, extruding, moulding, and handbuilding. It is, however, desirable to draw attention to the several methods that may be adopted to produce certain of them, and so some over-lapping and repetition are unavoidable.

10.9.2 PRODUCTS BASED ON SPREAD FABRICS

As indicated in Chapter 8, much spreading of fabrics is carried out as a preliminary to their incorporation in the products covered by the first four sections of this chapter. In addition, final products are made from fabrics, usually spread with natural or synthetic rubber on one side only, and used for their waterproof quality, e.g. macintoshes, groundsheets, and tarpaulin-type covers. Some interest was aroused several years ago in the production of waterproof garments incorporating specially compounded rubbers which allowed the permeation of water vapour but not liquid water. This process appeared to depend on the rather coarse dispersion of hydrophilic fillers, such as whiting (Wingfoot Corpn, 1946, 1949; Martin, Sell, and Habeck, 1950; Blow and Pike, 1952). The abrasion resistance of the film was not satisfactory, and little came of the idea.

The mechanical, electrical, automobile, and aircraft engineering industries have requirements for a great variety of gaskets and diaphragms which for strength and stiffness must have fabric reinforcement. These are frequently made by spreading fabrics of cotton, rayon, nylon, polyester, or glass fibre with rubber, all types of elastomer being used. The shape is punched from the fully cured sheet. Surface finish is important in many applications and determines the method of vulcanising the spread fabric. The finest smoothest surface is obtained by moulding, but often acceptable finish can be achieved by wrapping the fabric on itself on a large drum with an interleaf of fine textured cloth, special paper, or plastics film.

Flat punched diaphragms can be post formed to simple convoluted form by a short heat treatment when pressed between a metal and a rubber shaped jig (Fig. 10.37).

10.9.3 PRODUCTS MADE BY THE CALENDERING PROCESS

10.9.3.1 *Unsupported Sheeting*

For quantity production of unsupported rubber sheet, i.e. without fabric insertion, the calendering operation is most economic assuming that the vulcanisate requirements can be met by a mix which will calender smoothly. Unsupported sheeting is required for hospital use, for sticky-back repair patches, for gaskets and diaphragms, and for cutting into thread for garment and golf ball manufacture. Thread was for many years cut from a block of frozen rubber (Section 1.2.2). Calendered sheet is also required for built-up products, moulding, etc.—e.g. sheet may be wrapped on a mandrel to form a tube for subsequent parting off into washers.

10.9.3.2 *Supported Sheeting*

Many of the products referred to in Section 10.9.2 can equally well be produced by means of the calender. The choice of method depends on many factors: the ability of the rubber mix to friction or calender smoothly, the degree of adhesion required and obtainable by the various methods, the thickness required, and the uniformity of thickness needed (Fosgate, 1968).

The vulcanising methods available for the cure of calender produced sheeting and fabric reinforced sheeting are the same as for spread fabric.

10.9.4   EXTRUSIONS

10.9.4.1   *Continuous or Long Lengths*

It is hardly necessary to do more than mention the variety—tubing, sealing and trimming strips for automobiles and domestic appliances, shoe welting, etc.—supplied in long lengths to other industries and users to cut up as required. The techniques do not call for any comment; the vulcanising methods are either batch (Section 8.9.1) or continuous (Section 8.9.2), and both solid and cellular rubber sections are manufactured. Trade marks, names, sizes, and other details may be imprinted on the extrusion by allowing it to pass over an engraved or embossed roller as it leaves the machine.

10.9.4.2   *Precision Extrusions*

The compounder's and the diemaker's arts are usually adequate to ensure that extrusions meet the designer's requirements dimensionally and shape-wise, at least in the smaller size range. There are, however, large section extrusions where collapse and distortion occurring during handling and cure demand special treatment. Tubing may need to be blown on to a mandrel; other sections require support on specially made jigs. For fine dimensional tolerances or when the extruded surface is not perfect enough, grinding the outside of tubing is commonly practised. In other instances, bends and other shapes are made by putting the uncured extrusion on to bent or shaped mandrels before cure.

10.9.4.3   *Products Made from Cured Extrusions*

Scarf-Joining. Vulcanised or semi-vulcanised solid or hollow extrusions are scarf- or butt-joined to produce rings for such items as seals and driving bands. The techniques vary. The freshly cut surfaces are coated with a solution of the same mix, with the addition of a bonding agent in the case of solvent-resistant polymers. A further aid to a satisfactory bonding is to insert a 0·1–0·2 mm calendered film between the two solutioned faces of the joint. After making the joint, it is clamped under light pressure in a simple two-piece mould, heated electrically or placed between heated platens. A short cure is given to the joint, followed frequently by an open steam or oven cure to complete the vulcanisation of the whole component.

Parted-Off Components from Extrusions. Uncured tubing or other extrusion is cut into suitable lengths for use as moulding blanks, by parting-off in a lathe, often by automatic mechanisms. The extrusion is supported on mandrels coated with rubber or other non-metallic material to protect the knife and avoid contamination of the blanks. Fully vulcanised tubing is similarly cut into washers and gaskets. If dimensional tolerances are tight on the internal and external diameters, the extrusion will need to have been cured on a mandrel of the correct size, allowing for shrinkage (Section 8.6.7), and the outside diameter ground before the parting-off operation. This is only one way of producing a rectangular section washer; punching or trepanning from calendered or moulded sheet and moulding are alternatives. Stationers bands are made from extruded tube allowed to flatten, vulcanised, and cut to the required width by guillotine.

Endless Shaped Seals from Extrusions. For automobile and aircraft window and domestic appliance door seals, accurately shaped corners may be required. A widely used technique, to save the cost of complicated and large moulds which would be needed to make them in one piece, is the corner moulding of extrusions. Semi-vulcanised extrusion is prepared and cut to length, the cut ends being coated with solution. Two lengths are placed accurately in a simple corner mould made to the shape of the corner with a section corresponding to that of the extrusion. The mould is clamped between the heated platens of a press, and rubber is transferred to the cavity to form a corner piece integral with the two lengths of extrusion. The operation is repeated with the necessary other lengths of extrusion to produce an endless seal shaped to fit the window or door. Alternative methods include the scarf-joining of the extrusion into an endless ring and then by means of suitable mandrels or formers (which may be of a meltable or soluble substance) setting the corners to shape by means of a heat treatment. In this case, the extrusion must be in such a state of cure that it will assume the required permanent set.

10.9.5   MOULDED PRODUCTS

The general methods and principles of moulding have been dealt with in Section 8.6. The large number of general mechanical goods, as they are termed in the industry, manufactured by moulding, usually in compression-type moulds, do not call for any particular comment. There are a few special items which will be briefly discussed.

10.9.5.1   *Solid Tyres*

Rubber-tyred wheels, casters, and drive pulleys are manufactured in a variety of sizes and sections, and also solid rubber tyres for perambulators, push chairs, etc., as separate items for fixing to a grooved wheel.

The rubber-tyred product, in which the rubber is moulded and bonded to the metal or other rigid wheel, is made by applying an adhesive or ebonite layer to the metal surface followed by an extruded section or calendered strip; the assembled unit is then placed in a simple compression two- or three-piece mould and vulcanised under heat and pressure.

Solid rubber tyres for application to a wheel are moulded with an endless metal helix and/or fabric core on the inner diameter to give it rigidity and stiffness to hold on the rim; sufficient stretch is present to permit assembly on a grooved wheel.

10.9.5.2 *Hollow Articles*

Inner tubes for tyres of all sizes are nowadays moulded. A pre-moulded rubber-based valve stem is cemented to an extruded tube, the ends of which are cut, usually with a hot knife, and butt-joined. The tube is then inflated and moulded in watchcase-type presses. The previous method, still in use for cycle tubes, used a vulcanised extruded tube with overlapped joint, made by hand.

Under this heading are included also play balls, which may be moulded on the lines of tennis balls (Section 10.7.2), and hand-built and open steam or oven cured products (Section 8.8), such as bladders, fuel tanks, liquid storage bags, and dinghies. In addition, there are the moulded hollow components and articles—boots and gaiters for engineering joints, cell sacks, handle grips, and hot water bottles—for which the mould contains a core rigidly positioned and located but separate so that after cure the product can be removed. Careful blank sizing and shaping is necessary to ensure uniform and

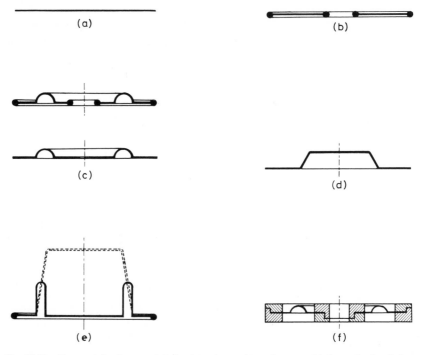

Fig. 10.37 Types of diaphragm: (a) flat (simple punchings from moulded or calendered sheet; the surface can be polished, cloth marked, or sueded); (b) beaded (essentially flat, but moulded with a bead on the inner and/or outer periphery to provide better sealing); (c) convoluted (formed from flat sheet or moulded with a bead); (d) dished (moulded with or without beads); (e) rolling type (usually made as a 'top hat' and assembled inverted to achieve true rolling action); (f) metal bonded (an elastomeric diaphragm of any shape is bonded or adhered to metal mating members to provide still better sealing and stability). Courtesy The Dunlop Co. Ltd

consistent wall thickness; if compression moulding is employed, the compound formulation and condition of the blank must be such that good knitting of the separate blanks occurs as they flow together. The vulcanisate, in the hot state as it leaves the mould, must have sufficient elongation to permit the extraction of the core, which, in the case of a hot water bottle, for example, is considerable. Transfer or injection moulding is necessary for many products of this type to eliminate uneven wall thickness, retracted spew (backrinding), air trapping, and other faults, and to save material.

10.9.5.3 *Diaphragms*

Reference has already been made to the production of simple flat or slightly convoluted diaphragms from unsupported and supported spread or calender proofed material. There are diaphragms with peripheral and/or centre beads and bosses for assembling and sealing purposes, and with deep convolutions which must be moulded (Fig. 10.37). No special techniques are called for, apart from accurate blank preparation for the reinforced diaphragms to ensure that during moulding the fabric is neither distorted, strained, nor damaged by the flow of rubber. Simple two-plate compression moulds with accurate dowelling or taper location are commonly used.

Top hat diaphragms, which are designed to roll in use giving long strokes with minimal resistance to movement, are the subject of patents (Taplin, 1956). The majority of such products have a fabric reinforcement which may

Table 10.13 BURST STRENGTHS OF DIAPHRAGM FABRICS: MULLENS TEST, 31·7 MM DIAMETER DISC

Fabric	*Burst strength* (kgf/cm^2)			
	60 g/m² fabric	*85 g/m² fabric*	*135 g/m² fabric*	*240 g/m² fabric*
Nylon ⎫ Polyester⎰ Glass	10 5·3	17·5 11	31 17·5	60 25

be square woven or knitted. The essential of the technique is the pre-forming of a top hat from the fabric and then moulding around it a rubber coating on one side of the fabric with the necessary beads or rims to provide sealing when assembled in the metal components.

It is common practice to produce diaphragms with the rubber on one side of the fabric only, which side faces the pressure; this ensures that no separation of fabric and rubber takes place as is liable to occur if the fabric side faces the pressure. Table 10.13 gives the burst strengths as a guide to the selection of fabrics for diaphragms.

10.9.5.4 *Seals*

Rubber seals are used to seal anything from a wrist watch to the freight door of an aircraft. The former may be an O-ring of bore 0·8 mm and cross-sectional

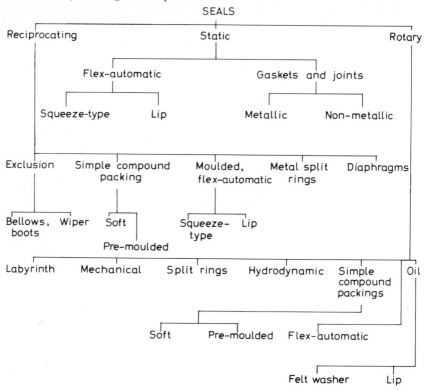

Fig. 10.38   *Seal types. From Goodsell (1967), courtesy Engineering, Chemical & Marine Press Ltd*

diameter 0·5 mm; the latter may have a section more than 40 mm across and a circumferential length of 15 m. Fig. 10.38 is a chart of seal types, not all of which are made wholly or partly of rubber. Some of these types have already been discussed as being manufactured by techniques other than moulding.

From the manufacturing point of view, those produced by moulding can be considered under three headings. Squeeze-type seals and lip seals are in common use, and within each type there are several section shapes (Figs 10.39 and 10.40). The former operate by distortion under radial or axial compressive

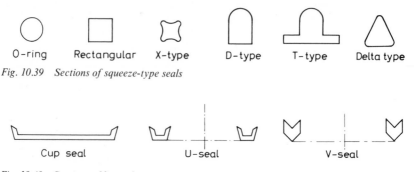

Fig. 10.39   *Sections of squeeze-type seals*

Fig. 10.40   *Sections of lip seals*

strain, as discussed below. The O-ring is very widely used, though some designers prefer the rectangular, D, or delta sections. The T-seal has the advantage that it can be clamped on assembly and does not therefore become easily displaced or twisted.

In the case of the lip seals, sealing is achieved by distortion of the lip assisted by the pressure in the system.

All-Rubber Seals. The majority of squeeze-type seals are all-rubber, and the same applies to a few of the simpler U-seals and cup-seals for light duty. Moulding is the method of production adopted for all except the smaller rectangular-section seals required in large quantities, which are made by punching or trepanning from sheet or by parting-off extrusion, and the large bore diameter ones, which are made by scarf-joining.

The moulds are two-plate compression type with single or multiple cavities. A substantial part of the technique lies in the mould design and manufacture to ensure perfect closure with thin flash and accurate match under the high-pressure moulding conditions that the product and material require. The blanks used are either cubes for small sizes or rectangular or cord section rings, prepared by parting-off from extruded tube, punching from sheet, or joining extruded cord. Their shape and size is important because most of the moulding faults, such as air trapping, retracted spew (back-rinding), flow marks, and 'not made', arise from incorrect blanks.

Considerable attention has been paid to transfer and injection of O-rings, and in particular to the production of flashless or 'less flash' products (Section 8.6.5.3).

Fabric-Containing Seals. Fabrics are incorporated in seals to increase modulus and stiffness and to provide strength and wear and tear resistance. Fabrics of all available fibres are used depending on the service use, in particular, temperature. Manufacture involves the thorough impregnation of the textile fabric with the rubber compound, either by spreading or frictioning, and then a careful building up to as near as possible the shape of the finished product before placing in the mould. During the moulding operation, minimal flow and distortion of the fabric must take place and trapping of fabric at the mould parting lines must be avoided as it detracts from the appearance of the seal, exposes it to ingress of liquid and other foreign matter, and generally reduces its service life.

Metal-Containing Seals. Many seals for rotary shaft applications contain metal stiffening, usually as a punched and formed cup to which the rubber is bonded (Fig. 10.41). Attention needs to be paid to the location of the metal

Metal-cased   Metal-cased with garter   Metal-cased bonded, with garter   Embedded metal case   Bonded metal-cased, spring loaded with tandem seal

Fig. 10.41   Sections of rotary shaft seals

during the moulding operation and choice of blank shape to ensure satisfactory and consistent products.

Finishing Seals. In lip seals, the sealing surface is the lip which must make uniform contact with the cylinder walls, against which it operates. Because it is not easy to mould this lip without a parting line near to its edge, it is common practice to make the mould parting line at the lip, which is made longer than required and knifed off to dimension in a lathe as a subsequent finishing operation. Automatic machines are often used to do this.

Flash removal from O-rings and moulded rectangular and similar section seals is carried out by high-speed felt and cotton mops and abrasive wheels.

Seal Design and Material Selection. The development and selection of rubber mixes for seals is dictated usually by three factors. First, the fluid with which the seal will be in contact; secondly, the temperature range of the environment in which it will operate; and thirdly, the mechanical environment including such factors as pressure, stressing, and abrasion. The fluid and temperature will restrict the choice of base polymer, usually to one or two; compounding will be directed to achieving the correct hardness and abrasion resistance and to adjusting the swelling behaviour of the rubber in the liquid. The designer will want to know the likely dimensional changes in the seal material due to uptake of liquid and differential thermal expansion.

Important in seal applications is the elastic property of the rubber vulcanisate. The distorted rubber exerts a pressure on the contacting surface to maintain the seal. Unfortunately, no rubber is perfectly elastic, and the stress in the rubber decays or relaxes with time (Section 3.3.3). This stress relaxation can be measured directly (Section 11.3.10) (Aston, Fletcher, and Morrell, 1969), or, more frequently, the degree of permanent compression set the rubber has taken up on removal of the stress is determined (Section 11.3.11).

There are two factors of importance in compression set testing. First, at the rubber–metal clamp interface, with the cylindrical test-piece of BS903, there can be any degree of slippage, and the amount of slippage affects the result. It is probably easiest to achieve the no-slip condition, but the 'recovered' test-piece in this case has concave top and bottom faces, which lead to measurement difficulties. This complication can be overcome by the use of an O-ring, or portion thereof, as a test-piece. There is, however, some evidence of a size effect here, probably related to the rate of cooling during the recovery step in the test, which is the second factor to influence the result.

At the conclusion of the heating period, the clamps and test-pieces can be allowed to cool before the strain is released; alternatively, the test-pieces can be removed from the clamps immediately and allowed to recover for the specified period of 30 min, either at the test temperature or at room temperature. The spread of results obtained by such variations of procedure depends on the magnitude of the viscous component of the elastomer's viscoelasticity. If high, not only is the spread high but one arbitrarily chosen method may give misleading and unreliable indications of the elastomer's behaviour as a seal. In the author's opinion, it is in such circumstances that stress relaxation observations are valuable as giving a more complete picture of what is taking place.

Other factors related to those mentioned above are the change of hardness on heating the rubber and on immersion in the liquid which the rubber is sealing. Too great a softening may lead to extrusion of the seal between the clearances, with ultimate rapid failure. Strain in the rubber affects liquid up-take, and diffusion rates of various liquids in rubbers of different types cover a wide range of values (Blow, Exley, and Southwart, 1968, 1969).

Nitrile rubber mixes are the workhorses of the seal industry, because of their resistance to mineral- and ester-based lubricants and to hydraulic liquids. The mineral oils offered vary in composition and in the swelling they produce in nitrile rubbers. It is important that the rubber and oil are matched. The publication by the Association of Hydraulic Equipment Manufacturers (1968) of their proposed 'seal compatibility index' method is of interest. The idea is to categorise oils by the volume change they produce in a standard nitrile rubber, stated as the SCI. Oils with similar SCI, within a tolerance, will be interchangeable; additionally from the SCI of an oil, seal manufacturers will be able to provide suitable rubbers. This scheme is based on the observed linear relation between the SCI value and the swelling of commercial nitrile rubbers in a range of 52 mineral-based oils from a number of oil suppliers.

10.9.6   HAND-BUILT PRODUCTS

The broad principles of the manufacture of the most important and typical products in this group have already been described (Section 8.8). All types of rubber are used, with a preference for natural rubber, if it meets the vulcanisate properties required, because of its building tack. Butyl rubber has become established as a reservoir-lining rubber and also for other water-storage containers which involve handwork. Nitrile, polyurethane, and chloroprene rubbers are used for fuel cells and storage containers on a base of nylon fabric.

ELEVEN

Testing Procedures and Standards

BY J. R. SCOTT

11.1 Introduction

It is not possible here to survey all rubber standards including product specifications, but a general picture of British and US test procedures is given, with reference to other national procedures containing interesting features, and some observations on specification making.

The International Organisation for Standardisation Technical Committee ISO/TC45—Rubber has since 1948 been preparing internationally agreed recommendations on many tests for raw and vulcanised rubbers, and more recently for products. These 'ISO Recommendations' are being widely adopted, to be used as the basis for the national standards issued by the standardising bodies of the member countries of ISO. Hence it will be convenient to refer primarily to the ISO Recommendations as the basic documents. It may be emphasised, however, that these are *recommendations*, not standards; product specifications should not quote them but should refer to the appropriate *national* standards.

11.2 Tests on Raw Rubbers and Unvulcanised Mixes

11.2.1 PLASTICITY

Standardised plasticity tests use either the rotation or the compression principle.

11.2.1.1 *Rotation Plastimeters*

Rotation plastimeters have two advantages over compression instruments:

1. They can give higher rates of shear, the upper limit being set only by the impossibility of dissipating the heat generated (even so, shear rates are

446

much lower—e.g. about 1 s $^{-1}$ in the Mooney viscometer—than those occurring in processing).

2. The test can be continued indefinitely under constant conditions, so as to measure scorch and cure characteristics.

The most widely used rotation instrument is the Mooney viscometer (Fig. 11.1) (ISO R289; BS 1673, Part 3; ASTM D1646). A disc-shaped rotor A turns inside a cylindrical chamber B (formed by two mating dies) filled with

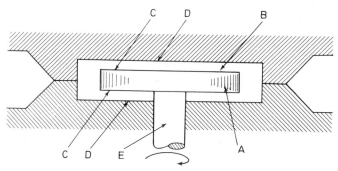

Fig. 11.1   Diagram of the Mooney viscometer

the rubber, thus producing shear between surfaces C and D. The resulting torque on shaft E gives a measure of the effective viscosity of the rubber.

It is necessary to control closely the rotor speed (normally 2 ± 0.02 rev/min), the closure force that holds the dies together, and especially the temperature (normally $100° \pm 0.5°C$); great care is needed to ensure that the *rubber* is at the temperature intended. For high molecular weight rubbers, lower speeds (Decker and Roth, 1953) or higher temperatures are more discriminating (thus, BS 4470 uses 125°C for high-viscosity butyl rubber). The reading time, i.e. running time before taking the reading, varies because different polymers take different times to reach a steady value; thus 4 min has been adopted for SBR (BS 3472; BS 3650), and 8 min for butyl rubber (BS 4470); ASTM D1646 specifies at least 2 min, although about 10 min is usually needed to reach equilibrium.

11.2.1.2   *Compression Plastimeters*

A small cylinder of the rubber is compressed axially between parallel plates under a definite force for a definite time, normally at elevated temperature, the compressed thickness being the 'plasticity number'. Such tests are simple and quick, but the shear rates produced are very low (usually below 0.1 s $^{-1}$), and scorch and cure rate cannot generally be assessed simply by continuing the test.

The original test (Williams, 1924), using *large* compression plates so that the rubber is pressed out into a disc, though now less used, is still included in ASTM D926 and in Soviet (GOST 415) standards.

By using *small* compression surfaces (approximately equal to the test-piece diameter) and suitably choosing other conditions, the test can be made very rapid. In this 'rapid plasticity test' (ISO R2007; BS 1673, Part 3) the test-piece, pre-compressed to 1 mm thickness, is pre-heated for 10 s, and the compressive force is then applied for 15 s, so that the test is completed in less than a minute. The test result, i.e. the thickness in mm/100, is the 'rapid plasticity number'. (Note that a high plasticity *number* denotes a stiff rubber, whereas a high *plasticity* generally denotes a soft rubber.) This test has been widely adopted for factory control of unvulcanised mixes, and more recently for the 'plasticity retention index' (PRI) test on plantation rubber (Section 11.2.3).

Instead of measuring the (variable) thickness under fixed compression conditions, the 'Defo' plastimeter (Odenwald and Baader, 1938; DIN 53514; GOST 10201) determines the force that compresses the rubber by a fixed amount in a fixed time. This force (in grams-force) is the 'Defo hardness' and is a measure of the effective viscosity of the rubber under the test conditions. The elastic recovery after removing the compressing force is often measured, as likewise in ASTM 926 and GOST 415; in the latter,

$$\text{Plasticity} = \frac{h_0 - h_2}{h_0 + h_1}$$

where h_0, h_1, and h_2 are the thicknesses before and after compression and after recovery respectively. However, it seems doubtful whether Defo hardness is more meaningful than the usual plasticity number; both are arbitrary figures, and the compression produced by a fixed force is as good as the force for a fixed compression.

11.2.2   SCORCH AND CURE RATE

Scorch was formerly assessed by plasticity tests on test-pieces previously heated at, say, 100°C or above, the increase in stiffness with heating time

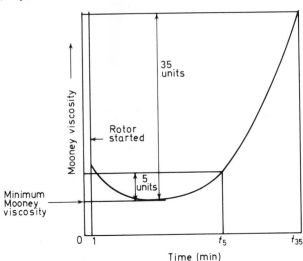

Fig. 11.2 Determination of scorch and cure characteristics by Mooney viscometer. From ISO R667, reproduced by courtesy of the British Standards Institution, 2 Park Street, London, W1A 2BS, from whom copies of the complete ISO Recommendation may be obtained

being observed. Rate of cure is still commonly assessed by vulcanising samples for various times (Section 11.2.4) and following the changes in tensile properties, hardness, set, swelling, etc.

These methods are time consuming, so, for scorch (probably almost entirely) and for rate of cure (to an increasing extent), methods are now used that show directly and continuously the increase in stiffness, i.e. crosslink density, of the rubber when held at a suitable temperature.

The Mooney viscometer, run continuously at, say, 120°C or 145°C, gives a curve (Fig. 11.2) on which 'scorch time' is the time t_5 for the reading to rise by (usually) 5 units above the minimum, while the upward slope indicates rate of cure; by convention, 'cure rate' is expressed (illogically) as the *time* for the reading to rise from 5 to 35 units above minimum (ISO R667; BS 1673, Part 3; ASTM D1646).

Such tests using *continuous* rotation cannot be run to the stage where the rubber is fully vulcanised. The more recent 'curemeters' use a small reciprocating motion, either rotary or linear, thus giving a continuous record of the

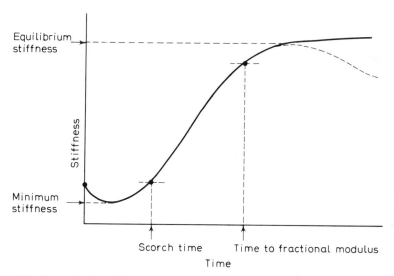

Fig. 11.3   Determination of plasticity, scorch, and cure characteristics by curemeter

stiffness of the rubber—at first low and largely viscous, but increasing and becoming elastic as vulcanisation progresses. From the resulting time–stiffness curve (Fig. 11.3) it is usual to deduce:

1. Minimum stiffness, representing the effective viscosity of the unvulcanised mix.
2. Scorch time, or incipient cure, being the time to a fixed small rise above the minimum.
3. Equilibrium stiffness, representing the elastic modulus of the fully vulcanised rubber; sometimes, the stiffness continues to increase slowly, so that an arbitrary 'equilibrium' or 'plateau' value must be selected. If the rubber shows reversion (broken line in Fig. 11.3), the maximum is taken.

4. Time to an arbitrary 'fractional modulus', commonly 90% of the equilibrium (or maximum) modulus, and representing approximately a technical cure.

It is possible also to deduce fundamental parameters of the vulcanisation process, namely induction period (initiation time) and order and rate constant of the reaction (DIN 53529).

Some instruments (e.g. the Shawbury Curometer) record the compliance of the rubber (inverse of stiffness); the resulting curve is then the inverse of Fig. 11.3, but can be analysed to give data corresponding to results 1–4 above.

Instruments working on the above principle (Section 5.5.2) have been devised: the Agfa Vulcameter (Gehman and Ogilby, 1965); the CEPAR apparatus (Conant and Claxton, 1965); Vulcanisation Tester No. 9011, Bayer System (Karl Frank G.m.b.H., 1969); the Monsanto rheometer (Wise and Decker, 1965); the Goodrich Viscurometer (Veith, 1965-2; Scott Testers Inc., 1965); the Vuremo apparatus (Bartha, Ször, and Ambrus, 1964); the Wallace–Shawbury Curometer (Norman, Hickman, and Payne, 1965; H. W. Wallace & Co. Ltd, 1969); the Zwick Elastometer (Zwick & Co. KG, 1966).

Standards for curemeter testing already include ASTM D2704, D2705, and D2706, DIN 53529, and GOST 12535 and are being drafted by ISO and BSI, as this type of test is rapidly replacing the older practice of making physical tests on vulcanised test slabs.

11.2.3   PLASTICITY RETENTION INDEX (PRI)

The plasticity retention test measures the resistance of raw natural (plantation) rubber to oxidation (BS 1673, Part 3). It is an ageing test in which small discs of the rubber are heated in air at $140° \pm 0.2°C$, and their rapid plasticity number (Section 11.2.1.2) is compared with that of unaged discs. The PRI is the percentage ratio of the aged to unaged plasticity numbers. Oxidation softens the rubber and hence lowers the PRI (Section 4.1.3).

11.2.4   MIXING AND VULCANISATION

Testing of raw rubbers and compounding ingredients involves preparing mixes which are generally vulcanised in either a curemeter (Section 11.2.2) or the more conventional platen press. Hence national standards for experimental mixing and vulcanising equipment and procedures have long existed, and ISO is drafting corresponding international recommendations. The further requirement of standard compounding ingredients has been met in the USA by making available selected and blended standard materials (Roth and Stiehler, 1959; listed also in ASTM D15), whilst other countries have prepared specifications for such materials (BS 4398; NF T43-001).

Essential features specified in mixing and vulcanising standards (e.g. BS 1674 and ASTM D15) are:

1. Two-roll mixing mill: roll diameter and speeds, control of temperature and nip, and (in BS 1674) safety guards and brakes.
2. Internal mixer (e.g. ASTM D15).

3. Vulcanising press: temperature control and mechanical pressure on the rubber during vulcanisation.
4. Vulcanising moulds.
5. Mixing procedure, general directions: accuracy of weighing, incorporation of ingredients, and manipulation of the batch on the rolls (roll temperature, nip, and order and time of adding ingredients depend on the mix formula and hence often appear in specifications for raw polymers, e.g. BS 3472, BS 3650, BS 4470).
6. Storage of mix before vulcanisation.
7. Vulcanisation procedure.

Unfortunately, even when following standard procedures, results can be disappointingly variable, especially as between different laboratories. The causes are not all evident, though they certainly include the personal factor in mill mixing, and the failure to achieve the correct vulcanising temperature (indeed, accurate temperature control in many forms of apparatus—Mooney viscometers, curemeters, vulcanising presses, etc.—is much more difficult than is commonly realised). Hence, in experimental mixing and vulcanising, e.g. as a basis for acceptance of raw materials, it is often preferable not to rely on absolute test figures, but to make direct comparative tests against standard materials.

11.3 Tests on Vulcanised Rubbers

11.3.1 SOME GENERAL CONDITIONS

11.3.1.1 *Temperature and Humidity Conditions*

The properties of vulcanised rubber vary with temperature and sometimes (e.g. electrical properties, cellular rubbers, rubber–textile products) with their moisture content also. Hence, to obtain consistent results, the test-pieces must be 'conditioned', i.e. brought to a predetermined temperature— normally the test temperature—and, where appropriate, into at least approximate equilibrium with air of a given relative humidity.

The ISO has for the time being (R471) accepted three alternatives for 'room' temperatures for conditioning and testing rubber: $20° \pm 2°C, 23° \pm 2°C,$ and $27° \pm 2°C$; the choice depends largely on climatic conditions, e.g. $20° \pm 2°C$ in the UK, and $23° \pm 2°C$ in the USA. Humidity, when controlled, is $65 \pm 5\%$ r.h. at 20°C or 27°C, but $50 \pm 5\%$ r.h. at 23°C—an awkward anomaly. Narrower tolerances ($\pm 1°C$; $\pm 2\%$ r.h.) may be used for sensitive materials. The usual minimum conditioning period is 3 hours for temperature only, or 16 hours if humidity also is involved (this time is often too short to reach moisture equilibrium, but this would take inconveniently long).

11.3.1.2 *Preferred Temperatures*

Tests are now often made at other than the 'room' temperatures mentioned above. To avoid the inconvenience that has arisen through haphazard choice of temperatures, the ISO rubber committee has agreed the following preferred list: $-75°$, $-55°$, $-40°$, $-25°$, $-10°$, $0°$, $20°$ ($23°$ and $27°$ as alternatives), $40°$, $50°$, $70°$, $85°$, $100°$, $125°$, $150°$, $175°$, $200°$, $225°$, and $250°C$. It is strongly urged that, in all standards, test temperatures should be selected from this list.

11.3.1.3   *Mechanical Conditioning*

In rubbers containing reinforcing fillers, there is a 'structure' which is more or less broken down by deformation, thereby greatly lowering the elastic modulus. As this lowered modulus determines the performance of rubbers subjected to repeated (dynamic) stressing in service, they should, strictly speaking, be tested after a suitable mechanical conditioning, or pre-stressing, to break down the structure. This procedure has not been applied consistently and is specified in few tests, e.g. compression stress–strain (BS 903, Part A4; ASTM D575) and rebound resilience and shear stress–strain (BS 903, Parts A8 and A14), though it is usual in testing cellular rubbers (Section 11.3.21). Dynamic tests (Section 11.3.13) involving repeated deformation cycles, of course, apply their own mechanical conditioning.

11.3.2   TENSILE STRESS–STRAIN

Tensile test-pieces in standard procedures are now usually those recommended by the ISO (R37; see also BS 903, Part A2; ASTM D412), i.e. two sizes of dumb-bell and the 'Schopper' ring with internal and external diameters 44·6 mm and 52·6 mm. The relative merits of dumb-bells and rings have been discussed (Scott, 1965, p. 62) and it will suffice here to note the following points:

1. Rings give tensile strengths as much as 33% below the true values as obtained (at least approximately) on dumb-bells. Elongations calculated on the *internal* circumference, as shown on the scales fitted to some ring-testing machines, give incorrect values of modulus, for which the elongation should be calculated on the *mean* circumference. However, these points are not important in control testing, which requires only comparative results.
2. The traditional method of measuring elongation on dumb-bells, i.e. watching two moving gauge marks against a scale, is inconvenient and liable to subjective errors. Hence it is gradually being replaced by automatic extensometers having, for example, light clamps in place of the gauge marks (Eagles and Payne, 1957). Alternatively, the elongation of the test length (between marks) can, with certain precautions, be deduced with sufficient accuracy from the distance between the grips (Khromov and Reznikovskiĭ, 1960).

Measurement of the force on the stretched test-piece has been much improved by eliminating the heavy pendulum dynamometer, which, because of its great inertia and its natural oscillation frequency, could give very inaccurate force indications. To an increasing extent, standard procedures now recommend or specify inertialess transducers which convert the force into (usually) an electrical signal; this has the further advantage that the amplification can be changed according to the force range involved, even during a single test (Payne and Smith, 1956).

All standard procedures follow the general technical practice of expressing tensile strength and modulus as force per unit area of the *initial* cross-section before stretching.

11.3.3   COMPRESSION STRESS–STRAIN

There are national standards (though not yet an ISO Recommendation) for determining the static compression stress–strain curve, usually on cylindrical test-pieces compressed axially (BS 903, Part A4; ASTM D575; GOST 265). To break down filler structure, it is usual to apply mechanical conditioning, i.e. a few deformation–retraction cycles, before making the test measurements. As in tensile tests, stress is calculated on the initial cross-section. Usually, the stress is not increased to a value where the test-piece breaks, as in a tensile test.

For solid (but not cellular) rubbers, the test on standard specimens appears to be relatively little used. This is probably because, in the absence of perfect lubrication at the end faces—which is difficult to achieve and in any case is not specified—the stress–strain curve depends on the shape of the test-piece, especially the ratio of height (axial dimension) to lateral dimensions, so the results are not directly applicable to a product having a different shape.

11.3.4   SHEAR STRESS–STRAIN

The fact that the shear stress–strain curve of vulcanised rubber is substantially linear affords a convenient means of measuring the elastic (in this case shear) modulus. A standard shear modulus test (ISO R1827; BS 903, Part A14) uses a quadruple bonded 'sandwich'. Here again, the rubber is mechanically conditioned by preliminary deformation–retraction cycles before the shear deformation produced by successively increasing forces is measured.

This test normally requires the preparation of special test-pieces from the unvulcanised mix, and hence is less generally applicable than tests of compression stress–strain or hardness, which also (directly or indirectly) measure an elastic modulus. On the other hand, the test could probably be combined with the measurement of the bond strength in shear (Section 11.3.20), which uses a similar test-piece and general procedure.

11.3.5   HARDNESS

Hardness tests on vulcanised rubber (e.g. BS 903, Part A26; ASTM D1415) now mostly follow the ISO Recommendation (R48). This is based on measuring the depth of indentation by a rigid ball under a dead load, the indentation being converted to international rubber hardness degrees (IRHD), the scale of which ranges from 0 (infinitely soft) to 100 (infinitely hard). The ISO Recommendation comprises a 'normal' test, for test-pieces at least 4 mm (preferably 8–10 mm) thick, and a 'micro' test, being an approximately 1:6 scaled-down version of this, for thinner (preferably 1·5–2·5 mm) or smaller test-pieces such as often have to be taken from finished products. The two forms of test give similar results if the ratio of normal to micro test-piece thickness is about 5:1 (Scott, 1965, p. 102)

A point still needing emphasis is that results of any indentation test depend on the thickness of the rubber, unless this is much greater than the depth of indentation; hence the specified standard thickness should be used wherever possible, and the thickness used should always be quoted.

The normal method described above is for the middle hardness range (30–85 IRHD) and uses a 2·50 mm diameter indenting ball. This is replaced by one of 5·0 mm diameter for low hardnesses (10–35 IRHD) or 1·0 mm diameter for high hardnesses (above 85 IRHD). These modifications, now being considered by the ISO, are included in BS 903, Part A26, which also gives special procedures for tests on the curved surfaces of rollers, O-rings, etc.

In British Standards for products, the above methods, using dead loads, are the only ones recognised for specifying hardness, and the ASTM makes a recommendation to the same effect. However, simpler spring-loaded 'pocket' hardness gauges are widely used for rapid checking of products. Examples are the Wallace hardness meter, with a hemispherical-ended indentor and a substantially constant spring force, calibrated to give readings in approximately IRHD; and the Shore durometers, with the indentor frustoconical (type A) or pointed conical (type D) and the spring force decreasing with increasing indentation, readings being in arbitrary degrees (those of type A are approximately equal to IRHD, though lower in the range below 30; type D has a more open scale intended for very hard materials).

Pocket-type hardness gauges are standardised in some countries (ASTM D2240; DIN 53505) but not by the BSI, for the reason given above. However, BS 2719 describes the precautions necessary for obtaining reliable readings; attention to these would avoid many of the difficulties that arise owing to differences between nominally identical instruments.

11.3.6   TEAR

Many forms of tear test-piece have been devised. Current standards use:

1. The crescent, Fig. 11.4(a), with one nick 0·50±0·08 mm deep; this is stretched lengthwise like a tensile dumb-bell (ISO R34; BS 903, Part A3; ASTM D624). A variant, Fig. 11.4(b), has five nicks (GOST 262).
2. The angle, Fig. 11.4(c), used with a nick (DIN 53515) or without a nick (ASTM D624); it is tested as the one-nick crescent form.
3. The trouser, Fig. 11.4(d), the two legs being pulled apart (DIN 53507).
4. The Delft, Fig. 11.4(e), with a central slit instead of a nick; this is useful because of its small size (ISO R816).

All the above are unsatisfactory in that part of the 'tearing' force is used in stretching the untorn part of the test-piece. This is avoided in a modified trouser test-piece which has a bonded-on layer of fabric (GOST 12014); a cut along the centre line is made through the fabric and part of the rubber, so as to leave 1 mm thickness of rubber uncut; this is then torn as in the normal trouser test-piece.

The tearing force is usually (ISO R34 and related standards) the highest

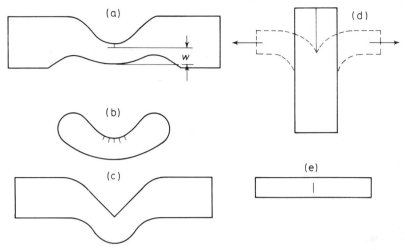

Fig. 11.4   Tear test-pieces: (a) crescent (w = initial untorn width); (b) crescent, original form, shown with five nicks; (c) angle; (d) trouser; (e) Delft. From Scott (1965), courtesy Maclaren

force recorded during tearing, though some kind of average may be used (DIN 53507; GOST 12014).

Tear resistance (or tear strength) is calculated in various ways:

1. The force per unit thickness of test-piece, on the assumption that force is proportional to thickness (though there is evidence to the contrary; Lukomskaya, 1960); this is applicable to all forms of test-piece.
2. The characteristic (or specific) tearing energy, i.e. the energy expended per unit area of torn surface formed; if stretching of the test-piece is eliminated (as in GOST 12014), this is numerically equal to the force per unit thickness.
3. The force to tear a test-piece of standard thickness and standard initial untorn width w. This assumes that the force is also proportional to w, which it is not, since the true tearing force would be independent of w. However, as the deviations from the standard width are usually small, no serious error arises (the British Standard omits the w correction from routine tests).

11.3.7   ABRASION

Standard abrasion tests depend on producing relative motion between the rubber and an abrasive surface (abradant) pressed together by a predetermined force.

Some national standards use the Du Pont (Williams or Grasselli) machine (NF T46-012; GOST 426), though the ISO Recommendation on this has been withdrawn because it does not correlate adequately with service wear—a criticism applicable to all other laboratory abrasion tests. Two rubber blocks are pressed against a rotating abrasive paper disc by either (a) a constant force, the same in all tests (constant load), or (b) a force that produces a constant frictional drag, and hence the same power input, in all tests (constant torque), this being claimed to correlate better with tyre wear.

Other machines in national standards include:

1. The Akron machine (BS 903, Part A9): a rubber test-disc is rotated so as to drive, by its edge, an abrasive wheel, the two being pressed together by a constant force; the abrasive action is produced by tilting the plane of the disc relative to that of the wheel.
2. The National Bureau of Standards machine (ASTM D1630): a rubber test-block is pressed, by constant force, against a rotating cylinder covered with abrasive paper.
3. The Conti machine (DIN 53516): this is similar to the National Bureau of Standards machine, but the test-block is traversed slowly along the length of the cylinder, so as not to pass repeatedly over the same abrasive surface; this procedure avoids loss of 'cutting power' and clogging of the abrasive with detritus—difficulties that arise in the Du Pont and National Bureau of Standards machines.
4. The Pico machine (ASTM D2228): this abrades by means of knives of controlled geometry and sharpness; standard rubbers are used for comparison and for checking the knives.

Owing to the lack of reproducibility of the abradant and to the loss of cutting power during use (at least with papers), the test result is usually expressed as the 'abrasion resistance index', i.e. the ratio of the abrasion loss of a standard (comparison) rubber to that of the test rubber; even this, however, does not entirely eliminate the effects of the variables mentioned. 'Abrasion loss' is generally the volume abraded in a given time (or sometimes over a given length of abrasion path); in the constant torque method this is also the volume abraded per unit of energy expended—indeed loss per unit energy is used in some other tests (NF T46-012; GOST 426).

Abraded rubber is generally removed by air jets (e.g. Du Pont) or brushing (e.g. Akron), but even so, some rubbers become sticky and clog the abradant. On the Akron machine, this can be minimised by feeding a powder (e.g. a mixture of an abrasive and Fullers' earth) between the rubber and the abradant.

11.3.8   FLEX CRACKING AND CUT GROWTH

Most national standard tests, based on the ISO Recommendations, use the De Mattia machine and measure the spontaneous development of cracks by repeated flexing (ISO R132; BS 903, Part A10), or the growth of a deliberately made cut through the test-piece (ISO R133; BS 903, Part A11; ASTM D813).

The test-piece is a flat strip, with a moulded transverse semicircular groove in which the strain is concentrated when the test-piece is bent by bringing its ends nearer together. Close tolerances on the test-piece thickness (overall, and below the groove) and on the maximum and minimum grip spacings ensure that the strain cycle at the groove base is exactly reproduced. The cycle frequency is not critical, but a special feature is the need to operate in an atmosphere free from ozone, which markedly affects the initiation of cracks.

Assessing flex-cracking resistance has proved difficult, and existing methods cannot be called ideal. The most common is to determine the number of flexing cycles required to reach each of several standard degrees of cracking defined verbally and/or by photographs; a difficulty is that rubbers differ in their pattern of cracking. There are methods (ISO R132) for combining the results into either a single measure of 'flex-cracking resistance', or measures of 'crack-initiation resistance' and 'crack-propagation resistance', though the latter distinction is omitted from the latest revision of the British Standard. However, refinements in assessing cracking tend to be negatived by the large variation between results of repeat tests, because of which it is often recommended that tests should be comparative.

The cut-growth test differs from that of flex cracking in that the test-piece is first punctured, at the base of the groove, by a chisel-shaped tool with a blade 2 mm wide. The length of the resulting cut (parallel to the groove axis) is measured before and after various periods of flexing, the test result being the numbers of flexing cycles for predetermined increases in this length. Again, variation between repeat tests is considerable.

Flex cracking can also be produced by rapidly repeated extension of, for example, a dumb-bell test-piece (e.g. ASTM D430).

11.3.9 FATIGUE

Fatigue tests have not so far been considered by the ISO. The ASTM tests (D623) use the Firestone and Goodrich flexometers (so called because they do not produce bending). In these, a test block or cylinder is compressed vertically by a constant force and is subjected to a rapidly alternating deformation, either in compression and of fixed amplitude [Goodrich, Fig. 11.5(a)] or by a rotary (eccentric) motion of the bottom face [Firestone, Fig. 11.5(b)]. The usual measurements are of temperature rise, permanent deformation (set or creep), and number of deformation cycles to failure.

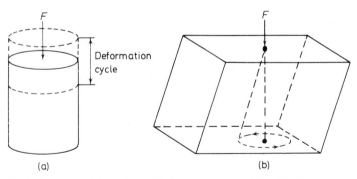

Fig. 11.5   Fatigue tests: (a) the Goodrich flexometer, and (b) the Firestone flexometer; F = constant compressive force. From Scott (1965), courtesy Maclaren

Another test (GOST 10952) uses a rubber cylinder which is rapidly rotated about its axis while bent through a fixed (but adjustable) angle, so as to produce alternating tension and compression; the test measures the number of cycles to failure [compare Fig. 11.8(c)].

These relatively complex modes of deformation are presumably chosen as being nearer to service conditions than repeatedly stretching a strip or dumb-bell till it fails (as, for example, in GOST 261).

11.3.10   CREEP AND STRESS RELAXATION

Creep and stress-relaxation tests have not been fully considered by the ISO, but national standards exist.

In the usual creep test (e.g. BS 903, Part A15), a constant force—tension, compression, or shear—is applied to a specified test-piece and the increase in deformation (beyond the initial 'instantaneous' value) is measured periodically, 'instantaneous' being in practice after stressing for, say, 10 min. In tension and compression, the cross-section (and hence the actual stress) change gradually but, as the test conditions and results are arbitrary, this is immaterial for practical purposes.

In stress-relaxation tests (BS 903, Part A15; ASTM D1390), the deformation is held constant and the reaction force exerted by the deformed test-piece is measured periodically.

For expressing the results, the proportionate increase in deformation (creep) or decrease in force (relaxation) may be either stated for one or more fixed test periods, or plotted against time, or preferably the logarithm of time, as this often gives an approximately straight line and hence a single 'rate' parameter.

11.3.11   SET

Standard methods of measuring set use either extension or compression.

Extension tests—not yet finally agreed by the ISO—use either parallel-sided strips, or dumb-bells or rings as used for tensile tests (e.g. BS 903, Part A5; ASTM D412). The test-piece is stretched to a specified elongation or stress (per unit initial cross-section) for a predetermined time, and the residual elongation (set) is measured after one or more periods of recovery.

More common are compression tests, using disc test-pieces, again under constant deformation (usually) or stress (ISO R815; BS 903, Part A6; ASTM D395). The use of compression has disadvantages because the result depends on the degree of slip between the rubber and the compression plates. A theoretical ideal would be complete slip, as the result would not depend on the test-piece thickness–diameter ratio. This, however, is difficult to achieve even with the lubrication sometimes specified, and it does not correspond to most service conditions. On the other hand, no-slip conditions necessitate standardising the 'shape-factor' (thickness–diameter ratio) and also cause the test-piece to have concave faces after recovery, thus complicating the measurement of recovered thickness and increasing the variability of test results. Indeed, compression set tests generally have proved troublesome owing to poor reproducibility, and there is a growing opinion that a test involving a more homogeneous type of deformation, e.g. extension, would give just as useful, and more reproducible, data.

However, in *all* set tests, the applied strain (or stress) and the deformation and recovery times are arbitrary, so the results have little absolute meaning and are useful only for comparative or control tests.

11.3.12   RESILIENCE

Standard tests mostly measure the rebound of a pendulum from the rubber test-piece. However, designs of pendulum vary considerably, and so do the corresponding test results. Hence ISO (R1767; see also BS 903 Part A8) has selected one type—the Lüpke pendulum—though the possibility of modifying the Schob pendulum to give similar results is being considered.

In the Lüpke pendulum, the striker is a steel bar suspended so as to remain horizontal [Fig. 11.6(a)]. It is released from a position (shown dotted) 100 mm above its rest position; the height of its rebound is measured (usually indirectly from the horizontal return swing) and expressed as percentage of the 100 mm fall, the result being the 'rebound resilience'.

Other forms of instrument use more conventional types of pendulum (Fig. 11.6): the Schob pendulum, Fig. 11.6(b) (DIN 53512); the Dunlop pendulum, Fig. 11.6(c); the Goodyear–Healey pendulum, Fig. 11.6(d) (ASTM D1054). In the Tripsometer, Fig. 11.6(e) (BS 903, Part A8), the

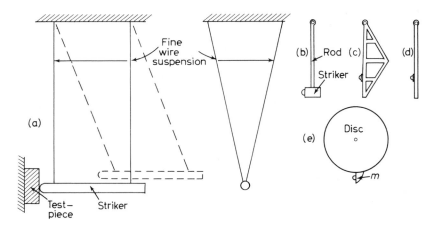

Fig. 11.6   Diagram of rebound resilience apparatus: (a) the Lüpke pendulum; (b) the Schob pendulum; (c) the Dunlop pendulum; (d) the Goodyear–Healey pendulum; (e) the Tripsometer (m = off-centre mass). From Scott (1965), courtesy Maclaren

striking energy is provided by a small off-centre mass *m* attached to a heavy disc rotating about its centre. Still other tests use a pendulum that strikes and rebounds from the centre of a rubber strip held horizontally under tension (GOST 10827), or a ball falling and rebounding vertically (Bashore resiliometer, ASTM D2632).

Some of the above instruments measure also the maximum deformation of the rubber under the impact, from which the dynamic modulus can be calculated.

Temperature must be closely controlled, and preferably variable over a wide range, as the change of resilience with temperature (especially cooling) is an important characteristic.

11.3.13   DYNAMIC TESTS

The term 'dynamic tests' conventionally denotes tests that measure the elastic moduli by subjecting rubber to rapid cycles of stress or strain, either a continuous train of constant-amplitude sinusoidal waves [Fig. 11.7(a)], or a decaying train [Fig. 11.7(b)], or a series of half-waves all in one direction

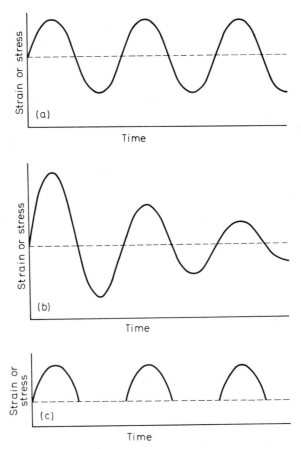

Fig. 11.7   Forms of stress or strain cycles in dynamic tests: (a) continuous constant-amplitude; (b) continuous decaying-amplitude; (c) successive half-waves (the intervals are not necessarily equal to the half-wave duration, as shown here). From Scott (1965), courtesy Maclaren

[Fig. 11.7(c)]. These deformations are usually superimposed on a fixed initial ('static' or 'mean') stress or strain.

Owing to the great variety of dynamic test machines and the necessarily wide range of test conditions, the British and US Standards (BS 903, Part A24; ASTM D2231; the ISO document is not yet finalised) do not specify a particular machine and procedure but give general guidance on the quantities that have to be measured, apparatus requirements and precautions for ensuring reliable results, the ranges of temperature, frequency, and amplitude to be covered, preferred values for these conditions, methods of interpreting and presenting results, and descriptions of suitable test machines given only as examples.

Standard machines using cycles as shown in Fig. 11.7(a) are of the 'non-resonant' type, i.e. they operate at frequencies away from the resonant (natural) frequency of the rubber test-piece (plus any other mass attached to it). Either a train of strain cycles is imposed on the test-piece and the resulting stress cycles recorded (e.g. the RAPRA sinusoidal-strain machine; Payne, 1956, 1961; BS 903, Part A24), or stress cycles are imposed and the strain cycles recorded (e.g. the Roelig machine, using mechanical drive, Roelig, 1938, 1943, 1945; DIN 53513; and 'transducer' machines, using higher frequency electromagnetic drive, BS 903, Part A24). In either case, the ratio of stress amplitude to strain amplitude gives the complex modulus (E^* or G^*), and the phase difference gives the loss angle δ. These and similar machines can operate in compression, shear, or tension.

In apparatus using cycles as shown in Fig. 11.7(b), an initial deformation starts a train of cycles that decay due to hysteresis; an example is the Yerzley oscillograph, Fig. 11.8(a) (Yerzley, 1940; ASTM D945), in which the oscillations of a beam are controlled by the rubber test-piece (in compression or shear).

In the 'rotary power loss' machine, Fig. 11.8(b) (Bulgin and Hubbard, 1958; BS 903, Part A24) successive half-cycle deformations [Fig. 11.7(c)] are produced by rotating a rubber-covered wheel against a drum. Owing to the phase difference between strain and stress, the distribution of stresses over the contact area is not symmetrical, and this sets up a torque on the drum, from which the loss angle is calculated. The complex modulus E^* is calculated from the compression of the rubber [d in Fig. 11.8(b)]. A somewhat similar arrangement is the Soviet 'roller' machine (GOST 10953), whilst GOST 10828 uses a rotating flexed cylinder [Fig. 11.8(c)] (compare GOST 10952; Section 11.3.9).

11.3.14 LOW-TEMPERATURE TESTS

Standard tests of low-temperature resistance depend on the fact that progressive cooling makes rubber stiff and ultimately glasslike (brittle), and causes recovery after deformation to be sluggish and incomplete.

Stiffening is the basis of torsion modulus tests using, for example, the Gehman apparatus (ISO R1432; ASTM D1053). The torsion (i.e. shear) modulus is measured by twisting a strip test-piece, at room and several reduced temperatures, to give a temperature–modulus curve. The result is commonly quoted as the temperatures at which the modulus is 2, 5, 10, . . . times its

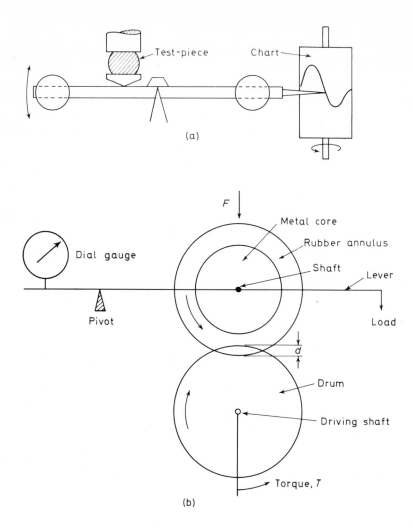

(a)

(b)

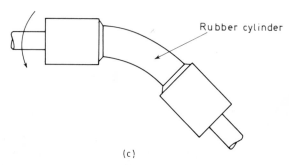

(c)

Fig. 11.8 Diagram of dynamic test apparatus: (a) the Yerzley oscillograph; (b) the rotary power loss machine; (c) method using rotating flexed cylinder. From Scott (1965), courtesy Maclaren

value at room temperature, but BS 903, Part A13 now prefers the temperature at which modulus increases to a predetermined value, e.g. corresponding to loss of technically useful flexibility—this is partly because measurements of the very low modulus at room temperature have proved unreliable.

Instead of torsion modulus, Young's modulus can be measured by bending a bar supported near the ends (ASTM D797), or a dynamic test (Section 11.3.13) can be used to measure the complex modulus (E^* or G^*), or its in-phase (E' or G') and out-of-phase (E'' or G'') components, over a range of temperatures.

Choice of the cooling medium is important; air (or other gas) involves long temperature equilibrium times, whereas a liquid, though avoiding this (and hence preferred in, for example, the British Standard) must be one that does not swell the rubber or extract plasticiser, etc., from it.

By using a long period at moderately low temperature, e.g. $-25°C$, the tendency of a rubber to crystallise can be assessed from the resulting gradual stiffening.

Brittleness temperature (brittle point) tests determine the temperature at which the rubber is *just not* brittle enough to break when a specified test-piece is hit with a striker (as tests are normally made at 2°C intervals, this is nearly enough the actual breaking temperature) (ISO R812; BS 903, Part A25; ASTM D746). The result is arbitrary, since it depends on the rate of strain produced, which in turn depends on the test-piece thickness and the striker velocity. Moreover, unlike the torsion modulus test, it fixes only one point on the temperature–stiffness curve. The choice of coolant involves the same problem as noted above.

A more fundamental measure is that of the glass transition temperature T_g ('glass–rubber transition' or 'second order transition'), conveniently determined as the temperature at which the slope of the temperature–volume curve changes abruptly, or at which (during gradual warming of a 'frozen' test-piece) the test-piece becomes deformable under load (GOST 12254).

Procedures depending on recovery from deformation comprise low-temperature set, temperature–retraction, and resilience tests.

Low-temperature set tests are made as described in Section 11.3.11 except that the test-piece is at a sub-zero temperature during compression and recovery (ISO R1653; ASTM D1229). In the ISO procedure, set is measured after recovery periods ranging from 10 s to 30 min (it is not clear why the normal-temperature set test specifies only 30 min).

Instead of measuring recovery at a fixed temperature, the temperature–retraction test measures the temperature at which a certain degree of recovery is reached (ASTM D1329). A dumb-bell is stretched to a specified elongation and cooled (e.g. to $-70°C$) to 'freeze' the elongation; the temperature is then slowly raised, and the temperatures corresponding to fixed recovery values are noted.

Resilience tests (Section 11.3.12) over a range of sub-zero temperatures show, by the rapid fall below a certain point, the loss of rubberlike properties, just as do measurements of elastic modulus; however, if the latter are made by the torsion modulus test (above), the two estimates of low-temperature resistance are not comparable because the deformation produced in the rebound resilience test is much more rapid.

Useful general guidance on low-temperature testing is given by ASTM D832.

11.3.15   RESISTANCE TO LIQUIDS (SWELLING)

Standard methods of assessing the effect of immersion in a liquid (e.g. ISO R1817; BS 903, Part A16; ASTM D471) include determinations of:

1. Change in volume, by hydrostatic weighing before and after immersion.
2. Changes in length, width, and thickness.
3. Amount of material extracted by the liquid.
4. Changes in tensile properties and hardness due to immersion.
5. Changes in tensile properties, hardness, and (in BS 903) volume after re-drying.
6. Change in weight of a rubber or rubber–textile sheet exposed to liquid on one side only.

Change in volume can be calculated from changes in length, width, and thickness, though not as accurately as by method 1; method 2 shows the anisotropic swelling produced by calender grain, which method 1 does not. A simple variant of method 2 measures only the change in length of a strip (ASTM D1460). Tests 3 and 5 are possible only if the liquid is volatile.

Change in volume ('swelling' if positive) is usually expressed as percentage of the initial volume of the test-piece. If absorption has reached approximate saturation, the result is independent of test-piece dimensions. Otherwise, it depends on these dimensions—in practice, on the thickness, as being much the smallest dimension; hence the thickness should be approximately constant, e.g. to $\pm 5\%$ in the British Standard. Indeed, if absorption is so slow as to be only superficial (e.g. water, viscous oils) the result is best expressed as volume of liquid absorbed per unit surface area of the test-piece, so as to be independent of thickness (BS 903, Part A16).

In high-temperature immersion tests, oxidation by atmospheric oxygen can markedly affect the action of the liquid; hence access of air should be restricted—a point not sufficiently appreciated.

Tests should, where possible, be made with the liquid encountered in service, especially when comparing different polymers as their relative liquid-resistance varies with the nature of the liquid. Otherwise, and for control purposes, use may be made of 'standard' liquids, some of which are pure chemical individuals or mixtures thereof, and some petroleum oils (originally 'ASTM oils'). These latter cannot be completely specified by the usual chemical and physical characteristics; hence uniform supplies can only be obtained from a particular crude oil source by a particular procedure, as is indicated or implied in the British and ASTM Standards.

11.3.16   PERMEABILITY

For permanent gases, standard methods measure the permeation of gas through a thin rubber test-piece under one of the following conditions:

1. With gas at constant high pressure on one side, and a small *constant volume* of (usually) the same gas, initially at substantially atmospheric

pressure, on the other; permeation increases the pressure on this side (ISO R1399; BS 903, Part A17).

2. As method 1, but with the gas on the second side kept at *constant pressure*, so that its volume increases.

3. With the test gas (e.g. hydrogen) on one side and another (e.g. air) on the other, both at atmospheric pressure; the concentration of the test gas that has diffused into the other gas is measured periodically, e.g. by the change in thermal conductivity of the latter (ASTM D815).

Permeability is calculated from the steady-state rate of pressure rise in method 1, or of volume increase in method 2, or of increase in test gas concentration in method 3.

For water vapour permeability, a thin disc of rubber is sealed around the rim of a shallow vessel containing a desiccant, and the whole kept in air of known relative humidity. Permeability is determined from the rate of weight increase of the assembly due to water vapour passing through the rubber and being absorbed by the desiccant (BS 3177).

11.3.17 ELECTRICAL TESTS

Tests on insulating rubbers, i.e. for volume and surface resistivity, dielectric strength, permittivity (dielectric constant), power factor (dissipation factor), and tracking, are made by methods used for insulating materials generally. Accordingly, the ISO is adopting the methods standardised by the International Electrotechnical Commission (IEC), the authoritative body in this field. Indeed IEC Recommendations 93 and 167 on resistivity measurements are already the basis of national standards (BS 903, Parts C1 and C2; ASTM D257); permittivity and power (dissipation) factor measurements are covered by IEC Recommendation 250 (also BS 903, Part C3; ASTM D150) and electric strength by BS 903, Part C4 and ASTM D149.

Conducting and antistatic rubbers call for special treatment in resistivity tests because the very high contact resistance between them and conventional electrode materials introduces serious errors (Norman, 1957). It is therefore necessary to use either (a) bonded brass electrodes [Fig. 11.9(a)], with which contact resistance is not high; (b) a measuring circuit in which this resistance is immaterial—e.g. the 'potential method' [Fig. 11.9(b)] in which a known current passing between electrodes C and C' produces a potential difference between E and E' measured by a system taking virtually no current; or (c) tin-foil electrodes adhered by colloidal graphite applied as an aqueous suspension [Fig. 11.9(c)]. [All three methods appear in BS 2044, and method (b) is in ASTM D991.] Moreover, conducting rubbers can be very sensitive to deformation, so that care is needed in handling test-pieces.

11.3.18 RESISTANCE TO DETERIORATION—AGEING

Tests so far considered by the ISO are (a) air oven, (b) oxygen pressure, (c) ozone cracking, and (d) light; the air pressure ('air bomb') method has not been considered and is not included in the British Standard.

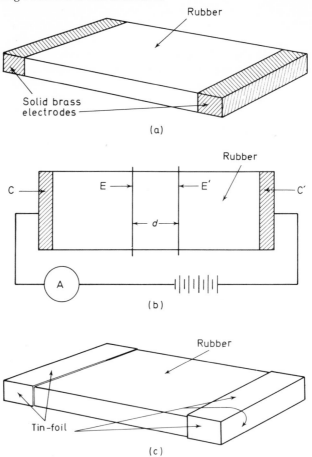

Fig. 11.9   Electrode systems for antistatic and conductive rubbers: (a) bonded brass electrodes; (b) potential method (C and C' are current electrodes; E and E' are measuring electrodes; A is a milliammeter; d is the test length); (c) adhered tin-foil electrodes. From Scott (1965), courtesy Maclaren

11.3.18.1   *Air Oven Ageing*

The air oven ageing test (ISO R188; BS 903, Part A19; ASTM D573) subjects prepared rubber test-pieces (*not* complete articles) to slowly circulating and renewed air at constant temperature. Developed from the original Geer oven at $70° \pm 1°C$, this now includes tests at higher temperatures, preferably $100° \pm 1°$, $125° \pm 1°$, $150° \pm 2°$, $175° \pm 2°$, $200° \pm 2°$, or $250° \pm 2°C$. Some Standards warn against using too high a temperature in an attempt to speed up a test intended to simulate prolonged normal-temperature ageing, the higher temperatures (e.g. 100°C or above) being regarded as valid only to simulate high-temperature service.

It has been objected that the $\pm 1°C$ and $\pm 2°C$ tolerances permit excessive variation in ageing rate, namely $\pm 10\%$ and $\pm 20\%$. So far as these tolerances refer to short-period fluctuations, the objection is not valid, as the average temperature could still be correct. However, the importance of ensuring the correct *average* is not generally made clear in standard procedures.

A refinement of the oven test—the 'cell oven' (ISO R188; BS 903, Part A19; ASTM D865)—was introduced because, if different rubbers are aged together, they affect each other through the volatile substances evolved (antioxidants, oxidation products, etc.). In the cell oven, each kind of rubber is contained in a separate cylindrical cell, each with its independent air circulation. The cells can conveniently be surrounded by a heating medium (liquid, saturated vapour, aluminium block) kept at the required temperature, so avoiding the trouble encountered, for example, with ovens having electrical units which heat the walls locally, causing radiation that overheats the test-pieces. Variants of the cell oven are 'test tube' ageing (ASTM D865) and a 'tubular' oven with forced ventilation (ASTM D1870).

11.3.18.2   *The Oxygen Pressure Method*

The oxygen pressure (oxygen bomb) method, using oxygen at 21 kgf/cm^2 and 70°C (ISO R188; BS 903, Part A19; ASTM D572), was intended to overcome a defect of the air oven test, namely non-uniform oxidation through the rubber, and also to accelerate the test still further. It was later found useful for detecting poor ageing due to traces of copper, manganese, etc., but appears now to be less used, possibly because the last effect is no longer such a serious problem. Although included in the ISO Recommendation and some national standards, it is now proposed to omit it from BS 903.

11.3.18.3   *The Air Pressure Method*

The air pressure (air bomb) method, using air at about 6 kgf/cm^2 and 120°–126°C, is a specialised test 'intended for estimating the probable life of rubber compounds designed for high-temperature service' (ASTM D454; also DIN 53508 and NF T46-006) and is not regarded as a general purpose ageing test.

11.3.18.4   *Ozone Cracking*

Ozone-cracking tests, in which stretched rubber strips are exposed to air of known ozone content at constant temperature, have proved difficult to standardise (ISO R1431; BS 903, Part A23; ASTM D1149). The main problems are:

1. Measuring the extremely low ozone concentrations used (down to 25–50 parts per 100 million)—hence the very detailed procedure specified.
2. Ensuring a uniform flow of ozonised air over the test-pieces (see, for example, the elaborate system in BS 903).

3. Assessing the degree of cracking—it is, indeed, generally recommended to record only the time of first appearance of cracks just visible with a lens (some workers find the naked eye more reliable); in the Soviet standard (GOST 6949), the time to complete rupture of the test-piece is also recorded, but this stage is generally regarded as too advanced to be of practical interest; no method of assessing intermediate stages of cracking has yet been found acceptable internationally.

11.3.18.5   *Light Ageing*

Light-ageing tests have likewise proved difficult to standardise. The main conclusions so far are: (a) the xenon lamp most nearly reproduces the effects of sunlight; (b) control of the test-piece temperature, especially avoidance of overheating, is essential; and (c) the light dosage (i.e. total light energy per unit area of test-piece) is best measured by simultaneously exposing standard blue-dyed wool specimens and comparing these with specimens faded to standard extents—a procedure well established in the textile industry.

11.3.18.6   *Weathering*

Weathering tests, using out-door exposure, have been standardised by, for example, the ASTM (D518; D1171) and are now being developed for British Standards.

11.3.18.7   *High-Energy Radiation Resistance*

Resistance to high-energy radiation is now the subject of standards, e.g. ASTM D1672 and D2309.

11.3.19   RUBBER-TO-TEXTILE ADHESION

Methods of assessing the adhesion of rubber to textile fabrics, as in hose, belting, etc., are mainly based on the ply-separation test of the ISO (R36; see also BS 903, Part A12; ASTM D413). This is a peeling (stripping) test in which the cover rubber (if any) and then the fabric plies are peeled off at constant rate. For flat products (e.g. belts), the test-piece is a 25 mm wide strip; for hose 50 mm or more in internal diameter, a 25 mm length is cut off and opened to form a strip; in ISO R36 the angle α of separation during peeling is uncontrolled [Fig. 11.10(a)], but in the BS and ASTM tests it is approx. 180°. For smaller hose, a 25 mm length is mounted on a mandrel and the plies are peeled off at approximately 90° [Fig. 11.10(b)].

The peeling force is recorded continuously over the test length. There is a difference of opinion as to whether the test result (force for standard 25 mm strip) should be the maximum, minimum, or an average; the ISO Recommendation uses 'the mean of the lowest 50% of peak values taken from the central 50% of the stripping trace'.

A simpler alternative to the above machine method is the dead-load method, used mainly in product specifications (also given in ASTM D413).

This stipulates that, under a given dead load, the ply under test must not separate at more than a certain rate, say 25 mm/min. Usually, also, steps are taken to ensure an approximately 180° angle of separation [Fig. 11.10(c)].

It has been shown (Borroff, Elliott, and Wake, 1951) that the result of a peeling test, though useful technically, is not a true measure of adhesion, for which the 'direct tension' test has been devised (BS 903, Part A27). Here the separating force is normal to the plane of the bond [Fig. 11.10(d)]. This necessitates the two components that are to be separated being adhered firmly to the end-pieces (metal cylinders) by which the force is applied; a cyanoacrylate adhesive (e.g. Eastman 910) is suitable. The end-pieces are

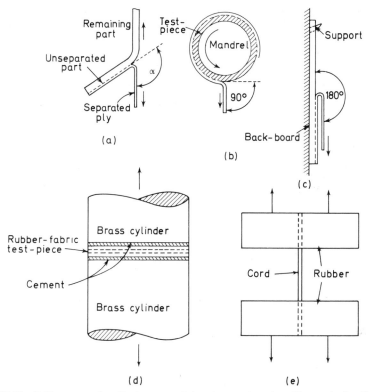

Fig. 11.10   Rubber-to-textile adhesion tests: (a) strip test-piece (machine method); (b) ring (hose) test-piece mounted on mandrel, (c) strip test-piece (dead-load method), supported against back-board to ensure approximately 180° separation; (d) direct tension test; (e) H-block test. From Scott (1965), courtesy Maclaren

pulled apart at constant rate till separation occurs; the test result is the force causing separation per unit area of the test-piece (compare Section 11.3.20, method 2).

For studying adhesion between rubber and textiles, ASTM D2630 describes a 'strap peel' test using a specially built-up and vulcanised multi-ply test-piece, tested before and/or after flexing; D430 covers resistance to ply separation, e.g. in belts and tyres, by repeated flexing.

Adhesion between rubber and individual tyre cords is measured by the H-block or H-pull test (ASTM D2138), or a variant of it. A cord (or, in practice, several) is vulcanised into a block of the rubber and then pulled out by an arrangement such as that shown is Fig. 11.10(e), the required force being measured.

11.3.20   RUBBER-TO-METAL BOND

Standard bond strength tests mostly follow the ISO, which has three methods:

1. With one metal plate (peeling test): a rubber strip, bonded by vulcanisation to a metal strip of the same width, is peeled off at constant rate, at approximately 90° [Fig. 11.11(a)]. In the ISO procedure (R813) the result is the maximum force recorded during separation; sometimes (e.g. BS 903, Part A21) the minimum also is recorded, while ASTM D429

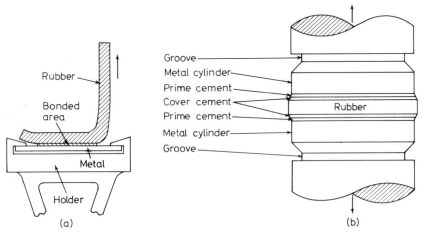

Fig. 11.11   Rubber-to-metal bond strength tests: (a) peeling test, and (b) button test; the grooves are for gripping in the test machine. From Scott (1965), courtesy Maclaren

and GOST 411 take a continuous record but do not say how it is evaluated; all these methods differ from the similar rubber–fabric peeling test (Section 11.3.19), for reasons not obvious.

2. With two metal plates (button test): a thin rubber disc is bonded to the faces of two metal cylinders, which are pulled apart at constant rate in a direction normal to the faces, and the force at failure noted; the result is the force per unit area of bonded face [Fig. 11.11(b)] (ISO R814; BS 903, Part A21; ASTM D429). With a very strong bond, failure may occur in the rubber or in one of the cement layers attaching it to the metal; the location of failure is therefore often recorded.

3. The quadruple shear test: this uses the test-piece and method of stressing described in Section 11.3.4 except that the 'rigid plates' are now metal, the bonding system is the one under investigation, and the force is increased to the point of failure. The result is the force per unit area of bond (BS 903, Part A28).

Other shear tests are used for measuring bond strength of rubber to wire, either single-strand or cord (ASTM D1871, D2229); these are analogous to the H-block test for tyre cord (Section 11.3.19).

11.3.21 CELLULAR RUBBERS

In the absence as yet of ISO Recommendations—though some are on the way—the present situation as regards standard tests for cellular rubbers is rather confused.

The most widely used and standardised tests on soft cellular rubbers are the following (details appear in BS 903, Parts F1–F9; BS 4443; also in BS 3093 and other product standards; BS 3667; ASTM D1055, D1056, and D1564):

1. Apparent density.
2. Stiffness under compression—often called 'indentation'.
3. Fatigue, i.e. change of thickness and loss of stiffness as a result of repeated compression.
4. Compression set.
5. Resilience.
6. Tensile and tear strengths.
7. Resistance to liquids.
8. Resistance to low temperatures.
9. Ageing.

For tests 2–4 two procedures are used: (a) compression is applied by a large circular plate (100–457 mm diameter) to a still larger test-piece—even a complete mattress; and (b) a small test-piece (29–100 mm in diameter or square, or a strip at least 20 mm wide) is compressed between larger plates. Mechanical conditioning (Section 11.3.1.3) is usually applied before the test. The test result in test 2 is usually the force(s) required to give predetermined compression(s). In test 4, the conditions are broadly as for solid rubber (Section 11.3.3), using either constant load [in procedure (a)] or constant deformation [in procedure (b)].

Tests 5–7 are made generally as for solid rubber, using: for resilience, rebound of a pendulum or a falling steel ball; for tensile properties, a large dumb-bell or (for elongation only) strip; for tear strength, a large angle or trousers test-piece (see Fig. 11.4); and for resistance to liquids, procedures modified because the cells are more or less accessible to the liquid.

Low-temperature resistance is assessed by the increase in stiffness after a cooling period. In the air oven, air pressure, and oxygen pressure ageing, the temperatures and pressures normally specified (Section 11.3.18) should not be exceeded—and may even have to be reduced—owing to the ready oxidisability of rubber in cellular form.

11.4 Product Specifications

It would be impossible, and not very helpful, to attempt here to discuss the multitude of specifications for rubber products; these are listed in the publications of national standardising bodies. As with test methods, the ISO Recommendations are intended as the basis for national standards, though as yet few Recommendations for rubber products have been issued.

It is therefore intended to deal rather with the coverage of existing specifications and with the philosophy of specification making, since readers of this book may well at some time be engaged in this task, whether for standards organisations, such as the BSI, or for Government departments or private companies.

Product specifications relating to rubber mostly fall into one of the groups given under the following headings.

11.4.1   RAW MATERIALS

Specifications for raw materials comprise those for natural and synthetic rubbers (BS 3472; BS 3650; BS 4396; BS 4470), latex (BS 4355), and compounding ingredients (BS 4398; NF T43-001).

11.4.2   VULCANISED RUBBERS

It has been found useful to standardise vulcanised rubbers having specified properties, so that users of rubber and designers of equipment incorporating it can often select a rubber that suits their purpose and is readily available from rubber manufacturers. Thus there are British Standards for vulcanisates made from natural rubber (BS 1154; BS 1155; BS 3106), styrene–butadiene rubber (BS 3515; BS 3629), butyl rubber (BS 3227), acrylonitrile–butadiene rubber (BS 2751; BS 3222), and chloroprene rubber (BS 2752), covering in each case a range of hardness. Only the properties are specified; the composition is left to the manufacturer, though formulae that have proved satisfactory are sometimes given for guidance.

Allied to this principle is the classification and coding of vulcanised rubbers according to properties. In ASTM D2000 this is combined with lists of actually available rubbers, in order 'to provide guidance to the engineer in the selection of practical, commercially available elastomeric materials, and further to provide a method for specifying these materials by the use of a simple "line call-out" designation'. The 'line-call out' or the code indicates the degree of heat and oil resistance and the range within which each of certain properties lies, namely tensile strength and hardness, plus such others as may be required—though excluding incompatible properties (Section 9.2.1).

11.4.3   MANUFACTURED RUBBER PRODUCTS

National standards for a great variety of rubber products have existed for many years. Hitherto, each country has produced these independently, with inevitable lack of uniformity in regard to dimensions and test requirements, but now that appropriate ISO Recommendations are coming forward, an increasing degree of agreement between different countries' standards may be expected.

11.4.4   MISCELLANEOUS

Useful British Standards are those giving the generally accepted dimensional tolerances for moulded and extruded rubber products (BS 3734), recommendations for conditions in which rubber products should be stored

(BS 3574), a glossary of terms used in the rubber industry (BS 3558), and a list of names and abbreviations in common use for rubbers and plastics (BS 3502; see also ASTM D1418).

11.5 Specification Making

Below are given some general observations which it is hoped will assist the user in obtaining and the manufacturer in supplying products suited to the application and of satisfactory quality, consistency, and cost. To this end, written specifications are usually an essential requirement, and it is important that they should be realistic, clear, and easy to operate.

The philosophy of specifications and the distinction between performance, buying, and control specifications are discussed in Section 2.11. The user should, unless exceptional circumstances exist, inform the manufacturer of the application and environment in which the rubber component will operate. There is, obviously, every advantage in using one of the standard vulcanised rubbers referred to above, and in many cases one of these can be chosen.

If documents have to be drawn up giving performance and properties, the difficult problem immediately arises of selecting appropriate tests. Ideally, the tests in a specification should guarantee that any product which passes them will give satisfactory service. A move towards this ideal is the increasing use of 'service' or 'performance' tests on the whole article; although this may well involve developing special tests and equipment for every product, the results are more directly informative than measurements of basic properties of the rubber, such as those described in this chapter. In any case, only a small proportion of the articles can usually be tested, and, for the results to be meaningful, the number tested and the method of selection must be based on statistical principles. An exception is safety equipment, where it may be necessary to test every individual article, e.g. electricians' safety gloves. Another procedure, especially for consumer goods, in the BSI Certification (Kite Mark) scheme, whereby the manufacture and control testing are subject to inspection and the marketed product to testing for compliance with the relevant British Standard.

Some properties have an obvious and direct relevance to service, e.g. hardness or elastic moduli for (say) an anti-vibration mounting, or resistance to liquids when these are encountered in service. Some others are relevant but do not give a *direct* indication of service, e.g. set, low-temperature tests, resilience, abrasion, flex cracking, ageing. Others again have little relevance to most uses, the 'classic' example being tests of tensile strength and breaking elongation; the long persistence of these is partly due to sheer inertia, partly to the belief that a strong stretchy rubber is 'good', but also because they are capable of showing any deterioration of quality by accidental or intentional contamination.

As to *methods* of test, these should be the national standards. Where the standard includes alternatives, the product specification must state clearly which alternative is to be used, e.g. form of test-piece for tensile or compression set tests, thickness of hardness test-pieces, oven ageing test conditions; this is often overlooked when specifications are being drafted.

Wherever possible, the necessary test-pieces should be taken from the

R

finished product—indeed, the micro-hardness test, the small tensile dumb-bell (even smaller types are being developed), and the Delft tear test-piece were introduced to help make this possible. The practice of allowing the tests to be made on special test-pieces separately vulcanised is unsatisfactory in that no attempt is usually made to check whether the product and test-piece are in fact similar, though this might be done, for example, by comparing their density and swelling.

Low or elevated temperatures, at which tests are to be carried out, should be selected from the list put forward by the ISO rubber committee (Section 11.3.1.2).

Fixing test limits is usually a matter of striking a balance between the manufacturer's view of what can reasonably be made and the user's view on what is required to ensure satisfactory service.

To this end, committees (e.g. of the BSI) include representatives of organisations concerned with these two aspects—though, with consumer goods, adequate presentation of the user's viewpoint is not always easy to obtain—as well as 'neutral' bodies such as research organisations. Wide circulation of draft standards for criticism before publication further helps to ensure that all points of view are taken into account.

In fixing limits, two common cases arise:

1. For established products known to give satisfactory service, the properties of these can form at least the basis for discussion.
2. If little information is available as to performance (as often with consumer goods), data may be obtained on the range of variation in properties among articles currently produced; the problem then is where, within this range, to draw the pass–fail dividing line—a point on which there is no 'golden rule'.

In either case, allowance must be made for the fact that, even with standard test methods, the manufacturer and user may get different results; hence, in fixing minimum or maximum test figures or tolerance ranges, the inevitable experimental error must be taken into account in accordance with statistical principles. Guidance on the application of these principles to rubber testing and specifications generally is being given by a committee recently set up by the BSI.

11.5.1   DIMENSIONS

Little need be said about fixing dimensions and their tolerances, beyond drawing attention to the advantages in uniformity that result from using dimensions based on one of the series of 'preferred numbers' (ISO R3; BS 2045) and to the existence of a schedule of dimensional tolerances for rubber products (BS 3734).

Finally, reference may be made to two useful documents issued by the former Federation of British Rubber and Allied Manufacturers (now British Rubber Manufacturers' Association), namely: *Guide to the Drafting of Specifications for Rubber Products* (FSAC/56, 1961), and *FSAC Questionnaires as a Basis for Rubber Product Standards* (FSAC/57, 1961).

TWELVE

Professional, Trade, Research, and Standards Organisations

BY C. M. BLOW

In this chapter, a brief description is given of the principal organisations in the UK connected with the rubber industry. In addition, some detail is given of a few of those in Europe and N. America. Neither list is claimed to be comprehensive but should nevertheless provide a starting point for any enquiries the reader may wish to make. The full names and addresses of the organisations referred to will be found in Section 12.5.

12.1 Professional Institutions

12.1.1 UNITED KINGDOM

The body that represents the interests of professional personnel employed in the industry and in firms supplying raw materials, equipment, and instruments to the rubber industry is the Institution of the Rubber Industry, founded in 1921. This body awards diplomas based on attainment of technical competence at three levels—licentiates, associates, and fellows. Local, national, and international conferences are organised at intervals, as well as regular technical meetings during the winter months at the main centres of the industry. Publications include the Journal, formerly Transactions and Proceedings, annual Reports on the Progress of Rubber Technology, Monographs, and Conference Proceedings.

Mention should be made of The Plastics Institute and of the Society of Chemical Industry which has a Plastics and Polymer Group, formed in 1940.

12.1.2 OVERSEAS

The Institution of the Rubber Industry has affiliated sections in Australasia, Ceylon, India, Malaysia, S. Africa, N. Ireland, and Pakistan. The professional body in the USA is the Division of Rubber Chemistry of the American Chemical Society, which holds symposia meetings twice a year and has many active Rubber Groups throughout the USA and Canada. *Rubber Chemistry and Technology* is published quarterly, with extra review issues, by the Rubber Division.

The Swedish Institution of Rubber Technology with a membership of rubber technicians and technologists organises an annual symposium and publishes proceedings.

12.2   Trade Associations

12.2.1   UNITED KINGDOM

The British Rubber Manufacturers' Association was formed in 1968 to become the unified body (replacing the Federation of British Rubber and Allied Manufacturers, the British Rubber and Allied Manufacturers' Associations, the Rubber Manufacturing Employers' Association, and the Tyre Manufacturers' Conference) and the organisation to represent the whole rubber industry in negotiations with government and other official bodies. Its aims, furthermore, are to promote the interests of manufacturers, processers, and distributors in the rubber industry, and to act as a reference point. It operates three divisions—Tyres, General Rubber Goods, and Personnel Services.

The National Joint Industrial Council for the Rubber Manufacturing Industry is concerned with encouraging sound employer–employee relations, fair wages and working conditions, safety regulations, and the maintenance of quality and output from the industry.

The Rubber Economic Development Committee of the National Economic Development Council was formed in 1965.

The producers of synthetic rubbers in the UK have formed the British Association of Synthetic Rubber Manufacturers, to promote co-operation in widening the market and increasing the usage of synthetic rubbers.

Two organisations concerned particularly with the natural rubber producing and marketing industries are respectively The Rubber Growers' Association (formed in 1907) and the Rubber Trade Association of London.

12.2.2   OVERSEAS

In the USA, the Rubber Manufacturers' Association (RMA) is the co-ordinator of the industry. It publishes the internationally accepted *Green Book— International Standards of Quality and Packing of Natural Rubber Grades*. The RMA *Rubber Handbook* gives specifications for many rubber products.

The International Institute of Synthetic Rubber Producers has its headquarters in New York. With a membership of 28, it aims to promote and further the interests both of the manufacturers of synthetic rubber polymers and of the general public in many aspects of the industry. Various committees cover technical aspects, nomenclature, packaging, and public relations. Research and development projects are sponsored by this Institute.

12.3 Research Organisations

12.3.1 UNITED KINGDOM

The industry in the UK is served by two research associations. The Rubber and Plastics Research Association at Shawbury, Shropshire, founded as the Research Association of British Rubber and Tyre Manufacturers in 1919, is the co-operative research organisation for the rubber and plastics industries. Its income comes from member companies' subscriptions, government grants, and sponsored work on a contract basis. Its Information Division, with its very extensive library, provides a service in answering technical and other queries and publishes monthly abstracts of current literature and patents. Research reports, technical reports, and translations also emanate from this organisation.

The Natural (formerly British) Rubber Producers' Research Association, formed in 1938 to carry out research and development for the benefit of the natural rubber production in the Far East, is now financed by the Malayan Rubber Fund Board. Its fundamental work on the chemistry and physics of natural rubber is of world-wide renown. Technical information about the processing and uses of natural rubber is issued in the form of technical reports and information sheets, as well as the quarterly *Rubber Developments*.

12.3.2 OVERSEAS

Also financed by the Malayan Rubber Fund Board is the Rubber Research Institute of Malaya at Kuala Lumpur, which is concerned with the cultivation of rubber trees, the collection of rubber, and its preparation for marketing. It publishes a Journal and *Planters' Bulletin*. There is a separate Rubber Research Institute of Ceylon.

In France, the Institut Français du Caoutchouc carries out research and development for the rubber-producing interests and provides a technical service for the industry. The IFC publishes *Revue Générale du Caoutchouc et des Plastiques*.

The Rubber Research Institute TNO, located at Delft, Holland, carries out research for rubber producers and rubber manufacturers, in addition to providing an advisory and information service.

12.4 Standards Institutes

The principal authorities in the various countries issuing standards for rubber materials and products are:

USA	American Society for Testing and Materials (ASTM)
UK	British Standards Institution (BSI)
Germany	Deutscher Normenausschuss (DIN)
USSR	State Committee for Standards, Measurements, and Measuring Instruments (GOST)
France	Association Française de Normalisation (NF)

The International Organisation for Standardisation (ISO) prepares internationally agreed Recommendations (see Chapter 11).

12.5   Addresses

American Chemical Society, Division of Rubber Chemistry, PO Box 123, University of Akron, Ohio 44304, USA

American Society for Testing and Materials, 1916 Race Street, Philadelphia, Pa., USA

British Association of Synthetic Rubber Manufacturers, 93 Albert Embankment, London SE1

British Rubber Manufacturers' Association, 9 Whitehall, London SW1

British Standards Institution, 2 Park Street, London W1A 2BS

Institut Français du Caoutchouc, 42 Rue Scheffer, Paris 16, France

Institution of the Rubber Industry, 4 Kensington Palace Gardens, London W8 4QR

International Institute of Synthetic Rubber Producers Inc., 54 Rockefeller Plaza, New York, NY 10020, USA

International Organisation for Standardisation, Geneva, Switzerland

Malayan Rubber Fund Board, Natural Rubber Building, PO Box 508, Kuala Lumpur, Malaysia

National Joint Industrial Council for the Rubber Manufacturing Industry:
   Employers: Trafalgar House, 9 Whitehall, London SW1
   Employees: National Union of General and Municipal Workers, Ruxley House, Ruxley Towers, Claygate, Near Esher, Surrey

Natural Rubber Producers' Research Association, 19 Buckingham Street, Adelphi, London WC2 6EJ
   Laboratories: 56 Tewin Road, Welwyn Garden City, Hertfordshire

The Plastics Institute, 11 Hobart Place, London SW1

Rubber and Plastics Research Association, Shawbury, Shrewsbury SY4 4NR

Rubber Economic Development Committee, National Economic Development Council, 21 Millbank Street, London SW1

The Rubber Growers' Association, Plantation House, 10/15 Mincing Lane, London EC3M 3DX

Rubber Manufacturers' Association, 444 Madison Avenue, New York, NY 10022, USA

Rubber Research Institute of Ceylon, Dartonfield, Agalawatta, Ceylon

Rubber Research Institute of Malaya, Pubat Penyelidekan Getah Tanah Malayu, Kuala Lumpur, Malaysia

Rubber Research Institute TNO, 178 Oostsingel, Delft, Holland, PO Box 15

Rubber Trade Association of London, Plantation House, 10/15 Mincing Lane, London EC3M 3DX

Society of Chemical Industry, Plastics and Polymer Group, 14 Belgrave Square, London SW1

Bibliography

General

Books covering the general subjects of polymer and rubber science and technology are given in the following list.

BATEMAN, L. (Ed.), *Chemistry and Physics of Rubber-like Substances*, Maclaren, London, 1963

CRAIG, A. S., *Rubber Technology: A Basic Course*, Oliver & Boyd, Edinburgh, 1963

DAVIS, C. C. AND BLAKE, J. T. (Eds), *Chemistry and Technology of Rubber*, Reinhold, New York, 1937

FISHER, H. L., *Chemistry of Natural and Synthetic Rubbers*, Reinhold, New York, 1957

HUKE, D. W., *Introduction to Natural and Synthetic Rubbers*, Hutchinson, London, 1961

KENNEDY, J. P. and TORNQUIST, E. G. M., *Polymer Chemistry of Synthetic Elastomers* (2 vols), Interscience, New York, 1967 and 1969

LEBRAS, J., *Rubber: Fundamentals of its Science and Technology*, Chemical Publishing Co., New York, 1957

MEMMLER, K., *Science of Rubber* (translated R. F. Dunbrook and V. N. Morris), Reinhold, New York, 1934

MORTON, M. (Ed.), *Introduction to Rubber Technology*, Chapman & Hall, London, 1959

MULLINS, L. (Ed.), *Proceedings of the NRPRA Jubilee Conference, Cambridge*, Maclaren, London, 1965

NAUNTON, W. J. S. (Ed.), *Applied Science of Rubber*, Arnold, London, 1961

STERN, H. J., *Rubber, Natural and Synthetic*, Maclaren, London, 1967

Specialised

In the following list, books relating to specific aspects of rubber technology and manufacture, for follow-up reading, are grouped and cross-referenced to the most appropriate section in the book.

CHAPTER 1

GOODYEAR, CHARLES, *Gum Elastic and its Varieties, with a detailed account of its applications and uses, and of the discovery of vulcanisation*, 1855
HANCOCK, THOMAS, *Personal Narrative of the Origin and Progress of the Caoutchouc or India Rubber Manufacture in England*, London, 1857
HOWARD, F. A., *Buna Rubber*, van Nostrand, New York, 1947
SCHIDROWITZ, P. and DAWSON, T. R. (Eds), *History of the Rubber Industry*, IRI, London, 1952

SECTION 2.3

SCOTT, J. R., *Ebonite: Its Nature, Properties and Compounding*, Maclaren, London, 1958

SECTION 2.4

BLACKLEY, D. C., *High Polymer Latices*, Maclaren, London, 1966
CARL, J. C., *Neoprene Latex*, Du Pont, Wilmington, Del., 1962
FLINT, C. F., *Chemistry and Technology of Rubber Latex*, Chapman & Hall, London, 1938
MADGE, E. W., *Latex Foam Rubber*, Maclaren, London, 1962
NOBLE, R. J., *Latex in Industry*, Rubber Age (Palmerton), New York, 1953
WINSPEAR, G. G., *The Vanderbilt Latex Handbook*, Vanderbilt, New York, 1954

SECTION 2.9

BUIST, J. M., *Ageing and Weathering of Rubber*, Heffer, Cambridge, 1956
CHARLESBY, A., *Atomic Radiation and Polymers*, Pergamon, Oxford, 1960
JAMES, D. I. (Ed.), *Abrasion of Rubber* (translated M. E. Jolley), Maclaren, London, 1967
NORMAN, R. H., *Conductive Rubber: Its Production, Application and Test Methods*, Maclaren, London, 1957

SECTION 2.10

ALLEN, P. W., LINDLEY, P. B. AND PAYNE, A. R., *Use of Rubber in Engineering*, Maclaren, London, 1967
BURTON, W. E., *Engineering with Rubber*, McGraw-Hill, New York, 1949
DAVEY, A. B. and PAYNE, A. R., *Rubber in Engineering Practice*, Maclaren, London, 1965
EDWARDS, W. H., *Rubber in Engineering Design*, Ministry of Aviation (Department of Materials, Tech. Memo. No. 7), London, 1960
LINDLEY, P. B., *Engineering Design with Natural Rubber*, NRPRA Tech. Bull., London, 1966

MCPHERSON, A. T. and KLEMIN, A. (Eds), *Engineering Uses of Rubber*, Chapman & Hall, London, 1956
MERNAGH, L. R. (Ed.), *Rubbers*, Morgan Grampian, West Wickham, 1969
PAYNE, A. R. and SCOTT, J. R., *Engineering Design with Rubber: The Properties, Testing, and Design of Rubber as an Engineering Material*, Maclaren, London, 1960

SECTION 2.11

HASLAM, J. and WILLIS, H. A., *Identification and Analysis of Plastics*, Iliffe, London, 1965
KLINE, G. M. (Ed.), *Analytical Chemistry of Polymers* (3 vols), Interscience, New York, 1959–1962
LEIGH-DUGMORE, C. H., *Microscopy of Rubber*, Heffer, Cambridge, 1961
SAUNDERS, K. J., *Identification of Plastics and Rubbers*, Chapman & Hall, London, 1966
WAKE, W. C., *Analysis of Rubber and Rubber-like Polymers*, Maclaren, London, 1958

CHAPTER 3

ANDREWS, E. H., *Fracture in Polymers*, Oliver & Boyd, Edinburgh, 1968
BINGHAM, E. C., *Fluidity and Plasticity*, McGraw-Hill, New York, 1923
FERRY, J. D., *Viscoelasticity of Polymeric Materials*, Wiley, New York, 1961
LEADERMAN, H., *Elastic and Creep Properties of Filamentous Materials*, Textile Foundation, Washington DC, 1943
LODGE, A. S., *Elastic Liquids*, Academic Press, New York, 1964
MASON, P. and WOOKEY, N. (Eds), *Rheology of Elastomers*, Pergamon, Oxford, 1958
SEVERS, E. T., *Rheology of Polymers*, Reinhold, New York, 1962
TRELOAR, L. R. G., *Physics of Rubber Elasticity*, Clarendon Press, Oxford, 1958
WAZER, J. R. VAN, LYONS, J. W., KIM, K. Y. and COLWELL, R. E., *Viscosity and Flow Measurement*, Interscience, New York, 1963

SECTION 4.1

BARAMBOIM, N. K., *Mechanochemistry of Polymers* (translated R. J. Moseley, Ed. W. F. Watson), Maclaren, London, 1964
BONNER, J. and VARNER, J. E., *Plant Biochemistry*, Academic Press, New York, 1965
BRYDSON, J. A. (Ed.), *Developments with Natural Rubber*, Maclaren, London, 1967
DAVIES, B. L. and GLAZER, J., *Plastics Derived from Natural Rubber*, The Plastics Institute, London, 1955
International Standards of Quality and Packing of Natural Rubber Grades (Green Book), Rubber Manufacturers' Association, New York

SECTION 4.2

BARRON, H., *Modern Synthetic Rubbers*, van Nostrand, New York, 1949
Description of Synthetic Rubbers and Latices, International Institute of
   Synthetic Rubber Producers, New York, 1968
PENN, W. S., *Synthetic Rubber Technology*, Maclaren, London, 1960
WHITBY, G. S., DAVIS, C. C. and DUNBROOK, R. F. (Eds), *Synthetic Rubber*,
   Chapman & Hall, London, 1954

SECTION 4.8

MURRAY, R. M. and THOMPSON, D. C., *The Neoprenes*, Du Pont, Wilmington,
   Del., 1963

SECTION 4.13

EABORN, C., *Organosilicon Compounds*, Butterworths, London, 1960
FORDHAM, G. (Ed.), *Silicones*, Newnes, London, 1961
FREEMAN, G. G., *Silicones*, Iliffe, London, 1962
MCGREGOR, R. R., *Silicones and their Uses*, McGraw-Hill, New York, 1954
NOLL, W., *Chemistry and Technology of Silicones*, Academic Press, New York,
   1968
ROCHOW, F. G., *An Introduction to the Chemistry of Silicones*, Chapman &
   Hall, London, 1961
STONE, F. G. A. and GRAHAM, W. A. G. (Eds), *Inorganic Polymers*, Academic
   Press, New York, 1968

SECTION 4.14

BUIST, J. M. and GUDGEON, H. (Eds), *Advances in Polyurethane Technology*,
   Maclaren, London, 1968
SAUNDERS, J. H. and FRISCH, K. C., *Polyurethanes*, Interscience, New York,
   1962
WRIGHT, P. and CUMMING, A. P. C., *Solid Polyurethane Elastomers*, Maclaren,
   London, 1969

SECTION 4.15

GAYLORD, N. G., *Polyethers, Part III: Polyalkylene Sulphides and Other
   Polythioethers*, Interscience, New York, 1962

CHAPTER 5

ALLIGER, G. and SJOTHUN, I. J. (Eds), *Vulcanization of Elastomers*, Chapman &
   Hall, London, 1964

HOFFMANN, W., *Vulcanization and Vulcanizing Agents*, Maclaren, London, 1967

SECTION 6.2

ALPHEN, J. VAN, SCHONLAU, W. I. K. and TEMPEL, M. VAN DEN, *Rubber Chemicals*, Elsevier, Amsterdam, 1956
Materials and Compounding Ingredients for Rubber and Plastics, Rubber World (Bill Bros.), New York, 1963
WILSON, B. J., *British Compounding Ingredients for Rubber*, Heffer, Cambridge, 1964

SECTION 6.3

Factice as an Aid to Productivity in the Rubber Industry, National College of Rubber Technology, London, 1962

SECTION 6.4

BALL, J. M. (Ed.), *Manual of Reclaimed Rubber*, Rubber Reclaimers' Association, New York, 1956
NOURRY, A. (Ed.), *Reclaimed Rubber: Its Development, Applications, and Future*, Maclaren, London, 1962

SECTION 6.5

SCOTT, G., *Atmospheric Oxidation and Antioxidants*, Elsevier, Amsterdam, 1965

SECTION 6.7

GORDON COOK, J., *Handbook of Textile Fibres* (4th edn, 2 vols), Merrow, Watford, 1968
MEREDITH, R., *Mechanical Properties of Textile Fibres*, North-Holland, Amsterdam, 1956

CHAPTER 7

DE BOER, J. H., *The Dynamical Character of Adsorption*, Clarendon Press, Oxford, 1953
KRAUS, G. (Ed.), *Reinforcement of Elastomers*, Interscience, New York, 1965
PARKINSON, D., *Reinforcement of Rubbers*, IRI and Lakeman, London, 1957
ROSS, S. and OLIVIER, J. P., *On Physical Adsorption*, Wiley, New York, 1964

CHAPTER 8

BUCHAN, S., *Rubber to Metal Bonding*, Lockwood, London, 1959
Machinery and Equipment for Rubber and Plastics, Rubber World (Bill Bros.), New York, 1963
PENN, W. S. (Ed.), *Injection Moulding of Elastomers*, Maclaren, London, 1969
WILLSHAW, H., *Calenders for Rubber Processing*, IRI and Lakeman, London, 1956

CHAPTER 9

COX, D. R., *Planning of Experiments*, Wiley, New York, 1958
DAVIS, O. L. (Ed.), *The Design and Analysis of Industrial Experiments*, Oliver & Boyd, Edinburgh, 1956
Education and Welfare, US Department of Health, Food and Drug Administration, Washington DC

SECTION 10.1

WOOD, E. C., *Pneumatic Tyre Design*, Heffer, Cambridge, 1955

SECTION 10.9

CARMICHAEL, C., *The Seals Book*, Penton, Cleveland, Ohio, 1964
MORSE, W. (Ed.), *Seals Handbook*, Morgan Grampian, West Wickham, 1969
WARRING, R. H., *Seals and Packings*, Trade & Technical Press, London, 1967
WHITE, C. M. and DENNY, D. F., *Sealing Mechanisms of Flexible Packings*, Her Majesty's Stationery Office, London, 1948

CHAPTER 11

KLUCKOW, P., *Rubber and Plastics Testing*, Chapman & Hall, London, 1963
Quality Control in the Rubber Industry: Symposium, Maclaren, London, 1967
Rubber Handbook: Specifications for Rubber Products, Rubber Manufacturing Association, New York
SCOTT, J. R., *Physical Testing of Rubbers*, Maclaren, London, 1965
SODEN, A. L., *A Practical Manual of Hardness Testing*, Maclaren, London, 1951

CHAPTER 12

YESCOMBE, E. R., *Sources of Information on the Rubber, Plastics and Allied Industries*, Pergamon, Oxford, 1968

Directories

Rubber Directory of Great Britain, Maclaren, London
Rubber Red Book: Directory of the Rubber Industry, Palmerton, New York

Trade Periodicals

Plastics & Rubber Weekly, Maclaren & Sons Ltd, Davis House, 69–77 High
  Street, Croydon, CR9 1QH
Plastics, Rubbers, Textiles (monthly), Rubber & Technical Press Ltd,
  Tenterden, Kent
Polymer Age (bimonthly), Rubber & Technical Press Ltd, Tenterden, Kent
Rubber Age (monthly), Palmerton Publishing Co. Inc., 101 West 31st Street,
  New York, NY 10001, USA
Rubber Journal (monthly), Maclaren & Sons Ltd, Davis House, 69–77 High
  Street, Croydon, CR9 1QH
Rubber World (monthly), Bill Brothers Publishing Co., 630 Third Avenue,
  New York, NY 10017, USA

References

Literature and Patent References—Name Index

All the publications and patents to which reference is made in the text are listed below in alphabetical order of the first author. Following the authors' name(s) is the year of publication, then the periodical name, volume, part, and page numbers, or the patent specification number. In the right-hand column, the section(s) of the book in which the reference appears is given.

Co-authors are also listed with a cross-reference to the first author.

ABOYTES, P. *See* Voet, A.	7.4.3.1
ABRAMS, W. J., (1962). *Rubb. Plast. Age*, **43**, 5, 451	4.10.5
ADAMEK, S. *See* Woodhams, R. T.	4.17
ALLEN, P. W., (1963). *The Chemistry and Physics of Rubber-like Substances* (Ed. L. Bateman), Maclaren, London, Ch. 5	4.1.6.2
ALLEN, P. W. and BLOOMFIELD, G. F., (1963). *The Chemistry and Physics of Rubber-like Substances* (Ed. L. Bateman), Maclaren, London, Ch. 1	4.1.7.1
ALLEN, P. W., LINDLEY, P. B. and PAYNE, A. R., (1967). *Use of Rubber in Engineering*, Maclaren, London	3.4.4
ALLEN, R. D. *See* Capito, J. E.	4.6.3
ALPHEN, J. VAN, (1954). *Proc. 3rd Rubb. Technol. Conf.*, 670	7.1
ALPHEN, J. VAN. *See* Houwink, R.	7.1
ALSTADT, D. M., (1955). *Rubb. Wld, N.Y.*, **133**, 2, 221	8.7.5, 10.5.5
AMBELANG, J. C., KLINE, R. H., LORENZ, O. M., PARKS, C. R., WADELIN, C. and SHELTON, J. R., (1963). *Rubb. Chem. Technol.*, **36**, 1497	6.5.3
AMBRUS, S. *See* Bartha, Z.	11.2.2
AMERICAN SOCIETY FOR TESTING AND MATERIALS (1965). *Spec. tech. Publs Am. Soc. Test. Mater.*, **383**	5.5.1, 5.5.2
AMERONGEN, G. J. VAN, (1964). *Rubb. Chem. Technol.*, **37**, 1065	2.9.3.10
ANDERSON, E. V., (1969). *Chem. Engng News*, **47**, July 14, 29	2.12
Anon. (1921). *India Rubb. J.*, **51**, 63	1.6
Anon. (1962). *Rubb. Plast. Age*, **43**, 1408	8.2.7
Anon. (1963). *Rubb. Plast. Age*, **44**, 1461	8.2.1
Anon. (1968-1). *Rubb. J.*, **150**, 5, 45	10.5.6
Anon. (1968-2). *Rubb J.*, **150**, 7, 21	6.1.5
Anon. (1968-3). *Rubb. J.*, **150**, 10, 25	8.6.6
Anon. (1968-4). *Rubb. Age, N.Y.*, **100**, 12, 60	8.10.1.4
Anon. (1968-5). *Rubb. Wld, N.Y.*, **159**, 2, 42	10.3.3
Anon. (1968-6). *Nucl. Engng, Lond.*, **13**, 146, 586	2.9.3.10
Anon. (1968-7). *Bundesgesundheitblatt*, **12**, 10, 160	9.4.10
Anon. (1969-1). *Chem. Engng News*, **47**, Jan. 13, 34	4.18.3

Anon. (1969-2). *Rubb. Dev.*, **22**, 3, 100   2.10.3
Anon. (1969-3). *Rubb. Age, N.Y.*, **101**, 6, 72   8.10.1.4
Anon. (1969-4). *Rubb. J.*, **151**, 10, 48   8.2.8
ARDLEY, S. M., (1963). *Rubb. Plast. Wkly.*, **145**, 8, 247   4.10.5
ARMSTRONG, R. T., LITTLE, J. R. and DOAK, K. W., (1944). *Ind. Engng. Chem.*, **36**, 628   5.2.2
ASSOCIATION OF HYDRAULIC EQUIPMENT MANUFACTURERS (1968). *Rubb. J.*, **150**, 9, 36;   2.9.3.5,
    J. Inst. Petrol., **54**, 530, 36   10.9.5.4
ASTON, M. W., FLETCHER, W. and MORRELL, S. H., (1969). *Proc. 4th int. Conf. Fluid*   10.9.5.4
    Sealing, 64
ATKIN, W. and NYE, H. F., (1969). *Jl IRI*, **3**, 16   5.5.4,
                                                                                         10.8.3.3
ATKINS, J. H. *See* Taylor, G. L.   7.4.2
AUER, E. E., DOAK, K. W. and SCHAFFNER, I. J., (1958). *Rubb. Chem. Technol.*, **31**,185 9.4.5

BAADER, T. *See* Odenwald, H.   11.2.1.2
BACHMAN, J. H., SELLERS, J. W., WAGNER, M. P. and WOLF, R. F., (1959). *Rubb. Chem.*   7.4.3.1
    Technol., **32**, 1286
BAHARY, W. S., SAPPER, D. I. and LANE, J. H., (1967). *Rubb. Chem. Technol.*, **40**, 1529 4.4.1,
                                                                                         4.4.3
BAKER, C. J. and RILEY, I. H., (1968). *Rubb. Plast. Age*, **39**, 577   4.13.4
BAKER, D., CHARLESBY, A. and MORRIS, J., (1968). *Polymer*, **9**, 437   4.13.5
BAKER, H. C., (1958). *Rubb. Dev.*, **11**, 2   4.1.4.4
BALL, J. M. and RANDALL, R. L., (1949). *Rubb. Age, N.Y.*, **64**, 718   6.4.5
BANBURY, F. H., (1916). US Pat. 1 200 070   1.8.1
BANKS, D. J. and WISEMAN, P., (1968). *J. appl. Chem., Lond.*, **18**, 9, 262   1.6
BARANANO, C. M. *See* Bruner, W. M.   10.5.2
BARNES, S., (1967). *Mach. Des.*, **39**, 1, 110   4.13.4
BARR, D. A. and HAZELDINE, R. W., (1955). *J. chem. Soc.*, 1881   4.18.1
BARTHA, Z., SZÖR, P. and AMBRUS, S., (1964). *Plaste Kautsch.*, **11**, 11, 670   11.2.2
BARTLETT, W. E., (1889). Brit. Pat. 11 800   1.9
BARTLETT, W. E., (1890). Brit. Pat. 16 348   1.9
BARTSCH, J. R., (1963). US Pats 3 313 545; 3 438 933   10.7.3.2
BASS, S. L., (1947). *Chemy Ind.*, **171**, 189   1.4.10.2
BAYER CO., THE, (1911). Germ. Pat. 265 221   1.6
BAYER CO., THE, (1912). Brit. Pat. 14 556   1.4.2
BAYER CO., THE, (1963). Germ. Pat. 1 135 909   1.6
BEATTY, J. R., (1964). *Rubb. Chem. Technol.*, **37**, 1341   9.4.5
BEATTY, J. R. *See* Juve, A. E.   8.6.7
BEAUMONT, R. A. *See* Hartman, N. E.   9.2.4
BEDFORD, C. W. *See* Sebrell, L. B.   1.6
BEERBOWER, A., PATTISON, D. A. and SAFFIN, G. D., (1964). *Rubb. Chem. Technol.*, **37**,   9.2.3.1,
    246   9.4.6.3
BENNETT, B. *See* Mueller, W. J.   9.3.6
BENNETT, D. A. and BURRIDGE, K. G., (1962). Brit. Pats 976 212; 976 213; 976 214   1.3
BERTSCH, P. F., (1961). *Rubb. Wld, N.Y.*, **144**, 3, 75   9.2.4
BEWLEY, H., (1845). Brit. Pat. 10 825   1.8.3
BIE, G. J. VAN DER, *et al.* (1968). *Eur. Congr. Plast. Elastomers*   4.7.2
BIGGS, B. S., (1958). *Rubb. Chem. Technol.*, **31**, 1051   6.5.3
BISHOP, E. T. *See* Holden, G.   4.7.2
BISHOP, E. T. and DAVISON, S., (1967). *Pacif. Conf. Chem. Spectrosc.*, California   4.7.2
BITTING, R. W., (1969). *Rubb. Age, N.Y.*, **101**, 6, 62   8.6.9
BLACKLEY, D. C., (1966). *High Polymer Latices, Vol. 2*, Maclaren, London   8.5.2.3
BLACKWELDER, B. R., (1924). US Pat. 1 520 925   1.9
BLOOMFIELD, G. F., (1968). *Jl IRI*, **2**, 287   4.1.5
BLOOMFIELD, G. F. *See* Allen, P. W.   4.1.7.1
BLOW, C. M., (1929). *IRI Trans.*, **5**, 417   2.6.2.1
BLOW, C. M., (1964). *Aircr. Engng*, **36**, 208   4.13.4
BLOW, C. M., (1968). *Tribology*, **1**, Mar., 81   4.13.4
BLOW, C. M., EXLEY, K. and SOUTHWART, D. W., (1968). *Jl IRI*, **2**, 282   10.9.5.4
BLOW, C. M., EXLEY, K. and SOUTHWART, D. W., (1969). *Jl IRI*, **3**, 22   10.9.5.4
BLOW, C. M. and HOPKINS, G., (1945). *J. Soc. chem. Ind., Lond.*, **64**, 316   10.5.5

488 *References*

BLOW, C. M. and PIKE, M., (1952). *Rubb. Dev.*, **5**, 61   10.9.2

BOGGS, C. R., (1919). US Pat. 1 296 469   1.6

BOLT, T. D., DANNENBERG, E. M., DOBBIN, R. E. and ROSSMAN, R. P., (1960). *Rubb. Plast. Age*, **41**, 1520   7.8

BONNER, J. and VARNER, J. E., (1965). *Plant Biochemistry*, Academic Press, New York, p. 685   4.1.1

BONNER, Z. D., (1969). *Rubb. Plast. Age*, **50**, 770   2.5.2

BOONSTRA, B. B. *See* Dannenberg, E. M.   7.4.1

BOONSTRA, B. B. *See* Polley, M.   7.4.3.2

BOONSTRA, B. B. *See* Renner, A.   7.1

BOONSTRA, B. B. and DANNENBERG, E. M., (1958). *Rubb. Age, N.Y.*, **82**, 838   7.8

BOONSTRA, B. B. and DANNENBERG, E. M., (1959). *Rubb. Chem. Technol.*, **32**, 825   3.4.5

BOONSTRA, B. B. and MEDALIA, A. I., (1963-1). *Rubb. Age, N.Y.*, **92**, 82; 892   7.7

BOONSTRA, B. B. and MEDALIA, A. I., (1963-2). *Rubb. Chem. Technol.*, **36**, 115   7.7

BOONSTRA, B. B. and TAYLOR, G. L., (1965). *Rubb. Chem. Technol.*, **38**, 943   7.6, 7.9

BOOTH, D. A. *See* Brzenk, F.   4.5.3

BORG WARNER (1961). Brit. Pats 873 359; 877 035; 877 923   10.5.3.2

BORROFF, E. M., ELLIOTT, R. and WAKE, W. C., (1951). *J. Rubb. Res.*, **20**, 5–6, 42   11.3.19

BORROFF, E. M. and WAKE, W. C., (1949). *IRI Trans.*, **25**, 190   8.5.2

BORROFF, E. M. and WAKE, W. C., (1951). *Ind. Engng Chem.*, **43**, 439   8.5.2.5

BOYER, R. F., (1963). *Rubb. Chem. Technol.*, **36**, 1303   4.13.4

BRADSHAW, W. D. and WELSBY, J. A., (1966). *IRI Proc.* **13**, 79   10.3.3

BREEN, A. W. VAN, *et al.* (1966). *Rubb. Plast. Age*, **47**, 1070   4.7.2

BRENNAN, J. J. *See* Dannenberg, E. M.   7.10.1

BRENNAN, J. J., DANNENBERG, E. M. and RIGBI, Z., (1967). *Proc. int. Rubb. Conf.*, 123   7.10.1

BRENNAN, J. J. and JERMYN, T. E., (1965), *J. appl. Polym. Sci.*, **9**, 2749   7.8

BRIGGS, G. J., EDWARDS, D. C. and STOREY, E. B., (1963). *Rubb. Chem. Technol.*, **36**, 621   10.8.2.3

BRIGGS, R. A. *See* Grüber, E. E.   4.15.2

BRISTOW, G. M. and WATSON, W. F., (1963). *The Chemistry and Physics of Rubber-like Substances* (Ed. L. Bateman), Maclaren, London, Ch. 14   4.1.7.2

BRITISH ASSOCIATION OF SYNTHETIC RUBBER MANUFACTURERS (1967). *Jl IRI*, **1**, 262; 319   4.10.4

BROKENBROW, B., SIMS, D. and STOKOE, A. L., (1969). *Rubb. J.*, **151**, 12, 30   2.10.3

BROOKE, T. A. and ROBINSON, P. M., (1964). Brit. Pat. 1 123 235   10.2.1.3

BROOKS, R. E., STRAIN, D. E. and MCALEVY, A., (1953). *India Rubb. Wld*, **127**, 791   4.9.1

BROOMAN, R. A., (1845). Brit. Pat. 19 852   1.8.3

BRUNER, W. M. and BARANANO, C. M., (1961). *Mod. Plast.*, **39**, 4, 97   10.5.2

BRUNI, G. and ROMANI, E., (1920). Ital. Pats 173 322; 173 364; Germ. Pat. 380 774   1.6

BRYANT, C. L., (1962). *Kautschuk Gummi*, **15**, 29   4.10.5

BRYDSON, J., (1969). *Rubb. Plast. Age*, **50**, 778   2.5.2

BRYDSON, J., (1970). *Polym. Age*, **1**, 1, 17   2.5.2

BRZENK, F. and BOOTH, D. A., (1968). *Jl IRI*, **2**, 217   4.5.3

BUCHAN, S., (1959). *Rubber to Metal Bonding* (2nd edn), Lockwood, London, p. 119   10.5.3.1

BUCHANAN, R. A., WEISLOGEL, O. E., RUSSELL, C. R. and RIST, C. E., (1968). *Ind. Engng Chem.—Product Res. Dev.* **7**, 2, 155   7.1

BUCKLER, E. J. and KRISTENSEN, I. M., (1967). *Jl. IRI*, **1**, 28   9.2.4

BUCKLEY, D. J., (1959). *Rubb. Chem. Technol.*, **32**, 1475   4.2.1.4, 4.5.2

BUECHE, A. M. *See* Safford, M. M.   4.13.5

BUECHE, F., (1959). *Rubb. Chem. Technol.*, **32**, 1269   3.3.5

BULGIN, D. and HUBBARD, G., (1958). *IRI Trans.*, **34**, 201   11.3.13

BURGESS, K. A., HIRSHFIELD, S. M. and STOKES, C. A., (1965). *Rubb. Age, N.Y.*, **97**, 9, 85   8.2.5

BURKE, O. W., (1965). *Reinforcement of Elastomers* (Ed. G. Kraus), Interscience, New York, p. 494   7.1

BURRELL, H., (1955). *Interchem. Rev.* **14**, 3   9.2.3.1

BURRIDGE, K. G. *See* Bennett, D. A.   1.3

CABLE, C. L. *See* Leeper, H. M.   7.4.2

CABOT CORPN (1953). Brit. Pat. 699 406   1.5

CAIN, M. E., KNIGHT, G. T., LEWIS, P. M. and SAVILLE, B., (1968). *Rubb. J.*, **150**, 11, 10   6.5.1

CALCOTT, W. S. and DOUGLASS, W. A., (1932). US Pat. 1 874 895   1.7

CALLAHAN, J. *See* Veatch, F.   4.10.2

CALLAN, J. E. *See* Hess, W. H.   7.10.3

CAPITO, J. E., INNES, F. and ALLEN, R. D., (1968). *Jl IRI*, **2**, 228   4.6.3

CARGAL, J. M., (1966). *IRI Proc.*, **13**, 232   8.2.7

CARLSON, R. H., (1961). *Am. Inst. elect. Engrs, 2nd Symp. on Butyl Rubber for Wire*   10.8.2.1
    and Cable Insulation

CHALFANT, D. L. *See* Schoenberg, E.   4.3.2

CHAPMAN, R., (1966). *Rubb. Plast. Age*, **47**, 7, 769   8.5.2.3

CHAPMAN, R., (1967). *Proc. int. Rubb. Conf.*, 405   6.7.2.2

CHAPMAN, R., (1968). *Jl IRI*, **2**, 233   2.7

CHARLESBY, A. *See* Baker, D.   4.13.5

CHEN, N. Y., (1966). *Jnl Polymer Sci.*, **B4**, 891   4.1.7.1

CHESNOKOV (1933). *Zh. rezin. Prom.*, **9**, 400   1.7

CLAXTON, W. E., (1959). US Pat. 2 904 994   5.5.2.3

CLAXTON, W. E. *See* Conant, F. S.   11.2.2

CLAYTON, R. E. and KOLEK, R. L., (1965). *Rubb. Wld, N.Y.*, **151**, 5, 95   2.7

CLIFTON, R. G., (1969). *Jl IRI*, **3**, 270   10.1.3

CLOUTIER, J. R., (1963). *Chemy. Can.*, **15**, 3, 15   10.8.2.1

COHAN, L. H., (1948). *Proc. 2nd Rubb. Technol. Conf.*, 365   7.8

COLLINS, E. A. and OETZEL, J. T., (1969). *Rubb. Age, N.Y.*, **101**, 5, 62   9.3.1

COLLINS, E. A. and OETZEL, J. T., (1970). *Rubb. Age, N.Y.*, **102**, 3, 64   9.3.1

COLLINS, J. M., JACKSON, W. L. and OUBRIDGE, P. S., (1965). *Rubb. Chem. Technol.*,   6.7.2.6
    38, 400

CONANT, F. S. and CLAXTON, W. E., (1965). *Spec. tech. Publs Am. Soc. Test. Mater.*,   11.2.2
    383, 36

COOK, F. R. *See* Voet, A.   7.4.3.1

CORISH, P. J., (1967). *Rubb. Chem. Technol.*, **40**, 326   9.2.3.1

COTTEN, G. R., (1966). *Rubb. Chem. Technol.*, **39**, 1553   7.8

COTTEN, G. R., (1969). *Kautschuk Gummi Kunststoffe*, **22**, 477   7.6

COX, D. R., (1969-1). *Rubb. J.*, **151**, 4, 49   9.5.1,
     10.5.2

COX, D. R., (1969-2). *Rubb. J.*, **151**, 5, 73   9.5.1,
     10.5.5

COX, D. R., (1969-3). *Rubb. J.*, **151**, 6, 36   9.5.1,
     10.5.3.2,
     10.5.6

CRANK, J., (1956). *Mathematics of Diffusion*, Oxford University Press, Oxford   3.4.3

CREASEY, J. R. and WAGNER, M. P., (1968). *Rubb. Age, N.Y.*, **100**, 10, 72   9.5.3

CRESPI, G. *See* Natta, G.   4.6.2,
     4.6.3

CROCKER, G. J., (1969). *Rubb. Chem. Technol.*, **42**, 30   8.5.2.1

CUNDIFF, R. R. *See* Weissert, F. C.   9.2.4

CUNNEEN, J. I., (1968). *Rubb. Chem. Technol.*, **41**, 182   6.5.3

CUTHBERT, C., (1954). *IRI Trans.*, **30**, 16   5.5.4

D'ADOLF, S. V., (1962). *Rubb. Wld, N.Y.*, **145**, 4, 72   6.4.3.4

D'ADOLF, S. V., (1963). *Rubb. Wld, N.Y.*, **149**, 3, 37   4.13.5

D'AMICO, J. J. *See* Leeper, H. M.   7.4.2

DANNENBERG, E. M., (1966). *Rubb. Age, N.Y.*, **98**, 9, 82; 10, 81   7.4.2

DANNENBERG, E. M. *See* Bolt, T. D.   7.8

DANNENBERG, E. M. *See* Boonstra, B. B.   3.4.5,
     7.8

DANNENBERG, E. M. *See* Brennan, J. J.   7.10.1

DANNENBERG, E. M. and BOONSTRA, B. B., (1955). *Ind. Engng Chem.*, **47**, 339   7.4.1,
     7.4.3.2

DANNENBERG, E. M. and BRENNAN, J. J., (1966). *Rubb. Chem. Technol.*, **39**, 597   7.10.1

DANNENBERG, E. M. and OPIE, W. H., (1958). *Rubb. Wld, N.Y.*, **138**, 85; 89   7.4.2

DARUWALLA, F. J. *See* Wood, J. O.   6.7.2.2

DAUBENBERGER, C. J., (1961). US Pat. 3 145 422   8.6.5.3

DAUGHERTY, J. P. and KEHN, J. T., (1968). *Rubb. Age, N.Y.*, **100**, 12, 47   10.3.3.2

DAVIS, C. C., (1951). *India Rubb. Wld*, **123**, 4, 433   1.6

DAVISON, S. *See* Bishop, E. T.   4.7.2

DE BOER, J. H., (1953). *The Dynamical Character of Adsorption*, Clarendon Press, Oxford   7.9

DE BOER, J. H., LINSEN, B. G. and OSINGA, T. J., (1965). *Jnl Catal.* **4**, 643   7.4.3.2

DEBRUYNE, N. A., (1953). *Research, Lond.*, **6**, 362   8.7.5

DECKER, G. E. *See* Wise, R. W.   11.2.2

DECKER, G. E. and ROTH, F. L., (1953). *India Rubb. Wld*, **128**, 339   11.2.1.1

DECKER, G. E., WISE, R. W. and GUERRY, D., (1962). *Rubb. Wld, N.Y.*, **147**, 68   5.5.2.4

DECREASE, W. M., (1960). *Rubb. Age, N.Y.*, **87**, 1013   10.5.5

DECREASE, W. M. and SCHAFER, J. H., (1966). US Pat. 3 282 883   10.5.3.2

DE FOURCROY, A. F., (1791). *Annln Chem.*, **11**, 225   1.3

DELGATTO, J., (1965). *Rubb. Wld, N.Y.*, **152**, 5, 69   4.13.4

DELGATTO, J., (1969). *Rubb. Wld, N.Y.*, **160**, 3, 41   2.8.3

DENECOUR, R. L. *See* Morton, M.   7.4.2

DESIENO, R. P., (1961). *Rubb. Wld, N.Y.*, **144**, 1, 75   4.13.5

DESIENO, R. P. and FUHRMAN, R. C., (1962). *Rubb. Age, N.Y.*, **91**, 84   4.13.4

D'IANNI, J. D., (1961). *Rubb. Chem. Technol.*, **34**, 367   4.3.2

DIBBO, A. A., (1966). *IRI Trans.*, **42**, 154   5.2.2

DICKIE, R. A. *See* Smith, T. L.   4.7

DIEM, H. E., TUCKER, H. and GIBBS, C. F., (1961). *Rubb. Chem. Technol.*, **34**, 191   4.3.2

DILLON, J. H. and JOHNSTON, N., (1933). *Physics*, **4**, 225   3.2.1

DIXON, S., REXFORD, D. R. and RIGG, J. S., (1957). *Ind. Engng Chem.*, **49**, 1687   1.4.10.1

DOAK, K. W. *See* Armstrong, R. T.   5.2.2

DOAK, K. W. *See* Auer, E. E.   9.4.5

DOBBIN, R. E. *See* Bolt, T. D.   7.8

DODWELL, R. C. *See* Le Nir, V. L.   10.8.3.3

DOEDE, C. M., (1961). *Rubb. Age, N.Y.*, **90**, 99   8.10.6

DOUGLASS, W. A. *See* Calcott, W. S.   1.7

DOYLE, G. M. and VAUGHAN, G., (1968). *Eur. Polym. Jnl*, **4**, 217   4.15.2

DRAKE, R. S. and MCGARTHY, W. J., (1968). *Rubb. Wld, N.Y.*, **159**, 1, 51   2.8.3

DUCK, E. W., (1968). *Chemy Ind.,* 219; 254; 286   2.8.3

DUCK, E. W. and LOCKE, J. M., (1968). *Jl IRI*, **2**, 223   4.4.2

DUMMER, W. *See* Scheele, W.   5.2.2, 5.3

DUNKER, H. C. L., (1952). Brit. Pat. 700 544   10.7.2

DUNLOP, J. B., (1888). Brit. Pat. 10 607   1.9

DUNLOP, J. B., (1889). Brit. Pat. 4115   1.9

DUNLOP RUBBER CO. LTD (1929). Brit. Pats 332 525; 332 526   1.3, 2.4.3.4

DUNLOP RUBBER CO. LTD (1967). Brit. Pat. Application 56105/67   10.2.1.3

DU PONT CO. LTD (1933-1). Brit. Pats 384 654; 395 131   1.4.5

DU PONT CO. LTD (1933-2). Brit. Pat. 387 340   1.4.5

DU PONT CO. LTD (1960). Brit. Pat. 838 281   1.4.10.1

DU PONT CO. LTD (1961). US Pat. 2 567 117   1.4.5

EAGLES, E. A. and PAYNE, A. R., (1957). *Rubb. Plast. Age*, **38**, 811   11.3.2

EATON, E. R. and MIDDLETON, J. S., (1965). *Rubb. Wld, N.Y.*, **152**, 3, 94   6.1.3.2

ECKER, R., (1959). *Kautschuk Gummi*, **12**, 351   9.4.1

ECKERT, F. J. *See* Scott, G. E.   4.2.1.4, 8.2.5

EDMONDSON, H. M., (1969). *IRI Conf. on rec. Dev. in Rubb. Compounding*, Manchester   8.2.1.2

EDWARDS, D. C. *See* Briggs, G. J.   10.8.2.3

EDWARDS, D. C. and STOREY, E. B., (1957). *Rubb. Chem. Technol.*, **30**, 122   9.4.1

EINHORN, S. C. and TURETZKY, S. B., (1964). *J. appl. Polym. Sci.*, **8**, 1257   3.2.1

EISENBERG, A. *See* Shen, M. C.   2.1

EISENSCHITZ, R., RABINOWITSCH, B. and WEISSENBERG, K., (1929). *Mitt. dt. Mater-PrüfAnst.*, **9**, 91   3.2.1

ELLIOTT, R. *See* Borroff, E. M.   11.3.19

ENDTER, F. W., (1952). *Kautschuk Gummi*, **5**, WT17   7.8

ENDTER, F. W., (1954). *Rubb. Chem. Technol.*, **27**, 1   7.8

ESSO RESEARCH (1960). Brit. Pat. 831 989   1.4.9

ESSO RESEARCH (1961). Brit. Pat. 868 022   1.4.9
ESTEVEZ, J. M. J., (1969). *Plast. Polym.*, **37**, 235   9.4.10
EVANS, B. B., (1953). *IRI Trans.*, **29**, 42   10.8.2,
                                                                              10.8.3.2
EVANS, C. W., (1968-1). *Proc. 3rd int. Lead Conf.*, Venice   10.3.3
EVANS, C. W., (1968-2). *Rubb. Wld, N.Y.*, **159**, 2, 41   10.3.3
EVANS, C. W., (1969-1). *Proc. S.G.F., Finl.*, **35**   10.3.2
EVANS, C. W., (1969-2). *Rubb. Chem. Technol.*, **42**, 984   10.3.3
EVANS, C. W., (1969-3). *Design Engng*, Oct., 133   10.3.3
EVANS, C. W., (1969-4). *Rubb. Age, N.Y.*, **101**, 9, 61   8.2.4
EVANS, C. W. and MELSOM, G. T., (1968). *Proc. Fluid Pwr int. Conf.*, London   10.3.3.2
EXLEY, K. *See* Blow, C. M.   10.9.5.4

FARBENINDUSTRIE, I. G., (1930). Brit. Pat. 339 225   1.4.2
FARBENINDUSTRIE, I. G., (1940). US Pat. 2 203 875   1.4.7
FARREL BIRMINGHAM (1953). Brit. Pat. 695 312   8.4.1.3
FARREL CORPN (1965). Brit. Pat. 998 725   8.3.2
FERRY, J. D., (1961). *Viscoelasticity of Polymeric Materials*, Wiley, New York   3.3.4
FERRY, J. D. *See* Williams, M. L.   3.3.4
FERRY, J. D., FITZGERALD, E. R., GRANDINE, L. D. and WILLIAMS, M. L., (1952). *Ind.*   3.3.4
   Engng Chem., **44**, 703
FETTES, E. M. and JORCZAK, J. S., (1950). *Ind. Engng Chem.*, **42**, 2217   1.4.6
FIRESTONE TIRE & RUBBER CO. (1942). US Pat. 2 271 834   1.6
FIRESTONE TIRE & RUBBER CO. (1947). US Pat. 2 415 029   1.6
FIRESTONE TIRE & RUBBER CO. (1949). US Pat. 2 459 736   1.6
FIRESTONE TIRE & RUBBER CO. (1957). Ital. Pats 559 160; 559 705; 560 200; 561 343   1.4.9
FISHER, H. L., (1957). *Chemistry of Natural and Synthetic Rubbers*, Reinhold, New   5.4.1
   York, p. 100
FISKE, R. L., (1961). *Adhes. Age*, **4**, 10, 33   10.5.2
FITZGERALD, E. R. *See* Ferry, J. D.   3.3.4
FLETCHER, W. *See* Aston, M. W.   10.9.5.4
FLEUROT, M., (1965). *Revue gén. Caoutch. Plast.*, **42**, 873   4.1.2.3
FLINT, C. F., (1962). *Proc. 4th Rubb. Technol. Conf.*, 107   9.4.1
FLORY, P. J., RABJOHN, N. and SHAFFER, M. C., (1949). *J. Polym. Sci.*, **4**, 225; 435   5.2.2,
                                                                              5.3
FLORY, P. J. and REHNER, J., (1943). *J. Chem. Phys.*, **11**, 512   3.3
FOLKMAN, D., LONG, D. and ROSENBAUM, R., (1966). *Science, N.Y.*, **154**, 148   4.13.4
FORD, F. P. *See* Gessler, A. M.   4.5.3,
                                                                              9.4.1
FOSGATE, C. M., (1968). *Rubb. Age., N.Y.*, **100**, 8, 58   10.9.3.2
FOUNTAIN, R. *See* Peer, S.   10.5.7
FROST, J. T. and LIVERPOOL ELECTRIC CABLE CO. (1939). Brit. Pat. 526 895   10.8.4
FUHRMAN, R. L. *See* DeSieno, R. P.   4.13.4

GAMMETER, J. R. *See* Johnson, J. T.   1.9
GARDNER, E. R., (1969). *Proc. IRI Conf. on Adv. Polym. Blends and Reinforcement*,   6.7.2.2
   Loughborough
GARVEY, B. S. and ROSTLER, K. S., (1968). *Rubb. Age, N.Y.*, **100**, 11, 59   9.2.4
GASKINS, F. H. *See* Philipoff, W.   3.2.2
GAUTIER, C. M., (1910). US Pat. 978 731   1.9
GAZZANA, P. *See* Marchesini, G.   10.8.2.2,
                                                                              10.8.2.3
GEHMAN, S. D. and OGILBY, S. R., (1965). *Spec. tech. Publs Am. Soc. Test. Mater.*, **383**,   11.2.2
   3
GELDART, D., (1969). *Chemy Ind.*, **11**, 311   8.9.2.4
GENERAL ELECTRIC CORPN (1953). Brit. Pat. 755 762   4.13.5
GENERAL TIRE & RUBBER CO., THE, (1955). Brit. Pat. 737 086   1.4.4
GENT, A. N., (1962). *J. appl. Polym. Sci.*, **6**, 433   3.3.3
GERMAN, R., VAUGHAN, G. and HANK, R., (1967). *Rubb. Chem. Technol.* **40**, 569   4.6.2
GESCHWIND, D. H., (1969). *Rubb. Age., N.Y.*, **101**, 1, 59   8.6.6
GESSLER, A. M., (1964). *Rubb. Age, N.Y.*, **94**, 598, 750   7.4.2

GESSLER, A. M., (1967). *Proc. int. Rubb. Conf.*, 249 — 7.4.3.1, 7.8

GESSLER, A. M., ZAPP, R. L., FORD, F. P. and REHNER, J., (1953). *Rubb. Age, N.Y.*, **74**, 59, 243; 397; 561 — 4.5.3, 9.4.1

GIBBS, C. F. *See* Diem, H. E. — 4.3.2

GIBSON, P. R. S. and PRATT, C. P., (1969). *Rubb. J.*, **151**, 9, 33 — 9.4.1

GILETTE, H. G., (1967). *Rubb. Wld, N.Y.*, **157**, 1, 67 — 8.6.5.3

GLAGOLER, V. A., IL'M, N. S. and KOSHELER, F. F., (1967). *Soviet Rubb. Technol.*, **26**, 2, 25 — 8.7.4

GOODMAN, L. T. *See* Woodcock, W. J. — 10.5.6

GOODRICH CO., B. F., (1933). US Pat. 1 919 361 — 8.3.2

GOODSELL, D. L., (1967). *Engng Mater. Des.*, **10**, 4, 537 — 10.9.5.4

GOODYEAR TIRE & RUBBER CO., (1961). Brit. Pat. 883 645 — 1.9

GOTTLOB, K. *See* Hofmann, F. — 1.6

GOWER, B. G. *See* Moore, R. A. — 2.8.3

GOY, R. S. *See* Wood, J. O. — 6.7.2.2

GOY, R. S. and MÖRING, P. L. E., (1964). *IRI Trans.*, **40**, T176 — 6.7.2.5

GRAHAM, D. J. and MORRIS, J. E., (1968). *Plrs' Bull. Rubb. Res. Inst. Malaya*, **99**, 200 — 4.1.2.3

GRANDINE, L. D. *See* Ferry, J. D. — 3.3.4

GRAY, C. H. and SLOPER, T., (1903). Brit. Pats 10 941; 10 942; 10 943 — 1.9'

GRAY, C. H. and SLOPER, T., (1913). Brit. Pat. 467 — 1.9

GRAY, H. *See* Winkelmann, H. A. — 1.7

GRAY, M., (1879). Brit. Pat. 5056 — 1.8.3

GREENSMITH, H. W., (1964). *J. appl. Polym. Sci.*, **8**, 1113 — 3.3.5.1

GREENSMITH, H. W. and THOMAS, A. G., (1955). *J. Polym. Sci.*, **18**, 189 — 3.3.5.1, 9.4.4.2

GREGORY, R., (1966). *Rubb. J.*, **148**, 9, 66; 10, 69 — 8.9.2

GRIFFIS, C. B. *See* Montermoso, J. — 4.18.1

GRIFFITH, A. A., (1921). *Phil. Trans. R. Soc.*, **A221**, 163 — 3.3.5.1

GRIJN, F. VAN DER. *See* Krol, L. H. — 9.3.1

GRIJN, J. VAN DER. *See* Krol, L. H. — 4.3.3

GRÜBER, E. E., BRIGGS, R. A. and MEYER, D. A., (1963). *Atti Congr. int. Plast. e Elastomer*, Turin, 315 — 4.15.2

GRÜBER, E. E., MEYER, D. A., SWART, G. H. and WEINSTOCK, K. V., (1964). *Ind. Engng Chem.—Product Res. Dev.*, **3**, 194 — 4.15.2

GUERRY, D. *See* Decker, G. E. — 5.5.2.4,

GUTH, E., (1948). *Proc. 2nd Rubb. Technol. Conf.*, 353 — 7.8

GUTH, E. and JAMES, H. M., (1943). *J. Chem. Phys.*, **11**, 455 — 3.3

GYSLING, H. and HELLER, H. J., (1961). *Kunststoffe*, **51**, 13 — 9.4.6.4

HABECK, B. W. *See* Martin, G. E. — 10.9.2

HAGMAN, J. F. *See* Thompson, D. C. — 9.4.6.2

HALBROOK, N. *See* Mueller, W. J. — 9.3.6

HALL, E. F., (1967). SATRA Memo. No. TM1353 — 10.4.3.4

HALPIN, J. C., (1964). *J. appl. Phys.*, **35**, 3133 — 3.3.5.2

HANK, R. *See* German, R. — 4.6.2

HANNSGEN, F. W., (1961). *Rubb. Plast. Age*, **42**, 166 — 4.3.3

HARKLEROAD, W. I., (1969). *Rubb. Chem. Technol.*, **42**, 666 — 10.3.3

HARLING, D. F. and HECKMAN, F. A., (1969). *Materie plast. ed Elastomeri*, **35**, 80 — 6.1.3.3, 7.4.3.2

HARRINGTON, R., (1957). *Rubb. Age, N.Y.*, **81**, 971; **82**, 461 — 2.9.3.9

HARRINGTON, R., (1958). *Rubb. Age, N.Y.*, **82**, 1003; **83**, 472 — 2.9.3.9

HARTMANN, N. E. and BEAUMONT, R. A., (1968). *Jl IRI*, **2**, 272 — 9.2.4

HARWOOD, J. A. C. *See* Mullins, L. — 7.10.1

HARWOOD, J. A. C. and PAYNE, A. R., (1968). *J. appl. Polym. Sci.*, **12**, 889 — 3.3.6

HARWOOD, J. A. C., PAYNE, A. R. and SMITH, J. F., (1969). *Kautschuk Gummi*, **10**, 548 — 3.3.2

HARWOOD, J. A. C., PAYNE, A. R. and WHITTAKER, R. W., (1969). *Proc. IRI Conf. on Adv. Polym. Blends and Reinforcement*, Loughborough — 3.3.6

HARWOOD, J. A. C. and SCHALLAMACH, A., (1967). *J. appl. Polym. Sci.*, **11**, 1835 — 3.3.2

HAZELDINE, R. W. *See* Barr, D. A. — 4.18.1

HEAL, C. J. A., (1963). *IRI Trans.*, **39**, 262 — 9.4.5

HEALY, J. C. *See* Morton, M. 7.4.2

HEAP. R. D. and NORMAN, R. H., (1966). RAPRA Tech. Rev. No. 36   5.5.4, 10.8.3.3

HECKMAN, F. A. *See* Harling, D. F.   6.1.3.3, 7.4.3.2

HECKMAN, F. A. and MEDALIA, A. I., (1969). *Jl IRI*, **3**, 66   7.4.1, 7.4.3.1

HEIDEMANN, W. *See* Peter, J. 5.5.2.1

HELLER, H. J. *See* Gysling, H. 9.4.6.4

HENTEN, K. VAN. *See* Krol, L. H. 9.3.1

HERCULES POWDER CO. LTD (1961). Brit. Pat. 857 938 1.4.9

HERCULES POWDER CO. LTD (1962). Brit. Pat. 902 385 1.4.9

HESS, W. H., SCOTT, C. E. and CALLAN, J. E., (1967). *Rubb. Chem. Technol.*, **40**, 371   7.10.3

HICKMAN, J. A. *See* Norman, R. H. 11.2.2

HICKS, A. E. and LYON, F., (1969). *Adhes. Age*, **12**, 5, 21 10.5.5

HILL, P., (1969). *Jl IRI*, **3**, 174   6.5.3, 9.4.6.1

HIRSCHFIELD, S. M. *See* Burgess, K. A. 8.2.5

HOFMANN, F. and GOTTLOB, K., (1914). US Pat. 1 149 580 1.6

HOFMANN, W., (1963). *Rubb. Chem. Technol.*, **36**, No. 5 4.10.1

HOFMANN, W., (1964). *Rubb. Chem. Technol.*, **37**, No. 2   4.10.1, 4.10.4

HOHENEMSER, W. *See* Meyer, K. H. 5.2.2

HOLDEN, G., BISHOP, E. T. and LEGGE, N. R., (1967). *Proc. int. Rubb. Conf.*, 287 4.7.2

HOPKINS, G. *See* Blow, C. M. 10.5.5

HOPPER, J. R., (1967). *Rubb. Chem. Technol.*, **40**, 463 3.2.5

HORN, J. B., (1968). *Rubb. J.*, **150**, 7, 13 6.1.6

HORN, J. B., (1969). *Rubb. Plast. Age*, **50**, 457   6.1.3.2, 6.1.6

HORNE, S. E., *et al.* (1956). *Ind. Engng Chem.*, **48**, 784   1.4.9, 4.2.2

HOUWINK, R. and ALPHEN, J. VAN, (1955). *J. Polym. Sci.*, **16**, 121 7.1

HOWARTH, H. *See* Thompson, C. W. 4.1.2.3

HSIEH, H. L., (1965). *Rubb. Plast. Age*, **46**, 394 4.2.2.1

HUBBARD, G. *See* Bulgin, D. 11.3.13

HUGHES, B. G., IZOD, D. A. W., MOAKES, R. C. W., SODEN, A. L. and WATSON, W. F., (1962). *Proc. 4th Rubb. Technol. Conf.*, 1   9.3.5, 10.8.3.3

HUNT, T. *See* Southwart, D. W.   2.6.2.1, 4.13.5

HURST, N. H. and JENKINS, D. L., (1954). Brit. Pat. 777 959 10.7.3.1

IDDON, M. I., (1965). *IRI Proc.*, **12**, 230 8.9.2

IDOL, J. D. *See* Veatch, F. 4.10.2

I. G. FARBENINDUSTRIE. *See* Farbenindustrie, I. G.   1.4.2, 1.4.7

IL'M, N. S. *See* Glagoler, V. A. 8.7.4

IMPERIAL CHEMICAL INDUSTRIES LTD, LAMBERT, A. and WILLIAMS, G. E., (1955). Brit. Pat. 723 838 1.7

INGLIS, C. E., (1913). *Trans. Instn nav. Archit.*, **60**, 219 3.3.5.1

INNES, F. *See* Capito, J. E. 4.6.3

IYENGAR, Y., (1963). *Rubb. Wld, N.Y.*, **148**, 6, 39 8.7.5

IYENGAR, Y., (1969). *J. appl. Polym. Sci.*, **13**, 2, 353 8.7.5

IZOD, D. A. W. *See* Hughes, B. G.   9.3.5, 10.8.3.3

IZOD, D. A. W. and MEARDON, J. T., (1960). RABRM Res. Rep. No. 100 8.5.2.5

JACKSON, W. L. *See* Collins, J. M. 6.7.2.6

JACOBS, H. L., (1969). *Am. chem. Soc. Rubb. Div. Meeting*, Los Angeles 8.2.8

JAMES, H. M. *See* Guth, E. 3.3

JANDIAL, A. P. and SPADE, R. L., (1967). *Proc. 7th NEMA–IEEE electl Insulation Conf.*, 226 10.8.2.2

494   References

JANSSEN, H. J. J. and WEINSTOCK, K. V., (1961). *Rubb. Chem. Technol.*, **34**, 1485   4.2.1.4

JENKINS, D. L. *See* Hurst, N. H.   10.7.3.1

JERMYN, T. E. *See* Brennan, J. J.   7.8

JOHNSON, J. T. and GAMMETER, J. R., (1916). US Pat. 1 177 112   1.9

JOHNSON, P. R. *See* Maynard, J. T.   5.4.1.2

JOHNSTON, N. *See* Dillon, J. H.   3.2.1

JONES, B. D., (1962). *Proc. 4th Rubb. Technol. Conf.*, 7   1.3

JONES, K. P., (1968). *Jl IRI*, **2**, 194   9.4.6.6

JORCZAK, J. S. *See* Fettes, E. M.   1.4.6

JURGELEIT, H. F., (1962). *Rubb. Age, N.Y.*, **90**, 5, 763   8.6.5.3

JURGELEIT, H. F., (1964). Brit. Pat. 1 022 084   8.6.5.3

JUVE, A. E. and BEATTY, J. R., (1954). *Rubb. Wld, N.Y.*, **131**, 62   8.6.7

JUVE, A. E. and BEATTY, J. R., (1955). *Rubb. Chem. Technol.*, **28**, 1141   8.6.7

JUVE, A. E., KARPER, P. W., SCHROYER, L. D. and VEITH, A. G., (1963). *Rubb. Wld, N.Y.*,   5.5.2.5
  149, 43

KARL FRANK, GMBH, (1969). Trade Lit.   11.2.2

KARPER, P. W. *See* Juve, A. E.   5.5.2.5

KEHN, J. T. *See* Daugherty, J. P.   10.3.3.2

KEILEN, J. J. and POLLAK, A., (1947). *Ind. Engng Chem.*, **39**, 480   7.1

KEMP, A. R. *See* Selker, M. L.   5.2.2

KHROMOV, M. K. and REZNIKOVSKII, M. M., (1960). *Physico-mechanical Testing of*   11.3.2
  Unvulcanised and Vulcanised Rubber (translated R. J. Moseley, R. A. Amos, and
  J. R. Scott), Maclaren, London, p. 109

KILLEFFER, D. H., (1962). *Banbury—The Master Mixer*, Palmerton, New York   1.8.1

KIPPING, F. S., (1937). *Proc. R. Soc.*, **A159**, 193   1.4.10.2

KIPPING, F. S., *et al.* (1923). *J. chem. Soc.*, **123**, 2590   1.4.10.2,
  4.13.1

KLEIN, P., (1923). Brit. Pats 223 188; 223 189   1.3

KLINE, R. H. *See* Ambelang, J. C.   6.5.3

KLÖTZER, E., (1969). *Jl IRI*, **3**, 129   9.5.3

KNIGHT, G. T. *See* Cain, M. E.   6.5.1

KOLEK, R. L. *See* Clayton, R. E.   2.7

KONKLE, G. M. *See* Youngs, D. C.   4.13.4

KONRAD, E., (1930). Brit. Pat. 360 821   4.10.1

KORZUN, E. A., (1962). *Rubb. Age, N.Y.*, **91**, 5, 777   4.13.5

KOSHELER, F. F. *See* Glagoler, V. A.   8.7.4

KOVASIC, P., (1955). *Ind. Engng Chem.*, **47**, 1090   4.8.2

KRAUS, G., (1965). *Reinforcement of Elastomers*, Interscience, New York, p. 232   3.2.5

KRISTENSEN, I. M. *See* Buckler, E. J.   9.2.4

KROL, L. H., (1968). *Rubb. J.*, **150**, 12, 20   4.2.2.1,
  4.4.3

KROL, L. H., GRIJN, F. VAN DER and HENTEN, K. VAN, (1964) *IRI Proc.*, **11**, 42   9.3.1

KROL, L. H., VERKERK, G. and GRIJN, J. VAN DER, (1963). *Rubb. Plast. Age*, **44**, 284   4.3.3

KUNCL, K. L. *See* Moore, R. A.   2.8.3

KUZMINSKII, A. S., (1966). *Rubb. Chem. Technol.*, **39**, 88   6.5.3

KYLE, J. W., (1968). *Jl IRI*, **2**, 133   10.2.1

LAAN, R. H. VAN DER, (1969). Hypalon Bull. No. 3A, Du Pont   4.9.3

LAKE, G. J. and LINDLEY, P. B., (1964). *Rubb. J.*, **146**, 10, 24; 79   10.5.6

LAMBERT, A. *See* Imperial Chemical Industries Ltd   1.7

LANDEL, R. F. *See* Williams, M. L.   3.3.4

LANDEL, R. F. and STEDRY, P., (1960), *J. appl. Phys.*, **31**, 1885   3.3.2

LANE, J. H. *See* Bahary, W. S.   4.4.1,
  4.4.3

LASMAN, H. R., (1965). *Encyclopaedia of Polymer Science, Vol. 2*, Interscience,   10.6.2.1
  New York, p. 532

LATOS, E. J. and SPARKS, A. K., (1969). *Rubb. J.*, **151**, 6, 18   6.5.1

LAWRENCE, R. *See* Mueller, W. J.   9.3.6

LEACH, W. R., (1966). *Rubb. Age, N.Y.*, **98**, 8, 71   4.15.1

LEBRAS, J. and PICCINI, I., (1950). *Bull. Soc. chim. Fr.*, **215**   7.1

LEBRAS, J. and PICCINI, I., (1951-1). *Ind. Engng Chem.*, **43**, 381   7.1

LEBRAS, J. and PICCINI, I., (1951-2). *Rubb. Chem. Technol.*, **24**, 649   7.1

LEEPER, H. M., CABLE, C. L., D'AMICO, J. J. and TUNG, C. C., (1956). *Rubb. Wld, N.Y.*,   7.4.2
    135, 413

LEGGE, N. R. *See* Holden, G.   4.7.2

LEHR, M. H., (1963). *Rubb. Chem. Technol.*, **36**, 1571   4.3.2

LE NIR, V. L. and DODWELL, R. C., (1968). *Wire. J.*, Oct., 90   10.8.3.3

LESER, W. H. *See* Weir, G. E.   4.13.4

LEVINE, N. B., (1969). *Rubb. Age, N.Y.*, **101**, 5, 45   4.18.2

LEWIS, J. T. *See* Turner, M. J.   4.13.4

LEWIS, P. M. *See* Cain, M. E.   6.5.1

LINDLEY, P. B. *See* Allen, P. W.   3.4.4

LINDLEY, P. B. *See* Lake, G. J.   10.5.6

LINSEN, B. G. *See* de Boer, J. H.   7.4.3.2

LITTLE, J. R. *See* Armstrong, R. T.   5.2.2

LIVERPOOL ELECTRIC CABLE CO. *See* Frost, J. T.   10.8.4

LOAN, L. D., MURRAY, R. W. and STORY, P. R., (1968). *Jl IRI*, **2**, 73   6.5.3

LOCKE, J. M. *See* Duck, E. W.   4.4.2

LOMBARDI, D. *See* Meyrick, C. I.   2.4.3.1

LONG, D. *See* Folkman, J.   4.13.4

LORD MANUFACTURING CO. (1958). Brit. Pat. 806 449   10.5.3.2

LORD MANUFACTURING CO. (1959). Brit. Pat. 822 725   10.5.3.2

LORENZ, O. *See* Parks, C. R.   7.9

LORENZ, O. *See* Scheele, W.   5.2.2,
    5.3

LORENZ, O. M. *See* Ambelang, J. C.   6.5.3

LUKOMSKAYA, A. I., (1960). *Physico-mechanical Testing of Unvulcanised and Vul-*   11.3.6
    canised Rubber (translated R. J. Moseley, R. A. Amos, and J. R. Scott),
    Maclaren, London, p. 19

LYON, F. *See* Hicks, A. E.   10.5.5

MAJOR, R. H. *See* Schoenberg, E.   4.3.2

MARCHESINI, G. and GAZZANA, P., (1962). *Proc. 4th Rubb. Technol. Conf.*, 162   10.8.2.2,
    10.8.2.3

MARK, H. *See* Meyer, K. H.   1.4.1

MARSH, P. A. *See* Voet, A.   7.4.3.1

MARSH, P. A. and VOET, A., (1968). *Rubb. Chem. Technol.*, **41**, 344   7.10.3

MARTIN, G. E., SELL, H. S. and HABECK, B. W., (1950). *Rubb. Age, N.Y.*, **66**, 409   10.9.2

MARTIN, G. M. *See* Wood, L. A.   2.9.3.11

MARTIN, G. M., ROTH, F. L. and STIEHLER, R. D., (1956). *IRI Trans.*, **32**, 189   3.3.2

MARZETTI, B., (1923). *India. Rubb. Wld*, **68**, 776   3.2.1

MASON, J., (1953). *IRI Trans.*, **29**, 148   10.8.3.3

MATSUI, J. *See* Takeyama, T.   8.5.2.1

MATTHEWS, F. E. and STRANGE, E. H., (1910). Brit. Pat., 24 790   1.4.2

MATUSIK, F. H., (1968). *Adhes. Age*, **11**, 12, 32   10.5.7

MAYNARD, J. T. and JOHNSON, P. R., (1963). *Rubb. Chem. Technol.*, **36**, 883; 963   5.4.1.2

MAYNARD, J. T. and MOCHEL, W. E., (1954). *J. Polym. Sci.*, **13**, 251   4.8.2

MAZZANTI, G., (1969). *Chemy Ind.*, **35**, Aug. 30, 1204   1.4.9

MAZZANTI, G. *See* Natta, G.   4.6.2

MCALEVY, A. *See* Brooks, R. E.   4.9.1

MCALPINE, A. *See* Thomson, R. N.   8.3.2

MCGARTHY, W. J. *See* Drake, R. S.   2.8.3

MCPHERSON, A. T., (1963). *Rubb. Chem. Technol.*, **36**, 1263   9.4.8,
    10.8.2.2

MEARDON, J. T. *See* Izod, D. A. W.   8.5.2.5

MEDALIA, A. I., (1961). *Rubb. Chem. Technol.*, **34**, 1134   7.7

MEDALIA, A. I., (1967). *Jnl colloid interface Sci. (U.S.)*, **24**, 393   7.4.3.1

MEDALIA, A. I., (1970). *Jnl colloid interface Sci. (U.S.)*, **32**, 115   7.4.3.1

MEDALIA, A. I. *See* Boonstra, B. B.   7.7

MEDALIA, A. I. *See* Heckman, F. A.   7.4.1,
    7.4.3.1

MEER, S. VAN DER, (1944). *Meded. RubbSticht.*, **47**, **48** 5.3
MEIER, D. J., (1967). *Am. chem. Soc. Symp. on Block Copolymers*, California 4.7.2
MELSOM, G. T. *See* Evans, C. W. 10.3.3.2
MEREDITH, R., (Ed.) (1956). *Mechanical Properties of Textile Fibres, North-Holland*, Amsterdam 6.7.2.6
MEYER, D. A. *See* Grüber, E. E. 4.15.2
MEYER, K. H. and HOHENEMSER, W., (1935). *Helv. chim. Acta*, **18**, 1061 5.2.2
MEYER, K. H. and MARK, H., (1928). *Ber. dt. chem. Ges.*, **61**, 1939 1.4.1
MEYER, K. H., SUSICH, G. VON and VALKO, E., (1932). *Kolloidzeitschrift*, **59**, 208 3.3
MEYRICK, C. I., VAUTIER, J. and LOMBARDI, D., (1969). *Rubb. Dev.*, **22**, 3, 92 2.4.3.1
MIDDLETON, J. S. *See* Eaton, E. R. 6.1.3.2
MILBERGER, E. C. *See* Veatch, F. 4.10.2
MOAKES, R. C. W. *See* Hughes, B. G. 9.3.5, 10.8.3.3
MOCHEL, W. E. *See* Maynard, J. T. 4.8.2
MOLONY, S. B., (1920). US Pat. 1 343 222 1.6
MONSANTO CO. (1940). US Pat. 2 191 657 1.6
MONSANTO CO. (1951). Brit. Pat. 655 668 1.6
MONTECATINI CO. (1960). Brit. Pats 856 736; 856 737 1.4.9
MONTERMOSO, J., GRIFFIS, C. B. and WILSON, A., (1962). *6th Joint Army, Navy and Air Force Conf. on Elastomers, Vol. 2*, 672 4.18.1
MOONEY, M., (1931). *J. Rheol.*, **2**, 210 3.2.1
MOONEY, M., (1934). *Ind. Engng Chem. analyt. Edn*, **6**, 147 3.2.3
MOONEY, M., (1936). *Physics*, **7**, 413 3.2.2
MOONEY, M., (1940). *J. appl. Phys.*, **11**, 582 3.3.2
MOONEY, M., (1958). *Rheology, Vol. 1* (Ed. F. R. Eirch), Academic Press, New York, Ch. 8 3.2.2
MOORE, C. G., (1964). *Proc. NRPRA Jubilee Conf.*, 167 5.2.2, 5.3, 5.6.1
MOORE, C. G. and WATSON, W. F., (1956). *J. Polym. Sci.*, **19**, 237 5.2.2
MOORE, R. A., KUNCL, K. L. and GOWER, B. G., (1969). *Rubb. Wld, N.Y.*, **159**, 5, 55 2.8.3
MORAN, A. L. *See* Radcliffe, R. R. 4.12.1
MORE, A. R., MORRELL, S. H. and PAYNE, A. R., (1959). *Rubb. J. int. Plast.*, **136**, 23, 858 5.5.2.2
MÖRING, P. L. E. *See* Goy, R. S. 6.7.2.5
MORRELL, S. H. *See* Aston, M. W. 10.9.5.4
MORRELL, S. H. *See* More, A. R. 5.5.2.2
MORRIS, J. *See* Baker, D. 4.13.5
MORRIS, J. E. *See* Graham, D. J. 4.1.2.3
MORTON, M., (1959). *Introduction to Rubber Technology*, Chapman & Hall, London 4.5.3
MORTON, M., HEALY, J. C. and DENECOUR, R. L., (1967). *Proc. int. Rubb. Conf.*, 175 7.4.2
MUELLER, N. N. *See* Thompson, D. C. 9.4.6.2
MUELLER, W. J., BENNETT, B., HALBROOK, N., SCHULLER, W. and LAWRENCE, R., (1969). *Rubb. Age, N.Y.*, **101**, 7, 43 9.3.6
MULLINS, L., HARWOOD, J. A. C. and PAYNE, A. R. (1965). *Jnl Polymer Sci.*, **B3**, 1119 7.10.1
MULLINS, L., HARWOOD, J. A. C. and PAYNE, A. R., (1966). *IRI Trans.*, **42**, T14 7.10.1
MULLINS, L. and WHORLOW, R. W., (1951). *IRI Trans.*, **27**, 55 2.6.2.1
MURRAY, R. M. and THOMPSON, D. C., (1964). *The Neoprenes*, Du Pont 4.8.3, 4.8.4, 5.4.2.1
MURRAY, R. W. *See* Loan, L. D. 6.5.3
NAIR, S. and SEKHAR, B. C., (1967). *Spec. Publs chem. Soc.*, **23**, 105 4.1.7.1
NATTA, G., CRESPI, G., VALVASSORI, A. and SARTORI, G., (1963-1). *Rubb. Chem. Technol.*, **36**, 1583 4.6.2, 4.6.3
NATTA, G., MAZZANTI, G., CRESPI, G., VALVASSORI, A. and SARTORI, G., (1963-2). *Rubb. Chem. Technol.*, **36**, 988 4.6.2
NAUNTON, W. J. S., (1930). *IRI Trans.*, **5**, 317 1.7
NAUNTON, W. J. S., (1961). *The Applied Science of Rubber*, Arnold, London 4.1.2.1, 4.1.7.2
NELMS, R. P. *See* Shaw, E. J. 10.8.3.3

NG, T. S. and SCHULZ, G. V., (1969). *Makromolek. Chem.*, **127**, 165 — 4.1.7.1

NORMAN, R. H., (1957). *Conductive Rubber*, Maclaren, London — 10.8.4, 11.3.17

NORMAN, R. H. *See* Heap, R. D. — 5.5.4, 10.8.3.3

NORMAN, R. H., HICKMAN, J. A. and PAYNE, A. R., (1965). *Spec. tech. Publs Am. Soc. Test. Mater.*, **383**, 19 — 11.2.2

NYE, H. F., (1967). *JI IRI*, **1**, 105 — 8.2.8, 10.8.3.1

NYE, H. F. *See* Atkin, W. — 5.5.4, 10.8.3.3

ODENWALD, H. and BAADER, T., (1938). *Proc. Rubb. Technol. Conf.*, 347 — 11.2.1.2

OETZEL, J. T. *See* Collins, E. A. — 9.3.1

OGILBY, S. R. *See* Gehman, S. D. — 11.2.2

OLIVIER, J. P. *See* Ross, S. — 7.9

OPIE, W. H. *See* Dannenberg, E. M. — 7.4.2

OSINGA, T. J. *See* de Boer, J. H. — 7.4.3.2

OSTROMISLENSKY, I. I., (1912). *Zh. russk. fiz.-khim. Obshch.*, **44**, 204 — 5.3

OSTROMISLENSKY, I. I., (1915). *Zh. russk. fiz.-khim. Obshch.*, **47**, 1892 — 1.6

OSTWALD, WO. and WA., (1908). Germ. Pat. 221 310 — 1.6

OUBRIDGE, P. S. *See* Collins, J. M. — 6.7.2.6

PAINTER, G. W., (1954). *Rubb. Age, N.Y.*, **74**, 701 — 4.13.4

PALMER, J. F., (1893). US Pat. 493 220 — 1.9

PARKES, A., (1840). Brit. Pat. 11 147 — 1.2.2, 5.3

PARKES, H. G., (1969). *J. natn. Cancer Inst.*, **43**, 1, 249 — 1.7

PARKS, C. R. *See* Ambelang, J. C. — 6.5.3

PARKS, C. R. and LORENZ, O., (1961). *J. Polym. Sci.*, **50**, 287 — 7.9

PARSHALL, A. M. and SAULINO, A. J., (1967). *Rubb. Wld, N.Y.*, **156**, 2, 78 — 8.2.7

PATTISON, D. A. *See* Beerbower, A. — 9.2.3.1, 9.4.6.3

PAYNE, A. R., (1956). *Revue gén. Caoutch.*, **33**, 10. 885 — 11.3.13

PAYNE, A. R., (1961). *Mater. Res. Stand.*, **1**, 12, 942 — 11.3.13

PAYNE, A. R., (1963). *Rubb. Plast. Wkly*, **144**, 293 — 5.5.2.2

PAYNE, A. R., (1966). *Rubb. Chem. Technol.*, **39**, 365; 915 — 7.4.3.1

PAYNE, A. R. *See* Allen, P. W. — 3.4.4

PAYNE, A. R. *See* Eagles, E. A. — 11.3.2

PAYNE, A. R. *See* Harwood, J. A. C. — 3.3.2, 3.3.6

PAYNE, A. R. *See* More, A. R. — 5.5.2.2

PAYNE, A. R. *See* Mullins, L. — 7.10.1

PAYNE, A. R. *See* Norman, R. H. — 11.2.2

PAYNE, A. R. and SMITH, J. F., (1956). *J. scient. Instrum.*, **33**, 11, 432 — 11.3.2

PAYNE, A. R. and WATSON, W. F., (1963). *IRI Trans.*, **39**, T125 — 7.4.3.1

PEACHEY, S. J. and SKIPSEY, A., (1921). *J. Soc. chem. Ind., Lond.*, **40**, 5 — 8.9.1.7

PEER, S. and FOUNTAIN, R., (1969). Brit. Pat. 1 166 427 — 10.5.7

PENN, W. S., (1969-1). *Rubb. J.*, **151**, 40 — 4.8.4

PENN, W. S., (1969-2). *Injection Moulding of Rubbers*, Maclaren, London — 8.6.6

PEREMSKY, R. (1963). *Kaucuk a plast. hmoty (Cz.)*, **2**, 37 — 7.10.1

PERKIN, W. H. JR., *et al.* (1910). *J chem. Soc.*, **97**, 1085 — 1.4.2

PETER, J. and HEIDEMANN, W., (1957). *Kautschuk Gummi*, **10**, 7, 168 — 5.5.2.1

PETER, J. and HEIDEMANN, W., (1958). *Kautschuk Gummi*, **11**, 6, 159 — 5.5.2.1

PETERSON, C. H., (1965). *Adhes. Age*, **8**, 7, 22 — 10.5.2

PHILIPOFF, W. and GASKINS, F. H., (1956). *J. Polym. Sci.*, **21**, 205 — 3.2.2

PICCINI, I. *See* LeBras, J. — 7.1

PICKLES, S. S., (1910). *J. chem. Soc.*, **97**, 1085 — 1.4.1

PIKE, M. *See* Blow, C. M. — 10.9.2

PIPER, G. H. and SCOTT, J. R., (1945). *J. scient. Instrum.*, **22**, 206 — 3.2.3

PIRELLI, SOC. PER AZIONI, (1966). Brit. Pat. 1 122 528 — 8.7.4

498   *References*

POLLAK, A. *See* Keilen, J. J.   7.1
POLLEY, M. and BOONSTRA, B. B., (1957). *Rubb. Chem. Technol.*, **30**, 170   7.4.3.2
PORTER, A. W. and RAO, O. A. M., (1926). *Trans. Faraday Soc.*, **23**, 311   3.2.3
PORTER, M. (1969). *Kautschuk Gummi Kunststoffe*, **22**, 8, 419   5.2.2
PRATT, C. P. *See* Gibson, P. R. S.   9.4.1

RABINOWITSCH, B. *See* Eisenschitz, R.   3.2.1
RABJOHN, N. *See* Flory, P. J.   5.2.2,
                                                                                     5.3
RADCLIFFE, R. R., RUGG, J. S., MORAN, A. L. and SMOOK, M. A., (1958). *Kautschuk Gummi*,   4.12.1
   11, WT242
RANDALL, R. L. *See* Ball, J. M.   6.4.5
RAO, O. A. M. *See* Porter, A. W.   3.2.3
REDMOND, G. B. *See* Wood, J. O.   6.7.2.3
REED, R. A., (1960). *Brit. Plast.*, **33**, 468   10.6.2.1
REHNER, J. *See* Flory, P. J.   3.3
REHNER, J. *See* Gessler, A. M.   4.5.3,
                                                                                     9.4.1
REISSINGER, S., (1960). *Kautschuk Gummi*, **13**, 195   9.4.1
RELLAGE, J. M., *et al.* (1969). *Int. Rubb. Conf.*, Moscow   4.3.3
RENAULT, REGIE NATIONALE DES USINES, (1958). Brit. Pat. 914 787   8.7.4
RENNER, A., BOONSTRA, B. B. and WALKER, D. F., (1969). *Proc. IRI Conf. on Adv. Polym.*   7.1
   Blends and Reinforcement, Loughborough
REVERTEX LTD (1931). Brit. Pat. 363 872   1.3
REXFORD, D. R. *See* Dixon, S.   1.4.10.1
REZNIKOVSKII, M. M. *See* Khromov, M. K.   11.3.2
RIGBI, Z., (1956). *Revue gén. Caoutch.*, **33**, 243   7.4.2
RIGBI, Z., (1957). *Bull. Res. Coun. Israel*, **6C**, 1, 67   7.4.2
RIGBI, Z. *See* Brennan, J. J.   7.10.1
RIGG, J. S. *See* Dixon, S.   1.4.10.1
RILEY, I. H., (1962). *Rubb. Plast. Wkly*, **143**, 1, 6; 2, 42; 3, 76   4.13.5
RILEY, I. H. *See* Baker, C. J.   4.13.4
RIST, C. E. *See* Buchanan, R. A.   7.1
RIVLIN, R. S., (1948). *Phil. Trans. R. Soc.*, **A241**, 379   3.3
RIVLIN, R. S. and SAUNDERS, D. W., (1951). *Phil. Trans. R. Soc.*, **A243**, 251   3.3.2
RIVLIN, R. S. and THOMAS, A. G., (1953). *J. Polym. Sci.*, **10**, 291   3.3.5.1
ROBINSON, P. M., (1960). Brit. Pat. 974 131   10.2.2
ROBINSON, P. M. *See* Brooke, T. A.   10.2.1.3
ROBINSON, R., (1928). Brit. Pat. 316 761   1.7
RODRIGUEZ, F. *See* Schwaber, D. M.   10.8.2.3
ROELIG, H., (1938). *Proc. Rubb. Technol. Conf.*, 821   11.3.13
ROELIG, H., (1943). *Kautschuk*, **19**, 47   11.3.13
ROELIG, H., (1945). *Rubb. Chem. Technol.*, **18**, 62   11.3.13
ROMANI, E. *See* Bruni, G.   1.6
ROSE, S. R., (1968). *Jnl Polymer Sci.*, **B6**, 857   4.18.3
ROSENBAUM, R. *See* Folkman, J.   4.13.4
ROSS, S. and OLIVIER, J. P., (1964). *On Physical Adsorption*, Wiley, New York   7.9
ROSSEM, A. VAN, (1958). *Rubber*, Servire, The Hague   7.2
ROSSMAN, R. P. *See* Bolt, T. D.   7.8
ROSTLER, K. S. *See* Garvey, B. S.   9.2.4
ROTH, F. L. *See* Decker, G. E.   11.2.1.1
ROTH, F. L. *See* Martin, G. M.   3.3.2
ROTH, F. L. and STIEHLER, R. D., (1959). *Proc. int. Rubb. Conf.*, 232   11.2.4
ROWLEY, T., (1881). Brit. Pat. 787   1.6
RUGG, J. S. *See* Radcliffe, R. R.   4.12.1
RUSSELL, C. R. *See* Buchanan, R. A.   7.1

SAAGEBARTH, K. A., (1968). *Jl IRI*, **2**, 185   5.2.1
SAFFIN, G. D. *See* Beerbower, A.   9.2.3.1,
                                                                                     9.4.6.3
SAFFORD, M. M. and BUECHE, A. M., (1953). US Pat. 2 710 290   4.13.5

SAPPER, D. I. *See* Bahary, W. S.   4.4.1, 4.4.3

SARTORI, G. *See* Natta, G.   4.6.2, 4.6.3

SAULINO, A. J. *See* Parshall, A. M.   8.2.7

SAUNDERS, D. W. *See* Rivlin, R. S.   3.3.2

SAVILLE, B. *See* Cain, M. E.   6.5.1

SAVILLE, B. and WATSON, A. A., (1967). *Rubb. Chem. Technol.*, **40**, 100   5.5.1

SCHAEFFER, W. D. and SMITH, W. R., (1955). *Ind. Engng Chem.*, **47**, 1286   7.4.2

SCHAFER, J. H. *See* DeCrease, W. M.   10.5.3.2

SCHAFFNER, I. J. *See* Auer, E. E.   9.4.5

SCHALLAMACH, A. *See* Harwood, J. A. C.   3.3.2

SCHEELE, W., LORENZ, O. and DUMMER, W., (1956). *Rubb. Chem. Technol.*, **29**, 1   5.2.2, 5.3

SCHIDROWITZ, P., (1914). Brit. Pat. 1 111   1.3

SCHIDROWITZ, P., (1921). Brit. Pat. 193 451   1.3

SCHIDROWITZ, P., (1922). Brit. Pat. 208 235   1.3

SCHMIDT, K. H., (1969). *Chem. Ind., Düsseld.*, **21**, 2, 94   4.10.2

SCHOENBERG, E., CHALFANT, D. L. and MAJOR, R. H., (1964). *Rubb. Chem. Technol.*, **37**, 103   4.3.2

SCHROYER, L. D. *See* Juve, A. E.   5.5.2.5

SCHULLER, W. *See* Mueller, W. J.   9.3.6

SCHULZ, G. V. *See* Ng, T. S.   4.1.7.1

SCHWABER, D. M. and RODRIGUEZ, F., (1967). *Rubb. Plast. Age*, **48**, 1081   10.8.2.3

SCHWARTZ, E. W., (1954). *Signal Corps Engng Lab., 3rd Annual Symp. on Tech. Progress in Communication Wires and Cables*   10.8.2.2

SCHWARTZ, E. W., (1955). *Kautschuk Gummi*, **8**, WT288   10.8.2.2

SCOTT, A. H., (1935). *Rubb. Chem. Technol.*, **8**, 401   2.9.3.11

SCOTT, C. E. *See* Hess, W. H.   7.10.3

SCOTT, G. E. and ECKERT, E. J., (1966). *Rubb. Chem. Technol.*, **39**, 533   4.2.1.4, 8.2.5

SCOTT, G. E. and ECKERT, E. J., (1967). *Jl IRI*, **1**, 99   8.2.5

SCOTT, J. R., (1965). *Physical Testing of Rubbers*, Maclaren, London   11.3.2, 11.3.5, 11.3.6, 11.3.9, 11.3.12, 11.3.13, 11.3.17, 11.3.19, 11.3.20

SCOTT, J. R. *See* Piper, G. H.   3.2.3

SCOTT TESTERS INC. (1965). Model GSV Goodrich–Scott Viscurometer, Tech. Leaflet   11.2.2

SEBRELL, L. B. and BEDFORD, C. W., (1925). US Pat. 1 544 687   1.6

SEKHAR, B. C., (1958). *Rubb. Chem. Technol.*, **31**, 425   4.1.5

SEKHAR, B. C., (1960). *Proc. NR Res. Conf.*, Kuala Lumpur, 512   4.1.5

SEKHAR, B. C., (1962). *Proc. 4th Rubb. Technol. Conf.*, 460   4.1.5, 4.1.7.1

SEKHAR, B. C., (1967). *Malaysian NR, New Presentation Methods*, Rubber Research Institute of Malaya, Kuala Lumpur   4.1.4.1, 4.1.5.1

SEKHAR, B. C. *See* Nair, S.   4.1.7.1

SELKER, M. L. and KEMP, A. R., (1945). *Ind. Engng Chem.*, **39**, 895   5.2.2

SELL, H. S. *See* Martin, G. E.   10.9.2

SELLERS, J. W. *See* Bachman, J. H.   7.4.3.1

SELLERS, J. W. and TOONDER, F. E., (1965). *Reinforcement of Elastomers* (Ed. G. Kraus), Ch. 13, p. 405   7.1

SEMPERIT (1937). Brit. Pat. 492 030   1.3

SHAFFER, M. C. *See* Flory, P. J.   5.2.2, 5.3

SHAW, E. J. and NELMS, R. P., (1967). *Rubb. J.*, **149**, 9, 17   10.8.3.3

SHELL INTERNATIONAL CHEMICAL CO. LTD (1961). Brit. Pats 881 119; 881 212   1.4.9

SHELL INTERNATIONAL CHEMICAL CO. LTD (1962). Brit. Pat. 907 579 — 1.4.9
SHELL INTERNATIONAL CHEMICAL CO. LTD (1965). Brit. Pat. 1 092 563 — 1.4.9
SHELTON, J. R., (1957). *Rubb. Chem. Technol.*, **30**, 1251 — 6.5.3
SHELTON, J. R. *See* Ambelang, J. C. — 6.5.3
SHEN, M. C. and EISENBERG, A., (1970). *Rubb. Chem. Technol.*, **43**, 95 — 2.1
SIBLEY, R. L., (1951). *Rubb. Chem. Technol.*, **24**, 211 — 5.3
SIMS, D., (1967). *Jl IRI*, **1**, 200 — 4.4.1
SIMS, D. *See* Brokenbrow, B. — 2.10.3
SINGLETON, R., (1964). *Rubb. J.*, **146**, 46 — 6.4
SINNOT, R. *See* Thomas, D. K. — 9.4.6.1
SKINNER, T. D. and WATSON, A. A., (1967). *Rubb. Age, N.Y.*, **99**, 11, 76 — 10.8.2
SKIPSEY, A. *See* Peachey, S. J. — 8.9.1.7
SLOPER, T. *See* Gray, C. H. — 1.9
SMALL, P. A., (1953). *J. appl. Chem., Lond.*, **3**, 71 — 9.2.3.1
SMIT, P. P. A. and VEGT, A. K. VAN DER, (1967). *Proc. int. Rubb. Conf.*, 99 — 4.3.3
SMIT, P. P. A. and VEGT, A. K. VAN DER, (1969). *Proc. int. Rubb. Conf.*, Moscow — 4.3.3
SMITH, D. R., (1967). *Adhes. Age*, **10**, 3, 25 — 10.5.2
SMITH, J. F. *See* Harwood, J. A. C. — 3.3.2
SMITH, J. F. *See* Payne, A. R. — 11.3.2
SMITH, M. G. *See* Thompson, C. W. — 4.1.2.3
SMITH, T. L., (1962). *Trans. Soc. Rheol.*, **6**, 61 — 3.3.2
SMITH, W. R., (1970). *Rubb. Chem. Technol.*, **43**, 960 — 7.4.3.2
SMITH, W. R. *See* Schaeffer, W. D. — 7.4.2
SMOOK, M. A. *See* Radcliffe, R. R. — 4.12.1
SODEN, A. L. *See* Hughes, B. G. — 9.3.5, 10.8.3.3

SOUTHERN, E., (1967). *Use of Rubber in Engineering*, Maclaren, London, Ch. 4 — 3.4.4
SOUTHERN, E. and THOMAS, A. G., (1967). *Trans. Faraday Soc.*, **63** 1913 — 3.4.3
SOUTHWART, D. W. *See* Blow, C. M. — 10.9.5.4
SOUTHWART, D. W. and HUNT, T., (1968). *Jl IRI*, **2**, 77; 79; 140 — 2.6.2.1, 4.13.5
SOUTHWART, D. W. and HUNT, T., (1969). *Jl IRI*, **3**, 249 — 2.6.2.1, 4.13.5
SOUTHWART, D. W. and HUNT, T., (1970). *Jl IRI*, **4**, 74; 77 — 2.6.2.1, 4.13.5
SPADE, R. L. *See* Jandial, A. P. — 10.8.2.2
SPALDING, A. G., (1963). Brit. Pat. 1 047 254 — 10.7.3.1
SPARKS, A. K. *See* Latos, E. J. — 6.5.1
SPARKS, W. J. *See* Thomas, R. M. — 1.4.7
SPRAGUE, G. R., (1968). *The Vanderbilt Rubber Handbook*, Vanderbilt, New York, p. 520 — 10.6.3.1
STAMBERGER, P., (1929). *The Colloid Chemistry of Rubber*, Oxford University Press, Oxford, p. 52 — 2.6.2.1
STANDARD OIL CO. (1935). Brit. Pat. 467 932 — 1.4.7
STAUDINGER, H., (1925). *Helv. chim. Acta*, **8**, 41 — 3.3
STAVELEY, F. W., *et al.* (1956). *Ind. Engng Chem.*, **48**, 778 — 1.4.9, 4.2.2
STEDRY, P. *See* Landel, R. F. — 3.3.2
STERN, H. J., (1945-1). *India Rubb. J.*, **108**, 615; 645; 648 — 1.2.2
STERN, H. J., (1945-2). *Rubb. Age Synth.*, **9**, 268; 295 — 1.5
STERN, H. J., (1967). *Rubber: Natural and Synthetic*, Maclaren, London — 1.2.1
STEVENS, H. P., (1931). Brit. Pat. 390 820 — 2.8.3
STIEHLER, R. D. *See* Martin, G. M. — 3.3.2
STIEHLER, R. D. *See* Roth, F. L. — 11.2.4
STOECK, P. J., (1969). *Rubb. Age, N.Y.*, **101**, 6, 64 — 9.2.1
STOKES, C. A. *See* Burgess, K. A. — 8.2.5
STOKOE, A. L. *See* Brokenbrow, B. — 2.10.3
STOREY, E. B. *See* Briggs, G. J. — 10.8.2.3
STOREY, E. B. *See* Edwards, D. C. — 9.4.1
STORY, P. R. *See* Loan, L. D. — 6.5.3
STRAIN, D. E. *See* Brooks, R. E. — 4.9.1

STRANGE, E. H. *See* Matthews, F. E.   1.4.2
STRAUB, R. M., (1969). Hypalon Bull. No. 1A, Du Pont   4.9.2
STUART, N., (1962). *Proc. 4th Rubb. Technol. Conf.*, 397   10.5.3.1
SUSICH, G. VON. *See* Meyer, K. H.   3.3
SWART, G. H. *See* Grüber, E. E.   4.15.2
SZÖR, P. *See* Bartha, Z.   11.2.2

TAKEYAMA, T. and MATSUI, J., (1969). *Rubb. Chem. Technol.*, **42**, 159   8.5.2.1
TALALAY, J. A., (1934). Brit. Pat. 445 138   1.3
TALALAY, J. A., (1942). *Rubb. Chem. Technol.*, **15**, 40   1.4.2
TALALAY, J. A., (1946). Brit. Pat. 619 619   2.4.3.4
TAPLIN, J. F., (1956). Brit. Pat. 855 348   10.9.5.3
TAYLOR, G. L. *See* Boonstra, B. B.   7.6,
                                                                            7.9
TAYLOR, G. L. and ATKINS, J. H., (1966). *J. phys. Chem., Ithaca*, **70**, 1678   7.4.2
TEW, H. J., (1956). *IRI Proc.*, **3**, 116   10.8.2,
                                                                            10.8.3.2
THOMAS, A. G. *See* Greensmith, H. W.   3.3.5.1,
                                                                            9.4.4.2
THOMAS, A. G. *See* Rivlin, R. S.   3.3.5.1
THOMAS, A. G. *See* Southern, E.   3.4.3
THOMAS, D. K. and SINNOTT, R., (1969). *Jl IRI*, **3**, 163   9.4.6.1
THOMAS, R. M. and SPARKS, W. J., (1939). US Pat. 2 356 128   1.4.7
THOMAS, R. M. and SPARKS, W. J., (1941). US Pat. 2 356 130; Brit. Pats 513 521; 523   1.4.7
    248; 543 308
THOMPSON, C. W., HOWARTH, H. and SMITH, M. G., (1966). *Rubb. J.*, **148**, 1, 24   4.1.2.3
THOMPSON, D. C. *See* Murray, R. M.   4.8.3,
                                                                            4.8.4,
                                                                            5.4.2.1
THOMPSON, D. C., HAGMAN, J. F. and MUELLER, N. N., (1958). *Rubb. Age., N.Y.*, **83**, 819   9.4.6.2
THOMPSON, J. and WATTS, J. T., (1966). *IRI Trans.*, **42**, 173   5.6.2
THOMPSON, R. W., (1845). Brit. Pat. 10 990   1.9
THOMSON, R. N. and MCALPINE, A., (1969). Brit. Pat. 1 138 362   8.3.2
TODD, R. V., (1962). *Rubb. Wld, N.Y.*, **146**, 2, 69   4.3.3
TOONDER, F. E. *See* Sellers, J. W.   7.1
TRAUBE, I., (1924). Brit. Pat. 226 440   1.3
TREIDA, G. F., (1969). *Rubb. Age, N.Y.*, **101**, 8, 52   8.6.6
TRELOAR, L. R. G., (1954). *Trans. Faraday Soc.*, **50**, 881   3.3.1
TRELOAR, L. R. G., (1958). *The Physics of Rubber Elasticity* (2nd edn), Oxford   2.1,
    University Press, Oxford   3.3.2
TUCKER, H. *See* Diem, H. E.   4.3.2
TUFTS, E., (1959). *Rubb. Age, N.Y.*, **84**, 936   4.12.1
TUNG, C. C. *See* Leeper, H. M.   7.4.2
TUNNICLIFFE, E., (1953). *IRI Trans.*, **29**, 55   10.8.3.3
TURETZKY, S. B. *See* Einhorn, S. C.   3.2.1
TURNER, M. J. and LEWIS, J. T., (1962). *Proc. 4th Rubb. Technol. Conf.*, 645   4.13.4
TWISS, D. F., (1925). *J. Soc. chem. Ind., Lond.*, **44**, 106T   2.6.2.1

UNION CARBIDE LTD (1960). Brit. Pats 851 002; 856 859   1.4.9
US RUBBER CO. (1954). Brit. Pat. 781 217   1.7
US RUBBER CO. (1957). Brit. Pat. 821 492   8.7.4
UTERMARK, W. L., (1923). Brit. Pat. 219 635   1.3

VALKO, E. *See* Meyer, K. H.   3.3
VALVASSORI, A. *See* Natta, G.   4.6.2,
                                                                            4.6.3
VANDENBERG, E. J., (1965). *Rubb. Plast. Age*, **46**, 1139   4.15.1
VARNER, J. E. *See* Bonner, J.   4.1.1
VAUGHAN, G. *See* Doyle, G. M.   4.15.2
VAUGHAN, G. *See* German, R.   4.6.2
VAUTIER, J. *See* Meyrick, C. I.   2.4.3.1

VEATCH, F., CALLAHAN, J. L., IDOL, J. D. and MILBERGER, E. C., (1962). *Petrol. Refiner*,   4.10.2
41, 11, 187
VEGT, A. K. VAN DER. *See* Smit, P. P. A.   4.3.3
VEITH, A. G., (1965-1). *Rubb. Chem. Technol.*, **38**, 700   9.4.4.2
VEITH, A. G., (1965-2). *Spec. tech. Publs Am. Soc. Test. Mater.*, **383**, 76   11.2.2
VEITH, A. G. *See* Juve, A. E.   5.5.2.5
VENUTO, L. J. *See* Wiegand, W. B.   1.5
VERKERK, G. *See* Krol, L. H.   4.3.3
VERVLOET, C., (1969). *Revue gén. Caoutch.*, **46**, 4, 474   4.4.3
VLIG, M., (1968). *Jl IRI*, **2**, 4   4.7.2
VOET, A. *See* Marsh, P. A.   7.10.3
VOET, A., ABOYTES, P. and MARSH, P. A., (1969). *Rubb. Age, N.Y.*, **101**, 78   7.4.3.1
VOET, A. and COOK, F. R., (1967). *Rubb. Chem. Technol.*, **40**, 1364   7.4.3.1
VOET, A. and COOK, F. R., (1968). *Rubb. Chem. Technol.*, **41**, 1207   7.4.3.1

WADELIN, C. *See* Ambelang, J. C.   6.5.3
WAGNER, M. P. *See* Bachman, J. H.   7.4.3.1
WAGNER, M. P. *See* Creasey, J. R.   9.5.3
WAKE, W. C., (1954). *Adhesion and Adhesives*, Society of Chemical Industry, London   6.7.3
WAKE, W. C., (1959). *Proc. int. Rubb. Conf.*, 406   7.10
WAKE, W. C. *See* Borroff, E. M.   8.5.2,
                                                                                       11.3.19
WALKER, D. F. *See* Renner, A.   7.1
WALL, F. T., (1942). *J. chem. Phys.*, **10**, 485   3.3
WALLACE & CO., H. W., (1969). The Wallace–Shawbury Curometer Mark VI, Tech.   11.2.2
Leaflet D4B
WATSON, A. A. *See* Saville, B.   5.5.1
WATSON, A. A. *See* Skinner, T. D.   10.8.2
WATSON, W. F., (1954). *Proc. 3rd Rubb. Technol. Conf.*, 553   7.8
WATSON, W. F., (1955). *Ind. Engng Chem.*, **47**, 1281   7.8
WATSON, W. F. *See* Bristow, G. M.   4.1.7.2
WATSON, W. F. *See* Hughes, B. G.   9.3.5,
                                                                                       10.8.3.3
WATSON, W. F. *See* Moore, C. G.   5.2.2
WATSON, W. F. *See* Payne, A. R.   7.4.3.1
WATTS, J. T. *See* Thompson, J.   5.6.2
WEINSTOCK, K. V. *See* Grüber, E. E.   4.15.2
WEINSTOCK, K. V. *See* Janssen, H. J. J.   4.2.1.4
WEIR, G. E., LESER, W. H. and WOOD, L. A., (1951). *Rubb. Chem. Technol.*, **24**, 366 4.13.4
WEISLOGEL, O. E. *See* Buchanan, R. A.   7.1
WEISS, M. L., (1922). US Pat. 1 411 231   1.6
WEISSENBERG, K., (1947). *Nature*, **159**, 310   3.2.1
WEISSENBERG, K. *See* Eisenschitz, R.   3.2.1
WEISSERT, F. C. and CUNDIFF, R. R., (1963). *Rubb. Age, N.Y.*, **92**, 881   9.2.4
WELCH, C. K., (1890). Brit. Pat. 14 563   1.9
WELSBY, J. A. *See* Bradshaw, W. D.   10.3.3
WHEELANS, M. A., (1966). *Rubb. J.*, **148**, 12, 26   8.6.6
WHEELANS, M. A., (1968). *Rep. Prog. Rubb. Technol.*   8.6.6
WHEELANS, M. A., (1969). *Injection Moulding of Elastomers* (Ed. W. S. Penn),   8.6.6
Maclaren, London, p. 82
WHITE, J. L., (1969). *Rubb. Chem. Technol.*, **42**, 281   3.2.5
WHITTAKER, R. W. *See* Harwood, J. A. C.   3.3.6
WHORLOW, R. W. *See* Mullins, L.   2.6.2.1
WIEGAND, W. B., (1920). *Can. chem. J.*, **4**, 160; *India Rubb. J.*, **60**, 379; 423; 453 7.2
WIEGAND, W. B. and VENUTO, L. J., (1929). Brit. Pat. 327 979   1.5
WIEGAND, W. B. and VENUTO, L. J., (1930). Can. Pat. 301 071   1.5
WIEGAND, W. B. and VENUTO, L. J., (1932). US Pat 1 889 429   1.5
WILDSCHUT, H. J. (1942). *Recl Trav. chim. Pays-Bas*, **61**, 898   7.1
WILLIAMS, G., (1860). *Proc. R. Soc.*, **10**, 516   1.4.1
WILLIAMS, G. E. *See* Imperial Chemical Industries Ltd   1.7

WILLIAMS, I., (1924). *Ind. Engng Chem.* **16**, 362 — 3.2.4, 11.2.1.2

WILLIAMS, M. L. *See* Ferry, J. D. — 3.3.4

WILLIAMS, M. L., LANDEL, R. F. and FERRY, J. D., (1955). *J. Am. chem. Soc.*, **77**, 3701 — 3.3.4

WILLSHAW, H., (1950). Brit. Pat. 638 290 — 8.4.1.3

WILSON, A. *See* Montermoso, J. — 4.18.1

WINGFOOT CORPN (1946). Brit. Pat. 649 094 — 10.9.2

WINGFOOT CORPN (1949). Brit. Pat. 677 556 — 10.9.2

WINKELMANN, H. A. and GRAY, H., (1924). US Pat. 1 515 642 — 1.7

WISE, R. W. *See* Decker, G. E. — 5.5.2.4

WISE, R. W. and DECKER, G. E., (1965). *Spec. tech. Publs Am. Soc. Test. Mater.*, **383**, 51 — 11.2.2

WISEMAN, P. *See* Banks, D. J. — 1.6

WOLF, R. F., (1960). *Rubb. Wld, N.Y.*, **142**, 2, 81 — 9.4.1

WOLF, R. F. *See* Bachmann, J. H. — 7.4.3.1

WOLFF, S., (1969). *Kautschuk Gummi Kunststoffe*, **22**, 367 — 7.6

WOLFF, S., (1970). *Kautschuk Gummi Kunststoffe*, **23**, 7 — 7.6

WOOD, B. B. J. *See* Woodhams, R. T. — 4.17

WOOD, J. O., GOY, R. S. and DARUWALLA, F. S., (1959). *Text. Res. J.* **29**, 669 — 6.7.2.2

WOOD, J. O. and REDMOND, G. B., (1965). *J. Text. Inst.*, **56**, T191 — 6.7.2.3

WOOD, L. A. *See* Weir, G. E. — 4.13.4

WOOD, L. A. and MARTIN, G. M., (1964). *J. Res. natn. Bur. Stand.*, **68A**, 259 — 2.9.3.11

WOOD, R. I., (1952). *J. Rubb. Res. Inst. Malaya*, **14**, 20 — 4.1.5

WOODCOCK, W. J. and GOODMAN, L. T., (1945). Brit. Pat. 572 261 — 10.5.6

WOODHAMS, R. T., ADAMEK, S. and WOOD, B. B. J., (1965). *Rubb. Plast. Age*, **46**, 1 — 4.17

WRAGG, A. L., (1967). *Aspects of Adhesion, Vol. 3* (Ed. D. J. Alner), p. 59 — 2.10.4

YERZLEY, F. L., (1940). *Rubb. Chem. Technol.*, **13**, 149 — 11.3.13

YOUNGS, D. C. and KONKLE, G. M., (1964). *Rubb. Age, N.Y.*, **95**, 736; 894 — 4.13.4

ZAPP, R. L. *See* Gessler, A. M. — 4.5.3, 9.4.1

ZIEGLER, K., (1959). *Brennst.-Chem.*, **40**, 7, 209 — 1.4.9

ZWICK & CO. KG, (1966). *Connaiss. Plast.*, **7**, 63, 20 — 11.2.2

Standards

Standards are denoted as follows:

ASTM. American Society for Testing and Materials, USA.
BS (British Standard). British Standards Institution, UK.
DIN (Deutsche Industrie Normen). Deutscher Normenausschuss, Germany.
GOST (State Standard). State Committee for Standards, Measurements, and Measuring Instruments, USSR.
IEC. International Electrotechnical Commission, Geneva, Switzerland.
ISO R. Recommendation of the International Organisation for Standardisation (ISO).
NF (Norme Française). Association Française de Normalisation, France.

Titles are given in English, where possible as on the original document, but translations or amendments are shown with square brackets. The ISO Recommendations marked with an asterisk (*) are still at the Draft Recommendation stage but will bear the same numbers when published as Recommendations.

ASTM D15-68a. Sample Preparation for Physical Testing of Rubber Products
ASTM D149-64. Tests for Dielectric Breakdown Voltage and Dielectric Strength of Electrical Insulating Materials at Commercial Power Frequencies
ASTM D150-68. Tests for A-C Loss Characteristics and Dielectric Constant (Permittivity) of Solid Electrical Insulating Materials
ASTM D257-66. Tests for D-C Resistance or Conductance of Insulating Materials

504 *References*

ASTM D297-67T. Chemical Analysis of Rubber Products

ASTM D395-67. Tests for Compression Set of Vulcanised Rubber

ASTM D412-68. Tension Testing of Vulcanised Rubber

ASTM D413-39 (1965). Tests for Adhesion of Vulcanised Rubber (Friction Test)

ASTM D429-68. Tests for Adhesion of Vulcanised Rubber to Metal

ASTM D430-59 (1965). Dynamic Testing for Ply Separation and Cracking of Rubber Products

ASTM D454-53 (1965). Air-pressure Heat Test of Vulcanised Rubber

ASTM D471-68. Test for Change in Properties of Elastomeric Vulcanisates Resulting from Immersion in Liquids

ASTM D518-61 (1968). Test for Resistance to Surface Cracking of Stretched Rubber Compounds

ASTM D572-67. Test for Accelerated Ageing of Vulcanised Rubber by the Oxygen-pressure Method

ASTM D573-67. Test for Accelerated Ageing of Vulcanised Rubber by the Oven Method

ASTM D575-67. Tests for Compression–Deflection Characteristics of Vulcanised Rubber

ASTM D623-67. Tests for Compression Fatigue of Vulcanised Rubber

ASTM D624-54. Test for Tear Resistance of Vulcanised Rubber

ASTM D746-64T. Test for Brittleness Temperature of Plastics and Elastomers by Impact

ASTM D797-64. Test for Young's Modulus in Flexure of Natural and Synthetic Elastomers at Normal and Subnormal Temperatures

ASTM D813-59 (1965). Test for Resistance of Vulcanised Rubber or Synthetic Elastomers to Crack Growth

ASTM D815-66. Test for Hydrogen Permeability of Rubber-coated Fabrics

ASTM D832-64. Recommended Practice for Conditioning of Elastomeric Materials for Low Temperature Testing

ASTM D865-62 (1968). Heat Ageing of Vulcanised Rubber by Test Tube Method

ASTM D926-67. Test for Plasticity and Recovery of Rubber and Rubber-like Materials by the Parallel Plate Plastometer

ASTM D945-59 (1968). Tests for Mechanical Properties of Elastomeric Vulcanisates Under Compressive or Shear Strains by the Mechanical Oscillograph

ASTM D991-68. Test for Volume Resistivity of Electrically Conductive Rubber and Rubber-like Materials

ASTM D1053-65. Measuring Low-temperature Stiffening of Rubber and Rubber-like Materials by Means of a Torsional Wire Apparatus

ASTM D1054-66. Test for Impact Resilience and Penetration of Rubber by the Rebound Pendulum

ASTM D1055-62. Specifications and Tests for Latex Foam Rubbers

ASTM D1056-68. Specifications and Tests for Sponge and Expanded Cellular Rubber Products

ASTM D1149-64. Test for Accelerated Ozone Cracking of Vulcanised Rubber

ASTM D1171-68. Test for Weather Resistance Exposure of Automotive Rubber Compounds

ASTM D1229-62 (1968). Test for Low-temperature Compression Set of Vulcanised Elastomers

ASTM D1329-60. Evaluating Low-temperature Characteristics of Rubber and Rubber-like Materials by a Temperature–Retraction Procedure (TR Test)

ASTM D1390-62 (1968). Test for Stress Relaxation of Vulcanised Rubber in Compression

ASTM D1415-68. Test for International Hardness of Vulcanised Natural and Synthetic Rubbers

ASTM D1416-67T. Chemical Analysis of Synthetic Elastomers (Solid Styrene–Butadiene Copolymers)

ASTM D1418-67. Nomenclature for Synthetic Elastomers and Latices

ASTM D1460-60 (1968). Test for Change in Length of an Elastomeric Vulcanisate Resulting from Immersion in a Liquid

ASTM D1564-64T. Testing Slab Flexible Urethane Foam

ASTM D1630-61 (1968). Test for Abrasion Resistance of Rubber Soles and Heels

ASTM D1646-68. Test for Viscosity and Curing Characteristics of Rubber by the Shearing Disk Viscometer

ASTM D1672-66. Recommended Practice for Exposure of Polymeric Materials to High Energy Radiation

ASTM D1765-67. Recommended Practice for Nomenclature for Rubber Grade Carbon Blacks

ASTM D1870-68. Test for Elevated Temperature Ageing Using a Tubular Oven

ASTM D1871-68. Tests for Adhesion of Vulcanised Rubber to Single-strand Wire

ASTM D2000-68. Classification System for Elastomeric Materials for Automotive Applications

ASTM D2138-67. Test for Static Adhesion of Textile Cord to Rubber (H-pull Test)

ASTM D2228-63T. Test for Abrasion Resistance of Rubber and Elastomeric Materials by the Pico Method

ASTM D2229-68. Test for Adhesion of Vulcanised Rubber to Wire Cord

ASTM D2230-63T. Extrudability of Unvulcanised Elastomeric Compounds

ASTM D2231-66. Recommended Practice for Forced Vibration Testing of Vulcanisates

ASTM D2240-68. Test for Indentation of Rubber and Plastics by Means of a Durometer

ASTM D2309-68. Testing Compression Set Induced in Vulcanised Rubber During Exposure to High Energy Nuclear Radiation

ASTM D2630-68T. Test for Strap Peel Adhesion of Textile Fabrics or Cords to Rubber

ASTM D2632-67. Test for Impact Resilience of Rubber by Vertical Rebound

ASTM D2704-68T. Measurement of Curing Characteristics with a Compression Modulus Cure Meter

ASTM D2705-68T. Measurement of Curing Characteristics with the Oscillating Disc Rheometer Cure Meter

ASTM D2706-68T. Measurement of Curing Characteristics with the Viscurometer Cure Meter

BS 903. Methods of Testing Vulcanised Rubber

Part A2 (1956). Determination of Tensile Stress–Strain Properties

Part A3 (1956). Determination of Tear Strength

Part A4 (1957). Determination of Compression Stress–Strain

Part A5 (1958) plus Amendment No. 1. Determination of Tension Set

Part A6 (1969). Determination of Compression Set

Part A8 (1963). Determination of Rebound Resilience

Part A9 (1957) plus Amendments Nos 1–2. Determination of Abrasion Resistance

Part A10 (1956). Determination of Resistance to Flex Cracking

Part A11 (1956) plus Amendments Nos 1–2. Determination of Resistance to Crack Growth

Part A12 (1968). Determination of Rubber-to-fabric Adhesion (Ply Separation)

Part A13 (1960) plus Amendments Nos 1–2. Determination of Resistance to Low Temperatures (Rigidity Modulus Test)

Part A14 (1970). Determination of Modulus in Shear (Bonded Quadruple Shear Test Piece)

Part A15 (1958). Determination of Creep and Stress Relaxation

Part A16 (1956). Determination of Swelling in Liquids

Part A17 (1960). Determination of Permeability to Gases (Constant-volume Method)

Part A19 (1956) plus Amendments Nos 1–4. Accelerated Ageing Tests

Part A21 (1961). Determination of Rubber-to-metal Bond Strength

Part A23 (1963). Determination of Resistance to Ozone Cracking under Static Conditions

Part A24 (1964). Dynamic Testing of Vulcanised Rubber

Part A25 (1968). Determination of Impact Brittleness Temperature

Part A26 (1969). Determination of Hardness

Part A27 (1969). Determination of Rubber-to-fabric Adhesion (Direct Tension)

Part A28 (1970). Determination of Adhesion of Vulcanised Rubber to Rigid Plates in Shear (Quadruple Shear Test)

Part C1 (1956). Determination of Surface Resistivity of Insulating Soft Vulcanised Rubber and Ebonite

Part C2 (1956). Determination of Volume Resistivity of Insulating Soft Vulcanised Rubber and Ebonite

Part C3 (1956) plus Amendments Nos 1–2. Determination of Permittivity and Power Factor of Insulating Soft Vulcanised Rubber and Ebonite

Part C4 (1957). Determination of Electric Strength of Insulating Soft Vulcanised Rubber and Ebonite

Parts F1–F9 (1956). Methods of Testing Soft Cellular Rubber

BS 1154 (1952) plus Amendments Nos 1–8. Vulcanised Natural Rubber Compounds

BS 1155 (1954) plus Amendments Nos 1–5. Vulcanised Extruded Natural Rubber Compounds and Tubing

BS 1673. Methods of Testing Raw Rubber and Unvulcanised Compounded Rubber

Part 3 (1969). Methods of Physical Testing

BS 1674 (1968). Equipment and Procedures for Mixing and Vulcanising Rubber Test Mixes

BS 2044 (1953) plus Amendment No 1. Laboratory Tests for Resistivity of Conductive and Antistatic Rubbers

BS 2045 (1965). Preferred Numbers

BS 2719 (1956). Methods of Use and Calibration of Pocket Type Rubber Hardness Meters
BS 2751 (1956) plus Amendments Nos 1–5. Vulcanised Butadiene/Acrylonitrile Rubber Compounds
BS 2752 (1956) plus Amendments Nos 1–5. Vulcanised Chloroprene Rubber Compounds
BS 3093 (1959) plus Amendments Nos 1–2. Latex Foam Rubber Hospital Mattresses
BS 3106 (1959). Non-silver-staining Natural Rubber Compounds
BS 3177 (1959) plus Amendment No. 1. Method for Determining the Permeability to Water Vapour of Flexible Sheet Materials used for Packaging
BS 3222 (1960) plus Amendment No. 1. Low Compression Set Butadiene/Acrylonitrile Vulcanised Rubber Compounds
BS 3227 (1960) plus Amendment No. 1. Vulcanised Butyl Rubber Compounds
BS 3472 (1963) plus Amendments Nos 1–3. Raw Styrene–Butadiene Rubbers (1500 series)
BS 3502 (1967). Schedule of Common Names and Abbreviations for Plastics and Rubbers
BS 3515 (1962). Vulcanised Styrene Butadiene Rubber (SBR) Compounds
BS 3558 (1968). Glossary of Terms Used in the Rubber Industry
BS 3574 (1963) plus Amendment No. 1. Recommendations for the Storage of Vulcanised Rubber
BS 3629 (1963). Vulcanised Extruded Styrene Butadiene Rubber (SBR) Compounds
BS 3650 (1963) plus Amendments Nos 1–2. Raw Oil-extended Styrene–Butadiene Rubbers (1700 series)
BS 3667. Methods of Testing Flexible Polyurethane Foam
  Parts 1–2 (1963) plus Amendment No. 1
  Parts 3–10 (1966).
BS 3734 (1964) plus Amendment No. 1. Schedule of Tolerances for Rubber Products in Solid Rubber and Ebonite
BS 4355 (1968). Centrifuged Ammonia-preserved Natural Rubber Latices
BS 4396 (1968). Raw Natural Rubber
BS 4398 (1969). Specification for Compounding Ingredients for Rubber Test Mixes
BS 4443 (1969). Methods of Test for Flexible Cellular Materials
BS 4470 (1969). Methods of Test for Raw Isobutylene–Isoprene Rubbers (IIR or Butyl)

DIN 53505 (1967). Testing of Rubbers; Shore A, C and D Hardness Test
DIN 53507 (1959). [Testing of Rubber; Tear Test with Strip Test-pieces]
DIN 53508 (1960). Testing of Rubber; Artificial Ageing of Soft Rubber
DIN 53512 (1965). Testing of Elastomers; Determination of Impact Resilience
DIN 53513 (1962). Testing of Rubber; Determination of the Visco-elastic Properties of Rubber under Forced Vibration beyond Resonance
DIN 53514 (1958). [Testing of Rubber; Determination of Plasticity by the Baader Hot Compression Test—Defo Test]
DIN 53515 (1962). Testing of Rubber and of Plastic Films; Tear Test using the Graves Angle Test Piece with Incision
DIN 53516 (1964). Testing of Rubber; Abrasion Test; Determination of Volume Loss
DIN 53529 (1969). Testing of Elastomers; Measurement of Plasticity, Determination of Reaction during Vulcanisation of Rubber Compounds by means of Vulcameters and Evaluation of Kinetic Reaction of Interlacing [Crosslinking] Isotherms

GOST 261-67. Rubber. Test Method of [by] Multiple Extension
GOST 262-53. [Tear Testing of Rubber]
GOST 265-66. Rubber. Method of Testing Short-time Static Compression
GOST 411-41. [Determination of Rubber-to-metal Bond Strength by Peeling]
GOST 415-53. [Determination of Plasticity of Raw Rubber and Unvulcanised Rubber Mixes by the Plastimeter]
GOST 426-66. Rubber. Method for the Determination of the Abrasion Resistance under Slipping
GOST 6949-63. Vulcanised Rubber. Test Method for the Determination of Failure in the Presence of Ozone under Static Deformation
GOST 10201-62. Rubber. Method of Determination of Stiffness of [Raw] Rubber and Unvulcanised Rubber Compounds
GOST 10827-64. Rubber. Method for the Determination of Dynamic Modulus and Internal Friction Modulus at the [by] Impact Extension
GOST 10828-64. Rubber. Method for the Determination of Dynamic Modulus and Internal Friction Modulus at [by] Flexing

GOST 10952-64. Rubber. Method for the Determination of Coefficient of Dynamic Fatigue at [by] Symmetrical Cycle[s] of Loading

GOST 10953-64. Rubber. Method for the Determination of Dynamic Modulus and Internal Friction Modulus at [by] Rolling

GOST 12014-66. Rubber. Method for the Determination of Specific Tearing Energy

GOST 12254-66. Synthetic and Natural Rubber. Method for the Determination of Glass-transition Temperature under Static Load

GOST 12535-67. Rubber. Method for the Determination of Kinetics of Vulcanisation

IEC Recommendation 93 (1958). Recommended Methods of Test for the Volume and Surface Resistivities of Electrical Insulating Materials (1st edn)

IEC Recommendation 167 (1964). Methods of Test for the Determination of the Insulation Resistance of Solid Insulating Materials (1st edn)

IEC Recommendation 250 (1969). Recommended Methods for the Determination of the Permittivity and Dielectric Dissipation Factor of Electrical Insulating Materials at Power, Audio and Radio Frequencies including Metre Wavelengths (1st edn)

ISO Recommendation R3 (1953). Preferred Numbers. Series of Preferred Numbers (1st edn)

ISO Recommendation R34 (1957). Determination of Tear Strength of Vulcanised Natural and Synthetic Rubbers (Crescent Test Piece) (1st edn)

ISO Recommendation R36 (1969). Determination of the Adhesive Strength of Vulcanised Rubbers to Textile Fabrics (2nd edn)

ISO Recommendation R37 (1968). Determination of Tensile Stress–Strain Properties of Vulcanised Rubbers (2nd edn)

ISO Recommendation R48 (1968). Determination of Hardness of Vulcanised Rubbers (2nd edn)

ISO Recommendation R132 (1959). Determination of Resistance to Flex Cracking of Vulcanised Natural or Synthetic Rubber (De Mattia Type Machine) (1st edn)

ISO Recommendation R133 (1959). Determination of Resistance to Crack Growth of Vulcanised Natural or Synthetic Rubber (De Mattia Type Machine) (1st edn)

ISO Recommendation R188 (1961). Accelerated Ageing or Simulated Service Tests on Vulcanised Natural or Synthetic Rubbers (1st edn)

ISO Recommendation R289 (1963) plus Amendment R289/A1. Determination of Viscosity of Natural and Synthetic Rubbers by the Shearing Disk Viscometer (1st edn)

ISO Recommendation R471 (1966). Standard Atmospheres for the Conditioning and Testing of Rubber Test Pieces (1st edn)

ISO Recommendation R667 (1968). Determination of Rate of Cure of Rubber Compounds by the Shearing Disk Viscometer (1st edn)

ISO Recommendation R812 (1968). Method of Test for Temperature Limit of Brittleness for Vulcanised Rubbers (1st edn)

ISO Recommendation R813 (1968). Preparation of Test Piece and Method of Test of the Adhesion of Vulcanised Rubber to Metal Where the Rubber is Assembled to One Metal Plate (1st edn)

ISO Recommendation R814 (1968). Preparation of Test Piece and Method of Test of the Adhesion of Vulcanised Rubber to Metal Where the Rubber is Assembled to Two Metal Plates (1st edn)

ISO Recommendation R815 (1968). Method of Test for the Compression Set under Constant Deflection at Normal and High Temperatures of Vulcanised Rubbers (1st edn)

ISO Recommendation R816 (1968). Determination of Tear Strength of Small Test Pieces of Vulcanised Rubbers (Delft Test Piece) (1st edn)

ISO Recommendation R1399*. Determination of the Permeability of Rubber to Gases (Constant-volume Method)

ISO Recommendation R 1431*. Testing Resistance of Vulcanised Rubbers to Ozone Cracking under Static Conditions

ISO Recommendation R1432*. Determination of the Stiffness of Vulcanised Rubbers (Gehman Test) at Low Temperature

ISO Recommendation R1653*. Determination of Compression Set of Vulcanised Rubbers under Constant Deflection at Low Temperatures

ISO Recommendation R1767*. Determination of Rebound Resilience of Vulcanised Rubber by the Lüpke Pendulum

ISO Recommendation R1817*. Method of Test for the Resistance of Vulcanised Rubber to Liquids

ISO Recommendation R1827*. Determination of Modulus in Shear of Rubber (Bonded Quadruple Shear Test Piece)
ISO Recommendation R2007*. Rapid Plasticity Test on Raw Rubber and Unvulcanised Compounded Rubber

NF T43-001 (1953). [Method of Preparing Vulcanised Specimens for the Specification or
Control Testing of Raw Rubber]
NF T46-006 (1966). [Rubber and Like Elastomers: Air Bomb Test at 125°C]
NF T46-012 (1946). [Rubber: Abrasion Test]

Index

In general, the names of individual rubbers follow the style of the common names used in Chapter 4. An exception to this is the use of 'ethylene–propylene rubbers' in place of either EPM or EPDM rubbers as the two polymer types receive parallel mention in most parts of the book. Synthetic isoprene rubber and its natural counterpart are listed separately as isoprene rubber and natural rubber, respectively. Very few entries occur under 'rubber' as the book is almost exclusively concerned with this topic.

T